Trends in Mathematics

Trends in Mathematics is a book series devoted to focused collections of articles arising from conferences, workshops or series of lectures.

Topics in a volume may concentrate on a particular area of mathematics, or may encompass a broad range of related subject matter. The purpose of this series is both progressive and archival, a context in which to make current developments available rapidly to the community as well as to embed them in a recognizable and accessible way.

Volumes of TIMS must be of high scientific quality. Articles without proofs, or which do not contain significantly new results, are not appropriate. High quality survey papers, however, are welcome. Contributions must be submitted to peer review in a process that emulates the best journal procedures, and must be edited for correct use of language. As a rule, the language will be English, but selective exceptions may be made. Articles should conform to the highest standards of bibliographic reference and attribution.

The organizers or editors of each volume are expected to deliver manuscripts in a form that is essentially "ready for reproduction." It is preferable that papers be submitted in one of the various forms of TEX in order to achieve a uniform and readable appearance. Ideally, volumes should not exceed 350-400 pages in length.

Proposals to the Publisher are welcomed at either:
Birkhäuser Boston, 675 Massachusetts Avenue, Cambridge, MA 02139, U.S.A.
math@birkhauser.com
or
Birkhäuser Verlag AG, PO Box 133, CH-4010 Basel, Switzerland
math@birkhauser.ch

International Symposium on Ring Theory

Gary F. Birkenmeier
Jae Keol Park
Young Soo Park

Editors

Birkhäuser
Boston • Basel • Berlin

Gary F. Birkenmeier
Department of Mathematics
University of Louisiana at Lafayette
Lafayette, LA 70504-1010
U.S.A.

Jae Keol Park
Department of Mathematics
Pusan National University
Pusan 609-735
Korea

Young Soo Park
Department of Mathematics
Kyungpook National University
Taegu 702-701
Korea

Library of Congress Cataloging-in-Publication Data

International symposium on ring theory / Gary F. Birkenmeier, Jae Keol Park, Young Soo Park, editors.
  p. cm – (Trends in mathematics)
  ISBN 0-8176-4158-0 (alk. paper) — ISBN 3-7643-4158-0 (alk. paper)
  1. Rings (Algebra)–Congresses. I. Birkenmeier, Gary F., 1946- II. Park, Jae Keol, 1947- III. Park, Young Soo, 1942- IV. Korea-China-Japan International Symposium on Ring Theory (3rd : 1999 : Kyongju-si, Korea) V. Series.

QA247.I58 2000
512'.4–dc21   00-051926

AMS Subject Classifications: 13A18, 13C14, 13C40, 13F30, 13N05, 16D50, 16D80, 16D90, 16D99, 16E10, 16E20, 16E30, 16E40, 16E50, 16G30, 16G60, 16L60, 16N80, 16P20, 16P50, 16P50, 16P60, 16P70, 16S36, 16S40, 16S90, 16W30

Printed on acid-free paper.
© 2001 Birkhäuser Boston   *Birkhäuser*

All rights reserved. This work may not be translated or copied in whole or in part without the written permission of the publisher (Birkhäuser Boston, c/o Springer-Verlag New York, Inc., 175 Fifth Avenue, New York, NY 10010, USA), except for brief excerpts in connection with reviews or scholarly analysis. Use in connection with any form of information storage and retrieval, electronic adaptation, computer software, or by similar or dissimilar methodology now known or hereafter developed is forbidden.
The use of general descriptive names, trade names, trademarks, etc., in this publication, even if the former are not especially identified, is not to be taken as a sign that such names, as understood by the Trade Marks and Merchandise Marks Act, may accordingly be used freely by anyone.

ISBN 0-8176-4158-0   SPIN 10748553
ISBN 3-7643-4158-0

Reformatted from editors' files by TEXniques, Inc., Cambridge, MA.
Printed and bound by Hamilton Printing Company, Rensselaer, NY.
Printed in the United States of America.

9 8 7 6 5 4 3 2 1

To all Ring Theorists—past, present, and future

Contents

Preface xi

Acknowledgments xiii

Participants xv

Generalized Deviations of Posets with Applications to
Chain Conditions on Modules
T. Albu and M. L. Teply 1

Stability Properties of Exchange Rings
P. Ara 23

Good Conditions for the Total
K. I. Beidar and F. Kasch 43

Semicentral Reduced Algebras
G. F. Birkenmeier, J. Y. Kim and J. K. Park 67

On Generalizations of Injectivity
J. Chen and N. Ding 85

Auslander-Gorenstein Rings for Beginners
J. Clark 95

The Flat Cover Conjecture and Its Solution
E. E. Enochs and O. M. G. Jenda 117

Some Results on Skew Polynomial Rings over a
Reduced Ring
J. Han, Y. Hirano and H. Kim 123

Derived Equivalences and Tilting Theory
D. Happel 131

CS-Property of Direct Sums of Uniform Modules
J. Kado, Y. Kuratomi and K. Oshiro 149

Generalized Principally Injective Maximal Ideals
J. Y. Kim, N. K. Kim and S. B. Nam 161

The Module of Differentials of a Noncommutative
Ring Extension
H. Komatsu 171

Dual Bimodules and Nakayama Permutations
Y. Kurata, K. Koike and K. Hashimoto 179

The Coinduced Functor and Homological Properties of
Hopf Modules
T. Li and Z. Wang 191

Hopf Algebra Coaction and Its Application to
Group-Graded Rings
G.-L. Liu 199

Non-Commutative Valuation Rings and Their Global Theories
H. Marubayashi 207

On the Maximal t-Corational Extensions of Modules
S. Morimoto 223

On Values of Cyclotomic Polynomials
K. Motose 231

Generalized Jordan Derivations
A. Nakajima 235

On Quasi-Frobenius Rings
W. K. Nicholson and M. F. Yousif 245

Theories of Harada in Artinian Rings and Applications to
Classical Artinian Rings
K. Oshiro 279

On Some Dimensions of Modular Lattices and Matroids
E. R. Puczyłowski 303

On Torsion-free Modules over Valuation Domains
K. M. Rangaswamy 313

Hecke Orders, Cellular Orders and Quasi-Hereditary Orders
K. W. Roggenkamp 329

Some Kind of Duality
M. Sato 355

On Inertial Subalgebras of Certain Rings
T. Sumiyama 365

Some Recent Results on Hopficity, Co-hopficity and
Related Properties
K. Varadarajan 371

Some Studies on QcF-coalgebras
M. Wang 393

Finitely Pseudo-Frobenius Rings
H. Yoshimura 401

Surjectivity of Linkage Maps
Y. Yoshino 421

Infinite Quivers and Cohomology Groups
P. Zhang 427

Open Problems 441

Preface

This volume is the Proceedings of the Third Korea-China-Japan International Symposium on Ring Theory held jointly with the Second Korea-Japan Joint Ring Theory Seminar which took place at the historical resort area of Korea, Kyongju, June 28–July 3, 1999. It also includes articles by some invited mathematicians who were unable to attend the conference. Over 90 mathematicians from 12 countries attended this conference.

The conference is held every 4 years on a rotating basis. The first conference was held in 1991 at Guilin, China. In 1995 the second conference took place in Okayama, Japan. At the second conference it was decided to include Korea, who hosted this conference of 1999.

During the past century Ring Theory has diversified into many subareas. This is reflected in these articles from over 25 well-known mathematicians covering a broad range of topics, including: Classical Ring Theory, Module Theory, Representation Theory, and the theory of Hopf Algebras. Among these peer reviewed papers are invited survey articles as well as research articles. The survey articles provide an overview of various areas for researchers looking for a new or related field to investigate, while the research articles give the flavor of current research. We feel that the variety of related topics will stimulate interaction between researchers. Moreover the Open Problems section provides guidance for future research. This book should prove attractive to a wide audience of algebraists.

Gary F. Birkenmeier, Lafayette, U. S. A.
Jae Keol Park, Pusan, Korea
Young Soo Park, Taegu, Korea
Editors

February, 2000

Acknowledgments

Our thanks go to all participants, contributors and referees for their participation and assistance in bringing out this volume in a timely manner. Financial support from the Korea Science and Engineering Foundation (KOSEF) and Kyungpook National University are greatly appreciated. Also our thanks go to a number of graduate students of Kyungpook National University and Pusan National University who spent a great deal of their time on various arrangements for the Conference, especially to Jaekyung Doh, Hai Lan Jin, Eun Jeong Kim, and Yeun Young Kim.

We appreciate Professors Yong Uk Cho, Juncheol Han, Chan Huh, Chan Yong Hong, Chol On Kim, Hongkee Kim, Jin Yong Kim, Nam Kyun Kim, Tai Kuen Kwak, Seog Hoon Rim and S. Tariq Rizvi for their advice and assistance which greatly contributed to the success of the Conference. We also appreciate Mr. Hogi Park and the staff of the Hyundai Hotel for their efficient arrangement of the facilities and accomodations and for providing an enjoyable environment for the Conference.

Finally, we must thank Ms. Youn Ju Park of Kyungpook National University for her excellent job of typing and uniformizing the manuscripts.

Gary F. Birkenmeier, Lafayette, U. S. A.
Jae Keol Park, Pusan, Korea
Young Soo Park, Taegu, Korea
Editors

February, 2000

Participants

Ricardo Alfaro, University of Michigan at Flint, Flint, MI 48502-1950, U. S. A.
Yoshitomo Baba, Osaka Kyoiku University, Kashiwara, Osaka 582-8582, Japan
K. Beidar, National Cheng Kung University, Tainan 701, Taiwan
Gary F. Birkenmeier, University of Louisiana at Lafayette, Lafayette, LA 70504-1010, U. S. A.
Jianlong Chen, Southeast University, Nanjing 210096, P. R. China
Miaosen Chen, Zhejiang Normal University, Jinhua, Zhejiang 321004, P. R. China
Zhizhong Chen, Northern Jiaotong University, Beijing 100044, P. R. China
Yong Uk Cho, Silla University, Pusan 617-736, Korea
Tae Ho Choe, McMaster University, Hamilton, ON L84 4K1, Canada
Jeong Ae Chun, Pusan National University, Pusan 609-735, Korea
J. Clark, University of Otago, P. O. Box 56, Dunedin, New Zealand
V. Dlab, Carleton University, Ottawa, Ontario K1S 5B6, Canada
Nanqing Ding, Nanjing University, Nanjing 210093, P. R. China
Xianneng Du, Anhui University, Heifei 230039, P. R. China
J. W. Fisher, University of Cincinnati, Cincinnati, OH 45221-0025, U. S. A.
Changlin Fu, Harbin University of Science and Technology, P. O. Box 603, Heilongjiang, P. R. China
Naoki Hamaguchi, Faculty of Environmental and Technical Science, Okayama University, Okayama 700, Japan
Juncheol Han, Kosin University, Pusan 606-701, Korea
Akihide Hanaki, Shinshu University, Matsumoto 390-8621, Japan
Dieter Happel, Fakultät fuer Mathematik, TU Chemnitz, D-09107 Chemnitz, Germany
Yasuyuki Hirano, Okayama University, Okayama 700, Japan
Chan Yong Hong, Kyung Hee University, Seoul 130-701, Korea
Osamu Iyama, Kyoto University, Kyoto 606-8502, Japan
S. K. Jain, Ohio University, Athens, OH 45702-2979, U. S. A.
Overtoun M. G. Jenda, Auburn University, Auburn, AL 36849-5307, U. S. A.
Young Cheol Jeon, Pusan National University, Pusan 609-735, Korea
Eric Jespers, Vrije University of Brussel, Pleinlaan 2, 1050 Brussel, Belgium

Jiro Kado, Osaka City University, Osaka 558-8585, Japan
F. Kasch, Universitaet München, D-80333 München, Germany
Boo Yoon Kim, Pusan National University, Pusan 609-735, Korea
Chol On Kim, Pusan National University, Pusan 609-735, Korea
Hongkee Kim, Gyeongsang National University, JinJu 660-701, Korea
Hwankoo Kim, Kyungpook National University, Taegu 702-701, Korea
Jin Yong Kim, Kyung Hee University, Suwon 449-701, Korea
Nam Kyun Kim, Kyung Hee University, Suwon 449-701, Korea
Shigeru Kobayashi, Department of Mathematics, Naruto University of Education, Naruto 772, Japan
Hiroaki Komatsu, Okayama Prefectural University, Okayama 719-1197, Japan
Yoshiki Kurata, Imaizumi 364-5, Hadano 257-0014, Japan
Yosuke Kuratomi, Yamaguchi University, Yamaguchi 753-8512, Japan
Mamoru Kutami, Yamaguchi University, Yamaguchi 753-8512, Japan
Dong J. Kwak, Kyungpook National University, Taegu 702-701, Korea
Tai Kuen Kwak, Daejin University, Pochun 487-800, Korea
Chong Yun Lee, Kyungnam University, Masan 631-701, Korea
Hyung Kyi Lee, Pusan National University, Pusan 609-735, Korea
Sang Bum Lee, Sangmyung University, Seoul 110-743, Korea
Yang Lee, Pusan National University, Pusan 609-735, Korea
Jae Kook Lee, Kyungpook National University, Taegu 702-701, Korea
Ya-nan Lin, Xiamen University, Xiamen 361005, P. R. China
Guilong Liu, China University of Geosciences, Beijing 100083, P. R. China
Hidetoshi Marubayashi, Naruto University of Education, Naruto 772, Japan
Jun Minamoto, Osaka City University, Osaka 558-8585, Japan
Shoji Morimoto, Hagi Koen Gakuin, Hagi, Yamaguchi 758-0047, Japan
Kaoru Motose, Hirosaki University, Hirosaki 036, Japan
Hiroshi Nagase, Osaka City University, Osaka 558-8585, Japan
Atsushi Nakajima, Okayama University, Okayama 700, Japan
Sang Bok Nam, Kyungdong University, Kosung 219-830, Korea
Kenji Nishida, Shinshu University, Matsumoto 390-8621, Japan
Tsunekazu Nishinaka, Okayama Shouka University, Okayama 700, Japan
Sei-Qwon Oh, Chungnam National University, Taejun 305-764, Korea
Kiyoichi Oshiro, Yamaguchi University, Yamaguchi 753-8512, Japan
J. Osterburg, University of Cincinnati, Cincinnati, OH 45221-0025, U. S. A.
F. Van Oystaeyen, University of Antwerp, Antwerp, Belgium

Participants

Jae Keol Park, Pusan National University, Pusan 609-735, Korea
Jeanam Park, Inha University, Incheon 402-751, Korea
Sangwon Park, Dong-A University, Pusan 604-714, Korea
Young Soo Park, Kyungpook National University, Taegu 702-701, Korea
Edmund Puczyłowski, University of Warsaw, 02-097 Warsaw, Banacha 2, Poland
K. M. Rangaswamy, University of Colorado, Colorado Springs, CO 80933-7150, U. S. A.
Seog Hoon Rim, Kyungpook National University, Taegu 702-701, Korea
S. Tariq Rizvi, Ohio State University at Lima, Lima, OH 45804-3576, U. S. A.
K. W. Roggenkamp, Universitaet Stuttgart, D-70550 Stuttgart, Germany
Masahisa Sato, Yamanashi University, Yamanashi 400-8511, Japan
Tae Young Seo, Pusan National University, Pusan 609-735, Korea
Keehong Song, Pusan National University, Pusan 609-735, Korea
Mirela Stefanescu, Ovidius University, 8700 Constanta, Romania
Chang Whan Suh, Pusan National University, Pusan 609-735, Korea
Takao Sumiyama, Aichi Institute of Technology, Toyota 470-03, Japan
Mark L. Teply, University of Wisconsin at Milwaukee, Milwaukee, WI 53201-0413, U. S. A.
H. Tsunashima, Yamaguchi University, Yamaguchi 753-8512, Japan
K. Varadarajan, University of Calgary, Calgary, Alberta T2N 1N4, Canada
Mingyi Wang, Southwest Jiaotong University, Chengdu 610031, P. R. China
Zhixi Wang, Capital Normal University, Beijing 100037, P. R. China
Hyun Sook Wu, Pusan National University, Pusan 609-735, Korea
K. Yamagata, Tokyo University of Agriculture and Technology, Tokyo 183-0054, Japan
Kenji Yokogawa, Okayama Science University, Okayama 700, Japan
Hiroshi Yoshimura, Yamaguchi University, Yamaguchi 753-8512, Japan
Yuji Yoshino, Okayama University, Okayama 700-8530, Japan
Suk Bong Yoon, Kyungpook National University, Taegu 702-701, Korea
M. F. Yousif, Ohio State University at Lima, Lima, OH 45804-3576, U. S. A.
Pu Zhang, University of Science and Technology of China, Hefei 230026, P. R. China

Yinhuo Zhang, Max-Planck Institute of Mathematics, Vivatsgasse 7, D-53111 Bonn, Germany

Alexander Zimmermann, Universite de Picardie, Jules Verne 33 rue Saint Leu, F-80039 France

Generalized Deviations of Posets with Applications to Chain Conditions on Modules

Toma Albu and Mark L. Teply

1. Introduction

The study of generalized deviations of a partially ordered set has its roots in the study of the Krull dimension of rings and modules. The concept of Krull dimension for commutative rings was developed by E. Noether and W. Krull in the 1920s. In 1923 E. Noether [27] explored the relationship between chains of prime ideals and dimensions of algebraic varieties. After five years, W. Krull [23] developed her idea into a powerful tool for arbitrary commutative rings satisfying the ascending chain condition for ideals. These rings are known today as Noetherian rings. Later, algebraists gave the name (classical) Krull dimension to the supremum of the length of finite chains of prime ideals in a ring.

Subsequently, the notion of the Krull dimension of a module as an object in an abelian category or poset emerged. In his famous Ph.D.Thesis [13] published in 1962, P. Gabriel introduced for objects in an abelian category an ordinal-valued dimension, which he named Krull dimension, by using a transfinite sequence of localizing subcategories. R. Rentschler and P. Gabriel [30] introduced in 1967 the deviation of an arbitrary poset, but only for finite ordinals, which agrees for commutative Noetherian rings with Gabriel's definition. The ordinal-valued version of the Rentschler-Gabriel definition was introduced in 1970 by G. Krause [21] for modules over arbitrary rings. His definition was extended to arbitrary posets in 1972 by B. Lemonnier [25], who introduced the ordinal-valued version of the Rentschler-Gabriel definition of the deviation of an arbitrary poset P, called in the sequel the Krull dimension of P.

An important step in the development of the Krull dimension for module theory was the AMS Memoir of Gordon and Robson [16]. They gave the name *Gabriel dimension* to Gabriel's original definition, after shifting the finite values by 1, and provided an incisive investigation of the Krull dimension of modules and rings. Roughly speaking, a module has Krull dimension $\leq \alpha$ in case any descending chain of submodules with "large" factors must terminate, where "large" means that the factors are not of Krull dimension $< \alpha$. Thus an Artinian module has Krull dimension ≤ 0, and the ring of integers, having no proper non-Artinian factors, has Krull dimension 1. More precisely, Gordon and Robson [16] gave the following inductive definition for a module M over a ring R with identity element to have Krull dimension. The *Krull dimension* of M, which will be denoted by $k(M)$, is defined by transfinite recursion as follows: if $M = 0$, $k(M) = -1$; if α is an ordinal and $k(M) \not< \alpha$, then $k(M) = \alpha$ provided there is no infinite descending chain $M = M_0 \supset M_1 \supset \ldots$ of submodules M_i such that, for $i = 1, 2, \ldots$, $k(M_{i-1}/M_i) \not< \alpha$. It is possible that there is no ordinal α such that $k(M) = \alpha$. In that case we say M has no Krull dimension. If R is a commutative Noetherian ring, the Krull dimension of the module R coincides with the classical Krull dimension of R defined by the supremum of the lengths of finite descending chains of prime ideals, in case the dimension is finite.

A considerable amount of work was done to relate the Krull dimension of a module to the length of chains of its submodules. In particular, Bass [8] worked on this question in the setting of commutative Noetherian rings. Then, Gulliksen [17] and Krause [22] extended Bass's work to the noncommutative ring situation.

Another way to view the Krull dimension of a module is that the Krull dimension measures how close a module is to being Artinian. This is done by seeing how many chains of order type ω^* (the order type of the negative integers) can be stuffed into a module. This viewpoint is the starting point for the general concept of deviation, where the order type ω^* is replaced by other order types or sets of order types.

Set theorists have studied order types for many years; e.g., see [14] and [33]. But the first to explore replacing ω^* by another order type and relating it to Krull dimension was Lemonnier [25]. He replaced ω^* by the order type ω (the order type of the positive integers) and formed a new dimension, which he called dual Krull dimension. He showed that a poset has Krull dimension if and only if it has dual Krull dimension; these dimensions may differ, and their relationship is still not understood very well.

Goodearl and Zimmermann-Huisgen [15] developed non-Noetherian results for their study of direct products of torsion modules over Ore domains. Their results are limited to the situation in which the Krull dimension of the module is countable. Examples are cited which indicate that their results cannot be extended when the Krull dimension is uncountable. To deal with the uncountable case, more general theories of Krull dimension were needed. These concepts were also developed by set theorists in the course of their study of deviation.

In [29] Pouzet and Zaguia defined the deviation of a partially ordered set (poset) P with respect to a linear order type γ (see Section 3 for details). When $\gamma = \omega^*$, this is precisely the deviation $k(P)$ of P, introduced by Lemonnier [25], and which will be called in the sequel the Krull dimension of P. Lemonnier [25] studied ω^*- and ω-deviation and characterized posets without deviation as the posets that have a chain of order type η of the rational numbers. Can this characterization be extended to arbitrary γ-deviation? Using the concept of a γ-dense set, Lau, Teply, and Boyle [24] answered this question for linearly ordered sets. The general answer for posets, or even modular lattices, is not known yet.

The poset setting for deviation (or generalized Krull dimension) has spawned a considerable number of investigations. In the poset setting, results on Krull dimension can be developed in their most natural setting and then be applied to other situations. In this exposition, we highlight two of our recent papers [6, 7] that study Krull dimension in the poset setting. Our results here are given without proof; the reader can consult [6, 7] for proofs.

Section 2 gives the terminology and notation that we use. Sections 3 and 4 are based on [6]; they give some existence and computational results for Krull dimension in general poset and modular lattice settings. Sections 5 and 6 are based on [7], which in turn was motivated by papers of Contessa [9, 10, 11], Osofsky [28], Karamzadeh and Motamedi [20], and Albu and Smith [4, 5]. Section 6 shows how our general results can be applied to study Grothendieck categories and torsion theories for modules over a ring with identity element.

2. Terminology and Notation

In this section we fix the notation used throughout this exposition and recall some basic definitions.

By a *poset* we mean a partially ordered set. Throughout this paper $\mathcal{P}$ will denote the class of all posets. If $x \leq y$ are elements in a poset $(P, \leq)$, then y/x will denote the interval $[x, y]$; i.e.,

$$y/x := \{\, a \in P \mid x \leq a \leq y \,\}.$$

Also, for any $z \in P$ we shall use the notation

$$(\leftarrow, z] := \{a \in P \mid a \leq z\}, \quad (\leftarrow, z) := \{a \in P \mid a < z\},$$

$$[z, \rightarrow) := \{a \in P \mid z \leq a\}, \quad (z, \rightarrow) := \{a \in P \mid z < a\}.$$

A poset is called *trivial* if it has no two distinct comparable elements (such posets are also known as *antichains*, cf. [31]). A *linearly ordered poset*, or a *chain*, is a poset $(C, \leq)$ having the following property: for any $x, y \in C$, $x \leq y$ or $y \leq x$. A poset P is called *dense* if it has at least two elements, and if for any $x < y$ in P there is a $z \in P$ such that $x < z < y$. The opposite poset of a poset P will be denoted by P^0.

Recall now some other basic definitions. Let P and Q be two posets, and let $f : P \to Q$ be a map. Then f is said to be *increasing* (resp. *strictly increasing*) if for any $x_1 \leq x_2$ in P one has $f(x_1) \leq f(x_2)$ (resp. if for any $x_1 < x_2$ in P one has $f(x_1) < f(x_2)$); f is an *isomorphism* in case f is (strictly) increasing, bijective and its inverse map f^{-1} is also increasing. The posets P and Q are called *isomorphic* if there exists an isomorphism $f : P \to Q$; we will designate this situation by $P \simeq Q$.

Two posets P and Q are said to have the *same (order) type* if $P \simeq Q$; i.e., P and Q are isomorphic posets.

By an *order type* we mean an "equivalence class" of $\mathcal{P}$ with respect to the "equivalence relation" $\simeq$ in $\mathcal{P}$. Throughout this exposition we employ Greek letters to denote linear order types. If α is an order type, then α^* will denote the order type of P^0, where P is any poset of order type α. If α and β are two order types, then $\alpha + \beta$ and $\alpha\beta$ are the usual sum and product of order types (see e.g., [19] or [31]). If α and β are ordinal numbers, then α^β denotes as usual the ordinal power (see e.g., [19]).

In the sequel we shall use the following traditional notation: ω for the order type of the set $\mathbb{N} = \{1, 2, \ldots\}$ of natural numbers, ζ for the order type of the set $\mathbb{Z}$ of all rational integers, η for the order type of the set $\mathbb{Q}$ of all rational numbers, and λ for the order type of the set $\mathbb{R}$ of all real numbers. The considered order in each of the four sets mentioned above is the usual one.

If α is an ordinal number, then a poset P is said to be *of type α* in case the posets P and $[0, \alpha)$ have the same type, where $[0, \alpha)$ denotes the set of all ordinal numbers β with $0 \leq \beta < \alpha$. The poset P is said to be *of type α^** if P and $[0, \alpha)^0$ have the same type. For any ordinal $\xi \geq 0$, we shall denote by ω_ξ the ξ-th initial ordinal. (See [18, Chapter 8]).

As in [29] or [31], a poset P is said to be *dispersed* or *scattered* in case P does not contain an isomorphic copy of $\mathbb{Q}$; we shall denote this situation by writing $\mathbb{Q} \not\leq P$. In general, if P and Q are two arbitrary posets, then the notation $P \leq Q$ will be used in case P is isomorphic to a subposet of Q endowed with the induced ordering; in this case one says that P "is sheltered in" Q (in French, P "s'abrite dans" Q, cf. [29]). For the contrary case we shall use the notation $P \not\leq Q$. Very frequently, an ordinal number α will be identified with the interval $[0, \alpha)$ of ordinal numbers.

A *linear order type* is an order type of a chain. So any ordinal number, being an order type of a well-ordered poset, is a linear order type, but not conversely.

If γ is a (linear) order type, then $|\gamma|$ will denote the cardinal number of any poset having order type γ.

Recall that a poset $(P, \leq)$ is *Noetherian* (or satisfies ACC) if there is no infinite strictly ascending chain $x_1 < x_2 < \cdots$ in P and *Artinian* (or satisfies DCC) if there is no infinite strictly descending chain $x_1 > x_2 > \cdots$ in P.

For all undefined notation and terminology on posets, the reader is referred to [18] or [31].

3. Generalized Deviation of Posets

The Krull dimension of a module is an important tool for the study of rings and their modules. Not only has a considerable amount of theory been developed for the Krull dimension and dual Krull dimension of modules (e.g., [16] or [26]), but also this dimension idea has given rise to a corresponding theory for partially ordered sets (e.g., see [1], [5], [6], [7], [24], [25], [29], or [34]). The most general definitions of this type are usually called Γ-*deviation* or Γ-*Krull dimension* in the literature, where Γ is a nonempty set of linear order types. The aim of this section is to study the properties of Γ-deviation and dual Γ-deviation.

We start by presenting the notion of Γ-*deviation* of an arbitrary poset $(P, \leq)$ in a slightly modified manner from the one given in [29].

Definition 3.1. Let $(P, \leq)$ be an arbitrary poset and Γ an arbitrary nonempty set of linear order types. The Γ-*deviation* of $(P, \leq)$, also called Γ-*Krull dimension* of $(P, \leq)$ and denoted in the sequel by $k_\Gamma(P)$, is an ordinal number defined recursively as follows:

$k_\Gamma(P) = -1$, where -1 is assumed to be the predecessor of zero, if and only if P is a trivial poset.

$k_\Gamma(P) = 0$ if and only if P is not trivial and contains no chain of order type γ for any $\gamma \in \Gamma$.

Let $\alpha > 0$ be an ordinal number and assume that we have already defined which posets have Γ-deviation β for any ordinal $\beta < \alpha$. Then we say that $k_\Gamma(P) = \alpha$ if and only if we do not have $k_\Gamma(P) = \beta$ for any ordinal $\beta < \alpha$, and for any $\gamma \in \Gamma$ and any chain C of P of type γ, there exist $a < b$ in C such that b/a, considered as an interval in P, has $k_\Gamma(b/a) = \beta$ for some $\beta < \alpha$.

In case $\Gamma = \{\gamma\}$, then instead of $k_{\{\gamma\}}(P)$, we shall simply write $k_\gamma(P)$, and call it the γ-*deviation* of P.

As noticed in [29, Proposition 2.4], the concept of Γ-deviation of a poset is a natural extension of the usual concept of Krull dimension. For a poset P we shall denote by $k(P)$ (resp. $k^0(P)$) the usual *Krull dimension* or *deviation* (resp. the usual *dual Krull dimension* or *codeviation*) of P (see e.g., [5] for the definitions of these notions).

Proposition 3.2. *Let $P \in \mathcal{P}$ and let α be an ordinal. Then the following statements hold:*

(1) *P has ω-deviation and $k_\omega(P) = \alpha \iff P$ has dual Krull dimension and $k^0(P) = \alpha$.*

(2) *P has ω^*-deviation and $k_{\omega^*}(P) = \alpha \iff P$ has Krull dimension and $k(P) = \alpha$.*

The *dual Γ-deviation* of a poset P, denoted by $k_\Gamma^0(P)$, is defined as being (if it exists !) the Γ-deviation $k_\Gamma(P^0)$ of the opposite poset P^0 of P. If we let
$$\Gamma^0 = \{\gamma^* \mid \gamma \in \Gamma\},$$
then it is clear that
$$k_\Gamma^0(P) = k_\Gamma(P^0) = k_{\Gamma^0}(P) \text{ and } k_{\Gamma^0}^0(P) = k_{\Gamma^0}(P^0) = k_\Gamma(P).$$

Note that in case Γ contains an order type γ with $|\gamma| \in \{0, 1, 2\}$, then the only (nonempty) posets having Γ-deviation are the trivial ones, and in that case their Γ-deviation is -1, by definition (see also [29, p.177]). So, to avoid trivialities, we shall assume in the sequel that $|\gamma| \geq 3$ for all $\gamma \in \Gamma$.

Sometimes the presence of a greatest element 1 and a least element 0 is important. If P fails to have them, then it is convenient to enlarge P by adjoining such elements to obtain a new poset P'. By [26, Lemma 6.1.15], if $k(P) \geq 0$, then $k(P) = k(P')$, and if $k(P) = -1$, then $k(P') = 0$. This property does not hold in general for Γ-deviation, as the following examples show:

Examples 3.3. (1) Let $P = \omega$ and $\gamma = \omega + 1$. Then P has no greatest element, and $P' = \omega + 1$; so $k_\gamma(P) = 0$, but $k_\gamma(P') = 1$.

(2) Let $P = \omega^*$ and $\gamma = (\omega + 1)^*$. Then P has no least element, and $P' = (\omega + 1)^*$; so $k_\gamma(P) = 0$, but $k_\gamma(P') = 1$.

(3) Let $P = \omega^* + \omega$ and $\gamma = (\omega + 1)^* + (\omega + 1)$. Then $P' = (\omega + 1)^* + (\omega + 1)$; so $k_\gamma(P) = 0$, but $k_\gamma(P') = 1$.

However, if Γ consists only of indecomposable linear order types we will see below that a result similar to [26, Lemma 6.1.15] holds. As in [31, Definition 10.5], a linear order type γ is said to be *indecomposable* if, whenever we partition a chain C of order type γ as $C = A \cup B$ with $A \cap B = \emptyset$, then either $\gamma \leq \alpha$ or $\gamma \leq \beta$, where α (resp. β) is the order type of A (resp. B). Notice that in [29] the term *impartible* is used for this concept. As in [31, Definition 10.1], a linear order type γ is called *additively indecomposable* if, whenever $\gamma = \alpha + \beta$ with α and β linear order types, then either $\gamma \leq \alpha$ or $\gamma \leq \beta$.

Clearly, any indecomposable order type is additively indecomposable, but not conversely; λ, the order type of $\mathbb{R}$, is additively indecomposable, but not indecomposable (see [31, p. 177]).

Proposition 3.4. *Assume that Γ consists only of indecomposable linear order types. Let P be an arbitrary poset which has neither a least nor a greatest element, and denote by P' the poset obtained by adjoining to P a least and a greatest element. If $k_\Gamma(P) \geq 0$, then $k_\Gamma(P) = k_\Gamma(P')$, and if $k_\Gamma(P) = -1$, then $k_\Gamma(P') = 0$.*

We are going now to examine when a poset P having Γ-deviation also has dual Γ-deviation. For $\Gamma = \{\omega\}$ or $\Gamma = \{\omega^*\}$, this always happens by [25, Corollaire 6]. For that, the next result is fundamental.

Lemma 3.5. [29, Proposition 2.2] *If P is a scattered poset, then P has Γ-deviation. The converse is true if Γ contains an order type γ with $|\gamma| \leq \aleph_0$.*

This result is the best one that could be expected. Indeed, if Γ fails to contain an order type γ with $|\gamma| \leq \aleph_0$, then it could be possible for a poset P to have Γ-deviation without being scattered. To see this, take $\Gamma = \{\lambda\}$ and $P = \mathbb{Q}$. Then P has Γ-Krull dimension 0, since $\lambda \not\leq \eta$, but $\mathbb{Q} \leq P$.

Proposition 3.6. *Suppose that Γ contains an order type γ with $|\gamma| \leq \aleph_0$. Then a poset P has Γ-deviation if and only if it has dual Γ-deviation.*

In case Γ does not contain any γ with $|\gamma| \leq \aleph_0$, then this result may fail, as the following remark from [29, p. 261] shows: for any ordinal $\xi > 0$ there exists a poset that has ω_ξ-deviation but does not have ω_ξ^*-deviation. We can ask if this happens also for any uncountable order type γ, or more generally, for any nonempty set Γ containing no finite or countable order type.

Corollary 3.7. [25, Corollaire 6] *Let P be an arbitrary poset. Then P has Krull dimension if and only if P has dual Krull dimension.*

From Lemma 3.5 it is clear that a poset P fails to have Krull dimension if and only if P contains a chain of order type η. Our next two results examine this direction further by using the density property of η.

Definition 3.8. A poset P is said to be Γ-*dense* if each of its proper intervals contains a chain of order type γ for some $\gamma \in \Gamma$.

The existence of Γ-deviation can be studied by using Γ-dense sets.

Proposition 3.9. *If a poset P contains a Γ-dense poset, then the Γ-deviation of P fails to exist.*

Proposition 3.10. [24, Theorem 1.4] *Let P be a linearly ordered set, and let γ be a linear order type. Then the γ-deviation of P fails to exist if and only if P contains a γ-dense chain.*

We conjecture that this result holds for any modular lattice with 0 and 1 and for any set Γ of linear order types instead of a single order type γ.

It seems very natural to ask about the relationship between the Γ-deviation $k_\Gamma(P)$ of a poset P and the γ-deviations $k_\gamma(P)$ for $\gamma \in \Gamma$. If P has Γ-deviation, then by [29, Proposition 2.6], $k_\gamma(P)$ exists for each $\gamma \in \Gamma$ and

$$\sup\{\,k_\gamma(P)\,|\,\gamma \in \Gamma\,\} \leq k_\Gamma(P)\,.$$

In order to examine the situation in which each $k_\gamma(P)$ exists ($\gamma \in \Gamma$), we need to introduce some terminology and notation.

Let $\{\delta_1, \delta_2, \ldots, \delta_n\}$ be a finite set of ordinals. Write each of these ordinals as a "Cantor sum", called also the *normal form*

$$\delta_i = \omega^{\alpha_{i1}}\beta_{i1} + \omega^{\alpha_{i2}}\beta_{i2} + \cdots + \omega^{\alpha_{it_i}}\beta_{it_i}\,,$$

where $\alpha_{i1} > \alpha_{i2} > \cdots > \alpha_{it} \geq 0$ and $0 \leq \beta_{ij} < \omega$ for each j. (See [19, p. 107] or [32, p. 320]). By using $\beta_{ij} = 0$ as necessary, we may assume that

$$\delta_i = \omega^{\alpha_1}\beta_{i1} + \omega^{\alpha_2}\beta_{i2} + \cdots + \omega^{\alpha_t}\beta_{it}\,,$$

where $\alpha_1 > \alpha_2 > \cdots > \alpha_t \geq 0$ and each $0 \leq \beta_{ij} < \omega$.

We define the *natural sum*, called also the *Hessenberg sum* (see [19, p. 109] or [32, p. 364]), $H(\delta_1, \delta_2, \ldots, \delta_n)$ of $\delta_1, \delta_2, \ldots \delta_n$ by

$$H(\delta_1, \delta_2, \ldots, \delta_n) = \omega^{\alpha_1}\left(\sum_{i=1}^{n}\beta_{i1}\right) + \omega^{\alpha_2}\left(\sum_{i=1}^{n}\beta_{i2}\right) + \cdots + \omega^{\alpha_t}\left(\sum_{i=1}^{n}\beta_{it}\right).$$

Now we are ready to state the result that determines how the existence of each $k_\gamma(P)$ affects the existence and magnitude of $k_\Gamma(P)$ when Γ is a finite set.

Theorem 3.11. *Let* $\Gamma = \{\gamma_1, \gamma_2, \ldots, \gamma_n\}$ *be a finite set of linear order types and let* P *be a poset such that* $k_{\gamma_i}(P) = \delta_i$ *for* $1 \leq i \leq n$. *Then* $k_\Gamma(P)$ *exists, and*

$$\max\{\delta_1, \delta_2, \ldots, \delta_n\} \leq k_\Gamma(P) \leq H(\delta_1, \delta_2, \ldots, \delta_n).$$

It is trivial to find examples in which $k_\Gamma(P)$ attains the lower bound in Theorem 3.11. For example, $k_{\{\omega,\omega^*\}}(\omega) = 1 = k_\omega(\omega) > k_{\omega^*}(\omega) = 0$.

Even in the finite case, $k_\Gamma(P)$ can attain the upper bound of Theorem 3.11. We can see this from the following examples.

Example 3.12. Let $P = (\omega^m)^* \times \omega^n$ be a poset with the lexicographical order, where $m, n < \omega$. Then $k_{\omega^*}(P) = m$, $k_\omega(P) = n$, and $k_{\{\omega,\omega^*\}}(P) = m + n$.

Example 3.13. Let $P = \omega^* \times \omega \times (\omega^\omega)^* \times \omega^\omega$ be a poset with the lexicographical order. Then $k_\omega(P) = \omega + 1$, $k_{\omega^*}(P) = \omega + 1$, and $k_{\{\omega,\omega^*\}}(P) = \omega 2 + 2$.

The method of construction in Example 3.13 can be used to construct examples of posets P that attain the maximum value of $k_\Gamma(P)$ in Theorem 3.11 in case $\Gamma = \{\omega, \omega^*\}$, $k_\omega(P) = \delta_1$, and $k_{\omega^*}(P) = \delta_2$, where δ_1 and δ_2 are arbitrary ordinals.

4. Γ-Deviation of Modular Lattices

The aim of this section is to investigate the Γ-deviation of modular lattices. We will show that in case the set Γ of linear order types consists only of indecomposable ones, then a basic property of the usual Krull dimension of posets holds also for Γ-deviation.

Throughout this section we shall use $\mathcal{M}$ to denote the class of all modular lattices with 0 and 1.

Proposition 4.1. *Let* $L \in \mathcal{M}$, *let* $a \in L$, *and assume that* Γ *contains an order type* γ *with* $|\gamma| \leq \aleph_0$. *Then* L *has* Γ-*deviation if and only if* $a/0$ *and* $1/a$ *both have* Γ-*deviation.*

Remarks 4.2. (1) We will see in Proposition 4.4 that the condition "Γ contains an order type γ with $|\gamma| \leq \aleph_0$" is not necessary in Proposition 4.1: it can be replaced by the condition "Γ *consists only of indecomposable linear order types*".

(2) Proposition 4.1 fails for a non-modular lattice, even for the usual Krull dimension, as the following simple example shows:

Let L be the poset that is the disjoint union of the set $\mathbb{Q}$ with a set $\{a_0, a, a_1\}$ having exactly three elements. We order the set L by using the usual order of $\mathbb{Q}$,

$$a_0 < a < a_1, \; a_0 < q < a_1, \; \forall q \in \mathbb{Q},$$

and no other distinct elements of L are comparable. With this order L becomes a lattice with greatest element a_1 and least element a_0, and L contains at least one pentagon (in fact many!). By [12, 3.2], it follows that L is not modular. We have $k(a/a_0) = 0$ and $k(a_1/a) = 0$, but $k(L)$ does not exist by Lemma 3.5, since $\mathbb{Q} \subseteq L$.

The next fact is needed to establish Proposition 4.4, which is the main result of this section.

Lemma 4.3. [29, Lemme 4.3] *Let Γ be a nonempty set of indecomposable linear order types, and let P_1 and P_2 be two posets. Then*

$$k_\Gamma(P_1 \times P_2) = \max\{\, k_\Gamma(P_1), k_\Gamma(P_2)\,\},$$

if either side exists, where $P_1 \times P_2$ is ordered componentwise.

Proposition 4.4. *Let Γ be an arbitrary nonempty set of indecomposable linear order types, $L \in \mathcal{M}$, and $a \in L$. Then*

$$k_\Gamma(L) = \max\{\, k_\Gamma(1/a), k_\Gamma(a/0)\,\},$$

if either side exists.

The condition that *all* of the order types of Γ are indecomposable is essential in Proposition 4.4. Even when Γ reduces to a single additively indecomposable order type γ, the result can fail. An example showing this is given in [6].

5. The Double Infinite Chain Condition and Balanced Krull Dimension

In this section we introduce and investigate the concepts of *double infinite chain condition* of an arbitrary poset and its Krull-like dimension extension, which we call *balanced Krull dimension*.

As particular cases we obtain the concepts of a DICC module [9], of a DICC object in an AB5 category [28] and of an α-DICC module [20]. Part of the results from [9], [20] and [28] follow more easily and more naturally from our more general setting.

Inspired by [9] and [28], we introduce the following definition.

Definition 5.1. An arbitrary poset P is said to be *DICC* or to have the *double infinite chain condition* if any chain of elements of P indexed by the integers $\mathbb{Z}$ stabilizes either to the right or to the left, or to both sides; that is, for any chain

$$\cdots \leq x_{-2} \leq x_{-1} \leq x_0 \leq x_1 \leq x_2 \leq \cdots$$

of elements of P, there exists $m \in \mathbb{Z}$ such that $x_{i+1} = x_i$ for all $i \geq m$ or $x_{i+1} = x_i$ for all $i \leq m$.

This concept can be related with those of an Artinian and a Noetherian poset as follows:

Proposition 5.2. *The following assertions are equivalent for a poset P:*
(1) P *is DICC.*
(2) *For any $x \in P$, the poset $(\leftarrow, x]$ is Artinian or the poset $[x, \rightarrow)$ is Noetherian.*
(3) *Any nonempty subset of P has either a minimal or a maximal element.*
(4) *The poset P does not contain a chain of order type ζ.*
(5) *The poset P does not contain a chain of order type $\omega^* + \omega$.*
(6) $k_\zeta(P) \leq 0$.
(7) $k_{\omega^*+\omega}(P) \leq 0$.

Remarks 5.3. (1) The definition of a DICC poset can be given by replacing chains of type

$$\cdots \leq x_{-2} \leq x_{-1} \leq x_0 \leq x_1 \leq x_2 \leq \cdots$$

with chains of type

$$\cdots \geq x'_{-2} \geq x'_{-1} \geq x'_0 \geq x'_1 \geq x'_2 \geq \cdots$$

by putting $x'_i = x_{-i}$ for each $i \in \mathbb{Z}$.

(2) A condition symmetric to that of (2) from Proposition 5.2, namely

For any $x \in P$, the poset $(\leftarrow, x]$ is Noetherian or the poset $[x, \rightarrow)$ is Artinian, is equivalent to

The poset P does not contain a chain of order type $\omega + 1 + \omega^*$,
or equivalently,

$$k_{\omega+1+\omega^*}(P) \leq 0.$$

(3) Condition (2) from Proposition 5.2 can be reformulated in terms of Krull dimension and dual Krull dimension as follows:

For any $x \in P$ the poset $(\leftarrow, x]$ has Krull dimension ≤ 0 or the poset $[x, \rightarrow)$ has dual Krull dimension ≤ 0.

This is related to the following result, which is a poset version of a module result [20, Proposition 4]:

Let P be an arbitrary poset. Then P has (dual) Krull dimension if and only if for each $x \in P$, either $(\leftarrow, x]$ or $[x, \rightarrow)$ has (dual) Krull dimension.

As in [9], a module M_R is said to be a *DICC module* in case any double infinite chain of submodules of M stabilizes either to the right or to the left, or to both sides; that is, for any chain

$$\cdots \leq X_{-2} \leq X_{-1} \leq X_0 \leq X_1 \leq X_2 \leq \cdots$$

of submodules of M, there exists $m \in \mathbb{Z}$ such that $X_{i+1} = X_i$ for all $i \geq m$ or $X_{i+1} = X_i$ for all $i \leq m$. This means precisely that M_R is a DICC module if the lattice $\mathcal{L}(M_R)$ of all submodules of M_R is a DICC poset. More generally, if $\mathcal{G}$ is a Grothendieck category, then an object $X \in \mathcal{G}$ is a *DICC object* (see [28]) if the lattice $\mathcal{L}(X)$ of all its subobjects is a DICC poset.

A Krull-like dimension extension of the concept of a DICC module has been provided in [20] by means of that of an α-DICC module. Below we introduce a more general concept for arbitrary posets. Then we shall relate this notion with that of the ζ-deviation of a poset.

Definition 5.4. Let P be an arbitrary poset and let $\alpha \geq 0$ be an ordinal number. Then P is said *to have balanced* (or *symmetric*) *Krull dimension* α, denoted as $bk(P) = \alpha$, or to be α–*DICC*, if α is the least ordinal such that for any double infinite chain

$$\cdots \leq x_{-2} \leq x_{-1} \leq x_0 \leq x_1 \leq x_2 \leq \cdots$$

of elements of P, there exists $m \in \mathbb{Z}$ such that $k^0(x_{i+1}/x_i) < \alpha$ for all $i \geq m$ or $k(x_{i+1}/x_i) < \alpha$ for all $i \leq m$.

By definition, $bk(P) = -1$ if and only if P is a trivial poset.

Clearly, the 0-DICC posets are the posets that are nontrivial and DICC. Also, the balanced (or symmetric) Krull dimension is indeed *symmetric*: for any poset P, $bk(P) = bk(P^0)$, if either side exists.

Proposition 5.5. *Let P be an arbitrary poset and let α be an ordinal number. Then $bk(P) = \alpha$ if and only if α is the least ordinal such that for all $x \in P$ one has $k((\leftarrow, x]) \leq \alpha$ or $k^0([x, \rightarrow)) \leq \alpha$.*

The next result was established in [28] for DICC objects in Grothendieck categories and in [20] for α-DICC modules. It can also be proved by a lattice-theoretic adaptation of the module-theoretic proof of [20, Theorem 1.2].

Proposition 5.6. *An upper continuous modular lattice L is α–DICC if and only if α is the least ordinal such that there exists $x \in L$ satisfying $k(x/0) \leq \alpha$, $k^0(1/x) \leq \alpha$, and $k^0(1/(x \wedge y)) \leq \alpha$ (or $k^0(x/(x \wedge y)) \leq \alpha$) for any $y \in L$ with $y \not\leq x$.*

Corollary 5.7. *An upper continuous modular lattice L is DICC if and only if there exists $x \in L$ such that $x/0$ is Artinian, $1/x$ is Noetherian, and $x/(x \wedge y)$ is Noetherian for any $y \in L$ with $y \not\leq x$.*

As known, if L is a modular lattice with 0 and 1 and if $a \in L$, then L is Artinian (resp. Noetherian) if and only if both $a/0$ and $1/a$ are Artinian (resp. Noetherian). This is also known to hold for the usual Krull dimension (resp. dual Krull dimension): one has

$$k(L) = \sup\{k(1/a), k(a/0)\} \; (\text{resp.} k^0(L) = \sup\{k^0(1/a), k^0(a/0)\})$$

if either side exists. (See [26] and Proposition 4.4).

However, a similar result does not hold for balanced Krull dimension, as the following example shows. Let L be the set $\mathbb{Z}$ of rational integers ordered in the usual way, enlarged by a greatest element g and by a least element l; then L is clearly a modular lattice, and for any $x \in \mathbb{Z}$, $bk(x/l) = 0$, $bk(g/x) = 0$, but $bk(L) = 1$.

The next result is a poset version of the corresponding one established in [20, Corollary 1.1].

Corollary 5.8. *The following conditions are equivalent for a poset P:*
  (1) *P has Krull dimension.*
  (2) *P has dual Krull dimension.*
  (3) *P has ζ-deviation.*
  (4) *P has balanced Krull dimension.*
In this case $k_\zeta(P) \leq bk(P) \leq \min\{k(P), k^0(P)\}$.

The first inequality in Corollary 5.8 can vary rather arbitrarily, as shown in the constructions below. For the reader's convenience we first list some known facts that will be used in our constructions. (See [32, p. 320], [25, Proposition 10], and [26, Proposition 6.1.9].)

Lemma 5.9. (1) *For each ordinal $\alpha \geq 0$,*
$$k([0, \omega^\alpha]^0) = k([0, \omega^\alpha)^0) = k((\omega^\alpha)^*) = \alpha.$$

(2) *Each ordinal δ can be uniquely written as a Cantor sum, called also the normal form*
$$\delta = \omega^{\alpha_1}\beta_1 + \omega^{\alpha_2}\beta_2 + \cdots + \omega^{\alpha_t}\beta_t,$$
where $\alpha_1 > \alpha_2 > \cdots > \alpha_t \geq 0$ and $0 < \beta_j < \omega$ for each j. Moreover, $k([0, \delta]^0) = k([0, \delta)^0) = k((\omega^{\alpha_1})^*) = \alpha_1$.

Let β be an ordinal, $\beta \geq 1$. Let H_β be the set of all formal "tuples"
$$h = (h_0, h_1, h_2, \ldots)$$
such that

(1) the entries (subscripts) of h are indexed by the ordinal interval $[0, \beta)$,

(2) each $h_i \in \mathbb{Z} \cup \{-\infty\}$,

(3) each "tuple" h has only finitely many entries different from $-\infty$, and

(4) each "tuple" h has at least one entry from $\mathbb{Z}$.

We order H_β with the reverse lexicographical order; this is the lexicographical order but *Hebrew-style*, from right to left. As usual, we consider that $-\infty < n$ for any $n \in \mathbb{Z}$.

Proposition 5.10. *For any $\beta \geq 1$,*
$$k_\zeta(H_\beta) = bk(H_\beta) = k(H_\beta) = k^0(H_\beta) = \beta.$$

To state Proposition 5.12, we need the following result.

Proposition 5.11. *For any ordinal $\alpha \geq 1$, consider the poset*
$$D_\alpha = (\omega^\alpha)^* + \omega^\alpha.$$
Then $k_\zeta(D_\alpha) = 1$ and $bk(D_\alpha) = \alpha$.

Let $1 \leq \beta < \alpha$ be ordinals. Define $P_{\alpha, \beta} = H_\beta + D_\alpha$; i.e., $P_{\alpha, \beta}$ is the linearly ordered set for which H_β and D_α have their previously defined orders and for which $h < d$ for every $h \in H_\beta$ and $d \in D_\alpha$.

Proposition 5.12. *For $1 \leq \beta < \alpha$, $k_\zeta(P_{\alpha,\beta}) = \beta$ and $bk(P_{\alpha,\beta}) = \alpha$.*

Notice that the ordinal ζ, which plays a fundamental role in the considerations above, has the property that $\zeta = \zeta^*$; i.e., it enjoys a sort of "symmetry" property. This leads us to the following definition.

Definition 5.13. An order type γ is said to be *symmetric* if $\gamma = \gamma^*$.

By [31, Exercise 1.45, p. 23], a linear order type γ is symmetric if and only if there exists another linear order type δ such that either $\gamma = \delta^* + \delta$ or $\gamma = \delta^* + 1 + \delta$. Clearly ζ, η and λ, the order type of the usually ordered set $\mathbb{R}$ of all real numbers, are symmetric order types. Moreover, $\zeta = \omega^* + \omega$, $\eta = \eta^* + 1 + \eta$, and $\lambda = \lambda^* + 1 + \lambda$.

Definition 5.14. Let Γ be a nonempty set of linear order types. The *symmetrization* of Γ is the set

$$\widetilde{\Gamma} = \{\gamma^* + 1 + \gamma \mid \gamma \in \Gamma\}.$$

In the sequel, for any order type γ we shall denote by $\widetilde{\gamma}$ the symmetric order type $\gamma^* + 1 + \gamma$. Clearly $\widetilde{\{\gamma\}} = \{\widetilde{\gamma}\}$ and $\widetilde{\omega} = \omega^* + 1 + \omega = \omega^* + \omega = \zeta$.

In this section we extend the concepts and results of the previous work by replacing the special ordinal ω with an arbitrary linear order type γ or, more generally, with a nonempty set Γ of linear order types.

Let C be a chain of order type $\widetilde{\gamma}$ of a poset P. Then C can be written as

$$C = C_- \cup \{c_0\} \cup C_+,$$

where C_- (resp. C_+) is a chain in P of order type γ^* (resp. γ), $c_0 \in P$, and

$$C_- < c_0 < C_+.$$

If X is a nonempty subset of P and $a \in P$, then we use the notation $X < a$ to mean that $x < a$ for all $x \in P$.

In the sequel we shall use this notation freely for any chain C of order type $\widetilde{\gamma}$ in a poset P.

Proposition 5.15. *The following assertions are equivalent for a poset P and a linear order type γ:*
  (1) $k_{\widetilde{\gamma}}(P) \leq 0$.
  (2) *For any $x \in P$, $k_\gamma^0((\leftarrow, x)) \leq 0$ or $k_\gamma((x, \rightarrow)) \leq 0$.*

Definition 5.16. Let P be an arbitrary poset, let γ be a linear order type, and let $\alpha \geq 0$ be an ordinal number. Then P is said to *have symmetric* (or *balanced*) γ-*deviation* α, denoted as $bk_\gamma(P) = \alpha$, if α is the least ordinal such that for any chain C of P of order type $\widetilde{\gamma}$ there exist $a_- < b_-$ in C_- with $k_\gamma^0(b_-/a_-) < \alpha$ or there exist $a_+ < b_+$ in C_+ with $k_\gamma(b_+/a_+) < \alpha$, where b_-/a_- and b_+/a_+ are considered as intervals in P.

The next result shows that the concept of symmetric ω-deviation is actually equivalent to that of balanced Krull dimension.

Proposition 5.17. *For any poset P one has $bk_\omega(P) = bk(P)$ if either side exists.*

Proposition 5.18. *Let P be an arbitrary poset, let α be an ordinal number, and let γ be a linear order type. Then $bk_\gamma(P) = \alpha$ if and only if α is the least ordinal such that for all $x \in P$ one has $k_\gamma^0((\leftarrow, x)) \leq \alpha$ or $k_\gamma((x, \rightarrow)) \leq \alpha$.*

Definition 5.19. Let P be an arbitrary poset, let Γ be a nonempty set of linear order types, and let $\alpha \geq 0$ be an ordinal number. Then P is said to *have symmetric* (or *balanced*) Γ-*deviation* α, denoted as $bk_\Gamma(P) = \alpha$, if α is the least ordinal such that for any $\gamma \in \Gamma$ and for any chain C of P of order type $\widetilde{\gamma}$, there exist $a_- < b_-$ in C_- with $k_\Gamma^0(b_-/a_-) < \alpha$ or there exist $a_+ < b_+$ in C_+ with $k_\Gamma(b_+/a_+) < \alpha$, where b_-/a_- and b_+/a_+ are considered as intervals in P.

Consequently, we can summarize our results as follows.

Proposition 5.20. *The following statements are equivalent for a poset P and a nonempty set Γ of linear order types that contains at least one order type γ with $|\gamma| \leq \aleph_0$:*
  (1) *P has (dual) Krull dimension.*
  (2) *P has Γ-deviation.*
  (3) *P has dual Γ-deviation.*
  (4) *P has $\widetilde{\Gamma}$-deviation.*
  (5) *P has symmetric Γ-deviation.*
In this case $k_{\widetilde{\Gamma}}(P) \leq bk_\Gamma(P) \leq \min\{k_\Gamma(P), k_\Gamma^0(P)\}$.

6. Applications to Modules and Grothendieck Categories

Throughout this section R will denote an associative ring with nonzero identity, Mod-R the category of all unital right R-modules, $\tau = (\mathcal{T}, \mathcal{F})$ a fixed hereditary torsion theory on Mod-R, and t the torsion radical associated with $(\mathcal{T}, \mathcal{F})$. The notation M_R will be used to emphasize that M is a right R-module. For any M_R we shall denote $\operatorname{Sat}_\tau(M) := \{N \mid N \leq M \text{ and } M/N \in \mathcal{F}\}$. Also, $\operatorname{Sat}_\tau(R) = \operatorname{Sat}_\tau(R_R)$.

It is known that for any M_R, $\operatorname{Sat}_\tau(M)$ is an upper continuous and modular lattice (e.g., see [35, Proposition 4.1, p. 207]).

For all undefined notation and terminology on categories and torsion theories the reader is referred to [2], [35].

Let $\mathcal{G}$ be a *Grothendieck category*, i.e., an abelian category with exact direct limits and with a generator. For any object X of $\mathcal{G}$ we shall denote by $\mathcal{L}(X)$ the lattice of all subobjects of X. The *Krull dimension* of $X \in \mathcal{G}$, denoted by $k(X)$, is defined as

$$k(X) := k(\mathcal{L}(X)).$$

In particular, for an R-module M_R, $k(M) := k(\mathcal{L}(M))$.

More generally, let Γ be a nonempty set of linear order types. Then the Γ-*deviation* of $X \in \mathcal{G}$, denoted by $k_\Gamma(X)$, is defined as

$$k_\Gamma(X) := k_\Gamma(\mathcal{L}(X)).$$

In particular, for an R-module M_R, $k_\Gamma(M) := k_\Gamma(\mathcal{L}(M))$.

In case $\Gamma = \{\gamma\}$, then instead of $k_{\{\gamma\}}(X)$, we shall simply write $k_\gamma(X)$ and call it the γ-*deviation* of X.

The τ-*Krull dimension* of M_R, denoted by $k_\tau(M)$, is defined in [3] as

$$k_\tau(M) := k(\operatorname{Sat}_\tau(M)).$$

Taking into account that the least element of the lattice $\operatorname{Sat}_\tau(M)$ is the torsion submodule $t(M)$ of M, we see that

$$k_\tau(M) = -1 \iff M \in \mathcal{T},$$

and $k_\tau(M) \leq 0$ if and only if M is τ-Artinian. Recall [2] that M is said to be τ-*Artinian* (resp. τ-*Noetherian*) if the lattice $\operatorname{Sat}_\tau(M)$ is Artinian (resp. Noetherian). Also, M is said to be τ-*finitely generated* if there exists a finitely generated submodule F of M such that $M/F \in \mathcal{T}$.

Note that $k_\tau(M)$ can be defined for M_R as

$$k_\tau(M) = -1 \iff M \in \mathcal{T},$$

and $k_\tau(M) = \alpha \geq 0$ if $k_\tau(M) \not< \alpha$ and if for any descending chain

$$M_1 \geq M_2 \geq \ldots \geq M_n \geq \ldots$$

in $\mathcal{L}(M)$, there exists an integer n_0 such that $k_\tau(M_n/M_{n+1}) < \alpha$ for all $n \geq n_0$. (See e.g., [3].)

More generally, the τ-Γ-*deviation* of a module M_R, denoted by $k_{\tau,\Gamma}(M)$, is defined as

$$k_{\tau,\Gamma}(M) := k_\Gamma(\operatorname{Sat}_\tau(M)).$$

In case $\Gamma = \{\gamma\}$, then instead of $k_{\tau,\{\gamma\}}(M)$, we shall simply write $k_{\tau,\gamma}(M)$, and call it the τ-γ-*deviation* of M.

In a similar way we can define for any Grothendieck category $\mathcal{G}$ and for any $X \in \mathcal{G}$ (resp. for any hereditary torsion theory τ on Mod-R and for any $M \in$ Mod-R) the *balanced Krull dimension* $bk(X)$ of X (resp. the τ-*balanced Krull dimension* $bk_\tau(M)$ of M) as

$$bk(X) := bk(\mathcal{L}(X)), \quad (\text{resp. } bk_\tau(M) = bk(\operatorname{Sat}_\tau(M))).$$

In case $bk_\tau(M) \leq 0$, we say that M is a τ-*DICC* module. The ring R is said to be a τ-*DICC* ring in case the right module R_R is τ-DICC.

The relative concepts of τ-balanced Krull dimension and τ-ζ-dimension of a module M can be also described inside the lattice $\mathcal{L}(M)$ of all submodules of M rather than inside that of the τ-saturated ones.

Proposition 6.1. *Let M_R be a module and let α be an ordinal. Then*

(1) $bk_\tau(M) = -1 \iff M \in \mathcal{T}$, and $bk_\tau(M) = \alpha \geq 0$ if α is the least ordinal such that for any double infinite chain

$$\cdots \leq M_{-2} \leq M_{-1} \leq M_0 \leq M_1 \leq M_2 \leq \cdots$$

of submodules of M, there exists $m \in \mathbb{Z}$ with $k_\tau^0(M_{i+1}/M_i) < \alpha$ for all $i \geq m$ or $k_\tau(M_{i+1}/M_i) < \alpha$ for all $i \leq m$.

(2) $k_{\tau,\zeta}(M) = -1 \iff M \in \mathcal{T}$, and $k_{\tau,\zeta}(M) = \alpha \geq 0$ if α is the least ordinal such that for any double infinite chain

$$\cdots \leq M_{-2} \leq M_{-1} \leq M_0 \leq M_1 \leq M_2 \leq \cdots$$

of submodules of M, there exists $m \in \mathbb{Z}$ with $k_{\tau,\zeta}(M_{i+1}/M_i) < \alpha$ for all $i \geq m$ or $k_{\tau,\zeta}(M_{i+1}/M_i) < \alpha$ for all $i \leq m$.

Most of the results established in the previous sections for posets or lattices can be applied to Grothendieck categories and to categories of modules equipped with hereditary torsion theories. We present below some of these applications.

Proposition 6.2. *Let $\mathcal{G}$ be a Grothendieck category. The following assertions are equivalent for an object $X \in \mathcal{G}$:*

(1) X has Krull dimension.

(2) X has dual Krull dimension.
(3) X has ζ-deviation.
(4) X has balanced Krull dimension.
In this case, $k_\zeta(X) \leq bk(X) \leq \min\{k(X), k^0(X)\}$.

Proposition 6.3. *Let τ be a hereditary torsion theory on* Mod-R. *The following assertions are equivalent for a module M_R:*
  (1) M has τ-Krull dimension.
  (2) M has dual τ-Krull dimension.
  (3) M has τ-ζ-deviation.
  (4) M has τ-balanced Krull dimension.
In this case, $k_{\tau,\zeta}(M) \leq bk_\tau(M) \leq \min\{k_\tau(M), k^0_\tau(M)\}$.

Proposition 6.4. *Let $\tau = (\mathcal{T}, \mathcal{F})$ be a hereditary torsion theory on* Mod-R *and let M_R be a module. Then $bk_\tau(M) = \alpha \geq 0$ if and only if α is the least ordinal such that there exists $N \leq M$ satisfying $k_\tau(N) \leq \alpha$, $k^0_\tau(M/N) \leq \alpha$, and $k^0_\tau(M/(N \cap P)) \leq \alpha$ (or $k^0_\tau(N/(N \cap P)) \leq \alpha$) for any $P \leq M$ with $P/(P \cap N) \notin \mathcal{T}$.*

For $\alpha = 0$ we obtain at once

Corollary 6.5. *Let $\tau = (\mathcal{T}, \mathcal{F})$ be a hereditary torsion theory on* Mod-R *and let M_R be a module. Then M is τ-DICC if and only if there exists $N \leq M$ such that N is τ-Artinian, M/N is τ-Noetherian, and $M/(N \cap P)$ (or $N/(N \cap P)$) is τ-Noetherian for any $P \leq M$ with $P/(P \cap N) \notin \mathcal{T}$.*

The next two results are relative versions of the corresponding results for Grothendieck categories [28].

Proposition 6.6. *Let $\tau = (\mathcal{T}, \mathcal{F})$ be a hereditary torsion theory on* Mod-R *and let M_R be a τ-DICC module which is not τ-Artinian. Then M is τ-finitely generated.*

Corollary 6.7. *Let $\tau = (\mathcal{T}, \mathcal{F})$ be a hereditary torsion theory on* Mod-R. *If R is τ-Noetherian, any τ-DICC right R-module is either τ-Artinian or τ-Noetherian.*

Recall that a ring R is called τ-*semiprime* if the τ-*closed prime radical* $\text{Rad}_\tau(R) = \bigcap \{P \mid P \text{ prime ideal with } R/P \in \mathcal{F}\}$ of R is zero. An ideal I of R is called τ-nilpotent if some power of I is τ-torsion.

We end this paper by presenting relative versions of some results from [20].

Proposition 6.8. *Let τ be a hereditary torsion theory on Mod-R, and assume that R is a τ-semiprime ring with τ-Krull dimension. Then*

$$bk_\tau(R) = \min(k_\tau(R), k_\tau^0(R)).$$

Corollary 6.9. *Any τ-semiprime τ-DICC ring is τ-Noetherian.*

Proposition 6.10. *Let R be a DICC ring that is not τ-Noetherian. If the τ-radical $\mathrm{Rad}_\tau(R)$ is τ-nilpotent, then $\mathrm{Rad}_\tau(R)$ is τ-Artinian.*

References

[1] T. Albu, *Classes of lattices (co)generated by a lattice and their global (dual) Krull dimension*, Discrete Math. **185** (1998), 1–18.

[2] T. Albu and C. Năstăsescu, *Relative Finiteness in Module Theory*, Marcel Dekker, New York, 1984.

[3] T. Albu and P. F. Smith, *Dual relative Krull dimension of modules over commutative rings*, in "Abelian Groups and Modules", edited by A. Facchini and C. Menini, Kluwer Academic Publisher, Dordrecht (1995), 1–15.

[4] T. Albu and P. F. Smith, *Localization of modular lattices, Krull dimension, and the Hopkins-Levitzki Theorem (I)*, Math. Proc. Cambridge Philos. Soc. **120** (1996), 87–101.

[5] T. Albu and P. F. Smith, *Localization of modular lattices, Krull dimension, and the Hopkins-Levitzki Theorem (II)*, Comm. Algebra **25** (1997), 1111–1128.

[6] T. Albu and M. L. Teply, *Generalized deviation of posets and modular lattices*, Discrete Math. **214** (2000), 1–19.

[7] T. Albu and M. L. Teply, *The double infinite chain condition and generalized deviations of posets and modules*, in "Algebra and its Applications", edited by D. V. Huynh, S. K. Jain and S. R. Lopez-Permouth, Contemporary Math. vol. 259, Amer. Math. Soc., Providence, 2000, 13–43.

[8] H. Bass, *Descending chains and the Krull ordinal of commutative noetherian rings*, J. Pure and Applied Algebra **1** (1971), 347–360.

[9] M. Contessa, *On rings and modules with DICC*, J. Algebra **101** (1986), 489–496.

[10] _____, *On DICC rings*, J. Algebra **105** (1987), 429–436.

[11] _____, *On modules with DICC*, J. Algebra **107** (1987), 75–81.

[12] P. Crawley and R. P. Dilworth, *Algebraic Theory of Lattices*, Prentice-Hall, Englewood Cliffs, New Jersey, 1973.

[13] P. Gabriel, *Des catégories abéliennes*, Bull. Soc. Math. France **90** (1962), 323–448.

[14] A. Gleyzal, *Order types and structure of orders*, Trans. Amer. Math. Soc. **48** (1940), 451–466.

[15] K. R. Goodearl and B. Zimmermann-Huisgen, *Length of submodule chains versus Krull dimension in non-noetherian modules*, Math. Z. **191** (1986), 519–527.

[16] R. Gordon and J. C. Robson, *Krull Dimension*, Memoirs Amer. Math. Soc. **133**, 1973.

[17] T. H. Gulliksen, *A theory of length for noetherian modules*, J. Pure and Applied Algebra **3** (1973), 159–170.

[18] K. Hrbacek and T. Jech, *Introduction to Set Theory*, Marcel Dekker, New York, 1984.

[19] E. Kamke, *Theory of Sets*, Dover, New York, 1950.

[20] O. A. S. Karamzadeh and M. Motamedi, *On α-DICC modules*, Comm. Algebra **22** (1994), 1933–1944.

[21] G. Krause, *On the Krull-dimension of left noetherian left Matlis-rings*, Math. Z. **118** (1970), 207–214.

[22] _____, *Descending chains of submodules and the Krull dimension of noetherian modules*, J. Pure and Applied Algebra **3** (1973), 385–397.

[23] W. Krull, *Primiidealketten in allgemeinen Ringbereichen*, Sitzungsber. Heidelberg Akad. Wissenschaft Math.-Natur. Kl. **7**. Abhandl. (1928), 3–14.

[24] W. G. Lau, M. L. Teply and A. K. Boyle, *The deviation, density, and depth of partially ordered sets*, J. Pure and Applied Algebra **60** (1989), 253–268.

[25] B. Lemonnier, *Déviation des ensembles et groupes abéliens totalement ordonnés*, Bull. Sci. Math. **96** (1972), 289–303.

[26] J. C. McConnell and J. C. Robson, *Noncommutative Noetherian Rings*, John Wiley & Sons, Chichester-New York, 1987.

[27] E. Noether, *Eliminationstheorie und allgemeine Idealtheorie*, Math. Ann. **90** (1923), 229–261.

[28] B. L. Osofsky, *Double Infinite Chain Conditions*, in "Abelian Group Theory", edited by R. Göbel and E. A. Walker, Gordon and Breach, New York-London (1987), 451–456.

[29] M. Pouzet and N. Zaguia, *Dimension de Krull des ensembles ordonnés*, Discrete Math. **53** (1985), 173–192.

[30] R. Rentschler and P. Gabriel, *Sur la dimension des anneaux et ensembles ordonnés*, C. R. Acad. Sci. Paris **265** (1967), 712–715.

[31] J. G. Rosenstein, *Linear Orderings*, Academic Press, New York, 1982.

[32] W. Sierpiński, *Cardinal and Ordinal Numbers*, Państwowe Wydawnictwo Naukowe, Warszawa, 1958.

[33] J. C. Shepherdson, *Well-ordered sub-series of general series*, Proc. London Math. Soc. **1**(3) (1951), 291–307.

[34] H. Simmons, *Generalized deviations of posets*, Discrete Math. **98** (1991), 123–139.

[35] B. Stenström, *Rings of Quotients*, Springer-Verlag, Heidelberg, New York, 1975.

Toma Albu
Faculty of Mathematics
Bucharest University
RO-70109 Bucharest
Romania
e-mail: Toma.Albu@imar.ro

Mark L. Teply
Department of Mathematics
The University of Wisconsin
Milwaukee, WI 53201-0413,
U. S. A.
e-mail: mlteply@csd.uwm.edu

Stability Properties of Exchange Rings

Pere Ara

Abstract

We survey some recent results on exchange rings, with emphasis on stability theorems in K-theory.

0. Introduction

It has been realized recently that the class of exchange rings has good stability properties in a K-theoretic sense. All known examples of exchange rings satisfy a weak cancellation condition called *separativity*. In the presence of this condition the stability properties are especially good.

Section 1 contains some basic facts on exchange rings. We also include a short discussion on non-unital exchange rings, which are important in the study of K-theory. Section 2 surveys some results concerning the stable rank of exchange rings, and Section 3 recalls the definition of separativity and gives large classes of exchange rings satisfying that condition. Section 4 deals with stability properties of exchange rings. The point here is that, in most cases, information at the K-theoretic level can be faithfully translated to ring theoretical information. We give a relative version of a recent result of Goodearl, O'Meara, Raphael and the author [AGOR] stating that the natural map $GL_1(R) \to K_1(R)$ is surjective for any separative exchange ring R. Also Perera has recently shown in [Per3] that, if I is a separative exchange ideal of a ring R, then a unit in R/I lifts to a unit in R if and only if $\delta([x]) = 0$, where $\delta : K_1(R/I) \to K_0(I)$ is the connecting map in K-theory.

1. Definition and Examples

We start by recalling the exchange properties for modules over unital associative rings. These properties were introduced by Crawley and Jónsson [CJ] for more general algebraic structures.

Partially supported by grants from the DGICYT (Spain) and the Comissionat per Universitats i Recerca de la Generalitat de Catalunya.

Definition 1.1. An R-module M has the exchange property if for every R-module A and any decompositions

$$A = M' \oplus N = \bigoplus_{i \in I} A_i$$

with $M' \cong M$, there exist submodules $A_i' \subseteq A_i$ such that

$$A = M' \oplus (\bigoplus_{i \in I} A_i').$$

It follows from the modular law that A_i' must be a direct summand of A_i for each i. If the above condition is satisfied whenever the index set I is finite, M is said to satisfy the *finite exchange property*. Obviously any finitely generated module with the finite exchange property satisfies the exchange property.

We refer the reader to [Fac, Chapter 2] for the basic theory of modules with the exchange property and its relation with the Krull-Schmidt-Remak-Azumaya Theorem. We highlight the following important fact:

Theorem 1.2. [Crawley and Jónsson, Warfield] (see [Fac, Theorem 2.8]) *The following conditions are equivalent for an indecomposable module M_R:*
  (a) *The endomorphism ring of M_R is local.*
  (b) *M_R has the finite exchange property.*
  (c) *M_R has the exchange property.*

It is still an open question whether the finite exchange property implies the exchange property for a general module M_R. An affirmative answer is known in various particular situations. For example, in a sequence of papers, it was shown for each module with an indecomposable decomposition; see [HI], [Ya] and [ZH-Z]. Also, Zimmermann-Huisgen [ZH] gave an affirmative answer for torsion modules over Dedekind domains, and Oshiro and Rizvi for quasi-continuous modules; see [OR] and [MMqc].

Following Warfield [W1], we say that a ring R is an *exchange ring* if R_R satisfies the (finite) exchange property. By [W1, Corollary 2], this definition is left-right symmetric.

The following characterization of exchange rings is very useful. It was obtained independently by Goodearl and Nicholson.

Theorem 1.3. [GW, p. 167], [Nic1, Theorem 2.1] *Let R be a unital ring. Then R is an exchange ring if and only if for every element $a \in R$ there exists an idempotent $e \in R$ such that $e \in aR$ and $1 - e \in (1 - a)R$.*

This theorem is very useful for checking that some rings are exchange rings. For example Stock used it to show that every π-regular ring is an exchange ring [Sto, Example 2.3]. Recall that a ring R is said to be π-regular in case for each $x \in R$ there are $y \in R$ and a positive integer n such that $x^n = x^n y x^n$. In particular all von Neumann regular rings are exchange rings. Another important class of algebras which can be seen to be exchange rings by using Theorem 1.3 is the class of C^*-algebras of real rank zero. These were introduced by Brown and Pedersen in 1991 [BP1]. In fact it was proved in [AGOP, Theorem 7.2] that the C^*-algebras with real rank zero are exactly the C^*-algebras which are exchange rings. This important connection opened the way for a transfer of technology between Ring Theory and Operator Algebras, which has been exploited already in both directions [AGOR, Per2, Per3]. For more examples where the characterization in Theorem 1.3 has been successfully applied, see [Bac] and [O'M2].

It is implicit in the proof of [Nic1, Theorem 2.1] that a module M_R satisfies the finite exchange property if and only if whenever $M \oplus B = A_1 \oplus A_2$ with $A_1 \cong A_2 \cong M$ there are $A_i' \subseteq A_i$ such that $M \oplus A_1' \oplus A_2' = A_1 \oplus A_2$. A similar statement holds for the $\aleph$-exchange property; see [ZH-Z, Proposition 3]. Since it is well known that the (finite) exchange property passes to direct summands and finite direct sums ([CJ, Lemma 3.10] or [Fac, Lemma 2.4]), we see that M has the finite exchange property if and only if the additive category add(M), whose objects are the R-modules which are isomorphic to direct summands of M^n for some $n \geq 1$, satisfies the exchange property (i.e., the exchange property in Definition 1.1 is satisfied whenever all the modules are in add(M) and the index set I is finite). In particular R is an exchange ring if and only if the additive category $FP(R)$ of finitely generated projective right R-modules satisfies the exchange property. Since for every right R-module M_R there is a categorical equivalence between the category $FP(\text{End}(M_R))$ and the category add(M_R) (see e.g., [Fac, Theorem 4.7]) we obtain a proof of Warfield's Theorem that a module M_R satisfies the finite exchange property if and only if End(M_R) is an exchange ring [W1, Theorem 2]. Also since the category $FP(R)$ is equivalent to the category $FP(R^{op})$ via duality, one obtains that R_R satisfies the exchange property if and only if $_R R$ does, which is the left-right symmetry of the exchange condition for rings. For a proof of this fact using the characterization of exchange rings given in Theorem 1.3, see [Nic2, Proposition].

The following result of Nicholson explains the role of the Jacobson radical $J(R)$ of a ring R in the theory of exchange rings.

Proposition 1.4. [Nic1, Proposition 1.5] *A ring R is an exchange ring if and only if $R/J(R)$ is an exchange ring and idempotents can be lifted modulo $J(R)$.*

It follows from Proposition 1.4 that the semilocal rings that are exchange rings coincide with the semiperfect rings. Also right self-injective rings and right continuous rings are exchange rings by [MoMu, Proposition 3.5 and Lemma 3.6].

Proposition 1.4 and various results concerning operator algebras (see [BP1], [Zha]) induced the author to think about the possibility of defining and working with non-unital exchange rings. This is particularly interesting when working with K-theory; see Section 4. The point here is that non-unital exchange rings I have good properties with respect to any unital ring R containing them as two-sided ideals. The definition is an adaptation of the characterization given in Theorem 1.3.

Definition 1.5. [A4] A non-unital ring I is an *exchange ring* if for every $x \in I$ there exist an idempotent $e \in I$ and elements $r, s \in I$ such that $e = xr = x + s - xs$.

In [A4] different characterizations of non-unital exchange rings are given. In particular, [A4, Theorem 1.2] gives a characterization in terms of the exchange property of suitable module decompositions. Every ideal of a unital exchange ring is an exchange ring, but there are non-unital exchange rings that cannot be embedded as ideals of any unital exchange ring [A4, Example 4, p. 412]. Note that the radical rings (i.e., those rings J which are the Jacobson radical of some unital ring R) are precisely the exchange rings without nonzero idempotents. The main result in [A4] is the following natural generalization of Proposition 1.4.

Theorem 1.6. [A4, Theorem 2.2] *Let I be an ideal of the (possibly non-unital) ring L. Then L is an exchange ring if and only if I and L/I are exchange rings and idempotents can be lifted modulo I.*

It is natural to ask whether the exchange condition for a non-unital ring I can be checked by looking at the corner rings eIe for the different idempotents $e \in I$. This is answered by the following result:

Theorem 1.7. [AGS, Theorem 3.3] *Let I be any ring. Let I_0 be the ideal of I generated by the idempotents of I. Then the following conditions are equivalent:*

(i) *I is an exchange ring.*
(ii) (a) *I/I_0 is a radical ring;*
    (b) *eIe is an exchange ring for all $e = e^2 \in I$.*

In fact it suffices to verify (b) for a family of idempotents which generates I_0 as an ideal [AGS, Theorem 3.2].

2. Stable Rank

For a ring R we denote by $V(R)$ the abelian monoid of isomorphism classes of objects from $FP(R)$, the category of finitely generated projective right R-modules. As we observed in Section 1, the fact that R is an exchange ring is stored in the category $FP(R)$, but unfortunately cannot be tested in $V(R)$ alone. For an easy example, consider the ring $\mathbb{Z}$ of integer numbers, which is not an exchange ring. However it follows easily from the exchange property that $V(R)$ is a refinement monoid when R is an exchange ring; see [AGOP, Corollary 1.3] or [Ha, Proposition 8.6.1]. Here an abelian monoid M is said to be a *refinement monoid* in case for each identity $a + b = x + y$ in M there are decompositions $x = x_1 + x_2$ and $y = y_1 + y_2$ such that $a = x_1 + y_1$ and $b = x_2 + y_2$. Some properties of rings which can be stated in terms of $FP(R)$ can also be stated in terms of $V(R)$ in case R is an exchange ring. In this section we present a representative example, the stable rank of a ring.

We recall briefly the definition of the (Bass) stable rank of a ring R. An n-row $\mathbf{a} = (a_1, \ldots, a_n) \in R^n$ is said to be *right unimodular* if $\mathbf{a}(^n R) = R$, where $^n R$ is the set of n-columns with coefficients in R. Given an $(n+1)$-row $(\mathbf{a}, b) \in R^n \times R$, we say that $(\mathbf{a}, b)$ is *reducible* in case there is an n-row $\mathbf{c}$ such that $\mathbf{a} + b\mathbf{c}$ is right unimodular. The *stable rank* of R, denoted $\mathrm{sr}(R)$, is the least positive integer n such that every right unimodular $(n+1)$-row is reducible, or ∞ if no such n exists.

Theorem 2.1. [W2, Theorem 1.6] *Let A be a finitely generated projective right module over a ring R. Then $\mathrm{sr}(\mathrm{End}(A)) \leq n$ if and only if whenever*

$$M = \left(\bigoplus_{i=1}^{n} A_i\right) \oplus H = A_0 \oplus Y$$

in $FP(R)$, with $A_i \cong A$ for all $i = 0, 1, \ldots, n$, there exist submodules $K \subseteq M$ and $L \subseteq \bigoplus_{i=1}^{n} A_i$ such that $M = L \oplus K = A_0 \oplus K$.

Note that Theorem 2.1 implies the following property for the class $a = [A] \in V(R)$ of a finitely generated projective R-module A with $\mathrm{sr}(\mathrm{End}(A)) \leq n$:

(1) Whenever $na + h = a + y$ for $a, h, y \in V(R)$, there exists $e \in V(R)$ such that $na = a+e$ and $y = e+h$. (Take $e = [E]$, where $\bigoplus_{i=1}^{n} A_i = L \oplus E$.)

This suggests the following definition of the stable rank of an element a of an abelian monoid M. Write $\mathrm{sr}_M(a) = n$ if n is the least positive integer such that (1) holds (with $V(R)$ replaced by M). By Theorem 2.1 we have $\operatorname{sr}_{V(R)}([A]) \leq \operatorname{sr}(\operatorname{End}_R(A))$, but in general the inequality is strict.

Now we are ready to obtain a result unifying [AGOP, Theorem 3.2] and [WuXu, Theorem 4] (the case $n = 1$ is due to Yu [Yu, Theorem 9]). The proof given here is new.

Theorem 2.2. *Let R be an exchange ring, let $A \in FP(R)$, and let $n \in \mathbb{N}$. Then the following conditions are equivalent:*
  (i) $\operatorname{sr}(\operatorname{End}_R(A)) \leq n$.
  (ii) $\operatorname{sr}_{V(R)}([A]) \leq n$.
  (iii) *If $nA \oplus B \cong A \oplus C$ for some $B, C \in FP(R)$, then B is isomorphic to a direct summand of C.*
  (iv) *Internal n-weak cancellation: If $nA \cong B_1 \oplus B_2$ and $A \cong B_1 \oplus B_3$, then B_3 is isomorphic to a direct summand of B_2.*

Proof. By Warfield's results (i)$\Rightarrow$(ii), as remarked before.

(ii)$\Rightarrow$(iii). Obvious.

(iii)$\Rightarrow$(iv). Suppose that $nA \cong B_1 \oplus B_2$ and $A \cong B_1 \oplus B_3$. Then $nA \oplus B_3 \cong A \oplus B_2$. By hypothesis B_3 is isomorphic to a direct summand of B_2.

(iv)$\Rightarrow$(i). Let $M = \bigoplus_{i=1}^n A_i \oplus H = A_0 \oplus Y$ be a decomposition in $FP(R)$ with $A_i \cong A$ for $i = 0, 1, \ldots, n$. By the finite exchange property of $\bigoplus_{i=1}^n A_i$ we have $M = \bigoplus_{i=1}^n A_i \oplus A_0' \oplus Y'$ where $A_0' \subseteq A_0$ and $Y' \subseteq Y$. By the modular law we can write $A_0 = A_0' \oplus A_0''$ and $Y = Y' \oplus Y''$. Let π denote the projection of M onto $\bigoplus_{i=1}^n A_i$ with kernel $A_0' \oplus Y'$. Set $Z = \pi(Y'')$. Then it is easily checked that $M = Z \oplus A_0'' \oplus A_0' \oplus Y'$. Since $nA \cong Z \oplus A_0''$ and $A \cong A_0' \oplus A_0''$, we obtain from (iv) a decomposition $Z = Z_1 \oplus Z_2$ such that $Z_1 \cong A_0'$. Since $Z_1 \cap A_0' = 0$ there is a module T such that $Z_1 \oplus A_0' = Z_1 \oplus T = A_0' \oplus T$. Note that

$$M = Z_1 \oplus Z_2 \oplus A_0'' \oplus A_0' \oplus Y' = T \oplus Z_2 \oplus A_0'' \oplus A_0' \oplus Y' = A_0 \oplus K,$$

where $K = T \oplus Z_2 \oplus Y'$. Write $\bigoplus_{i=1}^n A_i = Z \oplus L'$ for some L'. Then we have

$$\begin{aligned} M &= \bigoplus_{i=1}^n A_i \oplus A_0' \oplus Y' = Z_1 \oplus Z_2 \oplus L' \oplus A_0' \oplus Y' \\ &= Z_1 \oplus Z_2 \oplus L' \oplus T \oplus Y' = (Z_1 \oplus L') \oplus K = L \oplus K, \end{aligned}$$

where $L = L' \oplus Z_1 \subseteq \bigoplus_{i=1}^n A_i$. It follows from Theorem 2.1 that $\operatorname{sr}(\operatorname{End}(A)) \leq n$ as desired. □

It follows from Theorem 2.2 that $\operatorname{sr}(\operatorname{End}_R(A)) = \operatorname{sr}_{V(R)}([A])$ for any finitely generated projective module A over an exchange ring R. Of course this makes easier the computation of the stable rank of these modules. For some applications of this fact see [Par2] and [Per2].

3. Separativity

A class $\mathcal{C}$ of modules is called *separative* if for all $A, B \in \mathcal{C}$ we have

$$A \oplus A \cong A \oplus B \cong B \oplus B \Rightarrow A \cong B.$$

A ring R is *separative* if $FP(R)$ is a separative class. Separativity for rings was introduced in [AGOP]. However, separativity is an old concept in semigroup theory; see [CP]. A semigroup S is called *separative* if for all $a, b \in S$ we have $a + a = a + b = b + b \Rightarrow a = b$. Clearly a ring R is separative if and only if $V(R)$ is a separative semigroup. Separativity provides a key to a number of outstanding cancellation problems for finitely generated projective modules over exchange rings; see [AGOP].

Separativity can be tested in various ways.

Theorem 3.1. [AGOP, Section 2] *For a ring R the following conditions are equivalent*:
  (i) R is separative.
  (ii) For $A, B \in FP(R)$, if $2A \cong 2B$ and $3A \cong 3B$, then $A \cong B$.
  (iii) For $A, B \in FP(R)$, if there exists $n \in \mathbb{N}$ such that $nA \cong nB$ and $(n+1)A \cong (n+1)B$, then $A \cong B$.
  (iv) For $A, B, C \in FP(R)$, if $A \oplus C \cong B \oplus C$ and C is isomorphic to direct summands of both mA and nB for some $m, n \in \mathbb{N}$, then $A \cong B$.

In case R is an exchange ring, separativity is also equivalent to the condition
  (v) For $A, B, C \in FP(R)$, if $A \oplus 2C \cong B \oplus 2C$, then $A \oplus C \cong B \oplus C$.

Outside the class of exchange rings, separativity can easily fail. In fact it is easy to see that a commutative ring R is separative if and only if $V(R)$ is cancellative. Among exchange rings, however, separativity seems to be the rule. It is not known whether there are non-separative exchange rings. This is one of the major open problems in this area. We list below some classes of (exchange) rings which are separative.

(1) All rings with stable rank 1. In fact $V(R)$ is cancellative in this case [Ev, Theorem 2]. This includes all unit-regular rings as well as all strongly π-regular rings [A3, Theorem 4] and hence all algebraic algebras over a field.

(2) Any ring whose finitely generated projective modules enjoy n-cancellation ($nA \cong nB \Rightarrow A \cong B$) for some $n > 1$, in the terminology introduced by Lam [Lam], because of Theorem 3.1(iii). This includes all

right self-injective rings (e.g. [G1, Theorem 3]) and all right $\aleph_0$-continuous regular rings [A1, Theorem 2.13], as well as all AW^*-algebras—even all Rickart C^*-algebras (see [A2, Theorem 2.7]). We refer the reader to [Lam] for more information about the n-cancellation property.

(3) The class of separative exchange rings is closed under taking corners, finite matrix rings, arbitrary direct products, direct limits, and factor rings [AGOP, Proposition 2.2]. Moreover, it is closed under extensions [AGOP, Theorem 4.2]: If R is an exchange ring and I is an ideal of R, then R is separative if and only if I and R/I are separative. The latter is a fundamental distinction between separativity and stable rank one, since the class of exchange rings with stable rank one is not closed under extensions; see for example [G2, Example 4.26]. In fact extensions of exchange rings with stable rank one are *strongly separative*, meaning that $A \oplus B \cong A \oplus A \Rightarrow A \cong B$ for all $A, B \in FP(R)$; see [AGOP, Section 5]. Combining this result with [Bac] we see that all right semiartinian rings are strongly separative.

In (4)–(5) we consider two classes of exchange rings for which a proof of separativity has been obtained by using monoid-theoretic arguments.

(4) A ring R is said to be *strictly unperforated* if, for every positive integer n and for every $P, Q \in FP(R)$, if nP is isomorphic to a proper direct summand of nQ, then P is isomorphic to a proper direct summand of Q. This can be stated in monoid-theoretic terms. For $a, b \in V(R)$ set $a \leq^* b$ in case there is a nonzero $z \in V(R)$ such that $a + z = b$. Then R is strictly unperforated if and only if $na \leq^* nb$ implies $a \leq^* b$ in $V(R)$. It follows from the results in [AGPT] that a simple exchange ring satisfies strict unperforation if and only if it satisfies a seemingly much weaker condition called weak comparability, introduced by K. O'Meara in the context of regular rings [O'M1]. (A monoid-theoretic version of this result has been recorded in [Per1, Lemma 3.7].) By [APa], any simple exchange ring satisfying strict unperforation is separative.

(5) Let s be a positive integer and let M be an abelian monoid. Say that M satisfies s-comparability if for any $p, q \in M$ either $p \leq sq$ or $q \leq sp$, where $\leq$ is the algebraic pre-order on M. If T is any ring, say that T satisfies s-comparability provided $V(T)$ satisfies s-comparability. Of course *comparability* stands for 1-comparability. Extending previous work by O'Meara, Tyukavkin and the author [AOT], E. Pardo proved in [Par2, Theorem 2.2] that any exchange ring satisfying s-comparability is separative.

(6) If R is a regular ring, then every projective right R-module P satisfies the exchange property, as proved by Stock [Sto]. Consequently the ring $E = \text{End}(P)$ is an exchange ring. Let P be a countably generated

projective right R-module. It has been proved in [APP, Corollary 1.9] that, if R is unit-regular and strictly unperforated, then $\text{End}(P)$ is separative. Also, if R is a regular ring satisfying s-comparability for some $s \geq 1$, then the same conclusion holds [APP, Corollary 3.4]. However it seems that in general the separativity property will not be inherited by $\text{End}(P)$, and this construction opens up a possibility for constructing non-separative exchange rings.

(7) Following [BP2], we say that an element x of a C^*-algebra A is *quasi-invertible* in case there are orthogonal closed ideals I and J in A such that $x + I$ is left invertible in A/I and $x + J$ is right invertible in A/J. A C^*-algebra A is *extremally rich* if the set of quasi-invertible elements of A is dense in A; see [BP2, Section 3]. It has been proved by Brown and Pedersen in [BP3] that every extremally rich C^*-algebra of real rank zero is separative.

For some applications of separativity in other contexts we refer the reader to [Br] and [OV].

4. Stability Properties of Exchange Rings

In this section we survey some recent results on the low K-theory of exchange rings and we add some information. Usually we will consider a non-unital exchange ring I which is an ideal of a unital ring R. We will shorten this condition by saying that "I is an exchange ideal of R". However we remark that being an exchange ring is an intrinsic property of the ring I without unit (see Section 1).

Recall that the Grothendieck group $K_0(R)$ of a unital ring R is the Grothendieck (or universal) group associated to the abelian monoid $V(R)$.

Let $FP(R, I) = \{A \in FP(R) \mid A = AI\}$ and let $V(I)$ be the abelian monoid of isomorphism classes of objects from $FP(R, I)$. The monoid $V(I)$ does not depend on the enveloping ring R, as can be seen from the following alternative description. Let $M(I) = \varinjlim M_n(I)$, where for $n \leq m$ the map $M_n(I) \to M_m(I)$ is defined by $x \mapsto \begin{pmatrix} x & 0 \\ 0 & 0 \end{pmatrix}$. For idempotents $e, f \in M(I)$ we say that e is equivalent to f, written $e \sim f$, in case there are $x \in eM(I)f$ and $y \in fM(I)e$ such that $e = xy$ and $f = yx$. Then $V(I)$ is isomorphic to the abelian monoid of equivalence classes of idempotents from $M(I)$. In this picture the sum in $V(I)$ is given by $[e]+[f] = [\begin{pmatrix} e & 0 \\ 0 & f \end{pmatrix}]$.

For any unital ring R, the group $K_0(R)$ has a natural pre-order obtained by taking as a positive cone $K_0(R)^+$ the classes in $K_0(R)$ of the finitely generated projective R-modules.

Theorem 4.1. [A4] *Let I be an exchange ideal of a unital ring R. Let $\Psi : K_0(R) \to K_0(R/I)$ be the natural map.*
  (a) $\Psi(K_0(R)^+) = \Psi(K_0(R)) \cap K_0(R/I)^+$.
  (b) *Ψ is surjective if and only if the idempotents of $M_n(R/I)$ can be lifted to the idempotents in $M_n(R)$ for all $n \geq 1$.*

When every finitely generated projective R/I-module is isomorphic to a direct sum of cyclic R/I-modules, condition (b) in Theorem 4.1 is equivalent to the condition of lifting idempotents modulo I. This is the case when R/I is an exchange ring.

For a ring I without unit, $K_0(I)$ is defined as the kernel of the natural map $K_0(I^1) \to K_0(\mathbb{Z})$, where $I^1 = I \oplus \mathbb{Z}$ is the unitization of I. The next result, which is a K-theoretic version of Theorem 1.6, follows easily from Theorem 4.1.

Theorem 4.2. [A4, Theorem 3.5] *Let I be an ideal of the (possibly non-unital) ring L. Then L is an exchange ring if and only if I and L/I are exchange rings and the canonical homomorphism $K_0(L) \to K_0(L/I)$ is surjective.*

It follows from Theorem 4.2 that, if R is a semilocal ring, then R is semiperfect if and only if the canonical map $K_0(R) \to K_0(R/J(R))$ is an isomorphism. Also if I is an exchange ideal of a unital ring R and R/I is a *purely infinite* regular right self-injective ring, then R will be automatically an exchange ring, because $K_0(R/I) = 0$ by [G2, Proposition 15.6]. Here a regular right self-injective ring T is said to be purely infinite in case it contains no nonzero directly finite central idempotents, cf. [G2, p.116].

Let $G(I)$ be the Grothendieck group of $V(I)$. We have a natural map $G(I) \to K_0(I)$ sending the class of P in $G(I)$ to the corresponding class in $K_0(I)$. When this map is an isomorphism we write $G(I) = K_0(I)$. Goodearl showed that $G(I) = K_0(I)$ when I is a regular ring, see [MM1, Proposition 1.2]. A useful fact in that case is that any (non-unital) regular ring I is a *ring with local units*, that is, given $x_1, \ldots, x_n \in I$ there exists an idempotent $e \in I$ such that $x_i \in eRe$ for all i, cf. [FU]. In particular the set E of idempotents in I is directed and $I = \varinjlim_{e \in E} eIe$. (Recall that E is partially ordered by $e \leq f$ if and only if $e = ef = fe$, for $e, f \in E$.) For general exchange rings I we do not have that E is directed, but we have a substitute 'modulo equivalence'. Versions of the next lemma have been obtained in [AGOR, Lemma 2.1] and [Per3, Lemma 2.1]. Define $GL_n(R, I)$ to be the kernel of the natural map $GL_n(R) \to GL_n(R/I)$.

Lemma 4.3. *Let I be an exchange ideal of a unital ring R. Given idempotents $e_1, \ldots, e_n \in I$, there exist an idempotent $e \in I$ and $u_i \in GL_1(R, I)$ such that $u_i e_i u_i^{-1} \leq e$ for all i.*

Proof. Clearly it suffices to show the case $n = 2$. The result follows from (1) and (2) below:

(1) If e and f are idempotents in I and $fR \leq eR$, then there is $u \in GL_1(R, I)$ such that $ufu^{-1} \leq e$.

(2) There is an idempotent $g \in e_1R + e_2R$ and $v \in GL_1(R, I)$ such that $(ve_2v^{-1})R \leq gR$ and $e_1R \leq gR$. (In particular $g \in I$.)

Proof of (1). Setting $e' = f + (1-f)e(1-f)$, we obtain an idempotent e' such that $eR = e'R$ and $e'f = fe' = f$. Set $u = e' + 1 - e$, which is invertible in R with inverse $u^{-1} = e + 1 - e'$. It is easy to check that $ufu^{-1} = fe$ and so $ufu^{-1} \leq e$. On the other hand $u = 1 + (e' - e)$ and so clearly $u \in GL_1(R, I)$.

Proof of (2). (cf. [AGOR, Lemma 2.1]) Write $R = e_1R \oplus (1-e_1)R = e_2R \oplus (1-e_2)R$ and apply the exchange property to get $R = e_1R \oplus A \oplus B$, where $e_2R = A \oplus A'$ and $(1-e_2)R = B \oplus B'$. Let $\phi : R_R \to R_R$ be the automorphism such that $\phi = id$ on $A \oplus B$ and $\phi = \pi$ on $A' \oplus B'$, where $\pi : R \to R$ is the projection onto e_1R with kernel $A \oplus B$. Let g be an idempotent in I such that $gR = e_1R \oplus A$ and let $v \in GL_1(R)$ such that $\phi(x) = vx$ for all $x \in R$. Write $C = A \oplus B$ and $D = e_1R$. The matrix associated to ϕ with respect to the decomposition $R_R = C \oplus D$ is of the form $\begin{pmatrix} 1_C & a \\ 0 & 1_D \end{pmatrix}$, where $a \in I$. It follows that $v \in GL_1(R, I)$. Clearly $e_1R \leq gR$ and $ve_2v^{-1}R = ve_2R \leq gR$, as desired. □

The proof of the next proposition, due to Goodearl, is a simplification of the original argument of F. Perera and the author.

Proposition 4.4. *Let I be an exchange non-unital ring. Then the natural map $G(I) \to K_0(I)$ is surjective.*

Proof. Let $\pi : I^1 \to \mathbb{Z}$ be the canonical quotient map, so that $K_0(I) := \ker K_0(\pi)$. Let $x \in K_0(I)$, and write $x = [e] - [f]$ for some idempotents $e, f \in M(I^1)$. Then $[\pi(e)] - [\pi(f)] = 0$ in $K_0(\mathbb{Z})$, which implies that $\pi(e)$ and $\pi(f)$ have the same rank.

By [A4, proof of 3.4], we can assume that $e = (1_r - g) \oplus h$ and $f = (1_s - p) \oplus q$ for some idempotents $g, h, p, q \in M(I)$ with $g \leq 1_r$ and $p \leq 1_s$. Now $\pi(e) = 1_r$ and $\pi(f) = 1_s$ in $M(\mathbb{Z})$, so having the same rank implies $r = s$. Hence, $x = [e] - [f] = ([e] + [g] + [p]) - ([f] + [p] + [g]) = ([1_r] + [h] + [p]) - ([1_r] + [q] + [g]) = [h] + [p] - [q] - [g]$. But $[h], [p], [q], [g]$ are all in the image of the natural map $G(I) \to K_0(I)$, so x is in this image too. □

We have not been able to prove the injectivity of the map $G(I) \to K_0(I)$ for a general non-unital exchange ring I.

Now we move to K_1. For a unital ring R set $GL(R) = \varinjlim GL_n(R)$, the infinite general linear group. Let $E_n(R)$ be the subgroup of $GL_n(R)$ generated by the elementary matrices $e_{ij}(r)$ for $r \in R$ and $i \neq j$, and let $E(R) = \varinjlim E_n(R)$. Recall that $K_1(R) = GL(R)/E(R) = GL(R)^{\text{ab}}$.

Definition. [P. M. Cohn] We say that R is a GE_n-ring if $GL_n(R)$ is generated by $E_n(R)$ and the subgroup $D_n(R)$ of diagonal invertible matrices. We say that R is a GE-ring if it is a GE_n-ring for every $n \geq 1$.

If R is a GE_n-ring then $E_n(R)$ is a normal subgroup of $GL_n(R)$ and so $GL_n(R) = D_n(R)E_n(R) = E_n(R)D_n(R)$. Moreover, if R is a GE-ring, then the natural map $GL_1(R) \to K_1(R)$ is surjective.

Theorem 4.5. [AGOR, Theorem 2.8] *If R is a separative exchange ring, then R is a GE-ring, and so the natural homomorphism $GL_1(R) \to K_1(R)$ is surjective.*

Question 4.6. Determine the kernel of the natural map $GL_1(R) \to K_1(R)$ when R is a separative exchange ring. This has been solved for unit-regular rings and regular right self-injective rings by Menal and Moncasi [MM2] and for exchange rings with primitive factors artinian by Chen and Li [CL]. In all these cases, one gets $K_1(R) = GL_1(R)^{\text{ab}}$ provided that $\frac{1}{2} \in R$.

Recently F. Perera has shown in [Per3] that the index is the only obstruction to lifting units modulo a separative exchange ideal; see Theorem 4.11. By using the methods in [AGOR] and [Per3] we will prove a relative version of Theorem 4.5. Let I be an ideal of a unital ring R. Define $E_n(I)$ to be the subgroup of $E_n(R)$ generated by the elementary matrices $e_{ij}(a)$ with $a \in I$. Say that an idempotent g of a ring R is *full* in case $RgR = R$.

Proposition 4.7. *Let I be a separative exchange ideal of a unital ring R, and let $b \in I$ such that $bR = (1-p)R$ and $Rb = R(1-q)$ for some idempotents $1-p, 1-q \in I$. If p and q are full idempotents in R, then there exists $v \in GL_1(R, I)$ such that $(1-p)v = b = v(1-q)$.*

Proof. First we prove that there is an idempotent $e \in I$ and $u_1, u_2 \in GL_1(R, I)$ such that $u_1(1-p)u_1^{-1} \leq e$ and $u_2(1-q)u_2^{-1} \leq e$ and both $e - u_1(1-p)u_1^{-1}$ and $e - u_2(1-q)u_2^{-1}$ are full in eRe.

Since $RpR = R$, there are idempotents $p_1, \ldots, p_n \leq p$ such that $(1-p)R \cong p_1R \oplus \cdots \oplus p_nR$. Since $1 - p \in I$, we have $p_i \in I$ for all i. By Lemma 4.3 there are $v_i \in GL_1(pRp, pIp)$ and an idempotent $h_1' \in pIp$ such that $v_i p_i v_i^{-1} \leq h_1'$ for $i = 1, \ldots, n$. Write $h_1 = 1 - p + h_1' \in I$. Then

h_1 is an idempotent in I such that $1 - p \leq h_1$, and $h_1 - (1-p)$ is a full idempotent in $h_1 R h_1$. A similar argument yields an idempotent $h_2 \in I$ such that $1 - q \leq h_2$ and $h_2 - (1 - q)$ is a full idempotent in $h_2 R h_2$. By Lemma 4.3 there are $u_1, u_2 \in GL_1(R, I)$ and an idempotent $e \in I$ such that $u_i h_i u_i^{-1} \leq e$. Then $u_1(1-p)u_1^{-1} \leq e$ and $u_2(1-q)u_2^{-1} \leq e$ and both $e - u_1(1-p)u_1^{-1}$ and $e - u_2(1-q)u_2^{-1}$ are full idempotents in eRe.

Since $Rb = R(1-q)$ and $bR = (1-p)R$, there is $d \in (1-q)R(1-p)$ such that $bd = 1-p$ and $db = 1-q$. Note that $u_1 b u_2^{-1}$ and $u_2 d u_1^{-1}$ are in eRe and $(u_1 b u_2^{-1})(u_2 d u_1^{-1}) = u_1(1-p)u_1^{-1}$ and $(u_2 d u_1^{-1})(u_1 b u_2^{-1}) = u_2(1-q)u_2^{-1}$. Therefore $u_1(1-p)u_1^{-1}$ and $u_2(1-q)u_2^{-1}$ are equivalent idempotents in eRe. Since I is separative and $e \in I$, the unital ring eRe is separative. Using the fact that $e - u_1(1-p)u_1^{-1}$ and $e - u_2(1-q)u_2^{-1}$ are full in eRe, we conclude from the separativity of eRe that $e - u_1(1-p)u_1^{-1}$ and $e - u_2(1-q)u_2^{-1}$ are equivalent idempotents in eRe. Therefore there is $u' \in GL_1(eRe)$ such that

$$(u_1(1-p)u_1^{-1})u' = u_1 b u_2^{-1} = u'(u_2(1-q)u_2^{-1}).$$

Put $u = u' + 1 - e \in GL_1(R, I)$ and note that we can replace u' with u in the above formula. Set $v = u_1^{-1} u u_2 \in GL_1(R, I)$. Then $(1-p)v = b = v(1-q)$, as desired. $\square$

Remark 4.8. Note that, in the hypothesis and notation of Proposition 4.7, we have $b = bv^{-1}b$ so that b is unit-regular in R with $v^{-1} \in GL_1(R, I)$. This adds extra information to [Per3, Proposition 1.4]. We could say that b is *unit-regular in I*.

Let I be an ideal of a unital ring R. Let $E(R, I)$ be the smallest normal subgroup of $E(R)$ that contains all the elementary matrices $e_{ij}(a)$ with $a \in I$. By [Ros, Theorem 2.5.3], $E(R, I)$ is normal in $GL(R, I)$ and in $GL(R)$ and, by the definition, $K_1(R, I) = GL(R, I)/E(R, I)$. We can now obtain a relative version of Theorem 4.5.

Theorem 4.9. *Let I be a separative exchange ideal of a unital ring R. Then the natural map $GL_1(R, I) \to K_1(R, I)$ is surjective.*

Proof. By an easy inductive argument it suffices to prove that, given a matrix $\alpha \in GL_2(R, I)$, there is an invertible element $u \in GL_1(R, I)$ such that $[\alpha] = [u]$ in $K_1(R, I)$. Let $\alpha \in GL_2(R, I)$ and put $\alpha = \begin{pmatrix} a & b \\ c & d \end{pmatrix}$, with $a - 1, d - 1, b, c \in I$. By applying [Per3, Lemma 2.2(a)] to the first row of α, we get $\beta \in E_2(I)$ such that $\alpha \beta = \begin{pmatrix} a' & b' \\ c' & d' \end{pmatrix}$, where $b' \in Rb$,

$b'R = (1-h)R, a'R = hR$, and $RhR = R$ for some idempotent $h \in R$. By applying [Per3, Lemma 2.2(b)] to the second column of the matrix $\alpha\beta$, we get $\gamma \in E_2(I)$ such that $\gamma(\alpha\beta) = \begin{pmatrix} a'' & b'' \\ c'' & d'' \end{pmatrix}$ where $b'' \in b'R$, $Rb'' = R(1-q)$, $Rd'' = Rq$ and $RqR = R$ for some idempotent $q \in R$. Now note that since b'' is a von Neumann regular element of R, we have $b''R = (1-p)R$ for some idempotent $p \in R$ and, using that $b'' \in b'R$, we get $h(1-p)R = hb''R \le h(1-h)R = 0$. Since $RhR = R$, we obtain that $RpR = R$. Since $[\alpha] = [\gamma\alpha\beta]$ in $K_1(R,I)$ we can change notation and assume that the $(1,2)$-entry b of the matrix α satisfies the hypothesis of Proposition 4.7. Therefore there is $v \in GL_1(R,I)$ such that $b = (1-p)v = v(1-q)$ and so the matrix $\text{diag}(v^{-1}, 1)\alpha$ has an idempotent in its $(1,2)$-entry. Hence the result follows from the following analogue of [AGOR, Lemma 2.3]. $\square$

Lemma 4.10. *Let I be an ideal of a unital ring R and let $\alpha \in GL_2(R,I)$. If the $(1,2)$-entry of α is an idempotent, then there is $u \in GL_1(R,I)$ such that $[\alpha] = [u]$ in $K_1(R,I)$.*

Proof. Let $f = f^2$ be the $(1,2)$-entry of α. After right multiplication by a matrix in $E_2(I)$, we may assume that $\alpha = \begin{pmatrix} (1-f)a & f \\ b & c \end{pmatrix}$ for some $a, b, c \in R$ such that $a-1, b, c-1 \in I$. Since the row $((1-f)a, f)$ is unimodular, we get $(1-f)aR = (1-f)R$ so that there is $t \in R(1-f)$ such that $(1-f)at = 1-f$. Note that $t-1 \in I$. Now by performing similar changes to those in the proof of [Per3, Lemma 2.3], we arrive at the identity

$$\alpha = \begin{pmatrix} 1 & 0 \\ bt+c & 1 \end{pmatrix} \begin{pmatrix} 0 & 1 \\ -1 & 0 \end{pmatrix} \begin{pmatrix} u & 0 \\ 0 & 1 \end{pmatrix} \begin{pmatrix} 1 & 0 \\ (1-f)a & 1 \end{pmatrix} \begin{pmatrix} 1 & -t \\ 0 & 1 \end{pmatrix},$$

where $u = (bt+c)(1-f)a - b \in GL_1(R,I)$. Write $\gamma \equiv \delta$ in case the classes of γ and δ in $K_1(R,I)$ coincide. Replacing $\begin{pmatrix} 0 & 1 \\ -1 & 0 \end{pmatrix}$ by $e_{21}(-1)e_{12}(1)e_{21}(-1)$ in the above expression we get

$$\alpha = \begin{pmatrix} 1 & 0 \\ bt+c-1 & 1 \end{pmatrix} \begin{pmatrix} 1 & 1 \\ 0 & 1 \end{pmatrix} \begin{pmatrix} 1 & 0 \\ -1 & 1 \end{pmatrix} \begin{pmatrix} u & 0 \\ 0 & 1 \end{pmatrix} \begin{pmatrix} 1 & 0 \\ (1-f)a & 1 \end{pmatrix} \begin{pmatrix} 1 & -t \\ 0 & 1 \end{pmatrix}$$

$$\equiv \begin{pmatrix} 1 & 1 \\ 0 & 1 \end{pmatrix} \begin{pmatrix} u & 0 \\ 0 & 1 \end{pmatrix} \begin{pmatrix} 1 & 0 \\ -u & 1 \end{pmatrix} \begin{pmatrix} 1 & 0 \\ (1-f)a & 1 \end{pmatrix} \begin{pmatrix} 1 & -t \\ 0 & 1 \end{pmatrix}$$

$$\equiv \begin{pmatrix} u & 0 \\ 0 & 1 \end{pmatrix} \begin{pmatrix} 1 & 1 \\ 0 & 1 \end{pmatrix} \left[\begin{pmatrix} 1 & u^{-1} - 1 \\ 0 & 1 \end{pmatrix} \begin{pmatrix} 1 & 0 \\ -u + (1-f)a & 1 \end{pmatrix} \times \right.$$
$$\left. \begin{pmatrix} 1 & -t+1 \\ 0 & 1 \end{pmatrix} \right] \begin{pmatrix} 1 & -1 \\ 0 & 1 \end{pmatrix}$$

$$\equiv \begin{pmatrix} u & 0 \\ 0 & 1 \end{pmatrix}.$$

□

For an ideal I of a unital ring R, denote by $\delta : K_1(R/I) \to K_0(I)$ the index map in K-theory; see for example [Ros, 2.5].

Theorem 4.11. [Per3] *Let I be a separative exchange ideal of a unital ring R and let $\Phi : K_1(R) \to K_1(R/I)$ be the natural map.*

(a) *A unit $u \in GL_1(R/I)$ can be lifted to a unit in R if and only if $\delta([u]) = 0$.*

(b) *Φ is surjective if and only if the invertible matrices in $GL_n(R/I)$ can be lifted to $GL_n(R)$ for all $n \geq 1$.*

Theorem 4.11(b) is the exact analogue of Theorem 4.1(b), though now we need the hypothesis of separativity for I. If R/I is a separative exchange ring, then we need only to lift invertible elements from R/I to R to get that Φ is surjective, because of Theorem 4.5.

Acknowledgements. The author thanks Ken Goodearl and Francesc Perera for their useful comments.

Note added in Proof: Recently, the author has constructed an example of a non-unital exchange ring I such that the natural map $G(I) \to K_0(I)$ is not injective, cf. the comment after Proposition 4.4.

References

[A1] P. Ara, *Aleph-nought-continuous regular rings*, J. Algebra **109** (1987), 115–126.

[A2] P. Ara, *Left and right projections are equivalent in Rickart C^*-algebras*, J. Algebra **120** (1989), 433–448.

[A3] _____, *Strongly π-regular rings have stable range one*, Proc. Amer. Math. Soc. **124** (1996), 2293–2298.

[A4] _____, *Extensions of exchange rings*, J. Algebra **197** (1997), 409–423.

[AGOP] P. Ara, K. R. Goodearl, K. C. O'Meara and E. Pardo, *Separative cancellation for projective modules over exchange rings*, Israel J. Math. **105** (1998), 105–137.

[AGOR] P. Ara, K. R. Goodearl, K. C. O'Meara and R. Raphael, *K_1 of separative exchange rings and C^*-algebras with real rank zero*, Pacific J. Math. **195** (2000), 261–275.

[AGPT] P. Ara, K. R. Goodearl, E. Pardo, and D. V. Tyukavkin, *K-theoretically simple von Neumann regular rings*, J. Algebra **174** (1995), 659–677.

[AGS] P. Ara, M. Gómez Lozano and M. Siles Molina, *Local rings of exchange rings*, Comm. Algebra **26**(12) (1998), 4191–4205.

[AOT] P. Ara, K. C. O'Meara and D. V. Tyukavkin, *Cancellation of projective modules over regular rings with comparability*, J. Pure and Applied Algebra **107** (1996), 19–38.

[APa] P. Ara and E. Pardo, *Refinement monoids with weak comparability and applications to regular rings and C^*-algebras*, Proc. Amer. Math. Soc. **124** (1996), 715–720.

[APP] P. Ara, E. Pardo and F. Perera, *The structure of countably generated projective modules over regular rings*, J. Algebra **226** (2000), 161–190.

[APe] P. Ara and F. Perera, *Multipliers of von Neumann regular rings*, Comm. Algebra **28** (2000), 3359–3385.

[Bac] G. Baccella, *Right semiartinian rings are exchange rings*, Preprint.

[Br] G. Brookfield, *Direct sum cancellation of Noetherian modules*, J. Algebra **200** (1998), 207–224.

[BP1] L. G. Brown and G. K. Pedersen, *C^*-algebras of real rank zero*, J. Functional Analysis **99** (1991), 131–149.

[BP2] _____, *On the geometry of the unit ball of a C^*-algebra*, J. Reine Angew. Math. **469** (1995), 113–147.

[BP3] _____, *Non-stable K-theory and extremally rich C^*-algebras*, in preparation.

[CL] H. Chen and F.-U. Li, *Whitehead groups of exchange rings with primitive factors artinian*, preprint.

[CP] A. H. Clifford and G. B. Preston, *The Algebraic Theory of Semigroups, Vol 1*, Math. Surveys **7**, Amer. Math. Soc., Providence, 1961.

[CJ] P. Crawley and B. Jónsson, *Refinements for infinite direct decompositions of algebraic systems*, Pacific J. Math. **14** (1964), 797–855.

[Ev] E. G. Evans, Jr., *Krull-Schmidt and cancellation over local rings*, Pacific J. Math. **46** (1973), 115–121.

[Fac] A. Facchini, *Module Theory: Endomorphism rings and direct sum decompositions in some classes of modules*, Progress in Math. **167**, Birkhäuser, Basel, 1998.

[FU] C. Faith and Y. Utumi, *On a new proof of Litoff's theorem*, Acta Math. Acad. Sci. Hungar. **14** (1963), 369–371.

[G1] K. R. Goodearl, *Direct sum properties of quasi-injective modules*, Bull. Amer. Math. Soc. **82** (1976), 108–110.

[G2] _____, *Von Neumann Regular Rings*, Pitman, London, 1979, Second Ed., Krieger, Malabar, 1991.

[G3] _____, *Partially Ordered Abelian Groups with Interpolation*, Math. Surveys and Monographs, **20**, Amer. Math. Soc., Providence, 1986.

[GW] K. R. Goodearl and R. B. Warfield, Jr., *Algebras over zero-dimensional rings*, Math. Ann. **223** (1976), 157–168.

[Ha] M. Harada, *Factor Categories with Applications to Direct Decompositions of Modules*, Marcel Dekker, New York, 1983.

[HI] M. Harada and T. Ishii, *On perfect rings and the exchange property*, Osaka J. Math. **12** (1975), 483–491.

[Lam] T.-Y. Lam, *Modules with isomorphic multiples and rings with isomorphic matrix rings, a survey*, Monographie **35** de l'Enseignement Mathématique, Genève, 1999.

[MM1]   P. Menal and J. Moncasi, *Lifting units in self-injective rings and an index theory for Rickart C^*-algebras*, Pacific J. Math. **126** (1987), 295–329.

[MM2]   _____, *K_1 of von Neumann regular rings*, J. Pure and Applied Algebra **33** (1984), 295–312.

[MoMu] S. H. Mohamed and B. J. Mueller, *Continuous and discrete modules*, London Math. Soc. Lecture Note Series **147** Cambridge Univ. Press, Cambridge, 1990.

[MMqc] S. H. Mohamed and B. J. Mueller, *On the exchange property for quasi-continuous modules*, in Abelian groups and modules (Padova, 1994), 367–372, Math. Appl. **343**, Kluwer Acad. Publ., Dordrecht, 1995.

[Nic1] W. K. Nicholson, *Lifting idempotents and exchange rings*, Trans. Amer. Math. Soc. **229** (1977), 269–278.

[Nic2] _____, *On exchange rings*, Comm. in Algebra **25**(6) (1997), 1917–1918.

[O'M1] K. C. O'Meara, *Simple regular rings satisfying weak comparability*, J. Algebra **141** (1991), 162–186.

[O'M2] _____, *The exchange property for row and column-finite matrix rings*, J. Algebra, to appear.

[OV]   K. C. O'Meara and C. Vinsonhaler, *Separative cancellation and multi-isomorphism in torsion-free abelian groups*, J. Algebra **221** (1999), 536–550.

[Os]   K. Oshiro, *Projective modules over von Neumann regular rings have the finite exchange property*, Osaka J. Math. **20** (1983), 695–699.

[OR]   K. Oshiro and S. T. Rizvi, *The exchange property of quasi-continuous modules with the finite exchange property*, Osaka J. Math. **33** (1996), 217–234.

[Par1] E. Pardo, *Monoides de refinament i anells d'intercanvi*, Ph.D. Thesis, Universitat Autònoma de Barcelona, 1995.

[Par2] _____, *Comparability, separativity, and exchange rings*, Comm. Algebra **24**(9) (1996), 2915–2929.

[Per1]   F. Perera, *The structure of positive elements for C^*-algebras with real rank zero*, International J. Math. **8** (1997), 383–405.

[Per2]   _____, *Ideal structure of multiplier algebras of simple C^*-algebras with real rank zero*, Canad. J. Math., to appear.

[Per3]   _____, *Lifting units modulo exchange ideals and C^*-algebras with real rank zero*, J. Reine Angew. Math. **522** (2000), 51–62.

[Ros]   J. Rosenberg, *Algebraic K-Theory and Its Applications*, Grad. Texts in Math. **147**, Springer-Verlag, Heidelberg, New York, 1994.

[Sto]   J. Stock, *On rings whose projective modules have the exchange property*, J. Algebra **103** (1986), 437–453.

[W1]   R. B. Warfield, Jr., *Exchange rings and decompositions of modules*, Math. Ann. **199** (1972), 31–36.

[W2]   _____, *Cancellation of modules and groups and stable range of endomorphism rings*, Pacific J. Math. **91** (1980), 457–485.

[We]   F. Wehrung, *Injective positively ordered monoids I*, J. Pure and Applied Algebra **83** (1992), 43–82.

[WuXu]   Tongsuo Wu and Yonghua Xu, *On the stable range condition of exchange rings*, Comm. Algebra **25**(7) (1997), 2355–2363.

[Ya]   K. Yamagata, *On rings of finite representation type and modules with the finite exchange property*, Sci. Rep. Tokyo Kyoiku Daigaku Sect. A **13** (1975), 1–6.

[Yu]   H.-P. Yu, *Stable range one for exchange rings*, J. Pure and Applied Algebra **98** (1995), 105-109.

[ZH]   B. Zimmermann-Huisgen, *Exchanging torsion modules over Dede- kind domains*, Arch. Math. **55** (1990), 241–246.

[ZH-Z]   B. Zimmermann-Huisgen and W. Zimmermann, *Classes of modules with the exchange property*, J. Algebra **88** (1984), 416–434.

[Zha] S. Zhang, *Certain C^*-algebras with real rank zero and their corona and multiplier algebras, Part I*, Pacific J. Math. **155** (1992), 169–197.

Departament de Matemàtiques
Universitat Autònoma de Barcelona
08193 Bellaterra (Barcelona), Spain
e-mail: para@mat.uab.es

Good Conditions for the Total

K. I. Beidar and F. Kasch

Abstract

Let R be a ring with $1 \in R$. In Mod-R the total $\text{Tot}(M, N)$ is a semi-ideal, which contains the radical $\text{Rad}(M, N)$, the singular ideal $\Delta(M, N)$ and the cosingular ideal $\nabla(M, N)$. We study conditions on modules Q and P, which imply that $\text{Rad}(Q, N) = \Delta(Q, N) = \text{Tot}(Q, N)$ and $\text{Rad}(M, P) = \nabla(M, P) = \text{Tot}(M, P)$ for all M and N. We prove that these equalities hold if Q is injective, resp. P is semi-perfect and projective. To get further results and interesting topics, we consider the question: For which Q is $\Delta(Q, N) = \text{Tot}(Q, N)$ for all N? Here we study rings R such that the condition $\Delta(Q, N) = \text{Tot}(Q, N)$ for all N implies that Q is a direct sum of injective modules. We conjecture that such rings must be right Noetherian and prove that they (and all their homomorphic images) are right Goldie rings. Further, the conjecture is confirmed in a number of cases.

1. Introduction

Given a ring R (with unity), we denote by Mod-R the category of right R-modules. In what follows the term "module" will mean "right module". Given a module $M \in$ Mod-R, we denote by $\mathcal{E}_R(M)$ the injective hull of M and by 1_M the identity endomorphism of M. When the context is clear, we shall write $\mathcal{E}(M)$ for $\mathcal{E}_R(M)$. If $\emptyset \neq X \subseteq M$, then we set

$$\text{Ann}_R(X) = \{r \in R \mid Xr = 0\}.$$

Let Ω be a class of homomorphisms of R-modules and $M, N \in$ Mod-R. We set $\Omega(M, N) = \Omega \cap \text{Hom}_R(M, N)$. Recall that the class Ω is called *a semi-ideal* in Mod-R if for any $A, B, C, D \in$ Mod-R we have that $\Omega(B, C) \neq \emptyset$ and $\text{Hom}_R(C, D)\Omega(B, C)\text{Hom}_R(A, B) \subseteq \Omega(A, D)$. Further, the class Ω is

---

*The joint work on this paper was started during the Third Korea-China-Japan International Symposium on Ring Theory, and the authors express their deep gratitude to the organizers of the conference for the invitations and creating the stimulating atmosphere.

called *an ideal* in Mod-R if it is a semi-ideal in Mod-R and $\Omega(M,N)$ is a subgroup of the group $\text{Hom}_R(M,N)$ for all $M, N \in \text{Mod-}R$.

Given $M \in \text{Mod-}R$, we set $\text{Aut}_R(M)$ to be the group of automorphisms of the module M. If $N \subseteq M$ is a submodule of M, then we shall write $N \subseteq^* M$ whenever N is an essential submodule of M and $N \subseteq^\circ M$ whenever N is a small submodule of M. Next, we shall write $N \subseteq^\oplus M$ whenever N is a direct summand of the module M. Finally, given $g \in \text{Hom}_R(M,N)$, we denote by $\text{Ke}(g)$ the kernel of g.

We are now in a position to introduce the main objects of this paper.

Definition 1.1. (1) The radical of M and N is $\text{Rad}(M,N) = \{g \in \text{Hom}_R(M,N) \mid \text{ for all } f \in \text{Hom}_R(N,M), (1_M - fg) \in \text{Aut}_R(M)\}$.

(2) The singular ideal of M and N is $\Delta(M,N) = \{g \in \text{Hom}_R(M,N) \mid \text{Ke}(g) \subseteq^* M\}$.

(3) The cosingular ideal of M and N is $\nabla(M,N) = \{g \in \text{Hom}(M,N) \mid \text{Im}(g) \subseteq^\circ N\}$.

(4) The total of M and N is $\text{Tot}(M,N) = \{g \in \text{Hom}_R(M,N) \mid \text{ for any } f \in \text{Hom}_R(N,M) \text{ with } fg = (fg)^2 \text{ then } fg = 0\}$.

(5) An element $g \in \text{Hom}_R(M,N)$ is called partially invertible (pi) whenever there exists $f \in \text{Hom}_R(N,M)$ such that $fg = (fg)^2 \neq 0$.

We show that the definition of pi elements is symmetric. That is to say, the following two conditions are equivalent:

$$\text{There exists } f \in \text{Hom}_R(N,M) \text{ such that } 0 \neq fg = (fg)^2; \quad (1)$$
$$\text{There exists } h \in \text{Hom}_R(N,M) \text{ such that } 0 \neq gh = (gh)^2. \quad (2)$$

Indeed, by the symmetry it is enough to show that (1) implies (2). Let $f \in \text{Hom}_R(N,M)$ with $0 \neq fg = (fg)^2$. Set $e = fg$. Consider $d = gef$. We have $d^2 = ge(fg)ef = gef = d$ and $e = fdg \neq 0$. Therefore $0 \neq d = d^2$. Set $h = ef$. Clearly $h \in \text{Hom}_R(N,M)$ and $0 \neq gh = (gh)^2$.

Let $n > 0$, $M_1, M_2, \ldots, M_{n+1} \in \text{Mod-}R$ and $g_i \in \text{Hom}_R(M_i, M_{i+1})$, $i = 1, 2, \ldots, n$. Suppose that $g_1 g_2 \ldots g_n$ is pi. We claim that then each g_i is pi. To proceed by induction on n, we start with the case $n = 2$. If $g_1 g_2$ is pi, then there exist equations

$$e = f(g_1 g_2) = (fg_1)g_2 = e^2 \neq 0 \quad \text{and}$$
$$d = (g_1 g_2)h = g_1(g_2 h) = d^2 \neq 0,$$

which show that g_2 and g_1 are pi. In the inductive case $n \geq 3$, we assume that our statement is true for $n - 1$. Since

$$g_1 g_2 \ldots g_n = (g_1 g_2 \ldots g_{n-2})(g_{n-1} g_n),$$

we conclude that both $g_1 g_2 \ldots g_{n-2}$ and $g_{n-1} g_n$ are all pi. It follows from the induction assumption that each g_i is pi, which proves our claim. It now follows at once that Tot is a semi-ideal (see also [11]). Next, that Rad, Δ and ∇ are ideals in Mod-R is well known (and easy to check). Further, this definition of Rad is symmetric, since $1_M - fg$ is an automorphism of M if and only if $1_N - gf$ is an automorphism of N.

Obviously the pi elements of $\operatorname{Hom}_R(M, N)$ are exactly those which are not in $\operatorname{Tot}(M, N)$.

Lemma 1.2. *For $g \in \operatorname{Hom}_R(M, N)$ the following conditions are equivalent*:

(a) g is pi;

(b) *There exist submodules A, C and B, D of M and N respectively such that $A \oplus C = M$, $B \oplus D = N$, the map $A \ni a \mapsto g(a) \in B$ is an isomorphism, $g(C) \subseteq D$ and $A \neq 0$*;

(c) *There exists $h \in \operatorname{Hom}_R(N, M)$ such that $hgh = h \neq 0$.*

The proof of the lemma will be given in Section 2 in spite of the fact that this was essentially proved in [13]. But we need a slightly sharpened form of the lemma. Note that (c) shows that the pi elements are exactly those which occur in the "middle" of the definition of regular elements.

The Tot was defined in 1982 by F. Kasch and then studied in several papers by F. Kasch and W. Schneider (see [10, 11, 12, 13, 14, 18]). In the context of the radical theory it was studied in [3] and its applications to the structure of rings were given in [2]. We continue to study Tot in the category of modules. In general $\operatorname{Tot}(M, N)$ is not additively closed, but the following additive property is always true:

$$\text{if } h \in \operatorname{Rad}(M, N) \text{ and } g \in \operatorname{Tot}(M, N), \text{ then } h + g \in \operatorname{Tot}(M, N)$$

(see [11]). Here we are interested in conditions under which $\operatorname{Tot}(M, N)$ agrees with one (or two) of the ideals in Definition 1.1. This would imply that $\operatorname{Tot}(M, N)$ is additively closed. The goal of this paper is twofold. First we shall study conditions on modules Q and P which imply that

$$\operatorname{Rad}(Q, N) = \Delta(Q, N) = \operatorname{Tot}(Q, N) \text{ for all } N,$$
$$\operatorname{Rad}(M, P) = \nabla(M, P) = \operatorname{Tot}(M, P) \text{ for all } M.$$

Next we shall investigate rings such that the condition $\Delta(Q, N) = \operatorname{Tot}(Q, N)$ for all N implies that Q is a direct sum of injective modules.

Special Case $M = R$. In this case, we can apply on $\operatorname{Hom}_R(R, N)$ the isomorphism

$$\beta : \operatorname{Hom}_R(R, N) \to N, \quad g \xrightarrow{\beta} g(1) \in N, \quad g \in \operatorname{Hom}_R(R, N).$$

How do we change the ideals Rad, Δ, and Tot if we apply β taking into account their definitions?

"$1_R - fg$ is an automorphism" changes to "$(1_R - fg)(1) = 1 - f(g(1))$ is an invertible element of R". It is now easy to see that the radical $\mathrm{Rad}(N)$ of N is contained in $\beta(\mathrm{Rad}(R, N))$. Later we shall give an example of a nonzero module N such that $\mathrm{Rad}(N) = 0$ while $\beta(\mathrm{Rad}(R, N)) = N$.

If $g \in \mathrm{Hom}_R(R, N)$, then

$$\mathrm{Ke}(g) = \{r \in R \mid 0 = g(r) = g(1)r\} = \mathrm{Ann}_R(g(1)),$$

and so $\beta(\Delta(R, N)) = \mathrm{Sin}(N)$, the singular submodule of N. Taking into account this fact, we call Δ in general *the singular ideal* of Mod-R and the dual ∇ *the cosingular ideal*. If $g \in \mathrm{Hom}_R(R, N)$, then $\mathrm{Im}(g) = g(R) = g(1)R$, and hence

$$\beta(\nabla(R, N)) = \mathrm{Rad}(N).$$

Finally, note that $g \in \mathrm{Hom}_R(R, N)$ is pi if and only if there exists $f \in \mathrm{Hom}_R(N, R) = N^*$ such that $0 \neq fg = (fg)^2$. Setting $e = fg$, we see that

$$e(1) = e^2(1) = e(1 \cdot e(1)) = e(1)e(1) = e(1)^2 = (fg)(1) = f(g(1)).$$

That means that for $g(1)$ there exists $f \in N^*$ such that $f(g(1))$ is a nonzero idempotent in R. Therefore $g(1) \in \beta(\mathrm{Tot}(R, N))$ if and only if $N^*(g(1))$ contains no nonzero idempotents. The latter condition is equivalent to the following one: for any $f \in N^*$ the equality $f(g(1))^2 = f(g(1))$ implies that $f(g(1)) = 0$. It is now clear how to define "partially invertible element" and the total of a module N.

Definition 1.3. An element $a \in N$ is called partially invertible (pi) if there exists $f \in N^*$ such that $f(a)$ is a nonzero idempotent of R. Further, the subset $\mathrm{Tot}(N) = \{x \in N \mid x \text{ is not pi}\}$ of N is called the total of N.

2. Good Condition for the Total

The main result of this section is the following theorem.

Theorem 2.1. (a) *If $Q \in$ Mod-R is injective, then*

$$\mathrm{Rad}(Q, N) = \Delta(Q, N) = \mathrm{Tot}(Q, N) \ for\ all\ N \in Mod\text{-}R.$$

(b) *If $P \in Mod\text{-}R$ is projective and semi-perfect, then*

$$\mathrm{Rad}(M, P) = \nabla(M, P) = \mathrm{Tot}(M, P) \ for\ all\ M \in Mod\text{-}R.$$

Good Conditions for the Total

The proof will be given in a series of steps in which we prove somewhat more than these results.

For abbreviation, we use the following notation. If $g \in \text{Hom}_R(M, N)$, then (A, B) is said to be *a g-pair*, if $A \subseteq^\oplus M$, $B \subseteq^\oplus N$ and g induces an isomorphism from A to B. If $A \subseteq^\oplus M$, then by

$$i_A : A \to M \quad \text{and} \quad \pi_A : M \to A$$

we denote the inclusion of A in M and the canonical projection of M onto A respectively. Clearly $\pi_A i_A = 1_A$ and $i_A \pi_A$ is the projector of A (remark: the range of π_A is A while the range of $i_A \pi_A$ is M). The isomorphism induced by g on a g-pair (A, B) is then $\pi_B g i_A$. This is independent of the decomposition $N = B \oplus V$, since for a g-pair we have $g(A) = B$.

If, for example, $\Delta(M, N) \subseteq \text{Tot}(M, N)$ for all M and N, then we write simply $\Delta \subseteq \text{Tot}$. Prior we have some trivial inclusions.

Remark 2.2. Rad, Δ, $\nabla \subseteq \text{Tot}$.

Proof. Assume that $g \in \text{Hom}_R(M, N)$ and $g \notin \text{Tot}(M, N)$. Then g is pi and so there exists $f \in \text{Hom}_R(N, M)$ with $0 \neq fg = (fg)^2$. Therefore $(1 - fg)fg = 0$ and hence $1 - fg$ is not an automorphism of M. We see that $g \notin \text{Rad}(M, N)$. Further,

$$\text{Ke}(g) \subseteq \text{Ke}(fg) = (1 - fg)M$$

is not an essential submodule of M and hence $g \notin \Delta(M, N)$. Assume that $\text{Im}(g) \subseteq^\circ N$. Then

$$(fg)(M) = \text{Im}(fg) = f(\text{Im}(g)) \subseteq^\circ M,$$

contradicting $M = [(fg)M] \oplus [(1 - fg)(M)]$. Therefore $\text{Im}(g) \not\subseteq^\circ N$ and so $g \notin \nabla(M, N)$. The proof is now complete. □

Proof of Lemma 1.2. Assume that g is pi. Then there exists $f : N \to M$ with $0 \neq fg = (fg)^2$. Set $u = fg$ and $v = guf$. Then $fvg = u^3 = u \neq 0$ and $v^2 = gu(fg)uf = gu^3 f = guf = v$. Set $A = uM$ and $B = vN$. Clearly $A \subseteq^\oplus M$ and $B \subseteq^\oplus N$. We now have

$$g(A) = gu(M) = gu^2(M) = gufg(M) = vg(M) \subseteq vN \quad \text{and}$$
$$vN = guf(N) \subseteq gu(M) \subseteq g(A),$$

and so $g(A) = B$. Further, $\text{Ke}(g) \cap A \subseteq \text{Ke}(fg) \cap A = \text{Ke}(u) \cap uM = 0$. Therefore $A \ni a \mapsto g(a) \in B$ is an isomorphism. Write $N = B \oplus D$. Since $B \subseteq g(M)$, by the modular law $g(M) = B \oplus U$ where $U = g(M) \cap D$. Set

$C = g^{-1}(U)$. Clearly $g(C) \subseteq D$; to finish the proof for (a)$\Rightarrow$(b), we show $M = A \oplus C$. For $m \in M$, it follows that $g(m) = b+u$, $b \in B$, $u \in U$. Since $g(A) = B$, there exists $a \in A$ with $g(a) = b$. Then follows $g(m - a) = u$, hence $m - a \in C$ and then $m \in A + C$, therefore $M = A + C$. Assume that $a \in A \cap C$. Then $g(a) \in B \cap U = 0$ and since $\pi_B g i_A$ is an isomorphism, it follows that $a = 0$. Together we have $M = A \oplus C$.

Suppose that (b) is satisfied. Define $f : N \to M$ by the rule $f|_D = 0$, $f|_B = (\pi_B g i_A)^{-1}$. Clearly $fgf = f \neq 0$ and so $fg \neq 0$ and $(fg)^2 = (fgf)g = fg$. Therefore both (a) and (c) are fulfilled.

Finally, assume that (c) is satisfied and let h be as in (c). Then $hg \neq 0$ and $(hg)^2 = (hgh)g = hg$, which proves (a). The proof is now complete. $\square$

Theorem 2.3. (a) *If M is a module such that every monomorphism $M \to M$ splits, then*

$$\Delta(M, N) \subseteq \text{Rad}(M, N) \text{ for all } N \in \text{Mod-}R.$$

(b) *Let $Q \in \text{Mod-}R$. Then*

$$\Delta(Q, M) = \text{Tot}(Q, M) \text{ for all } M \in \text{Mod-}R$$

if and only if Q is an essential extension of a direct sum of injective submodules.

(c) *If Q is injective, then for all $N \in \text{Mod-}R$ and each $g \in \text{Hom}_R(Q, N)$ there exists $f \in \text{Hom}_R(N, Q)$ with $gfg - g \in \Delta(Q, N)$.*

Hint: The assumption in (a) is a special case of the condition (C_2) in the definition of continuous modules (see [6, 17]).

Proof. (a) Let $g \in \Delta(M, N)$ and $f \in \text{Hom}_R(N, M)$. Then $fg \in \Delta(M, M)$. Since $\text{Ke}(fg) \cap \text{Ke}(1_M - fg) = 0$ and $\text{Ke}(fg) \subseteq^* M$, we conclude that $\text{Ke}(1_M - fg) = 0$ and hence $1_M - fg$ is a monomorphism, which splits by assumption: $\text{Im}(1_M - fg) \subseteq^\oplus M$. As $\text{Ke}(fg) \subseteq \text{Im}(1_M - fg)$ and $\text{Ke}(fg) \subseteq^* M$, also $\text{Im}(1_M - fg) \subseteq^* M$. But a direct summand, which is essential in M, must be equal to M. Hence $1_M - fg$ is an automorphism and therefore $g \in \text{Rad}(M, N)$.

(b) Suppose that $\Delta(Q, M) = \text{Tot}(Q, M)$ for all $M \in \text{Mod-}R$. It follows from Zorn's lemma that Q contains a family $\{E_i \mid i \in I\}$ of injective submodules such that $\sum_{i \in I} E_i = \oplus_{i \in I} E_i$, and for any injective submodule G of Q we have that $G \cap \sum_{i \in I} E_i \neq 0$. Set $E = \sum_{i \in I} E_i$ and let F be a complement of E. We claim that $F = 0$. Assume to the contrary that $F \neq 0$ and set $M = \mathcal{E}(F)$. By assumption $\Delta(Q, M) = \text{Tot}(Q, M)$. Clearly $E + F = E \oplus F$ and so there exists a homomorphism $\alpha : E + F \to M$

such that $\alpha(E) = 0$ and $\alpha|_F$ is the canonical embedding $F \to \mathcal{E}(F) = M$. Since M is injective, we may assume that $\alpha : Q \to M$. As $\text{Ke}(\alpha) \cap F = 0$, $\alpha \notin \Delta(Q, M)$ and so it is pi. By Lemma 1.2 there exists an α-pair (A, B). Since M is injective, both B and A are injective as well. Clearly $A \cap \text{Ke}(\alpha) = 0$. Recalling that $E \subseteq \text{Ke}(\alpha)$, we conclude that $A \cap E = 0$, a contradiction to the choice of $\{E_i \mid i \in I\}$. Thus $\oplus_{i \in I} E_i \subseteq^* Q$ and (b) is satisfied.

Conversely, assume that $\oplus_{i \in I} E_i \subseteq^* Q$ where each E_i is injective. Since $\Delta(Q, M) \subseteq \text{Tot}(Q, M)$ for all M by Remark 2.2, it is enough to show that $\Delta(Q, M) \supseteq \text{Tot}(Q, M)$. Assume to the contrary that $\Delta(Q, M) \not\supseteq \text{Tot}(Q, M)$ and pick $f \in \text{Tot}(Q, M) \setminus \Delta(Q, M)$. Then $\text{Ke}(f) \not\subseteq^* Q$ and so there exists $i \in I$ such that $\text{Ke}(f) \cap E_i \not\subseteq^* E_i$. This is easy to see. If we assume that for all $i \in I$ $\text{Ke}(f) \cap E_i \subseteq^* E_i$, then by [9, 5.1.8],

$$K = \oplus_{i \in I}(\text{Ke}(f) \cap E_i) \subseteq^* \oplus_{i \in I} E_i.$$

By assumption $\oplus_{i \in I} E_i \subseteq^* Q$, so $K \subseteq^* Q$, and since $K \subseteq \text{Ke}(f)$ also $\text{Ke}(f) \subseteq^* Q$, contradicting $f \notin \Delta(Q, M)$. Further, let G be a complement of $\text{Ke}(f) \cap E_i$ in E_i. Then G is a nonzero injective submodule of E_i. Clearly $G \cong f(G)$ and hence $f(G)$ is an injective module. We have that $G \subseteq^\oplus Q$, $f(G) \subseteq^\oplus M$ and f induces an isomorphism $G \to f(G)$. By Lemma 1.2, f is pi, contradicting $f \in \text{Tot}(Q, M)$.

(c) If $g \in \Delta(Q, N)$, then for each $f \in \text{Hom}_R(N, Q)$ we have $gfg - g \in \Delta(Q, N)$. Therefore we can and we will assume that $g \notin \Delta(Q, N)$. Let C be a complement of $\text{Ke}(g)$ in Q. Then C is an injective module and g induces an isomorphism ϕ of C and $g(C)$. Therefore $g(C)$ is also injective and so $C \subseteq^\oplus Q$ and $g(C) \subseteq^\oplus N$. We set $f = i_C \phi^{-1} \pi_{g(C)}$. Since C is a complement of $\text{Ke}(g)$, we have $C + \text{Ke}(g) \subseteq^* Q$. We show that $C + \text{Ke}(g) \subseteq \text{Ke}(gfg - g)$. Indeed, let $c \in C$ and $a \in \text{Ke}(g)$. Then $gfg(c + a) = gfg(c) = g(c) = g(c + a)$, which implies that $gfg - g \in \Delta(Q, N)$. The proof is now complete. □

In the case $Q = N$, the proof of statement (c) of Theorem 2.3 shows that only the assumption of Q to be continuous is needed. If $S = \text{End}_R(Q)$, then it follows that

$$\text{Rad}(S) = \Delta(S) = \text{Tot}(S)$$

(the first equation is well known) and $S/\Delta(S)$ is regular. If we specialize to $R = Q = N$ with continuous R_R, then

$$\text{Rad}(R) = \text{Sin}(R) = \text{Tot}(R)$$

and $R/\text{Sin}(R)$ is regular. If $Q = R_R$ is injective, then

$$\beta(\text{Rad}(R, N)) = \text{Sin}(N) = \text{Tot}(N) \text{ for all } N.$$

Example. Let F be a field and $F[x]$ the polynomial algebra in x over F. Set $R = F[x]/(x^2 F[x])$ and $y = x + x^2 F[x]$. Clearly R is a local ring with radical $N = Fy$. It is easy to see that R is self-injective and N is a simple R-module. Therefore $\operatorname{Rad}(N) = 0$. We claim that $\beta(\operatorname{Rad}(R, N)) = N$. Indeed, given $0 \neq z \in N$, we define a homomorphism $g : R \to N$ by the rule $g(r) = zr$, $r \in R$. Then $g(1) = z$ and so $\beta(g) = z$. Let $f \in \operatorname{Hom}_R(N, R)$. Since N is simple and the socle of R is equal to N, we conclude that $\operatorname{Im}(f) \subseteq N = \operatorname{Rad}(R)$. Therefore $\operatorname{Im}(fg) \subseteq \operatorname{Rad}(R)$. Set $a = fg(1)$. Then $a \in \operatorname{Rad}(R)$ and so $1 - a$ is an invertible element of R. Given $r \in R$, we have

$$(1_R - fg)(r) = r - fg(r) = r - fg(1)r = r - ar = (1-a)r$$

and so $1_R - fg$ is an automorphism of R_R which forces $g \in \operatorname{Rad}(R, N)$. Since $\beta(g) = z$, we conclude that $z \in \beta(\operatorname{Rad}(R, N))$ and so we have $\beta(\operatorname{Rad}(R, N)) = N$.

Theorem 2.4. (a) *If N is a module such that every epimorphism $N \to N$ splits, then*

$$\nabla(M, N) \subseteq \operatorname{Rad}(M, N) \text{ for all } M.$$

(b) *If P is projective and semi-perfect, then*

$$\nabla(M, P) = \operatorname{Tot}(M, P) \text{ for all } M.$$

(c) *If P is projective and semi-perfect, then for all M and each $g : M \to P$ there exists $f \in \operatorname{Hom}_R(P, M)$ such that $gfg - g \in \nabla(M, P)$.*

Proof. (a) Assume that $g \in \nabla(M, N)$. Then for any $f \in \operatorname{Hom}_R(N, M)$ we also have $gf \in \nabla(N, N)$. Since $\operatorname{Im}(gf) + \operatorname{Im}(1_N - gf) = N$ and $\operatorname{Im}(gf) \subseteq^\circ N$, it follows that $\operatorname{Im}(1_N - gf) = N$. Therefore $1_N - gf$ is an epimorphism, which, by assumption, splits. It follows that $\operatorname{Ke}(1_N - gf) \subseteq^\oplus N$. But

$$\operatorname{Ke}(1 - gf) \subseteq \operatorname{Im}(gf) \subseteq^\circ N$$

forcing $\operatorname{Ke}(1_N - gf) = 0$. That is to say $1_N - gf$ is an automorphism of N and so $g \in \operatorname{Rad}(M, N)$.

(b) In view of Remark 2.2, we have only to prove that if $g \notin \nabla(M, P)$, then $g \notin \operatorname{Tot}(M, P)$. Therefore we assume for $g \in \operatorname{Hom}_R(M, P)$ that $\operatorname{Im}(g)$ is not small in P. Therefore there is a supplement $C \neq P$ of $\operatorname{Im}(g)$ in P such that $C \cap \operatorname{Im}(g) \subseteq^\circ P$ and there exists a supplement B of C in P with $B \subseteq \operatorname{Im}(g)$ (see [20, 42.3]). Therefore $D = B \cap C \subseteq^\circ B, C$ by [20, Section 41] and so $D \oplus D \subseteq^\circ B \oplus C$ by [20, 19.3]. Consider the map $f : B \oplus C \to P$ given by the rule $f(b, c) = b + c$, $b \in B$, $c \in C$. Clearly

Ke$(f) = \{(d, -d) \mid d \in D\} \subseteq D \oplus D \subseteq^\circ B \oplus C$, implying Ke$(f) \subseteq^\circ B \oplus C$. Since P is projective, f splits and so Ke$(f) = 0$. That is to say, $B \cap C = 0$ and $B + C = P$. We see that $P = B \oplus C$ and so B is projective. Since $C \cap \text{Im}(g) \subseteq^\circ P$ and Im(g) is not small in P, we conclude that $B \neq 0$. Now we consider $\pi_B g : M \to B$. Since $B \subseteq \text{Im}(g)$, $\pi_B g$ is surjective and therefore splits: $M = A \oplus \text{Ke}(\pi_B g)$. Then $A \ni a \mapsto \pi_B g(a) \in B$ is an isomorphism and therefore $\pi_B g$ is pi and then also g is pi; that is, $g \notin \text{Tot}(M, P)$.

(c) If $g \in \nabla(M, P)$, then $gfg - g \in \nabla(M, P)$ for all f. Assume that $g \notin \nabla(M, P)$. We shall use submodules of M and P from the proof of (b). We define $f \in \text{Hom}(P, M)$ by

$$f|_C = 0 \quad \text{and} \quad f|_B = (\pi_B g i_A)^{-1}.$$

Given $x \in \text{Ke}(\pi_B g)$ and $a \in A$, we have $g(x) = c \in C \cap \text{Im}(g)$ (because $\pi_B g(x) = 0$) and $g(a) = c_1 + b$ where $c_1 \in C$, $b \in B$. Then $\pi_B g(a) = b$ and so

$$gfg(x + a) = gf(c + c_1 + b) = gf(b) = g(a) = c_1 + b.$$

We now have that

$$(gfg - g)(x + a) = c_1 + b - (c + c_1 + b) = -c \in C \cap \text{Im}(g) \subseteq^\circ P$$

and hence $(gfg - g)(M) \subseteq C \cap \text{Im}(g) \subseteq^\circ P$ implying $gfg - g \in \nabla(M, P)$. The proof is now complete. □

Remarks. As the proofs of (b) and (c) show, in the case $M = P$ the assumption for P can be weakened to "P is discrete" (see [17]). If $T = \text{End}(P)$, then

$$\text{Rad}(T) = \nabla(T) = \text{Tot}(T).$$

If $M = R$, then Theorem 2.4 implies that $\text{Rad}(R, P) = \nabla(R, P) = \text{Tot}(R, P)$ and then, using $\nabla(R, P) = \text{Rad}(P)$, we get

$$\beta(\text{Rad}(R, P)) = \text{Rad}(P) = \text{Tot}(P).$$

If R_R is discrete, then it follows that

$$\text{Rad}(R) = \text{Tot}(R).$$

Proof of Theorem 2.1. (a) It follows from both Remark 2.2 and Theorem 2.3(a) and (b) that

$$\text{Tot}(Q, N) = \Delta(Q, N) \subseteq \text{Rad}(Q, N) = \text{Tot}(Q, N) \text{ for all } N,$$

which proves statement (a).

(b) Both Remark 2.2 and Theorem 2.4(a) and (b) imply that

$$\text{Tot}(M, P) = \nabla(M, P) \subseteq \text{Rad}(M, P) \subseteq \text{Tot}(M, P) \text{ for all } M,$$

so statement (b) is also proved. □

3. The Inverse Problem

Let

$$\Omega(R) = \{Q \in \text{Mod-}R \mid \Delta(Q, M) = \text{Tot}(Q, M) \text{ for all } M \in \text{Mod-}R\}$$

and

$$\Gamma(R) = \{Q \in \text{Mod-}R \mid Q \text{ is a direct sum of injective modules}\}.$$

When the context is clear, we shall write Ω for $\Omega(R)$ and Γ for $\Gamma(R)$.

We consider the following questions:
(a) If $Q \in \Omega$, is then Q injective?
(b) If $\Omega(R) = \Gamma(R)$, is the ring R right Noetherian?
(c) If $P \in \text{Mod-}R$ and $\nabla(M, P) = \text{Tot}(M, P)$ for all M, is then P projective (and semi-perfect)?

It follows from Theorem 2.3(b) that $\Omega(R) \supseteq \Gamma(R)$. Let R and N be as in the previous example. Then $i_N \notin \Delta(N, R)$ and $i_N \in \text{Tot}(N, R)$ because R is a local ring. Therefore in general $\text{Tot} \neq \Delta$.

A first step in our study of these questions is the following theorem.

Theorem 3.1. (a) *If $Q \in \text{Mod-}R$ has the a.c.c. for direct summands and $Q \in \Omega$, then Q is injective.*

(b) *If R is a right Noetherian ring and $Q \in \Omega$, then Q is injective; in particular $\Omega(R) = \Gamma(R)$.*

(c) *If $P \in \text{Mod-}R$ has the a.c.c. for direct summands and $\nabla(M, P) = \text{Tot}(M, P)$ for all M, then P is projective.*

First we prove the following useful lemma.

Lemma 3.2. *Let $M, N \in \text{Mod-}R$, $A \subseteq^\oplus M$ and $B \subseteq^\oplus N$. Then:*
(a) *If $\Delta(M, N) = \text{Tot}(M, N)$, then $\Delta(A, B) = \text{Tot}(A, B)$.*
(b) *If $\nabla(M, N) = \text{Tot}(M, N)$, then $\nabla(A, B) = \text{Tot}(A, B)$.*

Proof. We shall prove only (a); one can prove (b) analogously. In view of Remark 2.2, it is enough to show that $\text{Tot}(A, B) \subseteq \Delta(A, B)$. Let $h \in \text{Tot}(A, B)$. Since both Tot and Δ are semi-ideals, $i_B h \pi_A \in \text{Tot}(M, N) = \Delta(M, N)$ and then $\pi_B i_B h \pi_A i_A = h \in \Delta(A, B)$. □

As an immediate corollary to (a) we have the following result.

Good Conditions for the Total

Corollary 3.3. *Let $Q \in \Omega$ and suppose that $A \subseteq^{\oplus} Q$. Then $A \in \Omega$.*

Proof of Theorem 3.1. (a) By Theorem 2.3(b), $\oplus_{i \in I} E_i \subseteq^* Q$ for some nonzero injective submodules E_i, $i \in I$, of Q. If $|I| < \infty$, then $Q = \oplus_{i \in I} E_i$ and there is nothing to prove. Assume that $|I| = \infty$. We may assume that $\{1, 2, \dots\} \subseteq I$. Then

$$E_1 \subseteq E_1 \oplus E_2 \subseteq \cdots \subseteq \oplus_{t=1}^{n} E_t \subseteq \cdots$$

is an infinite increasing chain of direct summands of Q, a contradiction. Thus $|I| < \infty$ and so (a) is proved.

(b) Since the direct sum of a family of injective modules over a right Noetherian ring is an injective module [8, Theorem 20.1], it has no proper essential extensions and so the result follows from Theorem 2.3(b).

(c) Suppose we have already found submodules $P_1, P_2, \ldots, P_n, P'$ of P such that $P = P' \oplus (\oplus_{i=1}^{n} P_i)$ and each P_i is a nonzero projective module (it is understood that if $n = 0$, then $P' = P$). In view of our assumption we may assume that P' has no nonzero direct summand that is projective. Assume that $P' \neq 0$. By Lemma 3.2(b), $\nabla(M, P') = \mathrm{Tot}(M, P')$ for all M. Choose a free module F with epimorphism $g : F \to P'$. Clearly $g \notin \nabla(F, P') = \mathrm{Tot}(F, P')$ and so g is pi. By Lemma 1.2, there exists a g-pair (A, B). Since A is projective, B is also projective, contradicting $B \subseteq^{\oplus} P'$. Thus $P' = 0$ and so P is projective. □

Theorem 3.4. *Let $0 \neq M, N \in \mathrm{Mod}\text{-}R$. Assume that either M or N satisfies the a.c.c. on direct summands. Then:*

(a) If $\Delta(M, N) = \mathrm{Tot}(M, N)$, then each monomorphism $g : M \to N$ splits.

(b) If $\nabla(M, N) = \mathrm{Tot}(M, N)$, then each epimorphism $g : M \to N$ splits.

Proof. (a) Let $g : M \to N$ be a monomorphism. Since $\mathrm{Ke}(g) = 0$, we get $g \notin \Delta(M, N) = \mathrm{Tot}(M, N)$ and so g is pi. In the set of all g-pairs there exists a maximal one, which we denote by (A, B). If $M = A$, then $B = g(A) \subseteq^{\oplus} N$ and there is nothing to prove. Assume that $M = A \oplus C$ and $N = B \oplus D$ where C and D are as in Lemma 1.2 and $C \neq 0$. Clearly either C or D satisfies the a.c.c. on direct summands. Next, $\Delta(C, D) = \mathrm{Tot}(C, D)$ by Lemma 3.2. Setting $g' = \pi_D g i_C$, we see that $g'(c) = g(c)$ for all $c \in C$ (because $g(C) \subseteq D$) and so g' is a monomorphism. As above we conclude that there exists a g'-pair (A', B') and so $(A \oplus A', B \oplus B')$ is also a g-pair, a contradiction. Statement (b) can be proved analogously. □

Let $\phi : R \to S$ be a ring homomorphism and let M be an S-module. Clearly M is an R-module under $mr = m\phi(r)$ for all $m \in M$, $r \in R$.

Recall that ϕ is called *an epimorphism of rings* if for any ring T and any ring homomorphisms $\beta, \gamma : S \to T$, $\beta\phi = \gamma\phi$ implies that $\beta = \gamma$. Further, an epimorphism of rings $\phi : R \to S$ is said to be *flat* if S is a flat left R-module.

Lemma 3.5. *Let $\phi : R \to S$ be a flat epimorphism of rings and $M \in$ Mod-S.*

(a) If M is injective, then M_R is also injective.
(b) If $M_R = U_R \oplus V_R$, then $US = U$ and $VS = V$.
(c) If $N_S \subseteq^* M_S$, then $N_R \subseteq^* M_R$.

Proof. Let M be an injective S-module, L a right ideal of R and $f : L \to M$ a homomorphism of R-modules. Consider the canonical embedding $i : L \to R$. Since S is a flat left R-module, $i \otimes 1 : L \otimes_R S \to R \otimes_R S$ is a monomorphism. Define the map $g : L \otimes_R S \to M$ by the rule $g(l \otimes s) = f(l)s$, $l \in L$, $s \in S$. Clearly g is a well-defined homomorphism of S-modules and so there exists a homomorphism of S-modules $h : R \otimes_R S \to M$ such that $h(l \otimes 1) = g(l \otimes 1) = f(l)$ for all $l \in L$. Define the map $p : R \to M$ by the rule $p(r) = h(r \otimes 1)$, $r \in R$. Clearly p is an additive map and $p(l) = f(l)$ for all $l \in L$. Further,

$$p(r) = h(r \otimes 1) = h(1 \otimes \phi(r)) = h(1 \otimes 1)\phi(r) = p(1)r, \quad r \in R,$$

and so p is a homomorphism of R-modules. Therefore M_R is injective.

Next assume that $M_R = U_R \oplus V_R$. According to [19, Chapter XI, Theorem 2.1] for every $b \in S$ there exist $a_1, a_2, \ldots, a_n \in R$ and $b_1, b_2, \ldots, b_n \in S$ such that

$$\sum_{i=1}^{n} \phi(a_i)b_i = 1 \quad \text{and} \quad b\phi(a_i) \in \phi(R) \quad \text{for all } i = 1, 2, \ldots, n. \quad (3)$$

Suppose that $US \not\subseteq U$. Then there exist $x, u \in U$, $b \in S$, $v \in V$ such that $xb = u + v$ and $v \neq 0$. Let a_i, b_i, $i = 1, 2, \ldots, n$ be as in (3). Then $xb\phi(a_i) = x(b\phi(a_i)) \in U$ and so $v\phi(a_i) = 0$ for all $i = 1, 2, \ldots, n$. Therefore

$$0 = \sum_{i=1}^{n}(v\phi(a_i))b_i = v\sum_{i=1}^{n}\phi(a_i)b_i = v,$$

a contradiction. Therefore $US = U$. Analogously $VS = V$.

Finally, assume that $N_S \subseteq^* M_S$. Let $0 \neq x \in M$. By assumption there exists $b \in S$ such that $0 \neq xb \in N$. Let a_i, b_i, $i = 1, 2, \ldots, n$, be as in (3). Since $\sum_{i=1}^{n}\phi(a_i)b_i = 1$, there exists i such that $xb\phi(a_i) \neq 0$. Since $b\phi(a_i) \in \phi(R)$ and $xR = x\phi(R)$, we conclude that $xR \cap N \neq 0$ and so $N_R \subseteq^* M_R$. The proof is complete. □

Good Conditions for the Total

Theorem 3.6. *Let R be a ring such that $\Omega(R) = \Gamma(R)$. Then:*
(a) *If B is an ideal of R and $\overline{R} = R/B$, then $\Omega(\overline{R}) = \Gamma(\overline{R})$.*
(b) *If $\phi : R \to S$ is a flat epimorphism of rings, then $\Omega(S) = \Gamma(S)$.*

Proof. First we make several general observations. Let T be an $\overline{R}$-module. Then T is canonically an R-module and $TB = 0$. Given an R-module S, we set
$$\alpha(S) = \{x \in S \mid xB = 0\}.$$
Clearly $\alpha(S)$ is an $\overline{R}$-module canonically. Let E be an injective R-module. We claim that
$$\alpha(E) \text{ is an injective } \overline{R}\text{-module.} \tag{4}$$
Indeed, given a right ideal L of the ring $\overline{R}$ and a homomorphism of $\overline{R}$-modules $f : L \to \alpha(E)$, f is also a homomorphism of R-modules and so there exists a homomorphism of R-modules $h : \overline{R} \to E$ such that $h(x) = f(x)$ for all $x \in L$. We have $0 = h(\overline{R}B) = h(\overline{R})B$ and so $h(\overline{R}) \subseteq \alpha(E)$, which proves our claim.

Let T be an $\overline{R}$-module and $E = \mathcal{E}_R(T)$. Note that $T \subseteq \alpha(E)$. Clearly $T_{\overline{R}} \subseteq^* \alpha(E)_{\overline{R}}$ because $T_R \subseteq^* \alpha(E)_R$. It now follows from (4) that
$$\alpha(E_R(T)) \text{ is an injective hull of the } \overline{R}\text{-module } T. \tag{5}$$
Given a family $\{M_i \mid i \in I\}$ of R-modules, one can readily check that
$$\alpha(\oplus_{i \in I} M_i) = \oplus_{i \in I} \alpha(M_i). \tag{6}$$
Further, if U and V are submodules of an R-module W and $UB = 0$, then clearly
$$\alpha(U + V) = U + \alpha(V). \tag{7}$$

Now let $M \in \Omega(\overline{R})$. By Theorem 2.3 there exists a family $\{U_i \mid i \in I\}$ of injective submodules of the $\overline{R}$-module M such that $\oplus_{i \in I} U_i \subseteq^* M$. Note that $\oplus_{i \in I} U_i$ is an essential submodule of the R-module M as well. Setting $E = \mathcal{E}(M)$, we denote by V_i an essential closure of U_i in E. Since $\oplus_{i \in I} U_i \subseteq^* M$, we have $\oplus_{i \in I} V_i \subseteq^* E$ and so $\oplus_{i \in I} V_i \subseteq^* M + \oplus_{i \in I} V_i = N$ as well. Therefore $N \in \Omega(R)$ by Theorem 2.3 and hence $N \in \Gamma(R)$. That is to say, $N = \oplus_{j \in J} W_j$ where each W_j is an injective R-module. Set $W'_j = \alpha(W_j)$, $j \in J$. By (4) each W'_j is an injective $\overline{R}$-module. Further, (6) implies that $\alpha(N) = \oplus_{j \in J} W'_j$. On the other hand $MB = 0$ and so both (6) and (7) imply
$$\alpha(N) = \alpha(M + \oplus_{i \in I} V_i) = M + \alpha(\oplus_{i \in I} V_i) = M + \oplus_{i \in I} \alpha(V_i).$$

By (5) each $\alpha(V_i)$ is an injective hull of the $\overline{R}$-module U_i. Since U_i is an injective $\overline{R}$-module, we conclude that $\alpha(V_i) = U_i$, $i \in I$, and so

$$\oplus_{j \in J} W'_j = \alpha(N) = M + \oplus_{i \in I} U_i.$$

Thus $M \in \Gamma(\overline{R})$ and so $\Omega(\overline{R}) \subseteq \Gamma(\overline{R})$. Since $\Gamma \subseteq \Omega$ for any ring, (a) is proved.

We now prove (b). Let $\phi : R \to S$ be as in (b) and let $M \in \Omega(S)$. Setting $E = E_S(M)$, we see that there exists a family $\{E_i \mid i \in I\}$ of injective submodules of E_S such that $(\oplus_{i \in I} E_i)_S \subseteq^* M_S \subseteq^* E_S$. By Lemma 3.5, E and each E_i are injective R-modules and $(\oplus_{i \in I} E_i)_R \subseteq^* M_R \subseteq^* E_R$. Therefore $M \in \Omega(R) = \Gamma(R)$ and so there exists a family $\{U_j \mid j \in J\}$ of injective submodules of E_R such that $M = \oplus_{j \in J} U_j$. Clearly every U_j is a direct summand of the R-module E and so it is a direct summand of the S-module E by Lemma 3.5. Therefore each U_j is an injective S-module and so $M \in \Gamma(S)$ forcing $\Omega(S) = \Gamma(S)$. □

A module $M \in$ Mod-R is called radical if $M = \text{Rad}(M)$. Further, we denote by Soc(M) the socle of M.

Theorem 3.7. *Suppose that $\Omega(R) = \Gamma(R)$. Then:*

(a) Every finitely generated right R-module has finite Goldie dimension.

(b) Any injective right R-module is an essential extension of a direct sum of a family of indecomposable injective R-modules.

(c) A direct sum of any family of nonsingular injective right R-modules is injective.

(d) Given a family of injective right R-modules $\{E_i \mid i \in I\}$ with injective hull E of $\oplus_{i \in I} E_i$, we have that $\text{Rad}(E) = \oplus_{i \in I} \text{Rad}(E_i)$. Further, if $\Lambda = \text{End}_R(E)$, then $E = \Lambda(\oplus_{i \in I} E_i)$.

(e) A direct sum of any family of radical injective right R-modules is injective.

(f) A direct sum of any family of pairwise isomorphic indecomposable injective right R-modules is injective.

(g) A direct sum of any family of simple injective right R-modules is injective.

Proof. Let M be a finitely generated R-module. Suppose to the contrary that M contains nonzero submodules $\{M_i \mid i = 1, 2, \ldots\}$ such that $\sum_{i=1}^{\infty} M_i = \oplus_{i=1}^{\infty} M_i$. We may assume without loss of generality that each M_i is cyclic and so contains a maximal submodule, say L_i. Setting $L = \sum_{i=1}^{\infty} L_i$, we see that the socle of the factor module M/L has infinite Goldie dimension. Factoring out the complement of

Good Conditions for the Total

the socle, we reduce the proof to the case when M contains a family $\{S_i \mid i = 1, 2, \ldots\}$ of simple modules such that $\mathrm{Soc}(M) = \oplus_{i=1}^{\infty} S_i \subseteq^* M$. Let $E = \mathcal{E}(M)$ and let E_i be an essential closure of S_i, $i = 1, 2, \ldots$. Since $\mathrm{Soc}(M) \subseteq^* M \subseteq^* E$, $\mathrm{Soc}(M) \subseteq^* E$. Further, $\sum_{i=1}^{\infty} E_i = \oplus_{i=1}^{\infty} E_i \subseteq^* E$ because $\mathrm{Soc}(M) \subseteq \oplus_{i=1}^{\infty} E_i$. Therefore E is an essential extension of $\oplus_{i=1}^{\infty} E_i$. Suppose that $M \subseteq \oplus_{i=1}^{\infty} E_i$. Since M is finitely generated, there exists a positive integer n such that $M \subseteq \oplus_{i=1}^{n} E_i$ and so $\mathrm{Soc}(M) \subseteq \oplus_{i=1}^{n} E_i$, a contradiction. We see that $M \not\subseteq \oplus_{i=1}^{\infty} E_i$. It follows from Zorn's lemma that there exists a submodule N of E maximal with respect to the properties $\oplus_{i=1}^{\infty} E_i \subseteq N$ and $M \not\subseteq N$. By the choice of N,

$$E/N \text{ is a nonzero subdirectly irreducible module.} \quad (8)$$

Clearly $\oplus_{i=1}^{\infty} E_i \subseteq^* N$ and so $N \in \Omega$ by Theorem 2.3. Since $\Omega = \Gamma$ by our assumption, N is a direct sum of injective modules, say $N = \oplus_{j \in I} U_j$ where each U_j is a nonzero injective module. Since $\mathrm{Soc}(M) \subseteq^* E$, we have

$$\mathrm{Soc}(M) \cap U_j \subseteq^* U_j \text{ for all } j \in I. \quad (9)$$

If $|I| < \infty$, then N is an injective submodule of E containing $\mathrm{Soc}(M) \subseteq^* E$ and so $N = E$, contradicting $M \not\subseteq N$. Therefore $|I| = \infty$. Write $I = I_1 \cup I_2$ where $I_1 \cap I_2 = \emptyset$ and $|I_1| = \infty = |I_2|$. Set $V_1 = \oplus_{j \in I_1} U_j$ and $V_2 = \oplus_{j \in I_2} U_j$. Clearly $N = V_1 \oplus V_2$. Let W_i be an essential closure of V_i in E, $i = 1, 2$. Clearly each W_i is injective and $W_1 \oplus W_2 = E$ because $\mathrm{Soc}(M) \subseteq N \subseteq V_1 \oplus V_2$ and $\mathrm{Soc}(M) \subseteq^* E$. Let π_i be the canonical projection of $E = W_1 \oplus W_2$ onto W_i, $i = 1, 2$. Let $j \in I_1$. Since π_1 acts identically on $\mathrm{Soc}(M) \cap U_j$, we conclude that

$$\pi_1(M) \supseteq \mathrm{Soc}(M) \cap U_j \text{ for all } j \in I_1. \quad (10)$$

Suppose that $\pi_1(M) \subseteq V_1 = \oplus_{j \in I_1} U_j$. Since M is finitely generated, there exists a finite subset J of I_1 with $\pi_1(M) \subseteq \oplus_{j \in J} U_j$ and so (10) implies that $\mathrm{Soc}(M) \cap U_i = 0$ for all $i \in I_1 \setminus J$, contradicting (9). Therefore $\pi_1(M) \not\subseteq V_1$. Since $N = V_1 \oplus V_2$, we see that

$$\pi_i(N) = V_i \subseteq N, \quad i = 1, 2. \quad (11)$$

If $\pi_1(M) \subseteq N$, then $\pi_1(M) \subseteq \pi_1(N) = V_1$, which is impossible. Analogously, $\pi_2(M) \not\subseteq N$. Therefore

$$\pi_i(M) \not\subseteq N, \quad i = 1, 2. \quad (12)$$

Set $\overline{E} = E/N$ and $\overline{M} = (M + N)/N \subseteq \overline{E}$. Both (11) and (12) imply that π_1 and π_2 induce nonzero idempotent endomorphisms σ_1 and σ_2 of $\overline{E}$ such that $\sigma_1 \sigma_2 = 0 = \sigma_2 \sigma_1$, contradicting (8). Thus (a) is proved.

Clearly (b) follows directly from (a). We prove (c). Let $G = \oplus_{i\in I} E_i$ be a direct sum of nonsingular injective R-modules. Suppose that $H = \mathcal{E}(G) \neq G$ and choose $x \in H \setminus G$. Clearly $(xR) \cap G \subseteq^* xR$. By (a) there exists a finitely generated submodule M of $(xR) \cap G$ such that $M \subseteq^* (xR) \cap G$. Clearly $M \subseteq^* xR$. Since $M \subseteq G = \oplus_{i\in I} E_i$, there exists a finite subset J of I such that $M \subseteq \oplus_{j\in J} E_j = U$. Obviously U is injective and $xR \cap U \subseteq^* xR$. Let V be an essential closure of $\oplus_{i\in I\setminus J} E_i$. Then $H = U \oplus V$. Let π be the canonical projection of $U \oplus V$ onto V. Since $xR \cap U \subseteq^* xR$, we conclude that $\pi(xR)$ is a singular module and so $\pi(xR) = 0$. That is to say $x \in xR \subseteq U \subseteq G$, a contradiction. Therefore G is injective and (c) is proved.

Let $G = \oplus_{i\in I} E_i$ be a direct sum of injective modules and $E = \mathcal{E}(G)$. Let $x \notin G$. Since $\oplus_{i\in I} E_i = G \subseteq^* xR + G$, there exists a family $\{U_j \mid j \in J\}$ of injective submodules of $xR + G$ such that $xR + G = \oplus_{j\in J} U_j$. Clearly there exists a finite subset T of J with $xR \subseteq U = \oplus_{t\in T} U_t$. Note that U is an injective module and by the modular law we have $U = xR + U \cap G$ because $xR \subseteq U \subseteq xR + G$. Therefore $U/(U \cap G)$ is a nonzero cyclic module and so U has a maximal submodule M such that $x \notin M$. Let V be a complement of U in E. Then $E = U \oplus V$ and so $M \oplus V$ is a maximal submodule of E with $x \notin M \oplus V$. Thus $x \notin \mathrm{Rad}(E)$ and so $\mathrm{Rad}(E) \subseteq G = \oplus_{i\in I} E_i$. Let Q be a maximal submodule of E. Then either $Q \supseteq G$ or $Q \cap G$ is a maximal submodule of G. In both cases $\mathrm{Rad}(G) \subseteq Q$ and so $\mathrm{Rad}(G) \subseteq \mathrm{Rad}(E)$. Take any $x \in \mathrm{Rad}(E)$. Then there exists a finite subset J of I with $x \in \oplus_{j\in J} E_j = P$. Since P is a direct summand of E, $\mathrm{Rad}(P) = \mathrm{Rad}(E) \cap P$. By the same reasoning, $\mathrm{Rad}(P) = \mathrm{Rad}(G) \cap P$. Therefore $x \in \mathrm{Rad}(E) \cap P = \mathrm{Rad}(G) \cap P$ and so $\mathrm{Rad}(E) = \mathrm{Rad}(G)$. Since $G = \oplus_{i\in I} E_i$,

$$\mathrm{Rad}(E) = \mathrm{Rad}(G) = \oplus_{i\in I} \mathrm{Rad}(E_i).$$

Further, let $\Lambda = \mathrm{End}_R(E)$ and set $H = \Lambda G$. Clearly H is a quasiinjective module. Suppose to the contrary that $H \neq E$ and let $x \in E \setminus H$. Denote by F the essential closure of xR in E. Then $E = F \oplus F'$ for some submodule F' of E. Let $\pi : E \to F$ be a canonical projection. Since H is quasiinjective, $\pi(H) \subseteq H$ and so $\pi(H)$ is a direct summand of H. Clearly $\oplus_{i\in I} E_i \subseteq H$ and so $H \in \Omega = \Gamma$ (see Theorem 2.3). By Corollary 3.3, $\pi(H) \in \Omega = \Gamma$ as well. Clearly $\pi(H) = H \cap F$. As $G \subseteq H$ and $G \subseteq^* E$, we conclude that $\pi(H) \subseteq^* F$. Recalling that F is an essential extension of xR and the module xR has finite Goldie dimension by (a), we conclude that both F and $\pi(H)$ have finite Goldie dimension. The inclusion $\pi(H) \in \Gamma$ now yields that $\pi(H)$ is a finite direct sum of indecomposable injective modules and so $\pi(H)$ is an injective module. Since $\pi(H) \subseteq^* F$, we have

$\pi(H) = F$ and so $x \in \pi(H) \subseteq H$, a contradiction. Therefore $E = H = \Lambda G$ and (d) is proved.

Let $G = \oplus_{i \in I} E_i$ be a direct sum of radical injective modules, $E = \mathcal{E}(G)$ and $\Lambda = \operatorname{End}_R(G)$. Clearly $G \subseteq \operatorname{Rad}(E)$ and so (d) implies that $G = \operatorname{Rad}(E)$. Obviously $\Lambda \operatorname{Rad}(E) = \operatorname{Rad}(E)$ and hence $G = \Lambda G = E$. We see that G is an injective module and so (e) is proved.

We now prove (f). In view of [8, Proposition 20.3A], it is enough to show that a direct sum of countably many pairwise isomorphic indecomposable injective modules is injective. Let $G = \oplus_{i=1}^{\infty} E_i$ be a direct sum of pairwise isomorphic indecomposable injective modules and $E = \mathcal{E}(G)$. Suppose that $G \ne E$ and pick $x \in E \setminus G$. By Zorn's lemma there exists a submodule M of E maximal with respect to the properties $G \subseteq M$ and $x \notin M$. Clearly E/M is a nonzero subdirectly irreducible module. Next, $\oplus_{i=1}^{\infty} E_i = G \subseteq^* M$ and so $M \in \Omega = \Gamma$. Therefore $M = \oplus_{i \in I} U_i$ where each U_i is a nonzero injective module. Recalling that $x \notin M$ and $M \subseteq^* E$, we conclude that M is not injective and so $|I| = \infty$. Write $I = I_1 \cup I_2$ where $|I_1| = \infty = |I_2|$ and $I_1 \cap I_2 = \emptyset$ and set $V = \oplus_{i \in I_1} U_i$, $W = \oplus_{i \in I_2} U_i$. Clearly $M = V \oplus W$. Let V' and W' be essential closures in E of V and W respectively. Clearly $V' \oplus W' = E$ and so

$$E/M = (V' \oplus W')/(V \oplus W) \cong (V'/V) \oplus (W'/W).$$

Recalling that E/M is subdirectly irreducible, we conclude that either $V = V'$ or $W = W'$. Say, $V = V'$; that is, V is injective. It follows from (b) that each $U_i = U_i' \oplus U_i''$ where U_i'' is an indecomposable injective module. Clearly $U_i'' \cap G \subseteq^* U_i''$. Let $0 \ne y \in U_i'' \cap G$. Since U_i'' has Goldie dimension 1, yR is an essential uniform submodule of U_i''. Pick an integer n such that $y \in \oplus_{j=1}^{n} E_j$. Let $\pi_j : \oplus_{j=1}^{n} E_j \to E_j$, $j = 1, 2, \ldots, n$, be canonical projections. Clearly $\cap_{j=1}^{n} \{(yR) \cap \operatorname{Ke}(\pi_j)\} = 0$. Since yR is a uniform module, there exists an index j such that $(yR) \cap \operatorname{Ke}(\pi_j) = 0$ and so yR is isomorphic to a submodule of E_j. As both E_j and U_i'' are indecomposable injective modules, we conclude that $E_j \cong U_i''$. It now follows from our assumption that each $U_i'' \cong E_1$. We now have

$$V = \oplus_{i \in I_1} U_i = \oplus_{i \in I_1}(U_i' \oplus U_i'') = (\oplus_{i \in I_1} U_i') \oplus (\oplus_{i \in I_1} U_i'')$$

and so $\oplus_{i \in I_1} U_i''$ is an injective module. Since $|I_1| = \infty$, we conclude that G is a direct summand of $\oplus_{i \in I_1} U_i''$ and so is injective (in fact $G \cong \oplus_{i \in I_1} U_i''$). Therefore (f) is proved.

Now let $G = \oplus_{i \in I} E_i$ be the direct sum of simple injective modules E_i, $i \in I$. Clearly $G = \operatorname{Soc}(E) \subseteq^* E$. Suppose that $G \ne E = \mathcal{E}(G)$ and take $x \in E \setminus G$. Let H be an essential closure of xR in E. By (a) the module xR has finite Goldie dimension and so H has also finite Goldie dimension. Clearly $H \cap \mathrm{Soc}(E) \subseteq^* H$. Since H has finite Goldie dimension, $\mathrm{Soc}(E) \cap H$ is a direct sum of finitely many simple modules each of which is isomorphic to some E_i and so $\mathrm{Soc}(E) \cap H$ is an injective module, forcing $H = \mathrm{Soc}(E) \cap H$. Therefore $x \in \mathrm{Soc}(E) = G$, a contradiction. □

Theorem 3.8. *Let $\Omega(R) = \Gamma(R)$ and let S be a homomorphic image of the ring R. Then:*

(a) Every finitely generated right S-module M has finite Goldie dimension and S has the a.c.c. on right ideals of the form $\mathrm{Ann}_S(X)$, $\emptyset \neq X \subseteq M$.

(b) S is a right Goldie ring with nilpotent prime radical $\mathrm{rad}(S)$.

Proof. By Theorem 3.6, $\Omega(S) = \Gamma(S)$ and so it is enough to prove both (a) and (b) for $S = R$. Let M be a finitely generated R-module. Then M has finite Goldie dimension by Theorem 3.7(a). Therefore its injective hull $E = \mathcal{E}(M)$ is a direct sum of a finite number of indecomposable injective modules, say $E = \oplus_{i=1}^n E_i$. Set $E_i^j = E_i$ and $E^j = \oplus_{i=1}^n E_i^j$, $j = 1, 2, \ldots$. Then
$$H = \oplus_{j=1}^\infty E^j = \oplus_{i=1}^n [\oplus_{j=1}^\infty E_i^j]$$
and so H is an injective module in view of Theorem 3.7(f). It now follows from [8, Proposition 20.3A] that R satisfies the a.c.c. on right ideals of the form $\mathrm{Ann}_R(X)$, $\emptyset \neq X \subseteq E$. Since $M \subseteq E$, we conclude that (a) is proved.

By Theorem 3.7(a), R_R has finite Goldie dimension. Next, (a) yields that R satisfies the a.c.c. on right annihilators and so R is a right Goldie ring. By [15, Theorem 1], $\mathrm{rad}(R)$ is nilpotent. □

Lemma 3.9. *Suppose that every right primitive ideal of R contains an element that is not a left zero divisor. Then every injective right R-module is radical.*

Proof. Let E be an injective R-module. Suppose that a simple module S is a homomorphic image of E. Let $P = \mathrm{Ann}_R(S)$. Clearly P is a right primitive ideal and so by assumption contains an element a that is not a left zero divisor. Given $x \in E$, we define the map $f : aR \to E$ by the rule $f(ar) = xr$, $r \in R$. Since a is not a left zero divisor, f is a well-defined homomorphism of R-modules and so there exists $y \in E$ such that $f(ar) = yar$. In particular, $x = f(a) = ya$. We see that $Ea = E$ and so $Sa = S$. On the other hand, $Sa \subseteq SP = 0$, a contradiction. Therefore $E = \mathrm{Rad}(E)$. □

Theorem 3.10. [4, Theorem] *Let $M \in$ Mod-R. Then the following conditions are equivalent*:
   (a) *Every homomorphic image of M has finite Goldie dimension.*
   (b) *Every submodule N of M contains a finitely generated submodule L such that $\mathrm{Rad}(N/L) = N/L$.*

Theorem 3.11. *Suppose that $\Omega(R) = \Gamma(R)$ and one of the following conditions is satisfied*:
   (a) $\mathcal{E}(R)$ *is a radical module*;
   (b) $\mathrm{Rad}(R)$ *contains an element a which is not a left zero divisor*;
   (c) R *is a semiprime ring, no minimal prime ideal of which is right primitive*;
   (d) R *is a right V-ring*;
   (e) R *is semilocal*;
   (f) R *is commutative.*
Then R is right Noetherian.

Proof. Suppose that (a) is satisfied. Let $H = \mathcal{E}(R)$ and let $\{E_i \mid i \in I\}$ be a family of injective R-modules. Consider a module $G = H \oplus (\oplus_{i \in I} E_i)$ and set $E = \mathcal{E}(G)$. By Theorem 3.7(d), $\mathrm{Rad}(E) = \mathrm{Rad}(H) \oplus (\oplus_{i \in I} \mathrm{Rad}(E_i))$ and so $H \subseteq \mathrm{Rad}(E) \subseteq G$. Let $\Lambda = \mathrm{End}_R(E)$. Clearly $\Lambda \mathrm{Rad}(E) = \mathrm{Rad}(E)$ and hence $\Lambda H \subseteq G$. Since $R \subseteq H$, $\Lambda H = E$ and so $G = E$. That is to say, both G and $\oplus_{i \in I} E_i$ are injective and hence R is right Noetherian by [8, Theorem 20.1].

Suppose that (b) is fulfilled. Then every injective R-module is radical by Lemma 3.9. It now follows from Theorem 3.7(e) that a sum of injective R-modules is injective and hence R is right Noetherian.

Assume that (c) is satisfied. By Theorem 3.8(b), R is a semiprime right Goldie ring. Let $P_1, P_2, \ldots, P_n$ be all the minimal prime ideals of R (see [5, Lemma 1.16]). Clearly $\cap_{i=1}^n P_i = 0$. Next, let P be a right primitive ideal of R. If $P \not\subseteq^* R$, then there exists a nonzero ideal Q of R such that $PQ = 0$. By assumption $P \neq P_i$ and so $Q \subseteq P_i$ for all $i = 1, 2, \ldots, n$, forcing $Q \subseteq \cap_{i=1}^n P_i = 0$, a contradiction. Therefore P is an essential right ideal of R and so P contains an element which is not a zero divisor by [5, Theorem 1.10]. It now follows from Lemma 3.9 that every injective R-module is radical and hence R is right Noetherian by Theorem 3.7(e).

Suppose that (d) is fulfilled. Then it follows from Theorem 3.7(g) that every semisimple R-module is injective. The result now follows from [8, Corollary 20.3E].

Assume that R is semilocal. Then it has only a finite number of nonisomorphic simple modules, say $S_1, S_2, \ldots, S_n$. Set $E_i = \mathcal{E}(S_i)$ and $E = \oplus_{i=1}^n E_i$. Clearly E is an injective cogenerator. It now follows from

Theorem 3.7(f) that the direct sum of any set of copies of E is injective and so R is right Noetherian by [8, Corollary 20.3C].

Finally, suppose that R is commutative. Assume to the contrary that R is not Noetherian. Let N be an ideal of R which is not finitely generated. Let L be a submodule of N. Then $N/L \subseteq R/L$ and so N/L has finite Goldie dimension by Theorem 3.7(a). It now follows from Theorem 3.10 that N contains a finitely generated ideal L such that N/L is a nonzero radical module. Pick $x \in N \setminus L$ and let $K = \{r \in R \mid xr \in N\}$. Clearly K is a proper ideal of R and so there exists a maximal ideal M of R containing K. Set $S = R \setminus M$. Note that

$$xs \notin L \quad \text{for all } s \in S. \tag{13}$$

Consider the localization $A = S^{-1}R$ of R with respect to S. It is well known that A is a flat R-module (see, for example [1, Corollary 3.6]). It is easy to see that the canonical homomorphism $\phi : R \to A$, $r \mapsto r/1 \in A$, $r \in R$, is a flat epimorphism of rings and so $\Omega(A) = \Gamma(A)$ by Theorem 3.6(b). It is well known that A is a local ring with maximal ideal $\phi(M)A = S^{-1}M$ and ϕ induces an isomorphism of factor rings R/M and $A/(\phi(M)A)$. It now follows from (e) that A is Noetherian and so $\phi(N)A = S^{-1}N$ is a finitely generated A-module. If $\phi(L)A = \phi(N)A$, then $\phi(x) \in \phi(L)A = S^{-1}L$ and so there exists $s \in S$ with $xs \in L$, contradicting (13). Therefore $\phi(N)A \supset \phi(L)A$ and hence there exists a maximal submodule U of the A-module $\phi(N)A$ containing $\phi(L)A$. The factor module $V = [\phi(N)A]/U$ is a simple A-module isomorphic to $A/[\phi(M)A]$ and so V is a simple R-module. Let $\pi : \phi(N)A \to V$ be a canonical projection. Clearly π is a homomorphism of R-modules. Obviously, $\pi\phi : N \to V$ is a homomorphism of R modules mapping L to 0. By the choice of L, $\pi\phi(N) = 0$ and so $\phi(N) \subseteq U$ forcing $\phi(N)A \subseteq U$, a contradiction. Thus R is Noetherian. □

Proposition 3.12. *Let R be a ring with prime radical N. Suppose that $\Omega(R) = \Gamma(R)$ and R is not right Noetherian. Then one of the two following conditions is fulfilled:*

(a) There exists a minimal prime ideal P of R such that $\overline{R} = R/P$ is a right primitive ring, which is not right Noetherian, and $\Omega(\overline{R}) = \Gamma(\overline{R})$.

(b) The ideal $\overline{N} = N/N^2$ of the ring $\overline{R} = R/N^2$ is not finitely generated as a right ideal, $\overline{R}/\overline{N}$ is right Noetherian and $\Omega(\overline{R}) = \Gamma(\overline{R})$.

Proof. First of all the equality $\Omega(\overline{R}) = \Gamma(\overline{R})$ follows from Theorem 3.6(a). Suppose that R/P is right Noetherian for any minimal prime ideal P of R which is right primitive. Consider the factor ring $A = R/N$. By

Theorem 3.8(b), A is a semiprime right Goldie ring and so it has only finitely many minimal prime ideals, say $P_1, P_2, \ldots, P_n$ (see [5, Lemma 1.16]). Suppose that P_i is not right primitive. Then A/P_i is a prime ring which is not primitive and $\Omega(A/P_i) = \Gamma(A/P_i)$ by Theorem 3.6(a). It now follows from Theorem 3.11(c) that it is right Noetherian. Therefore A/P_j is a right Noetherian ring for all $j = 1, 2, \ldots, n$. Consider A/P_j as an A-module. Clearly A/P_j is a Noetherian A-module and so $\oplus_{j=1}^{n} A/P_j$ is a Noetherian A-module. Since $\cap_{j=1}^{n} P_j = 0$, A is isomorphic to a submodule of $\oplus_{j=1}^{n} A/P_j$ and so A is a right Noetherian ring. Set $\overline{R} = R/N^2$ and $\overline{N} = N/N^2$. Then $\overline{R}/\overline{N} \cong A$ and so it is right Noetherian. Next, assume that $\overline{N}$ is a finitely generated right $\overline{R}$-module. Then there exist elements $x_1, x_2, \ldots, x_m \in R$ such that $N = \sum_{i=1}^{m} x_i R + N^2$ and so $N^2 = \sum_{i=1}^{m} x_i R + N^3$. Therefore $N = \sum_{i=1}^{m} x_i R + N^3$. Continuing in this fashion we get that $N = \sum_{i=1}^{m} x_i R + N^k$ for any positive k. Since N is nilpotent by Theorem 3.8(b), we conclude that $N = \sum_{i=1}^{m} x_i R$. It now follows from [16, Lemma 1] that each N^k is a finitely generated right ideal of the ring R. Pick t such that $N^t = 0$. Consider the factor module N^k/N^{k+1}. Since N^k is a finitely generated R-module, N^k/N^{k+1} is a finitely generated module over R/N. By the above result $A = R/N$ is right Noetherian and so N^k/N^{k+1} is a Noetherian R/N-module. Therefore it is a Noetherian R-module. We have $N \supseteq N^2 \supseteq \ldots \supseteq N^t = 0$ and every factor module N^k/N^{k+1} is Noetherian. Therefore N is a Noetherian R-module. Since R/N is a right Noetherian ring, it is a Noetherian R-module. Therefore both R/N and N are Noetherian R-modules and hence R is right Noetherian, a contradiction. Thus $\overline{N}$ is not finitely generated. The proof is now complete. □

Conjecture. Let R be a ring such that $\Omega(R) = \Gamma(R)$. Then R is right Noetherian.

In view of Proposition 3.12 it is enough to check this conjecture in two cases: when R is right primitive and when $R/\mathrm{rad}(R)$ is right Noetherian, $\mathrm{rad}(R)^2 = 0$.

References

[1] M. F. Atyah and I. G. Macdonald, *Introduction to Commutative Algebra*, Addison-Wesley, New York, 1969.

[2] K. I. Beidar, *On rings with zero total*, Beiträge zur Algebra und Geom. **38** (1997), 233–238.

[3] K. I. Beidar and R. Wiegandt, *Radicals induced by the total of rings*, ibid., 149–159.

[4] V. P. Camilo, *Modules whose quotients have finite Goldie dimension,* Pacific. J. Math. **69** (1977), 337–338.

[5] A. W. Chatter and C. R. Hajarnavis, *Rings with Chain Conditions,* Pitman, Boston, 1980.

[6] N. V. Dung, D. V. Huynh, P. F. Smith and R. Wisbauer, *Extending Modules,* Pitman Research Notes in Math. Ser. **313**, Longman, 1994.

[7] C. Faith, *Algebra I. Rings, Modules, and Categories,* Springer-Verlag, Heidelberg, New York, 1976.

[8] _____, *Algebra II. Ring Theory,* Springer-Verlag, Heidelberg, New York, 1976.

[9] F. Kasch, *Moduln und Ringen,* Teubner, Stuttgart, 1977 (see also English, Russian and Chinese editions).

[10] _____, *Moduln mit LE-Zerlegung und Harada-Moduln,* Lecture Notes, München, 1982.

[11] _____, *Partiell invertierbare homomorphismen und das total,* Algebra Berichte **60**, Verlag R. Fisher, München, (1988), 1–14.

[12] _____, *The total in the category of modules,* General Algebra (1988), 129–137, Elsevier Sci. Pub., Amsterdam.

[13] F. Kasch and W. Schneider, *The total of modules and rings,* Algebra Berichte **69**, Verlag R. Fischer, München, (1992), 1–85.

[14] _____, *Exchange properties and the total,* Advances in Ring Theory, edited by S. K. Jain and S. Tariq Rizvi, Birkhäuser, Boston, 1997, 163–174.

[15] C. Lanski, *Nil subrings of Goldie rings are nilpotent,* Canad. J. Math. **21** (1969), 904–907.

[16] G. O. Michler, *Prime right ideals and right Noetherian rings,* in Ring Theory, edited by R. Gordon, Academic Press, New York, (1972), 251–255.

[17] S. H. Mohamed and B. J. Müller, *Continuous and Discrete Modules,* London Math. Soc. Lecture Notes **147**, Cambridge Univ. Press, Cambridge, 1990.

[18] W. Schneider, *Das Total von Moduln und Ringen,* Algebra Berichte **55**, Verlag R. Fischer, München, (1987), 1–59.

[19] B. Stenström, *Rings of Quotients,* Springer-Verlag, Heidelberg, New York, 1975.

[20] R. Wisbauer, *Foundations of Module and Ring Theory,* Gordon and Breach, New York, 1991.

K. I. Beidar
Department of Mathematics
National Cheng Kung University
Tainan 701, Taiwan
e-mail: `beidar@mail.ncku.edu.tw`

F. Kasch
Mathematisches Institut
Universität München
D-80333 München, Germany

Semicentral Reduced Algebras

Gary F. Birkenmeier, Jin Yong Kim and Jae Keol Park

Abstract

An idempotent e of an algebra R is *left semicentral* if $Re = eRe$. If 0 and 1 are the only left semicentral idempotents in R, then R is called *semicentral reduced*. Recent results on generalized triangular matrix algebras and semicentral reduced algebras are surveyed. New results are provided for endomorphism algebras of modules and for semicentral reduced algebras. In particular, semicentral reduced rings which are right FPF, right nonsingular, or left perfect are described.

0. Introduction

Throughout this paper, K denotes a commutative ring with unity, R an associative K-algebra with unity 1, and all modules are unital right R-modules, unless indicated otherwise.

This paper provides a survey of recent results on generalized triangular matrix algebras and semicentral reduced algebras. Moreover, new results are developed for endomorphism algebras of modules and semicentral reduced algebras. The motivation for the concept of a semicentral reduced algebra can be found in structure theory, in particular, in the theory of indecomposability.

Recall that an idempotent $e \in R$ is said to be *(centrally) primitive* if $\{0, e\}$ are the only (central) idempotents in eRe. If 1 is (centrally) primitive, then R is said to be *indecomposable* as an (algebra) R-module. A set of orthogonal idempotents $\{e_1, \ldots, e_n\}$ of R is called *complete* if $e_1 + \cdots + e_n = 1$. Observe that for a complete set of orthogonal idempotents $\{e_1, \ldots, e_n\}$ there is a K-algebra isomorphism $\phi : R \to A$, where A is the K-algebra of n-by-n matrices whose (i, j)-th position is of the form $e_i r_{ij} e_j$ where $r_{ij} \in R$, defined by

$$\phi(x) = \begin{pmatrix} e_1 x e_1 & \cdots & e_1 x e_n \\ \vdots & \cdots & \vdots \\ e_n x e_1 & \cdots & e_n x e_n \end{pmatrix}$$

for $x \in R$. Note that if $\{e_1, \ldots, e_n\}$ is a complete set of orthogonal (centrally) primitive idempotents, then each "diagonal" subalgebra $e_i R e_i$ is indecomposable as an (algebra) $e_i R e_i$-module.

We say R has a (*complete*) *generalized matrix representation of size* n if there exists an isomorphism

$$\theta : R \to \begin{pmatrix} R_1 & R_{12} & \cdots & R_{1n} \\ R_{21} & R_2 & \cdots & R_{2n} \\ \vdots & \vdots & \ddots & \vdots \\ R_{n1} & R_{n2} & \cdots & R_n \end{pmatrix},$$

where each R_i is an (indecomposable R_i-module) K-algebra and R_{ij} is a left R_i-right R_j-bimodule. We say that a generalized matrix representation for R is *trivial* if it has size one. Thus an idempotent e is primitive if and only if eRe has only the trivial generalized matrix representation.

In [2] an idempotent $e \in R$ is called *left* (*right*) *semicentral* if $Re = eRe$ ($eR = eRe$). This definition provides the key to developing an abstract theory of generalized triangular matrix representation. As is well known [10], a left semicentral idempotent induces a 2-by-2 generalized triangular matrix representation. We use $\mathcal{S}_\ell(R)$ and $\mathcal{S}_r(R)$ for the sets of all left and right semicentral idempotents, respectively. Again taking e to be an idempotent of R, observe that $\mathcal{S}_\ell(eRe) = \{0, e\}$ if and only if $\mathcal{S}_r(eRe) = \{0, e\}$; when this occurs e is said to be *semicentral reduced* [6]. If 1 is semicentral reduced, then R is said to be *semicentral reduced*. Clearly any algebra with only trivial idempotents is semicentral reduced. Every prime ring is semicentral reduced. The 2-by-2 upper triangular matrix F-algebra over a field F is an indecomposable F-algebra which is not semicentral reduced. Hence the concept of a semicentral reduced idempotent lies properly "between" the concepts of a primitive idempotent and a centrally primitive idempotent.

An ordered set $\{b_1, \ldots, b_n\}$ of nonzero distinct idempotents in R is called a set of *left triangulating idempotents* of R if all the following hold:
(i) $1 = b_1 + \cdots + b_n$;
(ii) $b_1 \in \mathcal{S}_\ell(R)$; and
(iii) $b_{k+1} \in \mathcal{S}_\ell(c_k R c_k)$, where $c_k = 1 - (b_1 + \cdots + b_k)$, for $1 \leq k \leq n-1$.

Similarly we define a set of *right triangulating idempotents* of R using (i), $b_1 \in \mathcal{S}_r(R)$, and $b_{k+1} \in \mathcal{S}_r(c_k R c_k)$. From part (iii) of the above definition, a set of left (right) triangulating idempotents is a set of pairwise orthogonal idempotents. Also from [6], R has a *generalized triangular matrix representation* if there exists a K-algebra isomorphism

$$\theta: R \to \begin{pmatrix} R_1 & R_{12} & \cdots & R_{1n} \\ 0 & R_2 & \cdots & R_{2n} \\ \vdots & \vdots & \ddots & \vdots \\ 0 & 0 & \cdots & R_n \end{pmatrix},$$

where each R_i is a K-algebra with unity and R_{ij} is a left R_i-right R_j bimodule for $i < j$. We say that R has a *complete generalized triangular matrix representation* if each R_i is semicentral reduced.

In Section 1, after providing basic properties of semicentral idempotents, we use results from [6] to show that R has a (complete) generalized triangular matrix representation if and only if R has a (complete) set of left triangulating idempotents. We then fully characterize algebras with a complete generalized triangular matrix representation. From this characterization we show that the algebras in many well-known classes have a complete generalized triangular matrix representation. Applications are made to the endomorphism rings of modules.

In Section 2 we show that the class of semicentral reduced rings is closed with respect to forming matrix algebras and trivial extensions.

Various criteria are developed in Section 3 to insure that a semicentral reduced ring is a prime ring. Moreover, semicentral reduced rings satisfying finiteness conditions are investigated.

Standard terminology and notation are adhered to as much as possible. Where there is conflict or confusion in the literature we define the term or notation as we plan to use it herein. We use $\mathbf{I}(R)$, $\mathbf{B}(R)$ and $\mathbf{N}(R)$ for the sets of idempotents, central idempotents, and nilpotents of R, respectively. Observe that $\mathcal{S}_r(R) \cap \mathcal{S}_\ell(R) = \mathbf{B}(R)$. If $\mathbf{N}(R) = 0$, then R is called *reduced*. We use $\mathbf{P}(R)$ for the prime radical of R (in the category of K-algebras), and for any nonempty subset X of R, we write $r(X)$ for $\{r \in R \mid Xr = 0\}$ and $\ell(X)$ for $\{r \in R \mid rX = 0\}$, which are called the right annihilator of X in R and the left annihilator of X in R, respectively. By prime ideal we mean a *proper* prime ideal. The right socle, the right singular ideal, and the right second singular ideal are denoted by $\mathrm{Soc}(R_R)$, $Z(R_R)$, and $Z_2(R_R)$, respectively. If X is a subset of R, then $X \subseteq_2 R$ and $X \subseteq_r R$ denote that X is an ideal of R and X is a right ideal of R, respectively. If $X, Y \subseteq_r R$, then $X \subseteq_r^{\mathrm{ess}} Y$ symbolizes that X is right essential in Y. We use $M_n(R)$ to denote the n-by-n full matrix algebra over R.

1. Preliminaries and Motivations

In this section we first provide the basic properties of semicentral idempotents. These properties have proved crucial in the development of the concepts of triangulating idempotents and generalized triangular matrix representations. We then provide a survey of results showing the equivalence of the (complete) set of left triangulating idempotents and (complete) generalized triangular matrix representation concepts. The application of these results to many well-known classes of algebras shows that, under mild finiteness conditions, their structure can be determined by that of the semicentral reduced algebras in the respective class.

Proposition 1.1. *Let $e \in \mathbf{I}(R)$. Then the following conditions are equivalent:*

(i) $e \in \mathcal{S}_\ell(R)$;
(ii) $1 - e \in \mathcal{S}_r(R)$;
(iii) $xe = exe$, for each $x \in R$;
(iv) $(1-e)Re = 0$;
(v) $(1-e)x = (1-e)x(1-e)$, for each $x \in R$;
(vi) eR is an ideal of R;
(vii) $R(1-e)$ is an ideal of R;
(viii) $eR(1-e)$ is an ideal of R and $eR = eR(1-e) \oplus Re$, as a direct sum of left ideals;
(ix) the function defined by $\phi(x) = \begin{pmatrix} exe & ex(1-e) \\ 0 & (1-e)x(1-e) \end{pmatrix}$ is a K-algebra isomorphism from R onto $\begin{pmatrix} eRe & eR(1-e) \\ 0 & (1-e)R(1-e) \end{pmatrix}$;
(x) $\{a \in R \mid Rea \subseteq Re\} = \{c \in R \mid ce = ec\}$;
(xi) $\ell(Re) = \ell(e)$.

Proof. Routine arguments establish the implications (i)$\Rightarrow$(ii)$\Rightarrow$(iii)$\Leftrightarrow$(iv)$\Leftrightarrow$(v), (iii)$\Rightarrow$(i), and (iii)$\Rightarrow$(vi). Assume (vi) and let $x \in R$. Then $xe = ey$ for some $y \in R$, and hence $(1-e)x = (1-e)[xe + x(1-e)] = (1-e)[ey + x(1-e)] = (1-e)x(1-e)$. Thus $R(1-e)x \subseteq R(1-e)$. So (vi) implies (vii) and (v). Similarly (vii) implies (vi). Now we have the equivalence of (i) through (vii). Assume these hold. Then (vi) and (vii) yield $eR(1-e)$ is an ideal. By (i), $Re = eRe$. Using the Peirce decomposition yields $R = R(1-e) \oplus Re$. So $eR = eR(1-e) \oplus eRe = eR(1-e) \oplus Re$ is a direct sum of left ideals. Thus (vii) implies (viii). Assume (viii), then eR is a left ideal. Hence (viii) implies (vi). The equivalence of (ix) and (iv) follows from a routine calculation. Therefore (i) through (ix) are equivalent. For the implication (i)$\Rightarrow$(x), let $A = \{a \in R \mid Rea \subseteq Re\}$ and $C = \{c \in R \mid ce = ec\}$. Clearly $C \subseteq A$. So let $a \in A$. Then $ea = xe$, for some $x \in R$. Then $ea = eae + ea(1-e) = eae + xe(1-e) = eae$. But $eae = ae$. So $a \in C$. Therefore $A = C$. Now assume (x), then $A = C$. Hence $(1-e)R \subseteq A = C$. So $(1-e)Re = 0$. Thus (x)$\Rightarrow$(iv).

Consequently, (i) through (x) are equivalent. Straightforward calculations show that (i)⇒(xi) and (xi)⇒(vii). □

Observe in Proposition 1.1(x) that $\{a \in R \mid Rea \subseteq Re\}$ is the idealizer of Re and $\{c \in R \mid ce = ec\}$ is the centralizer of e.

Corollary 1.2. (i) *If $e \in \mathbf{I}(R)$, then e is a left (right) semicentral idempotent in the idealizer of eR (Re).*
(ii) *If R is semiprime, then $\mathcal{S}_\ell(R) = \mathbf{B}(R) = \mathcal{S}_r(R)$.*

Proof. Part (i) follows from Proposition 1.1(vi), and part (ii) is a consequence of Proposition 1.1(iv) and (viii). □

For a nonempty subset X of R, let $\mathcal{S}_\ell(X)$ be the set of all $e = e^2 \in X$ such that $xe = exe$ for all $x \in X$.

Proposition 1.3. *Let $e \in \mathcal{S}_\ell(R)$ and $f \in \mathbf{I}(R)$. Then:*
(i) *$fR = A \oplus B$, as a direct sum of right ideals of R, and $B \subseteq eR$ with $A \cap eR = 0$;*
(ii) *R/eR is isomorphic to $(1-e)R = (1-e)R(1-e)$ both as K-algebras and as right R-modules;*
(iii) *every left ideal of eR is a left ideal of R and every right ideal of $(1-e)R(1-e)$ is a right ideal of R;*
(iv) *if L is a left ideal of R, then $\mathcal{S}_\ell(L) = L \cap \mathcal{S}_\ell(R)$ (in particular, $\mathcal{S}_\ell(Re) = \mathcal{S}_\ell(eRe) = eRe \cap \mathcal{S}_\ell(R))$;*
(v) *$f\mathcal{S}_\ell(R)f \subseteq \mathcal{S}_\ell(fRf)$;*
(vi) *$fe, fef \in \mathbf{I}(R)$;*
(vii) *if f is a primitive idempotent of R such that $fe \neq 0$, then $fef = f$ and efe is a primitive idempotent in eRe;*
(viii) *if $f \in \mathcal{S}_r(R)$ and X is an ideal of R, then eXf is an ideal of R;*
(ix) *$eR + fR = (e + f - ef)R$ and $e + f - ef \in \mathbf{I}(R)$;*
(x) *if $f \in \mathcal{S}_\ell(R)$, then $e + f - ef \in \mathcal{S}_\ell(R)$.*

Proof. (i) The proof of this part is a K-algebra analog of that given for Lemma 1(v) in [2] and uses Proposition 1.1(i) and (ii).

(ii) Write $R = eR \oplus (1-e)R$ as a direct sum of right ideals of R. Using Proposition 1.1, the desired result follows from a K-algebra analog of Lemma 1.1 in [2].

(iii) This part is an immediate consequence of part (ii) and a K-algebra analog of Lemma 1.1 in [2].

(iv) Let $v \in \mathcal{S}_\ell(L)$. Then $Lv \subseteq Rv = (Rv)v \subseteq Lv$. Hence $Rv = Lv$. So $vRv = vLv$. Therefore $Rv = Lv = vLv = vRv$. Consequently, $v \in \mathcal{S}_\ell(R)$. The containment $L \cap \mathcal{S}_\ell(R) \subseteq \mathcal{S}_\ell(L)$ is immediate. Using the equivalence of (i) and (iii) in Proposition 1.1, $Re = eRe$. Hence $\mathcal{S}_\ell(eRe) \subseteq \mathcal{S}_\ell(R)$.

(v) Let $b \in \mathcal{S}_\ell(R)$ and $r \in R$. Then it follows that $(fbf)(frf)(fbf) = (ff)(frf)(fbf) = (frf)(fbf)$. So $fbf \in \mathcal{S}_\ell(fRf)$.

(vi) A routine calculation using the equivalence of (i) and (iii) of Proposition 1.1 establishes this part.

(vii) First observe that $0 \neq fe = fefe$; so $fef \neq 0$. Then primitivity of f implies that the idempotent fef must be f. Let u be a nonzero idempotent in $(efe)(eRe)(efe)$. Routine calculations, making use of $e \in \mathcal{S}_\ell(R)$ and f an idempotent, yield that $ue = u$, $fu = u$, $uf = fuf$, and $(uf)(uf) = uf$. Since $uf = 0$ implies $u = ufe = 0$, we have that uf is a nonzero idempotent in fRf. Primitivity of f in R then yields $uf = f$. Then $u = ufe = fe = efe$. So efe is the only nonzero idempotent in eRe.

(viii) Observe that $R(eXf)R = eReXfRf \subseteq eXf$.

(ix) Let $s, t \in R$. Then $(e+f-ef)(es+ft) = es+fes-efes+eft+ft-eft = es+ft$. So $eR+fR \subseteq (e+f-ef)R$. Thus $eR+fR = (e+f-ef)R$. A routine calculation yields $e + f - ef \in \mathbf{I}(R)$.

(x) This part is shown by a calculation similar to that used in part (ix). □

Proposition 1.4. (i) *If h is a multiplicative homomorphism from R into a K-algebra, then $h(\mathcal{S}_\ell(R)) \subseteq \mathcal{S}_\ell(h(R))$.*

(ii) *Let $e \in \mathcal{S}_\ell(R) \cup \mathcal{S}_r(R)$ and $b \in \mathcal{S}_\ell(eRe) \cup \mathcal{S}_r(eRe)$. Then the function $h : R \to bRb$, defined by $h(r) = brb$, for each $r \in R$ is a K-algebra homomorphism.*

(iii) *The sum of any finite set of pairwise orthogonal, left semicentral idempotents is a left semicentral idempotent.*

(iv) *If S is any subalgebra of R, then $S \cap \mathcal{S}_\ell(R) \subseteq \mathcal{S}_\ell(S)$.*

(v) *Let $e \in \mathcal{S}_\ell(R)$ and $h : R \to eRe$ the K-algebra homomorphism defined by $h(x) = exe$ for all $x \in R$. Then h is a homomorphism from the lattice of right ideals of R into the lattice of right ideals of eRe.*

Proof. The proofs of parts (i), (iii) and (iv) are straightforward arguments. For (ii), observe that $b \in eRe$ implies $eb = b = be$. So for each $x, y \in R$, $bxyb = bexyeb$. Using $e \in \mathcal{S}_\ell(R) \cup \mathcal{S}_r(R)$ and $b \in \mathcal{S}_\ell(eRe) \cup \mathcal{S}_r(eRe)$, we have that $bxyb = bexeyeb = bexebeyeb = bxbyb$. So $h(xy) = bxb^2yb = h(x)h(y)$. For (v), let X, Y be right ideals of R. Then $e(X + Y)e = eXe + eYe$. Clearly $e(X \cap Y)e \subseteq eXe \cap eYe$. Let $exe = eye \in eXe \cap eYe$ where $x \in X$ and $y \in Y$. Then $exe = xe = ye = eye$. So $xe \in X \cap Y$. Hence $exe \in e(X \cap Y)e$. Thus $e(X \cap Y)e = eXe \cap eYe$. □

In general the function h in Proposition 1.4(v) is not a lattice homomorphism for left ideals. This can be seen by taking $R = \begin{pmatrix} F & F \\ 0 & F \end{pmatrix}$, $e =$

$\begin{pmatrix} 1 & 0 \\ 0 & 0 \end{pmatrix}$, $L_1 = \begin{pmatrix} F & 0 \\ 0 & 0 \end{pmatrix}$ and $L_2 = \{\begin{pmatrix} x & x \\ 0 & 0 \end{pmatrix} \mid x \in F\}$, where F is a field. Then $eL_1e \cap eL_2e = L_1 \neq e(L_1 \cap L_2)e = 0$.

Proposition 1.5. *Let M be a right R-module and $e = e^2 \in End_R(M)$. Then $e \in \mathcal{S}_\ell(End_R(M))$ if and only if eM is a fully invariant submodule of M.*

Proof. Let $E = End_R(M)$. Assume $e \in \mathcal{S}_\ell(E)$, and $f \in E$. By Proposition 1.1, $f(eM) = fe(M) = efe(M) \subseteq eM$. Hence eM is fully invariant.

Conversely, assume that eM is fully invariant and $h \in E$. Then $h(eM) = he(M) \subseteq eM$. Hence $ehe = he$. Therefore $e \in \mathcal{S}_\ell(E)$. □

Thus $End_R(M)$ is semicentral reduced if and only if M has no nontrivial fully invariant direct summand.

Example 1.6. Let M be a module with a simple submodule S. Let H be the homogeneous component of M containing S and H^c the essential closure of H in M. Then Proposition 1.5 yields that $End_R(H^c)$ is a semicentral reduced ring.

Proposition 1.7. *Let R be an indecomposable algebra in which every nonzero nilpotent ideal is either left essential in R or right essential in R. Then R is semicentral reduced.*

Proof. Assume $0 \neq e \in \mathcal{S}_\ell(R)$ such that $e \neq 1$. By Proposition 1.1, $eR = Re \oplus eR(1-e)$. Since $e \notin \mathbf{B}(R)$, then $0 \neq eR(1-e) \subseteq_2 R$. So $eR(1-e)$ is neither left nor right essential in R, a contradiction. □

The remaining results of this section (except Corollary 1.11) are from [6] and provide the motivation for studying semicentral reduced algebras.

By showing [6, Lemma 1.2] that a set of nonzero idempotents $\{b_1, \ldots, b_n\}$ is left triangulating if and only if it is a complete ordered set such that $b_j R b_i = 0$ (for all $i < j \leq n$), we obtain the following result via the K-algebra isomorphism from an algebra with a complete set of idempotents to a generalized triangular matrix algebra (as previously indicated).

Proposition 1.8. [6, Proposition 1.3] *R has a (complete) set of left triangulating idempotents if and only if R has a (complete) generalized triangular matrix representation.*

Thus if the set of left triangulating idempotents is complete, then the "diagonal" algebras are semicentral reduced. Also note that R is semicentral reduced if and only if R has no nontrivial generalized triangular matrix representation. The following result indicates the left-right symmetry in the triangulating idempotent concept.

Proposition 1.9. [6, Corollary 1.7] *The ordered set* $\{b_1, \ldots, b_n\}$ *is a (complete) set of left triangulating idempotents if and only if* $\{b_n, \ldots, b_1\}$ *is a (complete) ordered set of right triangulating idempotents.*

The next result fully characterizes algebras with a complete generalized triangular matrix representation and shows that this property is a mild finiteness condition.

Theorem 1.10. [6, Theorem 2.9] *The following conditions are equivalent*:
  (i) R *has a complete set of left triangulating idempotents*;
  (ii) $\{bR \mid b \in \mathcal{S}_\ell(R)\}$ *is a finite set*;
  (iii) $\{bR \mid b \in \mathcal{S}_\ell(R)\}$ *satisfies ACC and DCC*;
  (iv) $\{bR \mid b \in \mathcal{S}_\ell(R)\}$ *and* $\{Rc \mid c \in \mathcal{S}_r(R)\}$ *satisfy ACC*;
  (v) $\{bR \mid b \in \mathcal{S}_\ell(R)\}$ *and* $\{Rc \mid c \in \mathcal{S}_r(R)\}$ *satisfy DCC*;
  (vi) $\{bR \mid b \in \mathcal{S}_\ell(R)\}$ *and* $\{cR \mid c \in \mathcal{S}_r(R)\}$ *satisfy DCC*;
  (vii) R *has a complete set of right triangulating idempotents*;
  (viii) R *has a complete generalized triangular matrix representation.*

The following corollary applies Theorem 1.10 to Module Theory.

Corollary 1.11. *Let* M *be a right* R-*module and* $E = End_R(M)$. *The following conditions are equivalent*:
  (i) E *has a complete generalized triangular matrix representation*;
  (ii) M *has ACC and DCC on fully invariant direct summands*;
  (iii) M *has only finitely many distinct fully invariant direct summands.*

Proof. This result is a consequence of Proposition 1.5, Theorem 1.10 and the fact that if $h \in E$ and $e = e^2 \in E$, then $hM \subseteq eM$ if and only if $hE \subseteq eE$. □

The following result shows that the semicentral reduced "diagonal subalgebras" in a complete generalized triangular matrix representation are unique up to isomorphism.

Theorem 1.12. [6, Theorem 2.10] *Let* $\{b_1, \ldots, b_n\}$ *and* $\{c_1, \ldots, c_k\}$ *each be a complete set of left triangulating idempotents of* R. *Then* $n = k$ *and there exists an invertible element* $\alpha \in R$ *and a permutation* σ *on* $\{1, \ldots, n\}$ *such that* $b_{\sigma(i)} = \alpha^{-1} c_i \alpha$ *for each* i. *Thus for each* i, $c_i R \cong b_{\sigma(i)} R$, *as* R-*modules, and* $c_i R c_i \cong b_{\sigma(i)} R b_{\sigma(i)}$, *as* K-*algebras.*

From Theorem 1.10 and Theorem 1.12, we have that if R satisfies certain mild finiteness conditions, then R has a unique complete generalized triangular matrix representation. Moreover, by Proposition 1.4, the diagonal subalgebras of R (which are of the form eRe for some $e = e^2$) are

Semicentral Reduced Algebras

also homomorphic images of R. In the next two propositions we indicate that the study of many well-known classes of algebras and rings can be reduced to the investigation of semicentral reduced algebras and rings from the same respective class.

Proposition 1.13. [6, Proposition 2.14] *If R satisfies any of the following conditions, then R has a complete generalized triangular matrix representation in which each R_i is semicentral reduced and satisfies the same condition as R:*

(i) R has a complete set of primitive idempotents;
(ii) R has no infinite set of orthogonal idempotents;
(iii) R_R has Krull dimension;
(iv) R has DCC on (idempotent generated, principal, or finitely generated) ideals;
(v) R has DCC on (idempotent generated, principal, or finitely generated) right ideals;
(vi) R has ACC on (idempotent generated, principal, or finitely generated) ideals;
(vii) R has ACC on (idempotent generated, principal, or finitely generated) right ideals;
(viii) R has either ACC or DCC on right annihilators;
(ix) R is a semilocal ring;
(x) R is a semiperfect ring;
(xi) R is a semiprimary ring.

Proposition 1.14. [6, Proposition 2.16] *If R has a complete generalized triangular matrix representation and satisfies any of the following conditions, then each R_i is semicentral reduced and satisfies the same condition as R:*

(i) R_R has Gabriel dimension;
(ii) R is a Baer ring;
(iii) R is a right semihereditary ring;
(iv) R is a right hereditary ring;
(v) R is an I-ring (i.e., every non-nil right ideal contains a nonzero idempotent);
(vi) R is a π-regular ring;
(vii) R is a right semiartinian ring;
(viii) R is a PI-ring;
(ix) R is a right PP-ring;
(x) R is a semiregular ring;
(xi) R has bounded index of nilpotency;
(xii) R is right self-injective.

Observe from Theorem 1.12 and Proposition 1.9 that the number of elements in a complete set of left triangulating idempotents is unique for a given algebra R (which has such a set) and that this is also the number of elements in any complete set of right triangulating idempotents of R. This motivates the following definition: R has *triangulating dimension n*, written $\text{Tdim}(R) = n$, if R has a complete set of left triangulating idempotents with n elements. Note that R is semicentral reduced if and only if $\text{Tdim}(R) = 1$. In general the cardinality of a complete set of primitive idempotents of R may not be unique. However our next result shows that triangulating dimension provides a lower bound for this cardinality.

Proposition 1.15. [6, Proposition 2.18] *If $\{e_1, \ldots, e_n\}$ is a complete set of primitive idempotents for R, then $\text{Tdim}(R) \leq n$.*

Recall that R has a "block decomposition" if and only if R has a complete set of centrally primitive idempotents [15, Sections 21, 22]. Our next result shows that if R has a complete set of left triangulating idempotents, then R has a "block decomposition" and $\text{Tdim}(R)$ is greater than or equal to the cardinality of a complete set of centrally primitive idempotents.

Proposition 1.16. [6, Proposition 2.20] *Let $\{b_1, \ldots, b_n\}$ be a complete set of left triangulating idempotents for R.*

(i) $c \in \mathbf{B}(R) \setminus \{0, 1\}$ if and only if there exists $I \subsetneq \{1, \ldots, n\}$ such that $c = \sum_{i \in I} b_i$ and $b_i R b_j = b_j R b_i = 0$, for each $i \in I$ and $j \notin I$.

(ii) R has a complete set of centrally primitive idempotents.

(iii) $\{b_1, \ldots, b_n\} \subseteq \mathcal{S}_\ell(R)$ if and only if $\{b_1, \ldots, b_n\}$ is a complete set of centrally primitive idempotents.

2. Matrix Algebras

In this section we investigate matrix algebras and trivial extensions over semicentral reduced algebras. Let E_{ij} denote the matrix unit with 1 in the (i, j)-position and 0 elsewhere.

Lemma 2.1. *If $\alpha = [a_{ij}] \in \mathcal{S}_\ell(M_n(R))$, then $a_{ii} \in \mathcal{S}_\ell(R)$ for all $i = 1, \ldots, n$.*

Proof. Let $x \in R$ and $X = xE_{ii}$. Then $\alpha X \alpha = X\alpha$. Hence $a_{ii} x a_{ii} = x a_{ii}$. So $a_{ii} \in \mathcal{S}_\ell(R)$. □

Theorem 2.2. *For any positive integer n, R is semicentral reduced if and only if $M_n(R)$ is semicentral reduced.*

Proof. Assume that R is semicentral reduced. Let $\alpha = [a_{ij}] \in \mathcal{S}_\ell(M_n(R))$. Then by Lemma 2.1, $a_{11} \in \mathcal{S}_\ell(R)$. Thus $a_{11} = 0$ or $a_{11} = 1$.

Case 1. $a_{11} = 0$. Then $E_{11}\alpha$ is nilpotent. In fact, $(E_{11}\alpha)^n = 0$. Now $0 = (E_{11}\alpha)^n = E_{11}^n \alpha = E_{11}\alpha$, and hence $a_{1i} = 0$ for every $i = 2, \ldots, n$. Next from $E_{12}\alpha = \alpha E_{12}\alpha$, it follows that $a_{2i} = 0$ for all $i = 1, 2, \ldots, n$. For the k-th row of α, from $E_{1k}\alpha = \alpha E_{1k}\alpha$, we get that $a_{ki} = 0$ for all $i = 1, 2, \ldots, n$. Thus $\alpha = 0$.

Case 2. $a_{11} = 1$. Since $E_{11}\alpha = \alpha E_{11}\alpha$, it follows that $a_{i1} = 0$ for all $i \neq 1$. Again since $E_{21}\alpha = \alpha E_{21}\alpha$, it follows that $a_{22} = 1$ and $a_{i2} = 0$ for all $i \neq 2$. Continuing this procedure yields $\alpha = 1$.

Consequently, by Cases 1 and 2, $M_n(R)$ is semicentral reduced.

Conversely, assume $M_n(R)$ is semicentral reduced. Let $e \in \mathcal{S}_\ell(R)$. Then $eE_{11} + \cdots + eE_{nn} \in \mathcal{S}_\ell(M_n(R))$. Thus $e = 0$ or $e = 1$. So R is semicentral reduced. $\square$

Question 2.3. Is the semicentral reduced property a Morita invariant property?

For a ring R and a bimodule $_RM_R$, the *trivial extension* of R by M is the ring $T(R, M) = R \oplus M$ with the usual addition and multiplication $(r_1, m_1)(r_2, m_2) = (r_1 r_2, r_1 m_2 + m_1 r_2)$. This is isomorphic to the ring of all "matrices" $\begin{pmatrix} r & m \\ 0 & r \end{pmatrix}$ where $r \in R$ and $m \in M$ and the usual matrix operations are used.

Proposition 2.4. *Let T be the trivial extension of R by M.*
(i) If R is semicentral reduced, then T is semicentral reduced.
(ii) If M is an ideal of R and T is semicentral reduced, then R is semicentral reduced.

Proof. (i) Assume R is semicentral reduced. Let $E = (e, x) \in \mathcal{S}_\ell(T)$. Then $e = e^2$ and $ex + xe = x$. For any $a \in R$, we have $(a, 0)(e, x) = (e, x)(a, 0)(e, x)$. So $ae = eae$ and hence $e \in \mathcal{S}_\ell(R)$. Therefore $e = 0$ or $e = 1$. If $e = 0$, then $x = 0$ and so $E = 0$. If $e = 1$, then $x = x + x$ and so $x = 0$. Hence $E = 1$. Therefore T is semicentral reduced.

(ii) Assume T is semicentral reduced. Suppose that $e \in \mathcal{S}_\ell(R)$. Then for any $(a, x) \in T$, it follows that $ae = eae$ and $xe = exe$. So we have that $(a, x)(e, 0) = (e, 0)(a, x)(e, 0)$. Therefore $(e, 0) \in \mathcal{S}_\ell(T)$. Thus we have that $(e, 0) = 0$ or $(e, 0) = 1$. So $e = 0$ or $e = 1$. Therefore R is semicentral reduced. $\square$

3. Semicentral Reduced Rings with Additional Conditions

In this section R denotes a ring with unity (i.e., $K = \mathbb{Z}$). Propositions 1.13 and 1.14 impel us to consider semicentral reduced rings satisfying additional conditions. As we see in our first lemma and corollary, semicentral reduced rings exhibit behavior similar to that of prime rings. We develop various criteria which ensure that a semicentral reduced ring is prime.

Lemma 3.1. [6, Lemma 2.1] *The following are equivalent:*
   (i) R *is semicentral reduced;*
   (ii) $(1-e)Re$ *and* $eR(1-e)$ *are both nonzero for each nontrivial idempotent* e;
   (iii) *if e is a nontrivial idempotent of R and A and B are subsets of R that contain e and $1-e$, respectively, then neither ARB nor BRA can be zero;*
   (iv) *if X is a right ideal of R and $e \in \mathbf{I}(R)$ such that $1-e \in X$, then $XeR = 0$ implies $e = 0$ or $e = 1$.*

Corollary 3.2. [6, Corollary 2.2] *Let R be semicentral reduced. Then exclusively, either:*
   (i) R *is reduced and* $\mathbf{I}(R) = \{0, 1\}$; *or*
   (ii) *for each nonzero idempotent* $e \in R$, eR *and* Re *each contains nonzero nilpotent elements.*

Recall [11] that R is called *quasi-Baer* if the right annihilator of every right ideal is generated, as a right ideal, by an idempotent.

Proposition 3.3. [6, Lemma 4.2] R *is a prime ring if and only if R is quasi-Baer and semicentral reduced.*

In [9] and [16], there are right primitive rings R with $Z(R_R) \neq 0$. But observe that if R is a prime ring, then either $Z_2(R_R) = 0$ or $Z_2(R_R) = R$. In many of our succeeding results we see this behavior for semicentral reduced rings.

From [8], R is said to be right *FI-extending* if every ideal is right essential in an idempotent-generated right ideal. Note that the class of right FI-extending rings properly contains the class of right CS (or extending) rings (hence all right self-injective rings). We use $<R, R>$ to denote the ideal generated by $\{xy - yx \mid x, y \in R\}$.

Lemma 3.4. *If R is right FI-extending, then there exists $e \in \mathcal{S}_\ell(R)$ such that* $<R, R> \subseteq_r^{ess} eR$.

Proof. There exists $e = e^2$ such that $<R, R> \subseteq_r^{ess} eR$. Let $x \in eR$. Then $x(eR) = (ex + (1-e)x)eR \subseteq eR + (1-e)xeR$. Observe $(1-e)xe = xe - exe = (xe - ex)e \in <R, R> \subseteq eR$. Thus $eR \subseteq_2 R$. Hence, by Proposition 1.1, $e \in \mathcal{S}_\ell(R)$. □

Theorem 3.5. *If R is semicentral reduced and right FI-extending, then exactly one of the following conditions is true:*
 (i) *R is a prime ring.*
 (ii) *$R = Z_2(R_R)$, $\mathbf{P}(R) \cap I \neq 0$ for all nonzero ideals I of R, and either:*
  (a) *R is a commutative uniform ring; or*
  (b) *$<R, R> \subseteq_r^{ess} R$.*

Proof. From [4], either $R = Z_2(R_R)$ or R is a right nonsingular quasi-Baer ring. By Proposition 3.3, if R is quasi-Baer, then R is a prime ring. So assume $R = Z_2(R_R)$. If R is semiprime, then again R is prime by Proposition 3.3 and [3, Lemma 2.2]. So assume $\mathbf{P}(R) \neq 0$. Let $0 \neq I$ be an ideal of R. Suppose $\mathbf{P}(R) \cap I = 0$. Since R is right FI-extending, there exists $e = e^2 \in R$ such that $I \subseteq_r^{ess} eR$. Then $eR \cap \mathbf{P}(R) = 0$ and so $e\mathbf{P}(R) = 0$. Thus $\mathbf{P}(R) \subseteq_r (1-e)R$. Noting that R is semicentral reduced, $eR(1-e) \neq 0$ from Proposition 1.1. If $eR(1-e) \subseteq I$, then $[ReR(1-e)R]^2 = 0$. Hence $eR(1-e) \subseteq I \cap \mathbf{P}(R) = 0$, a contradiction. So there exists $x \in R$ such that $ex(1-e) \notin I$. Since $I \subseteq_r^{ess} eR$, there exists $y \in I$ such that $0 \neq ex(1-e)y \in I$. Let $Y = \{y \in R \mid ex(1-e)y \in I\}$. So $[R(ex(1-e)Y)]^2 = 0$. Hence $0 \neq R(ex(1-e)Y) \subseteq I \cap \mathbf{P}(R) = 0$, a contradiction. Thus $\mathbf{P}(R) \cap I \neq 0$. The remainder of the proof follows from Lemma 3.4. □

Recall that a ring is called *right bounded* if every essential right ideal contains a nonzero ideal. A ring R is called *right FPF* if every faithful finitely generated module is a generator in the category of right R-modules.

Proposition 3.6. *Let R be a semicentral reduced right FPF ring. Then either:*
 (i) *R is a right bounded prime left and right Goldie ring; or*
 (ii) *$R = Z_2(R_R)$ and $\mathbf{P}(R) \subseteq_r^{ess} R$.*

Proof. By [13, Theorem 5.1] $R = Z_2(R_R)$ or R is right nonsingular. Assume $Z(R_R) = 0$, then [13, Theorem 3.3] yields that R is semiprime. From [12, p. 168], R is a quasi-Baer ring. By Proposition 3.3, R is a prime ring. From [13, Theorem 4.7], R is right bounded and left and right Goldie. If $R = Z_2(R_R)$, [5, Proposition 3.11] yields that $\mathbf{P}(R) \subseteq_r^{ess} R$. □

Proposition 3.7. *Let R be a semicentral reduced regular ring with bounded index. Then R is simple Artinian.*

Proof. This result is a consequence of [14, Theorem 7.9, p. 74]. □

Proposition 3.8. [6, Proposition 2.3] *Let R be semicentral reduced with $\text{Soc}(R_R) \neq 0$. Then either R is a division ring or $\text{Soc}(R_R) = U \oplus V \oplus X$, as a direct sum of right ideals, where:*

(i) *U is an ideal of R and $U^2 = 0$;*

(ii) *V is a direct sum of minimal ideals V_j (of R), where $V_j = Re_j R$, $e_j \in \mathbf{I}(R)$, and each $e_j R$ is a minimal right ideal of R such that $e_j R \cap \mathbf{N}(R) \neq 0$;*

(iii) *X is a direct sum of the minimal right ideals X_i (of R), where each X_i is isomorphic as a right R-module to a nilpotent right ideal of R and each X_i is generated by an idempotent.*

In the next corollary we use the notation introduced in Proposition 3.8.

Corollary 3.9. [6, Corollary 2.4] *The following are equivalent:*

(i) *R is semicentral reduced with a complete set of primitive idempotents and $V \neq 0$;*

(ii) *R is isomorphic to a full n-by-n matrix algebra over a division ring, for some n.*

Lemma 3.10. *Let I be an ideal of R and $e = e^2$ such that $I \subseteq_r^{ess} eR$. Then:*

(i) *$(1-e)Re \subseteq Z(R_R)$;*

(ii) *if X is an ideal of R such that $(1-e)Re \subseteq X$, then $eR + X$ is an ideal of R; hence $eR + Z(R_R)$ is an ideal of R;*

(iii) *if $Z(R_R) \subseteq eR$ and $Y \subseteq_r R$ such that $I \subseteq_r^{ess} Y$, then $Y \subseteq eR$ and eR is an ideal of R.*

Proof. (i) Let $x \in R$. There exists an essential right ideal L of R such that $exL \subseteq I$. Then $(1-e)RexL \subseteq eR \cap (1-e)R = 0$. Thus $(1-e)ReR \subseteq Z(R_R)$.

(ii) For any $y \in R$, $ye = (1-e)ye + eye$, so (ii) follows.

(iii) Let $y \in Y$ such that $y \notin eR$. Then $(1-e)y \neq 0$. Hence there exists an essential right ideal J such that $yJ \subseteq I$. Then $(1-e)yJ = 0$. So $(1-e)y \in Z(R_p) \cap (1-e)R = 0$. Therefore $Y \subset eR$. □

Proposition 3.11. *Let R be right nonsingular. Then R is semicentral reduced if and only if no nonzero proper idempotent generated right ideal is a right essential extension of an ideal.*

Proof. Assume R is semicentral reduced. Let $0 \neq e = e^2 \in R$ such that there exists an ideal I of R with $I \subseteq_r^{ess} eR$. By Proposition 1.1 and Lemma 3.10(iii), $e \in \mathcal{S}_\ell(R)$. Hence $e = 1$, so eR is not proper in R.

Semicentral Reduced Algebras 81

Conversely, no nonzero proper idempotent generated right ideal is an ideal. Hence $\mathcal{S}_\ell(R) = \{0,1\}$. Thus R is semicentral reduced. □

Recall R is said to be *orthogonally finite* if R has no infinite set of orthogonal idempotents.

Lemma 3.12. *Let R be an orthogonally finite ring with $\mathrm{Soc}(R_R) \subseteq_r^{ess} R$. Then there exists $b \in \mathcal{S}_\ell(R)$ such that $Z(R_R) + X \subseteq_r^{ess} bR$, where X is the ideal generated by the square-zero minimal right ideals.*

Proof. If every minimal right ideal is square-zero, then we are done. So assume that not every minimal right ideal of R is square-zero. Let $e_1 R$ be a minimal right ideal of R such that $e_1^2 = e_1$. Now $e_1 R \cap Z(R_R) = 0$ because $Z(R_R)$ contains no nonzero idempotents, and $e_1 R \cap X = 0$ because X is a nil ideal. Hence $e_1 R \cap (Z(R_R) + X) = 0$ and so $e_1(Z(R_R) + X) = 0$. Then $Z(R_R) + X \subseteq (1-e_1)R$. If $Z(R_R) + X \subseteq_r^{ess} (1-e_1)R$, let $b = 1-e_1$. By Lemma 3.10, $b \in \mathcal{S}_\ell(R)$. So assume $Z(R_R) + X$ is not right essential in $(1-e_1)R$. Then there exists a minimal right ideal $e_2 R \subseteq (1-e_1)R$ such that $e_2^2 = e_2$ and $e_2 R \cap (Z(R_R) + X) = 0$. Thus $e_2(Z(R_R) + X) = 0$. So $Z(R_R) + X \subseteq (1-e_1)R \cap (1-e_2)R$. Note that $e_1 e_2 = 0$, so $e_1 + e_2 - e_2 e_1$ is an idempotent and $e_1 R \oplus e_2 R = (e_1 + e_2 - e_2 e_1)R$. Also $(1-e_1)R \cap (1-e_2)R = (1 - (e_1 + e_2 - e_2 e_1))R$. Thus there exist orthogonal idempotents c_1 and c_2 such that $c_1 R = e_1 R$, $c_2 R = e_2 R$, and $Z(R_R) + X \subseteq [1 - (c_1 + c_2)]R$. If $Z(R_R) + X \subseteq_r^{ess} [1 - (c_1 + c_2)]R$, let $b = 1 - (c_1 + c_2)$. By Lemma 3.10, $b \in \mathcal{S}_\ell(R)$. Otherwise continue this procedure. Since R is orthogonally finite, this procedure terminates in the desired $b \in \mathcal{S}_\ell(R)$ after a finite number of steps. □

Theorem 3.13. *Let R be a semicentral reduced ring. If R is orthogonally finite and $\mathrm{Soc}(R_R) \subseteq_r^{ess} R$, then either:*
  (i) *R is simple Artinian; or*
  (ii) *$[\mathrm{Soc}(R_R)]^2 = 0$ (hence $R = Z_2(R_R)$).*

Proof. If $\mathbf{P}(R) = 0$, then R is simple Artinian by Corollary 3.9. So assume $\mathbf{P}(R) \neq 0$. Observe that if $Z(R_R) \neq 0$, then there exists a family of square-zero minimal right ideals $\{I_i\}$ such that $\sum I_i \subseteq_r^{ess} Z(R_R)$. Now from Lemma 3.12, $\mathbf{P}(R) \subseteq_r^{ess} R$. Since $\mathrm{Soc}(R_R) \subseteq_r^{ess} R$, it follows that $Z(R_R) = \ell(\mathrm{Soc}(R_R))$. Assume $Z(R_R)$ is not right essential in R. Then there exists a minimal right ideal M such that $M \cap \ell(\mathrm{Soc}(R_R)) = 0$. Hence there exists a minimal right ideal Y such that $MY \neq 0$. But since $\mathbf{P}(R) \subseteq_r^{ess} R$, we have that $Y \subseteq \mathbf{P}(R)$, so $Y^2 = 0$. Hence $M = MY = (MY)Y = 0$, a contradiction. Thus $R = Z_2(R_R)$ and $\mathrm{Soc}(R_R) \subseteq Z(R_R)$. So $[\mathrm{Soc}(R_R)]^2 = 0$. □

There is an alternative proof for Theorem 3.13 for the case when $\mathbf{P}(R) \neq 0$. Indeed, if $\mathbf{P}(R) \neq 0$, then $\mathbf{P}(R) \subseteq_r^{ess} R$ as in the above proof of Theorem 3.13. Therefore $[\mathrm{Soc}(R_R)]^2 = 0$ by Proposition 3.8, and hence $R = Z_2(R_R)$. But the above proof of Theorem 3.13 stands alone without Proposition 3.8.

Corollary 3.14. *Let R be semicentral reduced with $\mathrm{Soc}(R_R) \subseteq_r^{ess} R$. If R is semilocal or has uniform dimension, then either:*
   (i) *R is simple Artinian; or*
   (ii) *$[\mathrm{Soc}(R_R)]^2 = 0$ (hence $R = Z_2(R_R)$).*

The following example illustrates Theorem 3.13(ii).

Example 3.15. [7] Let $R = \mathbb{Z}_3[S_3]$, the group algebra of the symmetric group S_3 on the three symbols $\{1, 2, 3\}$ over the field $\mathbb{Z}_3$ of three elements. Denote $\sigma = (123)$ and $\tau = (12)$ in S_3. Let $e_1 = 2 + \tau$ and $e_2 = 2 + 2\tau$ in R. Then since $e_1 R$ and $e_2 R$ are uniform, $\{e_1, e_2\}$ is a complete set of primitive idempotents. Now say that $b \in S_\ell(R)$. From Proposition 1.3(vii), either $bR = 0$, $bR = e_1 R$, $bR = e_2 R$, or $bR = R$. If $bR = e_1 R$, then $e_1 R$ is an ideal of R and so $e_1 \in S_\ell(R)$. Hence $\sigma e_1 = e_1 \sigma e_1$, which is a contradiction. Similarly, $bR \neq e_2 R$. Therefore $bR = 0$ or $bR = R$. So R is semicentral reduced. But $R/J(R) \cong \mathbb{Z}_3 \oplus \mathbb{Z}_3$, which is not simple Artinian as in [7]. Thus this example also shows that, in general, semicentral idempotents cannot lift modulo a nilpotent ideal.

Corollary 3.16. *R is left perfect if and only if R has a complete generalized triangular matrix representation where each diagonal ring R_i is left perfect and either simple Artinian or $[\mathrm{Soc}(R_{iR_i})]^2 = 0$.*

Proof. This result is a consequence of Proposition 1.13(v), Theorem 3.13, and [1, Proposition 28.11, p. 319]. □

Acknowledgements. The authors wish to thank the referee for helpful comments and suggestions for the improvement of the paper, especially for Theorem 3.5 and Lemma 3.12. The first author is grateful for the kind hospitality he received at Pusan National University and at Kyung Hee University. The second author was supported by the Academic Research Fund of the Ministry of Education, Korea, Project No. BSRI-97-1432, while the third author was supported in part by the Korea Research Foundation, Project No. 1998-001-D00006.

References

[1] F. W. Anderson and K. R. Fuller, *Rings and Categories of Modules*, Springer-Verlag, Heidelberg, New York, 1974.

[2] G. F. Birkenmeier, *Idempotents and completely semiprime ideals*, Comm. Algebra **11** (1983), 567–580.

[3] _____, *A generalization of FPF rings*, Comm. Algebra **17** (1989), 855–884.

[4] _____, *Decompositions of Baer-like rings*, Acta Math. Hung. **59** (1992), 319–326.

[5] _____, *When does a supernilpotent radical essentially split off?*, J. Algebra **172** (1995), 49–60.

[6] G. F. Birkenmeier, H. E. Heatherly, J. Y. Kim, and J. K. Park, *Triangular matrix representations*, J. Algebra **230** (2000), 558–595.

[7] G. F. Birkenmeier, J. Y. Kim, and J. K. Park, *A couterexample for CS-rings*, Glasgow Math. J. **42** (2000), 263–269.

[8] G. F. Birkenmeier, B. J. Müller, and S. T. Rizvi, *Modules in which every fully invariant submodule is essential in a direct summand*, Preprint.

[9] K. A. Brown, *The singular ideals of group rings*, Quart. J. Math. Oxford **28**(2) (1977), 41–60.

[10] S. U. Chase, *A generalization of the ring of triangular matrices*, Nagoya Math. J. **18** (1961), 13–25.

[11] W. E. Clark, *Twisted matrix units semigroup algebras*, Duke Math. J. **24** (1967), 417–423.

[12] C. Faith, *Injective quotient rings of commutative rings, Module Theory*, Lecture Notes in Math. **700**, Springer-Verlag, Heidelberg, New York (1979), 151–203.

[13] C. Faith and S. Page, *FPF Ring Theory: Faithful Modules and Generators of Mod-R,* London Math. Soc. Lecture Notes Series **88**, Cambridge Univ. Press, Cambridge, 1984.

[14] K. R. Goodearl, *Von Neumann Regular Rings* (2nd edition), Krieger, Malabar, 1991.

[15] T. Y. Lam, *A First Course in Noncommutative Rings*, Springer-Verlag, Heidelberg, New York, 1991.

[16] J. Lawrence, *A singular primitive ring*, Proc. Amer. Math. Soc. **45** (1974), 59–62.

Gary F. Birkenmeier
Department of Mathematics
University of Louisiana at Lafayette
Lafayette, LA 70504-1010, U. S. A.
e-mail: gfb1127@interval.usl.edu

Jin Yong Kim
Department of Mathematics
Kyung Hee University
Suwon 449-701, Korea

Jae Keol Park
Department of Mathematics
Pusan National University
Pusan 609-735, Korea
e-mail: jkpark@hyowon.cc.pusan.ac.kr

On Generalizations of Injectivity

Jianlong Chen and Nanqing Ding

Abstract

Let R be a ring. The following results are proven.

1. If R is semilocal, right Kasch and right simple-injective, then R is semiperfect, left and right finitely cogenerated.

2. If R is left perfect, right simple-injective and left (or right) pseudo-coherent, then R is QF.

3. If R is semilocal, right CF and right mininjective, then R is QF.

4. R is strongly regular if and only if R is a weakly right duo ring whose simple singular left R-modules are GP-injective.

1. Introduction

Throughout this paper R is an associative ring with identity and all modules are unitary. We write M_R to indicate that M is a right R-module and $N \subseteq M$ to mean N is a submodule of M. The left and right annihilators of a subset X of R are denoted by $l(X)$ and $r(X)$, respectively. Let M be a right R-module. For each subset A of R, the left annihilator of A in M is denoted by $l_M(A)$. As usual, J and $Soc(_RR)$ ($Soc(R_R)$) denote respectively the Jacobson radical and the left (right) socle of R.

Generalizations of injectivity have been studied in many papers such as [B], [CD], [DC], [G1], [H], [KNK], [NY1-3], [TC] and [X2]. The motivation of the present discussion is from [NY2], [NY3] and [KNK]. In this paper, we first prove that if R is semilocal, right Kasch and right simple-injective, then R is semiperfect, left and right finitely cogenerated. As a corollary, we have that R is a D-ring if and only if R is semilocal, right Kasch, left and right simple-injective. Then it is shown that, for a left perfect and right simple-injective ring R, if R is left (or right) pseudo-coherent, then R is QF. We also show that if R is semilocal, right CF and right mininjective, then R is QF. In particular, two earlier results are obtained as consequences. Finally, we have that a ring R is strongly regular if and only if R is a weakly right duo ring whose simple singular left R-modules are GP-injective.

2. Simple-injectivity

Let M_R and N_R be R-modules. Following Harada [H], M is said to be a *simple-N-injective* module if, for any submodule $X \subseteq N$ and any R-homomorphism $\gamma : X \to M$ such that $\text{im}(\gamma)$ is simple, there exists an R-homomorphism $\bar\gamma : N \to M$ such that $\bar\gamma|_X = \gamma$. We call a ring R *right simple-injective* if R_R is simple-R-injective; equivalently, if I is a right ideal of R and $\gamma : I \to R$ is an R-homomorphism with simple image, then γ is given by left multiplication by an element of R.

A ring R is called *right Kasch* if every simple right R-module embeds in R, equivalently $l(M) \neq 0$ for every maximal right ideal M of R.

Lemma 2.1. *If R is right Kasch, right simple-injective, then $l(J)$ is an essential left ideal of R.*

Proof. The proof is similar to that of [NY1, Lemma 2.3]. If $0 \neq b \in R$, choose M maximal in bR and let $\sigma : bR/M \to R_R$ be monic (for R is right Kasch). If $\alpha : bR \to R_R$ is defined by $\alpha(x) = \sigma(x + M)$ for $x \in bR$, then $\text{im}(\alpha) = \sigma(bR/M)$ is simple. Thus $\alpha = a\cdot$ is left multiplication by an element $a \in R$ since R is right simple-injective, and so $ab = \alpha(b) = \sigma(b + M) \neq 0$. But $abJ = a(bJ) = \alpha(bJ) = 0$ because $bJ \subseteq M$, so $0 \neq ab \in Rb \cap l(J)$, as required. □

Lemma 2.2. *If M_R is simple-R-injective, then $l_M(A \cap B) = l_M(A) + l_M(B)$, where A and B are right ideals of R and A is finitely generated semisimple.*

Proof. First we prove that, for any right ideal I of R and any homomorphism $\phi : I \to M$ such that $\text{im}(\phi)$ is finitely generated semisimple, then ϕ can be extended to a homomorphism $R_R \to M$. In fact, let $\text{im}(\phi) = \bigoplus_{i=1}^n S_i$, where each S_i is simple, and let $\pi_i : \text{im}(\phi) \to S_i$ be the projection. Then $\text{im}(\pi_i\phi) = S_i$ is simple, and so there exists $t_i \in M$ such that $\pi_i\phi(a) = t_i a$ for any $a \in I$. Put $t = \sum_{i=1}^n t_i$, then

$$\phi(a) = \sum_{i=1}^n \pi_i\phi(a) = \sum_{i=1}^n t_i a = ta$$

for $a \in I$. This shows that ϕ can be extended to a homomorphism from R_R to M.

Secondly, let $x \in l_M(A \cap B)$, where A is a finitely generated semisimple right ideal and B is a right ideal of R. Following a standard argument due to Ikeda and Nakayama we define a map $\phi : A+B \to M_R$ by $\phi(a+b) = xa$ for all $a \in A$ and $b \in B$. It is easy to see that ϕ is an R-homomorphism and im$(\phi) = xA$. Since $A \to xA \to 0$ is exact, xA is also finitely generated semisimple. By the first part of the proof, there exists $m \in M$ such that $xa = \phi(a+b) = m(a+b)$ for any $a \in A$ and $b \in B$. Let $a = 0$, then $m \in l_M(B)$, and let $b = 0$, then $x - m \in l_M(A)$. Thus $x = x - m + m \in l_M(A) + l_M(B)$, which implies $l_M(A \cap B) \subseteq l_M(A) + l_M(B)$. The reverse inclusion is clear. So the result follows. □

Recall that a ring R is called semilocal if R/J is semisimple Artinian. R is called right *P-injective* if every R-homomorphism from a principal right ideal of R into R extends to an endomorphism of R. R is called right *GPF* [NY1] if R is right P-injective, semiperfect, and $Soc(R_R)$ is essential as a right ideal. A ring R is said to be *right mininjective* [NY2] if every R-homomorphism from a simple right ideal of R into R extends to an endomorphism of R. Clearly, right simple-injective rings are right mininjective.

Theorem 2.3. *If R is semilocal, right Kasch and right simple-injective, then R is left GPF, left and right finitely cogenerated.*

Proof. Since R is semilocal, $J = \bigcap_{i=1}^n M_i$, where each M_i is a maximal right ideal. We may assume, without loss of generality, that

$$\bigcap_{i=1}^n M_i \neq \bigcap_{1 \leq j \leq n} M_j \, (j \neq k) \text{ for any } k, 1 \leq k \leq n.$$

By [CD, Lemma 2.7], $l(J) = \sum_{i=1}^n l(M_i)$. Since R is right simple-injective, R is right mininjective. By [NY2, Theorem 2.3], $l(M_i)$ is a minimal left ideal, and so $l(J)$ is a finitely generated left ideal. By Lemma 2.1, $l(J)$ is essential in $_RR$. But $Soc(_RR) = l(J)$ by [K, Theorem 9.3.5], and $Soc(_RR) = Soc(R_R)$ by [NY2, Lemma 4.2]. Thus R is left finitely cogenerated.

Let now $c_1R, c_2R, \ldots, c_nR$ be the representatives for the isomorphism classes of simple right R-modules, where $c_i \in R$ ($i = 1, 2, \cdots, n$). Then $Rc_1, Rc_2, \ldots, Rc_n$ are simple left R-modules by [NY2, Theorem 1.14(1)]. If $Rc_i \cong Rc_j$, then $c_iR \cong c_jR$ by [NY1, Theorem 1.1] (for R is left P-injective by [NY2, Lemma 4.2]), and so $i = j$. Therefore $Rc_1, Rc_2, \ldots, Rc_n$ are the representatives for the isomorphism classes of simple left R-modules. Consequently, R is left Kasch, left P-injective, and so R is right finitely cogenerated by [CD, Theorem 2.8] and $J = lr(J)$

by [CD, Theorem 2.3(3)]. But $r(J) = Soc(_RR)$ (for R is semilocal), and then $r(J)$ is a finitely generated semisimple right ideal. Thus $l(I \cap r(J)) = l(I) + lr(J) = l(I) + J$ for any right ideal I of R by Lemma 2.2, and hence idempotents can be lifted over J by [HN, Theorem 3.8]. Therefore R is semiperfect, whence R is left GPF. □

Corollary 2.4. *If R is left and right Kasch, right simple-injective, then R is left GPF, left and right finitely cogenerated.*

Proof. By [NY2, Lemma 4.2], $I = rl(I)$ for every right ideal I of R, and hence R is semilocal by [GG, Theorem 2.5]. So the result follows from Theorem 2.3. □

A ring R is called a *D-ring* [HN] if $I = lr(I)$ and $K = rl(K)$ for every left ideal I and right ideal K of R.

Corollary 2.5. *A ring R is a D-ring if and only if R is semilocal, right Kasch, left and right simple-injective.*

Proof. ($\Rightarrow$). R is semilocal by [HN, Theorem 3.4], and R is left and right simple-injective by [HN, Proposition 5.2]. Obviously, R is two-sided Kasch.

($\Leftarrow$). By the proof of Theorem 2.3, R is two-sided Kasch. Hence R is a D-ring by [NY2, Lemma 4.2]. □

Remark 1. A left perfect, left and right simple-injective ring is QF by [NY3, Proposition 3]. Hence a left perfect D-ring is QF by Corollary 2.5. This gives an affirmative answer to a question raised by Xue [X1, P. 754].

Recall that a ring R is called *left pseudo-coherent* [B] if $l(S)$ is finitely generated for every finite subset S of R. An idempotent e in a ring R is called *local* if eRe is a local ring. A ring R is called *right minfull* [NY2] if it is semiperfect, right mininjective and $Soc(eR) \neq 0$ for each local idempotent $e \in R$.

Nicholson and Yousif [NY3] conjectured that: A left perfect, right simple-injective ring is right self-injective. The following theorem is motivated by this conjecture.

Theorem 2.6. *Assume that R is left perfect and right simple-injective. If R is left (or right) pseudo-coherent, then R is QF.*

Proof. Since R is left perfect, $Soc(eR) \neq 0$ for every local idempotent $e \in R$. Thus R is a right minfull ring, and so R is left and right Kasch by [NY2, Theorem 3.7]. By the proof of Theorem 2.3, $S = Soc(_RR) =$

$Soc(R_R)$ is a finitely generated left and right ideal. Clearly, $J \subseteq l(S)$. Since R is left and right Kasch, $J = l(S) = r(S)$. By hypotheses, J is a finitely generated left (or right) ideal. Since R is left perfect, R is left (or right) Artinian by [O, Lemma 11]. But R is left and right mininjective by [NY2, Lemma 4.2], and hence R is QF by [NY2, Corollary 4.8]. □

Remark 2. Recall that a ring R is called left *f*-injective [G1] (= *f.g.* injective in [B]) if every homomorphism from a finitely generated left ideal of R into R extends to an endomorphism of R. Björk [B, Theorem 4.3] proved that if R is a left f-injective ring which is also left perfect and right pseudo-coherent, then R is a QF ring. In relation with the preceding theorem, we may pose the following question: if R is a left simple-injective ring which is also left perfect and right (or left) pseudo-coherent, is R a QF ring?

3. Mininjectivity

Recall that a ring R is called a right *CF* ring [G2] if every cyclic right R-module embeds in a free module, or equivalently, every right ideal of R is a right annihilator of a finite subset of R.

Theorem 3.1. *Let R be a semilocal ring. If R is right CF and right mininjective, then R is QF.*

Proof. Since R is right CF, R is right Kasch and left P-injective. Let now $c_1 R, c_2 R, \ldots, c_n R$ be the representatives for the isomorphism classes of simple right R-modules, where $c_i \in R$ ($i = 1, 2, \ldots, n$). Then $Rc_1, Rc_2, \ldots, Rc_n$ are simple left R-modules by [NY2, Theorem 1.14(1)]. If $Rc_i \cong Rc_j$, then $c_i R \cong c_j R$ by [NY1, Theorem 1.1], and so $i = j$. Hence $Rc_1, Rc_2, \ldots, Rc_n$ are the representatives for the isomorphism classes of simple left R-modules. Thus R is left Kasch, and so R is right Artinian by [GG, Corollary 2.6]. Clearly, R is left and right mininjective, and so R is QF by [NY2, Corollary 4.8]. □

Corollary 3.2. [X2, Theorem 7] *or* [RS, Theorem 3.5] *Let R be a semiperfect ring. If R is right CF and right mininjective, then R is QF.*

Recall that a ring R is called right *A-injective* [TC] if R satisfies the following two conditions:
(1) $l(I \cap K) = l(I) + l(K)$ for each pair of right ideals I and K of R.
(2) $lr(a) = Ra$ for each $a \in R$.

Clearly, right A-injective rings are right P-injective, and hence right mininjective. So we have the following corollary.

Corollary 3.3. [TC, Theorem 7] *Let R be a semilocal ring. If R is right CF and right A-injective then R is QF.*

Corollary 3.4. *Let R be left finite dimensional. If R is right CF and right mininjective then R is QF.*

Proof. Since R is right CF, R is left P-injective. Thus R is semilocal by [NY1, Theorem 3.3], and so R is QF by Theorem 3.1. □

4. GP-injectivity

A right R-module M is called *GP-injective* [KNK] (= *YJ-injective* in [Y3]) if, for any $0 \neq a \in R$, there exists a positive integer n such that $a^n \neq 0$ and any right R-homomorphism from $a^n R$ to M extends to one from R_R to M. A ring R is called *weakly right duo* (abbreviated WRD) if for any $a \in R$, there exists a positive integer n such that $a^n R$ is an ideal of R. R is said to be *right quasi-duo* if every maximal right ideal is an ideal. R is abelian if every idempotent element of R is central.

Recently, N. K. Kim, S. B. Nam and J. Y. Kim [KNK] proved the following:

Theorem. *The following statements are equivalent.*

(1) *R is strongly regular.*

(2) *R is a WRD ring whose simple singular right R-modules are GP-injective.*

(3) *R is an abelian right quasi-duo ring whose simple singular right R-modules are GP-injective.*

We note that strong regularity is a left-right symmetric property. It is natural to ask whether the condition "simple singular right R-modules are GP-injective" in the above-mentioned theorem can be replaced by "simple singular left R-modules are GP-injective". The answer is "Yes". To answer this question, we first prove the following lemma which may be viewed as the intensification of [DC, Theorem 3.4].

Lemma 4.1. *Let R be a ring whose simple singular left R-modules are GP-injective. If R is abelian or semiprime, then for each $0 \neq a \in R$, there exists a positive integer n with $a^n \neq 0$ such that $RaR + l(a^n) = R$. In particular, $a^n \in RaR a^n$ and $J = 0$.*

Proof. First we prove that if $RaR + l(a) \subseteq M$ for some maximal left ideal M of R, then M is essential. In fact, if M is not essential, then $M \cap Rb = 0$ for some $0 \neq b \in R$. Thus $M \oplus Rb = R$, and so $M = Re$, where $e^2 = e$. If R is abelian, since $a \in RaR \subseteq M = Re$, $a = ae = ea$. Thus $1 - e \in l(a) \subseteq M = Re$, and so $1 \in M$, a contradiction. If R is semiprime, since $aR(1-e) \subseteq R(1-e) \cap RaR \subseteq R(1-e) \cap M = 0$, $aR(1-e) = 0$. Thus $(R(1-e)a)^2 = 0$, and so $(1-e)a = 0$, which leads to a contradiction by the foregoing proof.

The following proof is similar to that in [X2, Proposition 2]. Here we prove it for the sake of completeness.

Next we show that if $0 \neq a \in R$ with $a^2 = 0$, then $RaR + l(a) = R$. In fact, if $RaR + l(a) \neq R$, then $RaR + l(a) \subseteq M$ for some maximal left ideal M of R, and so M is essential by the first part of the proof. Hence R/M is GP-injective. Now define $\phi : Ra \to R/M$ by $\phi(ra) = r + M$ for $r \in R$. It is clear that ϕ is an R-homomorphism. Thus there exists $c \in R$ such that $1 + M = a(c + M)$ by the GP-injectivity of R/M, and hence $1 - ac \in M$. But $ac \in RaR \subseteq M$, and so $1 \in M$, a contradiction. Therefore $RaR + l(a) = R$ for $0 \neq a$ with $a^2 = 0$.

Finally, let $0 \neq a \in R$. If a is nilpotent, then there exists n such that $a^n \neq 0$ and $a^{n+1} = 0$. Thus $(a^n)^2 = 0$, and so $Ra^n R + l(a^n) = R$ by the second part of the proof, whence $RaR + l(a^n) = R$.

If a is not nilpotent, then $a^m \neq 0$ for any positive integer m. Let $I = \sum_{i=1}^{\infty}(Ra^i R + l(a^i))$. If $I \neq R$, then $I \subseteq N$ for some maximal left ideal N. Thus $RaR + l(a) \subseteq I \subseteq N$, and so N is essential, whence R/N is GP-injective. Therefore there exists a positive integer n such that any left R-homomorphism from Ra^n to R/N extends to an R-homomorphism from R to R/N. Now define $f : Ra^n \to R/N$ via $f(ra^n) = r + N$ for $r \in R$. It is easy to see that f is a left R-homomorphism. Thus there exists $c \in R$ such that $1 + N = a^n(c + N)$, and so $1 - a^n c \in N$. But $a^n c \in I \subseteq N$, and hence $1 \in N$, a contradiction. So $I = R$. Since ${}_RR$ is finitely generated, $R = \sum_{i=1}^{m}(Ra^i R + l(a^i))$ for some positive integer m. It follows that $R = RaR + l(a^m)$. Consequently, for each $0 \neq a \in R$, there exists a positive integer n such that $a^n \neq 0$ and $R = RaR + l(a^n)$. Thus $Ra^n = RaRa^n + l(a^n)a^n = RaRa^n$, and hence $a^n \in RaRa^n$. If $0 \neq a \in J$, then $a^n = ba^n$, where $b \in RaR \subseteq J$. Since $1 - b$ is invertible, $a^n = 0$. This is a contradiction. So $J = 0$. This completes the proof. □

Theorem 4.2. *The following statements are equivalent for a ring R.*

(1) R is strongly regular.

(2) R is a WRD ring whose simple singular right R-modules are GP-injective.

(3) R is a WRD ring whose simple singular left R-modules are GP-injective.

Proof. (1)⇔(2) by [KNK, Theorem 7]. (1)⇒(3) is clear.

(3)⇒(1). Since R is WRD, R is right quasi-duo by [Y2, Proposition 2] and abelian by [Y1, Lemma 4]. Hence R/J is reduced by [R, Proposition 4.4]. But $J = 0$ by Lemma 4.1, and so R is reduced. Thus R is weakly regular by [KNK, Corollary 5], and hence R is strongly regular by [R, Proposition 4.7] (for R is right quasi-duo). □

Theorem 4.3. *The following statements are equivalent for a ring R.*
  (1) *R is strongly regular.*
  (2) *R is a semiprime (or abelian) right quasi-duo ring whose simple singular right R-modules are GP-injective.*
  (3) *R is a semiprime (or abelian) right quasi-duo ring whose simple singular left R-modules are GP-injective.*

Proof. The proof is the same as that of Theorem 4.2. □

Acknowledgements. This work was partially supported by the National Natural Science Foundation of China (No.19701008 & 19771046), and supported by Hwa-Ying Culture and Education Foundation.

References

[AF] F. W. Anderson and K. R. Fuller, *Rings and Categories of Modules,* Springer, Heidelberg, New York, 1973.

[B] J. E. Björk, *Rings satisfying certain chain conditions,* J. Reine Angew. Math. **245** (1970), 63–73.

[CD] J. L. Chen and N. Q. Ding, *On general principally injective rings,* Comm. Algebra **27**(5) (1999), 2097–2116.

[DC] N. Q. Ding and J. L. Chen, *Rings whose simple singular modules are YJ-injective,* Math. Japon. **40**(1) (1994), 191–195.

[G1] R. N. Gupta, *On f-injective modules and semihereditary rings,* Proc. Nat. Inst. Sci. India Part A **35** (1969), 323–328.

[G2] J. L. Gómez Pardo, *Embedding cyclic and torsion-free modules in free modules,* Arch. Math. **44** (1985), 503–510.

[GG] J. L. Gómez Pardo and P. A. Guil Asensio, *Torsionless modules and rings with finite essential socle,* Lecture Notes in Pure and Appl. Math. **201** (1998), 261–278.

[H] M. Harada, *On almost relative injective of finite length,* preprint.

[HN] C. R. Hajarnavis and N. C. Norton, *On dual rings and their modules*, J. Algebra **93** (1985), 253–266.

[K] F. Kasch, *Modules and Rings*, Academic Press, London, New York, 1982.

[KNK] N. K. Kim, S. B. Nam and J. Y. Kim, *On simple singular GP-injective modules*, Comm. Algebra **27**(5) (1999), 2087–2096.

[NY1] W. K. Nicholson and M. F. Yousif, *Principally injective rings*, J. Algebra **174** (1995), 77–93.

[NY2] _____, *Mininjective rings*, ibid. **187** (1997), 548–578.

[NY3] _____, *On perfect simple-injective rings*, Proc. Amer. Math. Soc. **125** (1997), 979–985.

[O] B. L. Osofsky, *A generalization of quasi-Frobenius rings*, J. Algebra **4** (1966), 373–388.

[R] M. B. Rege, *On von Neumann regular rings and SF-rings*, Math. Japon. **31**(6) (1986), 927–936.

[RS] J. Rada and M. Saorin, *On two open problems about embedding of modules in free modules*, preprint.

[TC] H. D. Tang and J. L. Chen, *On left A-injective rings*, (in Chinese), J. Nanjing Univ. Math. Biq. **7**(2) (1990), 163–169. Zbl. Math. 735: 16002.

[X1] W. M. Xue, *A note on perfect self-injective rings*, Comm. Algebra **24**(2) (1996), 749–755.

[X2] _____, *A note on YJ-injectivity*, Riv. Mat. Univ. Parma **1**(6) (1998), 31–37.

[Y1] X. Yao, *Weakly right duo rings*, Pure and Appl. Math. Sci. **21** (1985), 19–24.

[Y2] H. P. Yu, *On quasi-duo rings*, Glasgow Math. J. **37** (1995), 21–31.

[Y3] R. Yue Chi Ming, *On regular rings and Artinian rings (II)*, Riv. Mat. Univ. Parma **11**(4) (1985), 101–109.

Jianlong Chen
Department of Mathematics
Harbin Institute of Technology
Harbin 150001, P. R. China
and
Department of Applied Mathematics
Southeast University
Nanjing 210096, P. R. China
e-mail: jlchen@seu.edu.cn

Nanqing Ding
Department of Mathematics
Nanjing University
Nanjing 210093, P. R. China
e-mail: nqding@netra.nju.edu.cn

Auslander-Gorenstein Rings for Beginners

John Clark

Abstract

We discuss past and current research on Noetherian rings which satisfy a condition introduced by Auslander involving the homological grade of modules. These rings include quasi-Frobenius rings, many commutative Noetherian rings and some non-commutative Noetherian rings arising in the theory of quantum groups.

1. Introduction

While the title of this paper is probably too ambitious, I will try to give beginners like myself a feel for the theory of Auslander-Gorenstein rings by illustrating the ideas involved with down-to-earth examples (which to the expert will be trivial) or by referring to the theory's commutative origins.

In what follows, R denotes an associative ring with identity. For homological concepts used below, we refer the reader to the texts by Jans [Ja] or Rotman [Ro]. The texts by McConnell and Robson [MR] and Goodearl and Warfield [GW] naturally serve Noetherian needs.

Definition 1.1. Let R be a ring. The *grade* $j(M)$ of a (left) R-module M is defined by

$$j(M) = \min\{\, i \mid \operatorname{Ext}_R^i(M, R) \neq 0\}$$

or ∞ if no such i exists.

Examples. (1) Clearly $j(0) = \infty$ and $j(F) = 0$ if F is a nonzero free R-module, since $\operatorname{Hom}_R(F, R) \neq 0$.

1991 *Mathematics Subject Classification.* 16E10.

Key words and phrases. Auslander condition, injective resolution, Auslander regular ring, quasi-Frobenius, pure module.

I am grateful to the organizers of the Conference for the invitation to talk and to submit this paper, based on my talk, to the Proceedings and for financial support. Special thanks go to Jae Keol Park and Gary Birkenmeier.

(2) Recall that a ring R is *quasi-Frobenius* (QF) if it is left and right artinian and left and right self-injective. Thus if R is QF and M is an R-module we have $\operatorname{Ext}_R^n(M, R) = 0$ for all $n \geq 1$. Moreover, if $M \neq 0$ one can show that $M^* = \operatorname{Hom}_R(M, R) \neq 0$ and so $j(M) = 0$. (To see this, first note that every simple R-module is isomorphic to a minimal left ideal of R; see, e.g., Jans [Ja, pp. 78–79].)

(3) Regarding the nonzero finitely generated abelian group M as a $\mathbb{Z}$-module, we have $\operatorname{Hom}_\mathbb{Z}(M, \mathbb{Z}) = M/t(M)$ and $\operatorname{Ext}_\mathbb{Z}^1(M, \mathbb{Z}) = t(M)$, where $t(M)$ is the torsion subgroup of M, so that $j(M) = 0$ if M is free and $j(M) = 1$ otherwise.

(4) It is noted by Björk and Ekström in [BjE] that, as a consequence of work by Levasseur [Le1], if R is left and right Noetherian and has finite left and right self-injective dimension, then for any f.g. R-module M there is an $i \geq 0$ for which $\operatorname{Ext}^i(M, R) \neq 0$. More generally, Fuller and Wang [FuWa] show that if R is left Noetherian of finite right injective dimension n, then for any f.g. left R-module M there is an $i \leq n$ for which $\operatorname{Ext}^i(M, R) \neq 0$.

(5) If N is a submodule of the R-module M, then we have the long exact sequence

$$0 \to \operatorname{Hom}(M/N, R) \to \operatorname{Hom}(M, R) \to \operatorname{Hom}(N, R) \to \operatorname{Ext}^1(M/N, R)$$
$$\to \cdots \to \operatorname{Ext}^n(N, R) \to \operatorname{Ext}^{n+1}(M/N, R) \to \operatorname{Ext}^{n+1}(M, R) \to \cdots$$

and from this it follows that $j(M) \geq \min\{j(N), j(M/N)\}$. Further properties of the grade function are presented in section 3 of [LiVV].

Notation. Simplifying notation we set $\operatorname{E}^i(M) = \operatorname{Ext}_R^i(M, R)$ for any right or left module M and for any $i \geq 0$ (so that, in particular, $\operatorname{E}^0(M) = \operatorname{Hom}_R(M, R)$). If M is a left (right) R-module, then right (left) multiplication in R induces a right (left) R-module structure on $\operatorname{E}^i(M)$.

For each $i, j \geq 0$ we denote $\operatorname{E}^i(\operatorname{E}^j(M))$ by $\operatorname{E}^{i,j}(M)$.

Definition 1.2. We say that a ring R
- satisfies the *Auslander condition* if for every Noetherian left or right R-module M and for all $i \geq 0$, $j(N) \geq i$ for all submodules $N \subseteq \operatorname{E}^i(M)$;
- is *Auslander-Gorenstein* (AG) if R is left and right Noetherian, satisfies the Auslander condition, and has finite left and right injective dimension;
- is *Auslander regular* if it is Auslander-Gorenstein and has finite global dimension.

It is important to note here that Zaks [Za] has shown that if R is both left and right Noetherian, then the left injective dimension of R equals its right injective dimension if both are finite.

A couple of introductory examples may help:

(1) If R is QF, then it is AG since $j(M) = 0$ for all nonzero Noetherian R-modules M. However a QF ring is Auslander regular only if it is semisimple artinian.

(2) Bass [Ba] proves that if R is a commutative Noetherian ring of finite self-injective dimension n, then R has Krull dimension n and this is equivalent to the condition that $j(E^i(M)) \geq i$ for all $i \geq 0$ and all f.g. R-modules M. Such rings are called *Gorenstein* and Bass assures us of their ubiquity. His arguments show that these are precisely the commutative Auslander-Gorenstein rings.

If moreover R has finite global dimension, then R is a regular ring (of Auslander-Buchsbaum-Serre fame, see Bruns and Herzog [BH] or Kaplansky [Ka]) and so these are precisely the commutative Auslander regular rings.

We now present perhaps the easiest example of a Noetherian ring which does not satisfy Auslander's condition. Its failure was noted by I. Reiten in her thesis and it features in [Bj3, p. 138] and [ASZ1, Example 5.4].

(3) Let V be a finite-dimensional vector space over the field K of dimension at least 2. Then the triangular matrix ring

$$R = \begin{bmatrix} K & V \\ 0 & K \end{bmatrix}$$

is hereditary Artinian but does not satisfy Auslander's condition, since if J denotes the radical of R, then $E^{0,1}(M) \neq 0$ for $M_R = R/J$. By contrast, Stafford [St1] notes that any hereditary Noetherian *prime* ring is Auslander regular.

2. Filtered Rings and Graded Rings

We define filtered rings and graded rings. These have proved invaluable in the construction and study of AG rings and in this role have been used extensively by many authors including Björk and Ekström in [Bj1–3] and [BjE1–2], Levasseur, Stafford and Zhang in [Le1–2], [LSt1–3], [StZh], and [Zh1–2], and Li and Van Oystaeyen in [Li1–3] and [LiO1–5].

For example, following earlier work by Roos [Ro], Björk employs these tools in [Bj1] to show that, for any field k of characteristic 0 and any $n \geq 1$, the Weyl algebra $A_n(k)$ is Auslander regular of global dimension n. ($A_n(k)$ is the k-algebra on $2n$ generators $x_1, \ldots, x_n, y_1, \ldots, y_n$ subject to the relations $x_i y_j - y_j x_i = \delta_{ij}$ and $x_i x_j - x_j x_i = y_i y_j - y_j y_i = 0$.)

Definition 2.1. A family $\mathcal{F}(R) = \{F_n : n \in \mathbb{Z}\}$ of additive subgroups of the ring R is called a *filtration* on R if
1. $1 \in F_0$,
2. for $i < j$, $F_i \subseteq F_j$,
3. for each i, j, $F_i \cdot F_j \subseteq F_{i+j}$, and
4. $\bigcup_{n \in \mathbb{Z}} F_n = R$.

If $F_n = 0$ for $n < 0$ then $\mathcal{F}(R)$ is called *positive*. (Note that the third condition implies that F_0 is a subring of R.)

For example, for any ring A, any ring endomorphism σ on A and any σ-derivation δ on A, the skew polynomial ring $R = A[x; \sigma, \delta]$ in the indeterminate x has a filtration given by $F_n = \{f(x) \in R : \deg f(x) \leq n\} \cup \{0\}$ for $n \geq 0$ and $F_n = 0$ for $n < 0$.

The first Weyl algebra $A_1(k)$ over the field k can be thought of as the skew polynomial ring over $k[x]$ in the indeterminate y with the identity as the endomorphism and the partial derivative $\partial/\partial x$ as the derivation. As such, $A_1(k)$ has the filtration defined above but also a positive filtration (known as the *Bernstein filtration*) where, for $j \geq 0$, F_j is the k-subspace of $A_1(k)$ generated by all $x^m y^n$ such that $m + n \leq j$. (See [Bj1].)

Definition 2.2. A family $\mathcal{G}(R) = \{G_n : n \in \mathbb{Z}\}$ of additive subgroups of the ring R is called a *grading* of R if
1. for each i, j, $G_i \cdot G_j \subseteq G_{i+j}$ and
2. $R = \bigoplus_{n \in \mathbb{Z}} G_n$ is an abelian group.

In this case, we also say that R is a *graded ring*. If $G_n = 0$ for $n < 0$, then $\mathcal{G}(R)$ is called *positive*.

If $\mathcal{F}(R)$ is a filtration on R, defining $G_n = F_n/F_{n-1}$ for each n and setting $S = \bigoplus_{n \in \mathbb{Z}} G_n$ gives a graded ring S with multiplication induced from defining $(a + F_{n-1})(b + F_{m-1}) = ab + F_{m-n-1} \in G_{m+n}$ for all $a \in F_n, b \in F_m$.

S is called the *associated graded ring* and is denoted by $\operatorname{gr} R$.

Roughly speaking, $\operatorname{gr} R$ is an approximation of R which is often easier to study and its properties can often be lifted back to R. For example, it is well known that if R is positively filtered and $\operatorname{gr} R$ is left Noetherian, then R is left Noetherian (see [MR, p. 27]). Of interest to us here are the following two results proved by Ekström [Ek1] using results of Björk [Bj3].

Theorem 2.1. *If R is AG, respectively Auslander regular, then for any ring automorphism σ on R and any σ-derivation δ, the skew polynomial ring $R[x; \sigma, \delta]$ is also AG, respectively Auslander regular.*

Recall that an element a of a ring R is called *normal* if $aR = Ra$. The proof of the next result uses the filtration on R given by $F_n = R$ for $n \geq 0$ and $F_n = (Ra)^n$ for $n < 0$.

Theorem 2.2. *Let a be a regular normal element of R with $a \in J(R)$, the Jacobson radical of R. If the factor ring R/Ra is AG, respectively Auslander regular, then so too is R.*

Conversely, Levasseur [Le2] notes that the Auslander-Gorenstein property is preserved when factoring out by a normal regular element. In a similar vein, it is shown in [LiVV] that if x is a regular central element of the Noetherian ring R and both R_x (the localization of R at the monoid generated by x) and R/Rx are Auslander regular, then so is R.

We can manufacture a second graded ring Gr R, known as the *Rees ring* of R. Here Gr $R = \bigoplus_{n \in \mathbb{Z}} F_n$ and multiplication is defined in Laurent polynomial fashion as $(a_n)(b_n) = (c_n)$ where $c_n = \sum_{i+j=n} a_i b_j$. There is then an interplay between R, gr R and Gr R. For example, Ekström [Ek1] and Li, Van den Bergh and Van Oystaeyen [LiVV] have shown

Theorem 2.3. *Let R be a filtered ring such that its associated Rees ring Gr R is Noetherian. Then if both R and gr R are AG, respectively Auslander regular, then so is Gr R.*

Of particular usefulness in property transfer are Zariskian filtrations which include the positive filtrations. Our definition below follows [LiO1] where they are characterized and exemplified. (See also [LiO5].)

Definition 2.3. A filtration $\mathcal{F}(R)$ on the ring R is called *left Zariskian* and R is called a *left Zariski ring* if
1. $F_{-1} \subseteq J(F_0)$ (the Jacobson radical of the subring F_0) and
2. the associated Rees ring Gr R is left Noetherian.

The terminology comes from commutative ring theory where a Noetherian ring R with ideal I is said to be Zariski with respect to the I-adic topology if $I \subseteq J(R)$ (see, e.g., [Ma]).

Björk [Bj3] and Li and Van Oystaeyen [LiO2] establish the following shortcut for Theorem 2.3 which was used in the proofs of Theorems 2.1 and 2.2 above. Forerunners appear in [Bj2] and [Le1].

Theorem 2.4. *Let R be a left and right Zariski ring. If its associated graded ring gr R is AG, respectively Auslander regular, then so too is R.*

Levasseur [Le2] has shown that if k is a field and R is a positively graded Auslander regular k-algebra in which $G_0 = k$, then R is a domain. Stafford continued the study of (not necessarily graded) Auslander regular k-algebras R in [St1], establishing that many such R are domains which are maximal orders in their division ring of quotients Q, i.e., if R is a subring of the ring S such that $aSb \subseteq R$ for some nonzero $a, b \in R$, then $R = S$. Then, using Theorem 2.4, Stafford and Zhang [StZh] showed the next important result.

Theorem 2.5. *Let R be a fully bounded Noetherian ring of finite injective dimension n. If $\{G_n \mid n \in \mathbb{Z}\}$ is a positive grading on R such that $G_0 = k$ is a central subfield of R and each G_n is a finite-dimensional k-vector space for $n > 0$, then R is Auslander regular of Krull dimension n.*

Stafford and Zhang's paper also contains a number of interesting examples and the result that local Noetherian PI rings of finite global dimension are Auslander regular.

This result was generalized by Teo [Te1] as follows.

Theorem 2.6. *Let R be a local fully bounded Noetherian ring of finite global dimension. Then R is Auslander regular.*

3. Minimal Injective Resolutions and Cohen-Macaulay Rings

Let
$$0 \longrightarrow R \longrightarrow I^0 \longrightarrow I^1 \longrightarrow \cdots \longrightarrow I^d \longrightarrow \cdots \quad (\star)$$
be a minimal injective resolution of R as a left R-module.

If R is commutative Noetherian, then Bass [Ba] (see also Bruns and Herzog [BH, Prop. 3.2.9]) has shown that, for each $i \geq 0$, we have
$$I^i \cong \bigoplus_{\mathfrak{p} \in \operatorname{Spec} R} E(R/\mathfrak{p})^{\mu_i(\mathfrak{p})},$$
where $\operatorname{Spec} R$ is the set of prime ideals of R, $E(-)$ denotes an injective hull and the multiplicity $\mu_i(\mathfrak{p})$, known as the *ith Bass number* with respect to $\mathfrak{p}$, is given by $\dim_{k(\mathfrak{p})} \operatorname{Ext}^i_{R_\mathfrak{p}}(k(\mathfrak{p}), R_\mathfrak{p})$ where $k(\mathfrak{p})$ denotes the residue field at $\mathfrak{p}$.

As a simple example, for $R = \mathbb{Z}$ we have the minimal injective resolution
$$0 \longrightarrow \mathbb{Z} \longrightarrow \mathbb{Q} \longrightarrow \mathbb{Q}/\mathbb{Z} \longrightarrow 0$$

and $\mathbb{Q} = E(R/0)$, $\mathbb{Q}/\mathbb{Z} = \bigoplus_{p \text{ prime}} E(\mathbb{Z}/p\mathbb{Z})$, noting that $\mu_i(p\mathbb{Z}) = 1$ for $i = 1$ and 0 otherwise, while $\mu_i(0) = 1$ for $i = 0$ and 0 otherwise.

Similarly, if R is a commutative quasi-Frobenius ring, then
$$0 \longrightarrow R \longrightarrow R \longrightarrow 0$$
is a minimal injective resolution of R. Since R is the direct sum of finitely many local rings, each having simple essential socle, we have $R \cong \bigoplus_{\mathfrak{p} \in \text{Spec } R} E(R/\mathfrak{p})$.

If moreover the commutative Noetherian ring R has finite self-injective dimension n, then n is the Krull dimension of R and every $E(R/\mathfrak{p})$ occurs at least once in the resolution $(\star)$. (In this regard, extending results of Hoshino [Ho2] and Iwanaga [Iw], Miyachi [Mi1] has shown that if R is left Noetherian, right coherent with finite left and right self-injective dimension, then every indecomposable injective left R-module appears as a summand of some I^i. Brown [Br1] and Iwanaga and Sato [IwS1] also examine the occurrence of indecomposable injectives as summands of the I^i.)

If R is local and Gorenstein, then $\mu_i(\mathfrak{p})$ is 1 if height $\mathfrak{p} = i$, and zero otherwise. It follows that each $E(R/\mathfrak{p})$ appears precisely once in $(\star)$ and I^i is the direct sum of the $E(R/\mathfrak{p})$ where $\mathfrak{p}$ runs through the set of prime ideals of height i. This homogeneity of I^i is referred to as *purity* and may be rephrased as saying that every non-zero finitely generated submodule of I^i has Krull dimension equal to $n - i$.

A natural question to ask is if this purity can be extended to the non-commutative setting. To set the scene for this we first have the following:

Definition 3.1. (See McConnell and Robson [MR, p. 210]) A *dimension function* ∂ on the left Noetherian ring R is a function which assigns a value $\partial(M)$ to each f.g. left R-module M, the possible values being $-\infty$, all non-negative reals and all infinite ordinals, and which satisfies the following:

(i) $\partial(0) = -\infty$,

(ii) if $0 \longrightarrow M' \longrightarrow M \longrightarrow M'' \longrightarrow 0$ is an exact sequence of f.g. left R-modules then $\partial(M) \geq \max\{\partial(M'), \partial(M'')\}$ with equality if the sequence splits and

(iii) if $M\mathfrak{p} = 0$ for some $\mathfrak{p} \in \text{Spec } R$ and M is a torsion module over $R/\mathfrak{p}$, then $\partial(M) + 1 \leq \partial(R/\mathfrak{p})$.

If in condition (ii) equality always hold, then ∂ is called *exact*.

The standard example of an exact dimension function is Krull dimension (Kdim) in the sense of Rentschler-Gabriel (see [MR, Chapter 6]). For algebras over a field, Gelfand-Kirillov dimension (GKdim) is often exact (see [KL] or [MR, Chapter 8]).

At this stage we should define another important class of Noetherian rings. Our definition is taken from Ajitabh, Smith and Zhang [ASZ1].

Definition 3.2. Let ∂ be a dimension function on the Noetherian ring R. We say that R is *Cohen-Macaulay with respect to* ∂, briefly ∂-*CM*, if

$$j(M) + \partial(M) = \partial(R) < \infty$$

for every nonzero Noetherian module M.

When R is an algebra over a field, we say that R is *Cohen-Macaulay* if it is so with respect to GKdim.

Cohen-Macaulay rings are prominent in the homological theory of commutative rings, as evidenced in the text by Bruns and Herzog [BH]. However, while every commutative Auslander regular ring is Cohen-Macaulay the same is not true when commutativity is dropped, as illustrated by the ring of 2×2 upper triangular matrices over a field. Much of the recent research on AG rings has also involved the Cohen-Macaulay condition but here we have given the Auslander condition centre stage.

If R is Noetherian and has injective dimension $n < \infty$, we define

$$\delta(M) = n - j(M)$$

for all Noetherian R-modules M.

Although δ is *not* a dimension function in general, it is not difficult to see that it is exact whenever it is a dimension function. If R is AG, then Levasseur [Le2, Prop. 4.5] has shown that δ is an exact dimension function, called the *canonical dimension function*.

The next definition is also taken from [ASZ1].

Definition 3.3. Let R be a Noetherian ring with finite injective dimension n. We say that a nonzero R-module M is
- *s-pure* (or $(n\text{-}s)$-*homogeneous*) if $\delta(N) = s$ for all nonzero Noetherian submodules $N \subseteq M$;
- *essentially s-pure* if it has an essential submodule which is s-pure;
- *s-critical* if it is s-pure and $\delta(M/N) < s$ for all nonzero submodules $N \subset M$.

Of course, we say that M is *pure* if it is s-pure for some s.

For example, if R is QF then every nonzero R-module is 0-pure and the 0-critical R-modules are precisely the simple R-modules. Also, if $R = \mathbb{Z}$, then the 1-pure R-modules are the free modules, the 0-pure R-modules are the torsion modules, $\mathbb{Z}$ is the only 1-critical module while again the 0-criticals are the simples. Levasseur (loc. cit.) shows that every Noetherian module over an AG ring R has a critical submodule and in fact a critical composition series.

Definition 3.4. If R is Noetherian of finite injective dimension n, then we say that the resolution $(\star)$ is
- *pure* if each I^i is $(n-i)$-pure,
- *essentially pure* if each I^i is essentially $(n-i)$-pure.

Note that our previous remarks show that if R is commutative local Gorenstein, then a minimal injective resolution of R is pure.

Ajitabh, Smith and Zhang in [ASZ1] and [ASZ2] investigate hypotheses under which an arbitrary Noetherian ring of finite injective dimension has a pure or essentially pure minimal injective resolution. In [ASZ1] they show that, under a mild assumption, the existence of an essentially pure resolution forces the ring R to be AG. However, they also exhibit examples due to Artin and Stafford to show that not every AG ring has such a resolution. On the positive side, they prove many AG rings with small injective dimension have pure or essentially pure resolutions. Moreover, they show

Theorem 3.1. Let R be an AG ring satisfying a polynomial identity.
1. If R is grade-symmetric, i.e., $j(_RM) = j(M_R)$ for every R-R-bimodule M finitely generated on both sides, then R has a pure resolution.
2. If R is Cohen-Macaulay with respect to either Gelfand-Kirillov or Krull dimension, then R is grade-symmetric.

Localization plays a key role in proving some of the results in [ASZ2]. As the authors note, Goodearl and Jordan in [GJ1-2] show that injectivity and injective dimension are usually fragile under localization unless one localizes at normal elements. However, it is shown in [ASZ2, Proposition 2.1] that if R is an AG ring and S is a multiplicatively closed Ore set of regular elements in R, then the localization RS^{-1} is also AG. If moreover R is also Cohen-Macaulay and the elements of S are central, then RS^{-1} is also CM.

4. Catenarity

Definition 4.1. A ring R is said to be *catenary* if, for any two prime ideals P and Q of R with $P \subset Q$, any two saturated chains of prime ideals between P and Q have the same length.

It is well known among commutative ring theorists that any commutative Cohen-Macaulay ring is catenary (see, e.g., [BH, Theorem 2.1.12]).

One of the early successes of Auslander-Gorenstein theory was its use by O. Gabber in establishing catenarity for enveloping algebras of finite dimensional solvable Lie algebras. This work was reported in [Ga] but can be found more easily by combining [LSt1, Appendix A] and [KL, Chapter 9]. A key ingredient in Gabber's argument was a property of pure submodules, now known as *Gabber's maximality principle* and described in the following theorem, which he established for Auslander regular rings. In [Bj3, Theorem 1.14] Björk generalized the principle to AG rings and with Ekström provided more information on pure modules over filtered AG rings R in [BjE] using the interplay between R, gr R and Gr R described in Section 2 and a technique borrowed from Fossum [Fo].

Theorem 4.1. [Gabber's maximality principle] *Let R be an AG ring and let M be a finitely generated s-pure module. Suppose that N is an R-module containing M and every finitely generated submodule of N is pure. Then there is a unique largest finitely generated submodule X of N containing M with the property that $j(X/M) \geq s + 2$.*

Using this principle and the essence of previously used methods, Goodearl and Lenagan [GL] prove the following theorem and apply it to several quantum groups. Here Spec R is called *normally separated* if for any pair of prime ideals $P \subset Q$ in R the factor Q/P contains a normal element of R/P. Also we say that *Tauvel's height formula* holds for the algebra R if

$$\text{height}(P) + \text{GKdim}(R/P) = \text{GKdim}(R)$$

for all P in Spec R. (Tauvel established this useful equality for enveloping algebras of solvable Lie algebras in [Tau].)

Theorem 4.2. *Let R be an affine AG Cohen-Macaulay algebra over a field k with finite Gelfand-Kirillov dimension. If Spec R is normally separated, then R is catenary. Moreover, if R is prime Tauvel's formula holds.*

Theorem 4.2 has been used further by Oh [Oh] to establish catenarity for more quantum groups. Moreover, if one assumes at the outset that R is a Noetherian k-algebra of finite injective dimension and R is positively graded as $R = \bigoplus_{i \geq 0} G_i$ with $G_0 = k$, then, in the spirit of Theorem 2.5, Zhang [Zh2] has shown that normal separation of Spec R implies that R is AG, Cohen-Macaulay and of finite Gelfand-Kirillov dimension and so, by Theorem 4.2, catenary.

In [YeZh], Yekutieli and Zhang are able to generalise Theorem 4.2 using what they call *Auslander dualizing complexes*.

5. n-Gorenstein Rings

Definition 5.1. A two-sided Noetherian ring R is said to be n-*Gorenstein* for $n \geq 1$ if, in a minimal injective resolution

$$0 \longrightarrow {}_R R \longrightarrow I^0 \longrightarrow I^1 \longrightarrow \cdots \longrightarrow I^n \longrightarrow \cdots$$

of R as a left R-module, the flat dimension $\mathrm{fd}(I^k)$ of I^k is at most k for each $0 \leq k \leq n-1$.

The following characterization of n-Gorenstein rings displays their symmetry (cf. the result of Zaks mentioned above). It is due to Auslander and appears in a categorical setting as Theorem 3.7 of [FGR].

Theorem 5.1. *Let R be a two-sided Noetherian ring. Then the following conditions are equivalent*:
 (a) R *is n-Gorenstein*;
 (b) *for any f.g. left R-module M, any integer $i \leq n$ and any submodule N of $\mathrm{E}^i(M)$, we have $j(N) \geq i$*;
 (c) *in a minimal injective resolution $0 \longrightarrow R_R \longrightarrow Q^0 \longrightarrow Q^1 \longrightarrow \cdots \longrightarrow Q^n \longrightarrow \cdots$ of R as right R-modules, the flat dimension $\mathrm{fd}(Q^k)$ of Q^k is at most k for each $0 \leq k \leq n-1$*;
 (d) *the right-left analogue of* (b).

It is clear from condition (b) of the Theorem that if R is n-Gorenstein for all $n \geq 1$, then R satisfies the Auslander condition. As noted in Section 1, the left and right self-injective dimensions of a Noetherian ring R coincide if they are both finite. Similarly, Auslander and Reiten proved in [AR1] that these two dimensions *always* coincide if R is an Artinian n-Gorenstein algebra for all $n \geq 1$ and they conjectured that such an algebra has finite self-injective dimension. This conjecture is discussed by Fuller and Iwanaga in [FuIw] and Iwanaga and Sato in [IwS2]. They show that if R is an n-Gorenstein ring with self-injective dimension n, then R is Auslander-Gorenstein. Hoshino [Ho3] investigates the conjecture further.

1-Gorenstein rings are also called QF-3 rings and as such were studied by Hoshino in [Ho1] and in Tachikawa's monograph [Ta]. Hoshino provides the following characterization.

Theorem 5.2. *A Noetherian ring R is 1-Gorenstein if and only if the R-double dual ( )** preserves monomorphisms of finitely generated (left) R-modules.*

Fuller and Iwanaga (loc. cit.) investigate 2- and 3-Gorenstein rings which are serial, proving that the latter include the former but that there are 3-Gorenstein rings which are not 4-Gorenstein.

Further to our discussion of minimal injective resolutions in Section 4, we have the following result due to Iwanaga and Miyachi [IwM] and generalizing an earlier result in [IwS2].

Theorem 5.3. *Let R be an AG ring of self-injective dimension n and with minimal injective resolution of $_RR$ as above. Then any indecomposable injective left R-module of flat dimension n is isomorphic to a direct summand of I^n and is the injective hull of some simple left R-module. Moreover if M is a left R-module of injective dimension n, then the nth injective in any minimal injective resolution of M has an essential socle.*

Iwanaga and Wakamatsu in [IwW] prove that, for any $n, k \geq 1$, the ring of upper triangular $k \times k$ matrices over an Artinian n-Gorenstein ring is also n-Gorenstein.

Auslander and Reiten prove in [AR1] the following variant of Auslander's Theorem 5.1 with a weakening of the grade and flat dimension conditions.

Theorem 5.4. *Let R be a two-sided Noetherian ring and n be a positive integer. Then the following conditions are equivalent:*

(a) if $0 \longrightarrow {}_RR \longrightarrow I^0 \longrightarrow I^1 \longrightarrow \cdots \longrightarrow I^k \longrightarrow \cdots$ is a minimal injective resolution of the left R-module R, then the flat dimension $\mathrm{fd}(I^k)$ of each I^k is at most $k+1$ for each $0 \leq k \leq n-1$;

(b) for any f.g. right R-module M, any integer $i \leq n$ and any submodule N of $\mathrm{E}^{i+1}(M)$, we have $j(N) \geq i$.

The class of rings described in Theorem 5.4 are further characterized by Auslander and Reiten in [AR1–2] using extension closure of syzygy modules and by Huang in [Hu] using extension closure of i-torsionfree modules as defined in [AR2].

Interestingly, the left-right symmetry enjoyed by n-Gorenstein rings fails for this new class of rings, as witnessed by an example due to Hoshino (see [Hu] or [Ho2]).

6. Duality

It is well known that if R is a quasi-Frobenius ring, then the dual functors $\mathrm{Hom}_R(-, R)$ define a duality between the categories of left and right f.g. R-modules and under this correspondence simples correspond to simples. In [Fo] Fossum showed that if R is any commutative Noetherian ring and M is an f.g. R-module with $j(M) \geq n$, then there is a natural homomorphism $M \longrightarrow \mathrm{E}^{n,n}(M)$. Later Reiten and Fossum [RF, Prop. 5] proved that a commutative ring R is $(n+1)$-Gorenstein if and only if, for each $0 \leq i \leq n$ and for each f.g. module M of grade at least i, we have $\mathrm{E}^{i,i}(\mathrm{Ext}^i(M,R)) \cong \mathrm{Ext}^i(M,R)$.

In this section we look at a similar duality over AG rings discussed by Ajitabh, Smith and Zhang in [ASZ1] and Iwanaga in [Iw].

Definition 6.1. Let R be an AG ring of self-injective dimension n. A Noetherian R-module M is called *holonomic* if $j(M) = n$.

Note that every nonzero f.g. module over a QF ring is holonomic while the holonomic $\mathbb{Z}$-modules are the f.g. torsion groups.

Iwanaga (loc. cit.) proves the following:

Theorem 6.1. *Let R be an AG ring and M be an f.g. left R-module such that $\mathrm{E}^i(M) = 0$ for any $i \neq j(M)$. Then there is a canonical isomorphism*

$$\sigma_M : M \longrightarrow \mathrm{E}^{j(M),j(M)}(M)$$

and the correspondence $M \longrightarrow \mathrm{E}^{j(M)}(M)$ gives a bijection between f.g. left and right R-modules with $\mathrm{E}^i(M) = 0$ for all $i \neq j(M)$. In particular, this correspondence gives a bijection between left and right holonomic modules. Moreover, if I^n (Q^n) denotes the last injective in a left (right) minimal injective resolution of R, then simple submodules of I^n and Q^n are holonomic and correspond bijectively.

Ajitabh, Smith and Zhang extend Theorem 6.1 in [ASZ1] by showing that if R is an AG ring of self-injective dimension n, then the categories of Noetherian left and right R-modules are in $(n+1)$-step duality. Here, categories $\mathfrak{C}$ and $\mathfrak{D}$ are said to be in $(n+1)$-*step duality* if there are dense subcategories

$$0 = \mathfrak{C}_0 \subset \mathfrak{C}_1 \subset \cdots \subset \mathfrak{C}_{n+1} = \mathfrak{C} \text{ and } 0 = \mathfrak{D}_0 \subset \mathfrak{D}_1 \subset \cdots \subset \mathfrak{D}_{n+1} = \mathfrak{D}$$

such that for each $i = 1, \ldots, n+1$ there are contravariant functors $F : \mathfrak{C}^i \longrightarrow \mathfrak{D}^i$ and $G : \mathfrak{D}^i \longrightarrow \mathfrak{C}^i$ such that $FG \simeq \mathrm{Id}_{\mathfrak{D}^i}$ and $GF \simeq \mathrm{Id}_{\mathfrak{C}^i}$, where $\mathfrak{C}^i$ and $\mathfrak{D}^i$ are the quotient categories $\mathfrak{C}_i/\mathfrak{C}_{i-1}$ and $\mathfrak{D}_i/\mathfrak{D}_{i-1}$ respectively.

As a consequence of this duality we have the following

Theorem 6.2. *If R is an AG ring of injective dimension n, then, for any f.g. R-module M, we have $\mathrm{Kdim}(M) \leq \delta(M) = n - j(M)$. In particular, the left and right Krull dimensions of R are bounded above by n and any holonomic R-module is artinian.*

7. Quantum Group Examples

We briefly mentioned earlier how some of the general theory on AG rings could be applied to establish their presence in the topical area of quantum groups. More instances of when the Auslander property is held by quantum groups are presented by Ajitabh, Smith and Zhang in [ASZ1–2], by Artin, Tate and Van den Bergh in [ATV] and [TV], by Brown, Goodearl and Lenagan in [Br2], [BrG] and [GL], by Giaquinto, Yekutieli and Zhang in [GiZh], [Ye] and [YeZh], and by Levasseur, Smith and Stafford in [LeSm], [LeSt1–3] and [St2].

We outline one of the more prominent examples.

Let k be an algebraically closed field of characteristic not 2 and let $\alpha, \beta, \gamma \in k$ satisfy

$$\alpha + \beta + \gamma + \alpha\beta\gamma = 0.$$

The *4-dimensional Sklyanin algebra* $S = S(\alpha, \beta, \gamma)$ is the graded k-algebra with generators x_0, x_1, x_2, x_3 subject to the six relations

$$x_0 x_1 - x_1 x_0 = \alpha(x_2 x_3 + x_3 x_2), \quad x_0 x_1 + x_1 x_0 = x_2 x_3 - x_3 x_2,$$

$$x_0 x_2 - x_2 x_0 = \beta(x_1 x_3 + x_3 x_1), \quad x_0 x_2 + x_2 x_0 = x_3 x_1 - x_1 x_3,$$

$$x_0 x_3 - x_3 x_0 = \gamma(x_1 x_2 + x_2 x_1), \quad x_0 x_3 + x_3 x_0 = x_1 x_2 - x_2 x_1.$$

When $\{\alpha, \beta, \gamma\} \cap \{0, \pm 1\} = \emptyset$, Artin, Tate and Van den Bergh [ATV] and Smith and Stafford [SmSt] (with a little help from Levasseur [Le2]) show that S is an Auslander-Gorenstein domain.

Now there are Sklyanin algebras for any dimension $n \geq 4$ — their definition is complicated, see [TV] — but the methods used in these two papers were not available beyond $n = 4$. However, Tate and Van den Bergh in [TV] show that all the n-dimensional Sklyanin algebras are AG domains of global dimension n and that they are also Cohen-Macaulay and maximal orders. Their proof uses results of Stafford and Zhang [StZh] and they suggested in an appendix to the paper that an alternative proof may be available using skew polynomial rings (recall Theorem 2.1). Such a proof was established by Teo in [Te2]. His approach was to show that the Sklyanin algebra S could be embedded "nicely" as a graded k-algebra in a ring R which is an iterated skew polynomial extension of a field extension K of k. He then established the desired properties for R and was able to pull these back to S. Interestingly, the properties for R were obtained by using an exact dimension function modelled on Gelfand-Kirillov dimension.

8. Tailpiece

Space and time limitations have meant that a number of important associated topics have not been covered in the above. In particular we have not discussed Artin-Schelter (AS) algebras or recent investigations of the Auslander condition using complexes. Our references include papers on these and other topics suffering inattention.

References

[ASZ1] K. Ajitabh, S. P. Smith and J. J. Zhang, *Auslander-Gorenstein rings*, Comm. Algebra **26** (1998), 2159–2180.

[ASZ2] _____, *Injective resolutions of some regular rings*, J. Pure and Applied Algebra **140** (1999), 1–21.

[ATV] M. Artin, J. Tate and M. Van den Bergh, *Modules over regular algebras of dimension* 3, Invent. Math. **106** (1991), 335–389.

[AB] M. Auslander and M. Bridger, *Stable Module Theory*, Mem. Amer. Math. Soc. **94** (1969).

[AR1] M. Auslander and I. Reiten, *k-Gorenstein algebras and syzygy modules*, J. Pure and Applied Algebra **92** (1994), 1–27.

[AR2] _____, *Syzygy modules for Noetherian rings*, J. Algebra **183** (1996), 167–185.

[Ba] H. Bass, *On the ubiquity of Gorenstein rings*, Math. Z. **82** (1963), 8–28.

[Bj1] J.-E. Björk, *Rings of Differential Operators*, North-Holland Math. Library **21**, Amsterdam-New York, 1979.

[Bj2] _____, *Filtered Noetherian rings*, Noetherian rings and their applications (Oberwolfach, 1983), Math. Surveys Monographs **24**, Amer. Math. Soc., Providence, RI, 1987, 59–97.

[Bj3] _____, *The Auslander condition on noetherian rings*, Séminaire Dubreil-Malliavin 1987–88, Lect. Notes Math. **1404**, Springer-Verlag, Heidelberg, New York, (1989), 137–173.

[BjE] J.-E. Björk and E. K. Ekström, *Filtered Auslander-Gorenstein rings*, Operator algebras, unitary representations, enveloping algebras, and invariant theory (Paris, 1989), Progr. Math. **92**, Birkhäuser, Boston, 1990, 425–448.

[Br1] K. A. Brown, *Fully bounded noetherian rings of finite injective dimension,* Quart. J. Math. (Oxford) **41** (1990), 1–13.

[Br2] _____, *Representation theory of Noetherian Hopf algebras satisfying a polynomial identity,* Contemp. Math. **229** (1998), 49–79.

[BrG] K. A. Brown and K. R. Goodearl, *Homological aspects of Noetherian PI Hopf algebras and irreducible modules of maximal dimension,* J. Algebra **198** (1997), 240–265.

[BH] W. Bruns and J. Herzog, *Cohen-Macaulay Rings,* Cambridge Univ. Press, Cambridge, 1994.

[Ek1] E. K. Ekström, *The Auslander condition on graded and filtered noetherian rings,* Séminaire Dubreil-Malliavin 1987–88, Lect. Notes Math. **1404**, Springer-Verlag, Heidelberg, New York, (1989), 220–245.

[Ek2] _____, *Homological properties of some Weyl algebra extensions,* Compos. Math. **75** (1990), 231–246.

[Fo] R. M. Fossum, *Duality over Gorenstein rings,* Math. Scand. **26** (1970), 165–176.

[FGR] R. M. Fossum, P. A. Griffith and I. Reiten, *Trivial Extensions of Abelian Categories,* Lect. Notes Math. **456**, Springer-Verlag, Heidelberg, New York, 1975.

[FuIw] K. R. Fuller and Y. Iwanaga, *On n-Gorenstein rings and Auslander rings of low injective dimension,* Proceedings of the Sixth International Conference on Representations of Algebras (Ottawa, ON,1992), Carleton-Ottawa Math. Lecture Note Ser. **14**, Carleton Univ., Ottawa, 1992, 12 pp.

[FuWa] K. R. Fuller and Y. Wang, *Redundancy in resolutions and finitistic dimensions of noetherian rings,* Comm. Algebra **21** (1993), 2983–2994.

[Ga] O. Gabber, *Equidimensionalité de la variété caractéristique 0,* Exposé de O. Gabber redigé par T. Levasseur, Université de Paris VI, 1982.

[GiZh] A. Giaquinto and J. J. Zhang, *Quantum Weyl algebras,* J. Algebra **176** (1995), 861–881.

[GJ1] K. R. Goodearl and D. A. Jordan, *Localizations of injective modules,* Proc. Edinburgh Math. Soc. **28** (1985), 289–299.

[GJ2] K. R. Goodearl and D. A. Jordan, *Localizations of essential extensions,* Proc. Edinburgh Math. Soc. **31** (1988), 243–247.

[GL] K. R. Goodearl and T. H. Lenagan, *Catenarity in quantum algebras,* J. Pure and Applied Algebra **111** (1996), 123–142.

[GW] K. R. Goodearl and R. B. Warfield, Jr., *An Introduction to Noncommutative Noetherian Rings,* London Math. Soc. Student Texts **16**, Cambridge Univ. Press, Cambridge, 1989.

[Ho1] M. Hoshino, *On dominant dimension of Noetherian rings,* Osaka J. Math. **26** (1989), 275–280.

[Ho2] ———, *On self-injective dimensions of Artinian rings,* Arch. Math. **54** (1990), 18–24.

[Ho3] ———, *On self-injective dimensions of Artinian rings,* Tsukuba J. Math. **18** (1994), 1–8.

[Hu] Huang Zhaoyong, *Extension closure of k-torsionfree modules,* Comm. Algebra **27** (1999), 1457–1464.

[Iw] Y. Iwanaga, *Duality over Auslander-Gorenstein rings,* Math. Scand. **81** (1997), 184–190.

[IwM] Y. Iwanaga and J. Miyachi, *Modules of the highest homological dimension over a Gorenstein ring,* Contemp. Math. **229** (1998), 193–199.

[IwS1] Y. Iwanaga and H. Sato, *Minimal injective resolutions of Gorenstein rings,* Comm. Algebra **18** (1990), 3835–3856.

[IwS2] ———, *On Auslander's n-Gorenstein rings,* J. Pure and Applied Algebra **106** (1996), 61–76.

[IwW] Y. Iwanaga and T. Wakamatsu, *Auslander-Gorenstein property of triangular matrix rings,* Comm. Algebra **23** (1995), 3601–3614.

[Ja] J. P. Jans, Jr., *Rings and Homology,* Holt, Rinehart and Winston, New York, 1964.

[Jo] P. Jørgensen, *Properties of AS-Cohen-Macaulay algebras,* J. Pure and Applied Algebra **138** (1999), 239–249.

[Ka] I. Kaplansky, *Commutative Rings,* Allyn and Bacon, Boston, 1970.

[KL] G. Krause and T. H. Lenagan, *Growth of algebras and Gelfand-Kirillov Dimension,* Pitman, Boston, 1985.

[Le1] T. Levasseur, *Complexe bidualisant en algèbre non commutative,* Séminaire d'Algèbre Paul Dubreil et Marie-Paule Malliavin, Paris 1983–84, Lect. Notes Math. **1146**, Springer-Verlag, Heidelberg, New York, (1985), 270–287.

[Le2] _____, *Some properties of non-commutative regular graded rings,* Glasgow Math. J. **34** (1992), 277–300.

[LeSm] T. Levasseur and S. P. Smith, *Modules over the 4-dimensional Sklyanin algebra,* Bull. Soc. Math. France **121** (1993), 35–90.

[LSt1] T. Levasseur and J. T. Stafford, *Rings of differential operators on classical rings of invariants,* Mem. Amer. Math. Soc. **412** (1989).

[LSt2] _____, *Differential operators commuting with invariant functions,* Comm. Math. Helv. **72** (1997), 426–433.

[LSt3] _____, *The quantum coordinate ring of the special linear group,* J. Pure and Applied Algebra **86** (1993), 181–186.

[Li1] Li Huishi, *Note on pure module theory over Zariskian filtered ring and the generalized Roos theorem,* Comm. Algebra **19** (1991), 843–862.

[Li2] _____, *Rees rings of grading filtrations and an application to Weyl algebras,* Comm. Algebra **21** (1993), 2967–2972.

[Li3] _____, *Lifting Ore sets of Noetherian filtered rings and applications,* J. Algebra **179** (1996), 686–703.

[LiVV] Li Huishi, M. Van den Bergh and F. Van Oystaeyen, *Global dimension and regularity of Rees rings for non-Zariskian filtrations,* Comm. Algebra **18** (1990), 3195–3208.

[LiO1] Li Huishi and F. Van Oystaeyen, *Zariskian filtrations,* Comm. Algebra **17** (1989), 2945–2970.

[LiO2] _____, *Global dimension and Auslander regularity of Rees rings,* Bull. Soc. Math. Belg. **43** (1991), 59–87.

[LiO3] _____, *Dehomogenization of gradings to Zariskian filtrations and applications to invertible ideals,* Proc. Amer. Math. Soc. **115** (1992), 1–11.

[LiO4] _____, *Sign gradations on group ring extensions of graded rings,* J. Pure and Applied Algebra **85** (1993), 311–316.

[LiO5] _____, *Zariskian Filtrations,* Kluwer Academic Publ., Dordrecht, 1996.

[MR] J. C. McConnell and J. C. Robson, *Non-commutative Noetherian Rings,* Wiley-Interscience, Chichester, 1987.

[Ma] H. Matsumura, *Commutative Ring Theory,* Cambridge Univ. Press, Cambridge, 1986.

[Mi1] J. Miyachi, *Duality for derived categories and cotilting bimodules,* J. Algebra **185** (1996), 583–603.

[Mi2] _____, *Injective resolutions of Noetherian rings and cogenerators,* Proc. Amer. Math. Soc. **128** (2000), 2233–2242.

[NV] C. Năstăcescu and F. Van Oystaeyen, *Graded Ring Theory,* North-Holland, Amsterdam, 1982.

[Oh] Oh Sei Qwon, *Catenarity in a class of skew polynomial rings,* Comm. Algebra **25** (1997), 37–49.

[Pr] K. L. Price, *Homological properties of color Lie superalgebras,* Advances in ring theory (Granville, OH, 1996), 287-293, Trends Math., Birkhäuser, Boston, (1997), 287–293.

[RF] I. Reiten and R. M. Fossum, *Commutative n-Gorenstein rings,* Math. Scand. **31** (1972), 33–48.

[Ro] J.-E. Roos, *Compléments à l'étude des quotients primitifs des algèbres de Lie semi-simples,* C. R. Acad. Sci. Paris **276** (1973), 447–450.

[R] J. J. Rotman, *An Introduction to Homological Algebra,* Academic Press, New York, 1979.

[Sa] H. Sato, *Note on holonomic modules over Gorenstein rings,* Bull. Fac. Ed. Wakayama Univ. Natur. Sci. **47** (1997), 7–15.

[SmSt] S. P. Smith and J. T. Stafford, *Regularity of the 4-dimensional Sklyanin algebra,* Compos. Math. **83** (1992), 259–289.

[St1] J. T. Stafford, *Auslander-regular algebras and maximal orders,* J. London Math. Soc. **50** (1994), 276–292.

[St2] _____, *Regularity of algebras related to the Sklyanin algebras,* Trans. Amer. Math. Soc. **341** (1994), 895–916.

[StZh] J. T. Stafford and J. J. Zhang, *Homological properties of (graded) Noetherian PI rings,* J. Algebra **168** (1994), 988–1026.

[Ste] D. R. Stephenson, *Artin-Schelter regular algebras of global dimension three,* J. Algebra **183** (1996), 55–73.

[SteZ] D. R. Stephenson and J. J. Zhang, *Growth of graded Noetherian rings,* Proc. Amer. Math. Soc. **125** (1997), 1593–1605.

[Ta] H. Tachikawa, *Quasi-Frobenius Rings and Generalizations. QF-3 and QF-1 Rings,* Lect. Notes Math. **351**, Springer-Verlag, Heidelberg, New York, 1973.

[TV] J. Tate and M. Van den Bergh, *Homological properties of Sklyanin algebras,* Invent. Math. **124** (1996), 619–647.

[Tau] P. Tauvel, *Sur les quotients premiers de l'algèbre enveloppante d'un algèbre de Lie résoluble,* Bull. Soc. Math. France **106** (1978), 177–205.

[Te1] K.-M. Teo, *Homological properties of Sklyanin algebras,* Comm. Algebra **24** (1996), 3027–3035.

[Te2] _____, *Homological properties of fully bounded Noetherian rings,* J. London Math. Soc. **55** (1997), 37–54.

[VHL] F. Van Oystaeyen, M. Houssein and Li Huishi, *Residual properties of graded algebras over domains,* Comm. Algebra **25** (1997), 3577–3586.

[Wi] L. Willaert, *Schematic algebras and the Auslander-Gorenstein property,* Algèbre non-commutative, groupes quantiques et invariants (Reims, 1995), Soc. Math. France, Paris, (1997), 149–156.

[Ye] A. Yekutieli, *The residue complex of a noncommutative graded algebra,* J. Algebra **186** (1996), 522–543.

[YeZh] A. Yekutieli and J. J. Zhang, *Rings with Auslander dualizing complexes,* J. Algebra **213** (1999), 1-51.

[Yi] Yi Zhong, *Injectively homogeneous Noetherian rings that are integral over their centers,* (in Chinese) Chinese Ann. Math. Ser. A **18** (1997), 477–482.

[Za] A. Zaks, *Injective dimension of semiprimary rings,* J. Algebra **13** (1969), 73–89.

[Zh1] J. J. Zhang, *Twisted graded algebras and equivalences of graded categories,* Proc. London Math. Soc. **72** (1996), 281–311.

[Zh2] _____, *Connected graded Gorenstein algebras with enough normal elements,* J. Algebra **189** (1997), 390–405.

Department of Mathematics and Statistics
University of Otago
PO Box 56
Dunedin, New Zealand
e-mail: jclark@maths.otago.ac.nz

The Flat Cover Conjecture and Its Solution

Edgar E. Enochs and Overtoun M. G. Jenda

Abstract

We will give a brief history of the developments that led to the flat cover conjecture and its positive solution.

Baer in [2] proved that every abelian group can be embedded in a divisible group and proved that divisible groups were injective in the category of $\mathbb{Z}$-modules (although he didn't use that terminology). He also proved that the abelian group A can be embedded in a divisible group D in a minimal fashion, i.e., if $A \subset D' \subset D$ with D' divisible, then $D' = D$. Then he showed that such a minimal embedding is unique up to isomorphism. More generally he proved that any module (over any ring) can be embedded in an injective module. We note that his proof of this fact was very set-theoretic.

In their elegant paper [5], Eckmann and Schöpf defined an essential extension of a module and proved that every module M can be embedded as an essential submodule of an injective module E (so E is an injective envelope of M). Then they proved that the injective envelope of M is unique up to isomorphism.

At about this time categorical notions were becoming clarified and so every categorical notion had a dual counterpart. The dual notion to that of an essential monomorphism (in an abelian category) is that of an epimorphism with a superfluous kernel. This raised the question of the existence of the dual of an injective envelope, i.e., the existence of a projective cover. We note that from the definition one can prove that a projective cover is also unique up to isomorphism.

The first examples of these covers were obtained by using Nakayama's lemma. This lemma says that if $J \subset R$ is the Jacobson radical of R and if M is a finitely generated (left) R-module, then JM is superfluous in M. Hence if P is a finitely generated projective module and if $S \subset JP$, then $P \to P/S$ is a projective cover. These notions were used by Eilenberg [6] and Eilenberg and Nakayama [7].

In the late 1950s Bass attacked the problem of projective covers full force [3]. We recall that Serre in an appendix to his GAGA article ([16],

1956) had defined flat modules. Bass considered the rings R such that every R-module has a projective cover. He called these rings left perfect (using left modules). Among the several characterizations of such rings he noted that they are precisely the rings for which a module is projective if and only if flat.

In the decades 1960–80 various other kinds of so-called envelopes and covers made their appearance. One example of such envelopes is the pure injective envelopes of Fuchs [13] and Warfield [17]. The definition of these envelope is closely modeled on that of the injective envelope. Torsion free covers also appeared in this period. And there was the familiar uniqueness result for both pure-injective envelopes and for torsion free covers.

Slowly it became clear that in these various envelopes and covers the important common element was the uniqueness result which followed from their various definitions. And so it seemed better to define envelopes and covers with this uniqueness as part of the definition – and then to see what conditions would guarantee their existence. So we now give the definition.

Definition. [11] If $\mathcal{C}$ is any category and if $\mathcal{F}$ is a class of objects of $\mathcal{C}$ and if X is an object of $\mathcal{C}$, then by an $\mathcal{F}$-cover of X we mean a morphism $\phi : F \to X$ such that $F \in \mathcal{F}$; such that $\text{Hom}\,(G, F) \to \text{Hom}\,(G, X)$ is surjective for any $G \in \mathcal{F}$ and such that any $f : F \to F$ such that $\phi \circ f = \phi$ is an automorphism of F.

It is clear from the definition that an $\mathcal{F}$-cover of X is unique up to isomorphism. The dual notion is that of an $\mathcal{F}$-envelope.

$\mathcal{F}$-covers and envelopes are often named after the class $\mathcal{F}$. For example, if $\mathcal{F}$ is the class of flat modules, an $\mathcal{F}$-cover is just called a flat cover. And so an injective envelope is an $\mathcal{E}$-envelope where $\mathcal{E}$ is the class of injective modules.

If $\phi : F \to X$ ($F \in \mathcal{F}$) as above has only the property that $\text{Hom}\,(G, F) \to \text{Hom}\,(G, X)$ is surjective, then $\phi : F \to X$ is called an $\mathcal{F}$-precover. So $\phi : F \to X$ would be a flat precover in the situation above. Preenvelopes are defined in a dual manner.

These definitions were given close in time by Auslander and Smalø in [1] and by Enochs in [11]. However Auslander and Smalø used the terminology of approximations (minimal left and minimal right) instead of covers and envelopes.

Using this terminology, Bass proved that if a ring is such that every module has a projective cover, then every module over that ring has a flat cover. So implicitly this raised the possibility that while projective covers are rare, modules might always have flat covers. This was explicitly stated as a conjecture in ([12], pp. 196).

Since torsion free is equivalent to being flat over a Prüfer domain, the existence of flat covers over these rings was guaranteed by the existence of torsion free covers [10]. From the period 1980–95 a few other classes of rings whose modules have flat covers were discovered. The biggest breakthrough though was the work of Xu [18]. He proved that modules over commutative noetherian rings of finite Krull dimension have flat covers (This important class of rings includes all coordinate rings of affine algebraic varieties). Xu then wrote his monograph "Flat Covers of Modules" [19]. This well written work stirred up interest in the flat cover conjecture.

And now we go back to 1979 when Salce wrote his paper "Cotorsion theories for abelian groups" [15]. Salce's definition of a cotorsion theory was in the category of abelian groups, but it applies to any abelian category.

If $\mathcal{A}$ is an abelian category and $\mathcal{F}, \mathcal{C}$ are classes of objects of $\mathcal{A}$, then $(\mathcal{F}, \mathcal{C})$ is called a cotorsion theory for $\mathcal{A}$ if an object F is in $\mathcal{F}$ if and only if $\mathrm{Ext}^1(F, C) = 0$ for all $C \in \mathcal{C}$ and if $C \in \mathcal{C}$ if and only if $\mathrm{Ext}^1(F, C) = 0$ for all $F \in \mathcal{F}$. Salce then says $(\mathcal{F}, \mathcal{C})$ has enough injectives if for every object A of $\mathcal{A}$ there is an exact sequence $0 \to A \to C \to F \to 0$ with $C \in \mathcal{C}$ and $F \in \mathcal{F}$. We note that if $D \in \mathcal{C}$, then the exact sequence $\mathrm{Hom}\,(C, D) \to \mathrm{Hom}\,(A, D) \to \mathrm{Ext}^1(F, D) = 0$ shows that $A \to C$ is in fact a $\mathcal{C}$-preenvelope. The cotorsion theory $(\mathcal{F}, \mathcal{C})$ is said to have enough projectives if for all A there is an exact sequence $0 \to C \to F \to A \to 0$, again with $C \in \mathcal{C}$ and $F \in \mathcal{F}$ (and so $F \to A$ will be an $\mathcal{F}$-precover). Then Salce raised the question of whether every cotorsion theory has enough injectives and projectives.

If $\mathcal{A}$ is the category of R-modules and $\mathcal{F}$ the class of flat modules, then there is a $\mathcal{C}$ so that $(\mathcal{F}, \mathcal{C})$ is a cotorsion theory on $\mathcal{A}$. Then using [11, Theorem 3.1], which says that if a module has a flat precover then it has flat cover, we see that the flat cover conjecture is equivalent to the requirement that this cotorsion theory $(\mathcal{F}, \mathcal{C})$ has enough projectives.

In [14], Göbel and Shelah studied abelian groups A such that $\mathrm{Ext}^1(A, A) = 0$ (the so-called "splitters"). A construction they used inspired a theorem of Eklof and Trlifaj which we give.

Given a cotorsion theory $(\mathcal{F}, \mathcal{C})$ in the category of left R-modules, a class $X \subset \mathcal{F}$ is said to cogenerate $(\mathcal{F}, \mathcal{C})$ if any module C is in $\mathcal{C}$ if and only if $\mathrm{Ext}^1(F, C) = 0$ for all $F \in X$.

Then Eklof and Trlifaj's result is:

Theorem. [9] *If the cotorsion theory $(\mathcal{F}, \mathcal{C})$ is cogenerated by a set $X \subset \mathcal{F}$ (and so not just a class), then $(\mathcal{F}, \mathcal{C})$ has enough injectives.*

Salce in [15] had shown how to argue that enough injectives for a cotorsion theory in a category of modules guarantees that the cotorsion theory has enough projectives.

From this result and the remarks above we see that the flat cover conjecture would be settled (positively) if it could be proved that the cotorsion theory $(\mathcal{F}, \mathcal{C})$ with $\mathcal{F}$ the class of flat modules is cogenerated by a set.

We will indicate how this follows from:

Lemma. *If R is any ring, there exists a cardinal number λ such that if $x \in M$ for any module M, there is a pure submodule $S \subset M$ with $x \in S$ and with $\operatorname{card}(S) \leq \lambda$.*

If we apply this lemma to a flat module F and use (essentially [8, Theorem 1.2]) we see that $(\mathcal{F}, \mathcal{C})$ is cogenerated by a set $X \subset \mathcal{F}$ of flat modules G with $\operatorname{card}(G) \leq \lambda$.

So we see that the flat cover conjecture is true.

But at almost exactly the same time this solution to the flat cover conjecture was found, El Bashir proved the next result.

Theorem. [4] *If R is any ring, then for each cardinal number λ there exists a cardinal number κ such that for any submodule $L \subset M$ of any module M with $\operatorname{card}(M) \geq \kappa$ and $\operatorname{card}(M/L) \leq \lambda$ (so M is "big" and M/L is "small") there is a nonzero submodule $S \subset L$ which is pure in M.*

Then El Bashir and Bican noted that this theorem (applied to flat modules) quickly gives that every module has a flat precover. And so this is another way to settle that flat cover conjecture.

These two solutions to the flat cover conjecture can be seen in [4], or in Chapter 7 of [12].

It seems that both methods can have (perhaps very different) uses in proving the existence of (pre)envelopes and (pre)covers in various settings.

References

[1] M. Auslander, and S. O. Smalø, *Preprojective modules over artin algebras,* J. Algebra **66** (1980), 61–122.

[2] R. Baer, *Abelian groups which are direct summands of every containing group,* Bull. Amer. Math. Soc. **46** (1940), 800–806.

[3] H. Bass, *Finitistic dimension and a homological generalization of semiprimary rings*, Trans. Amer. Math. Soc. **95** (1960), 466–488.

[4] L. Bican, R. El Bashir, and E. Enochs, *The existence of flat covers*, Preprint.

[5] B. Eckmann and A. Schöpf, *Über injective Moduln*, Arch. Math. **4** (1953), 75–78.

[6] S. Eilenberg, *Homological dimension and syzygies*, Ann. Math. **64** (1956), 328–336.

[7] S. Eilenberg and T. Nakayama, *On the dimension of modules and algebras V*, Nagoya Math. J. **11** (1957), 9–12.

[8] P. C. Eklof, *Homological algebra and set theory*, Trans. Amer. Math. Soc. **227** (1977), 207–225.

[9] P. C. Eklof and J. Trlifaj, *How to make Ext vanish*, Preprint.

[10] E. Enochs, *Torsion free covering modules*, Proc. Amer. Math. Soc. **14** (1963), 884–889.

[11] _____, *Injective and flat covers, envelopes, and resolvents*, Israel J. Math. **39** (1981), 33–38.

[12] E. Enochs and O. M. G. Jenda, *Relative Homological Algebra*, de Gruyter, in press.

[13] L. Fuchs, *Algebraically compact modules over noetherian rings*, Indian J. Math. **9** (1967), 357–374.

[14] R. Göbel and S. Shelah, *Cotorsion theories and splitters*, Trans. Amer. Math. Soc., to appear.

[15] L. Salce, *Cotorsion theories for abelian groups*, Symposia Math. **23** (1979), 11–32.

[16] J.-P. Serre, *Géométrie algébrique et géométrie analytique*, Ann. Inst. Fourier **6** (1956), 1–42.

[17] R. Warfield, *Purity and algebraic compactness for modules*, Pacific J. Math. **28** (1969), 699–719.

[18] J. Xu, *The existence of flat covers over noetherian rings of finite Krull dimension*, Proc. Amer. Math. Soc. **123** (1995), 27–32.

[19] _____, *Flat Covers of Modules,* Lecture Notes in Math. **1634**, Springer-Verlag, Heidelberg, New York, 1996.

Edgar F. Enochs
Department of Mathematics
University of Kentucky
Lexington, KY 40506-0027, U. S. A.
e-mail: enochs@ms.uky.edu

Overtoun M. G. Jenda
Department of Discrete Mathematics
120 Mathematics Annex
Auburn University
Auburn, AL 36489-5307, U. S. A.

Some Results on Skew Polynomial Rings over a Reduced Ring

Juncheol Han, Yasuyuki Hirano and Hongkee Kim

0. Introduction

Throughout this paper, R denotes an associative ring with unity. Let σ be an endomorphism of a ring R. A skew polynomial ring $R[x;\sigma]$ is the ring of polynomials in x over R with the usual addition and with new multiplication defined by $xa = \sigma(a)x$ for each $a \in R$ (see [3], [4]). A left (right) annihilator of a subset U of R is defined by $l_R(U) = \{a \in R \mid aU = 0\}$ ($r_R(U) = \{a \in R \mid Ua = 0\}$). Recall that R is a *reduced* ring if it has no nonzero nilpotent elements.

A ring R is called (*quasi-*) *Baer* if the right annihilator of every (right ideal) nonempty subset of R is generated by an idempotent. In [3], a ring R is called *right* (resp. *left*) *p.q.-Baer* if the right (resp. left) annihilator of a principal right (resp. left) ideal is generated by an idempotent.

A ring R is called a *right* (resp. *left*) *p.p.-ring* if the right (resp. left) annihilator of any element of R is generated by an idempotent.

In [3], the following results were proved:

(1) A ring R is a right p.q.-Baer ring if and only if the polynomial ring $R[x]$ is a right p.q.- Baer ring.

(2) A ring R is a quasi-Baer ring if and only if the polynomial ring $R[x]$ over R is a quasi-Baer ring.

In this paper, first we will consider when the skew polynomial ring $R[x;\sigma]$ is a reduced ring. Next we try to extend the above results (1) and (2) to the skew polynomial ring $R[x;\sigma]$.

1. Some Conditions for $R[x;\sigma]$ to be Reduced

Recall that R is an *abelian* ring if every idempotent of R is central. We can easily observe that every reduced ring is abelian and in a reduced ring R, $l_R(U) = r_R(U)$ for any subset U of R.

There exists an example of a skew polynomial ring $R[x;\sigma]$ which is not a reduced ring even though R is a reduced ring with an automorphism σ.

Example 1.1. Let F be a field and $R = F \times F$ with an automorphism σ given by $\sigma(a, b) = (b, a)$ for all $(a, b) \in R$. Then R is a reduced ring. In this case, the skew polynomial ring $R[x; \sigma]$ is not a reduced ring because $(1, 0)x (\neq 0) \in R[x; \sigma]$ but $(1, 0)x(1, 0)x = 0$.

Proposition 1.2. *Let R be a ring with an endomorphism σ. Then the following are equivalent:*
  (1) $R[x; \sigma]$ *is reduced.*
  (2) R *is reduced and for* $a \in R$, $a\sigma(a) = 0$ *implies* $a = 0$.
  (3) R *is reduced and, for any* $f = \sum_{i=0}^{n} a_i x^i$ *and any* $g = \sum_{j=0}^{m} b_j x^j$ *in* $R[x; \sigma]$, $fg = 0$ *if and only if* $a_i b_j = 0$ *for all i and j, $0 \leq i \leq n, 0 \leq j \leq m$.*

Proof. (1)$\Rightarrow$(3). Suppose that $a_i b_j = 0$ for all i and j ($0 \leq i \leq n, 0 \leq j \leq m$). Then $a_i R[x; \sigma] b_j = 0$ because $R[x; \sigma]$ is reduced. In particular, $a_i x^k b_j = a_i \sigma^k(b_j) x^k = 0$, and so $a_i \sigma^k(b_j) = 0$. Now $fg = 0$ is obvious.

Conversely, suppose that $fg = 0$ for $f = \sum_{i=0}^{n} a_i x^i$ and $g = \sum_{j=0}^{m} b_j x^j$ in $R[x; \sigma]$. Then $a_n x^n b_m = 0$. Since $R[x; \sigma]$ is reduced, this implies $a_n b_m = 0 = b_m a_n$. Then $(\sum_{i=0}^{n-1} b_m a_i x^i)(\sum_{j=0}^{m} b_j x^j) = b_m fg = 0$. By induction on $m + n$, we obtain that $b_m a_i b_j = 0$ for all i and j. In particular, we have $b_m a_i b_m = 0$ for all i. Since R is reduced, we have $a_i b_m = 0$ and so $a_i \sigma(b_m) = 0$ because $a_i R[x, \sigma] b_m = 0$. Therefore $f(\sum_{j=0}^{m-1} b_j x^j) = 0$. By induction hypothesis, we obtain $a_i b_j = 0$ for all i and j ($0 \leq i \leq n, 0 \leq j \leq m$).

(3)$\Rightarrow$(2). Let $a \in R$ with $a\sigma(a) = 0$. If we take $f = ax$ and $g = a$, then $fg = a\sigma(a)x = 0$. Then $a^2 = 0$ by (3). Since R is reduced, this implies $a = 0$.

(2)$\Rightarrow$(1). First we claim that, for $a \in R$ and any positive integer n, $a\sigma^n(a) = 0$ implies $a = 0$. So assume that $a\sigma^n(a) = 0$. Since R is reduced, $ab\sigma^n(a) = 0$ for any $b \in R$. Hence $a\sigma^{n-1}(a)\sigma(a\sigma^{n-1}(a)) = a\sigma^{n-1}(a)\sigma(a)\sigma^n(a) = 0$. By hypothesis, this implies $a\sigma^{n-1}(a) = 0$. Continuing this process, $a\sigma(a) = 0$ and so $a = 0$.

Now let $f = a_n x^n + \cdots + a_0 \in R[x; \sigma]$ with $a_n \neq 0$. Then $f^2 = a_n \sigma^n(a_n) x^{2n} + \cdots + a_0^2$. By the above claim, $f^2 \neq 0$. □

Corollary 1.3. *Let R be a reduced ring with a monomorphism σ. Assume that $\sigma(P) \subseteq P$ for any minimal prime ideal P in R. Then R is reduced if and only if $R[x; \sigma]$ is reduced.*

Proof. ($\Leftarrow$). This is clear.
($\Rightarrow$). By Proposition 1.2, it is sufficient to show that for any $a \in R$, $a\sigma(a) = 0$ implies $a = 0$. Let $a\sigma(a) = 0$ for $a \in R$. Then $a\sigma(a) \in P$ for

any minimal prime ideal P of R. Since R is reduced, P is a completely prime ideal of R. So $a \in P$ or $\sigma(a) \in P$. If $a \in P$, then $\sigma(a) \in \sigma(P) \subseteq P$ by hypothesis. Hence $\sigma(a) \in P(R) = 0$, where $P(R)$ is the prime radical of R, and so $a = 0$ because σ is a monomorphism. □

Corollary 1.4. *Let R be a ring with an endomorphism σ. Suppose that R has only finitely many minimal prime ideals. Then the following are equivalent:*
 (1) *$R[x;\sigma]$ is reduced.*
 (2) *R is reduced, σ is monic and $\sigma(P) \subseteq P$ for any minimal prime ideal P in R.*

Proof. (1)⇒(2). Let $P_1, \ldots, P_n$ be the minimal prime ideals of R. If there exists $0 \neq a \in P_2 \cap \cdots \cap P_n$ such that $\sigma(a) \in P_1$, then $a\sigma(a) = 0$. By Proposition 1.2, this is impossible. Hence $0 = P_1 \cap \sigma(P_2 \cap \cdots \cap P_n) = P_1 \cap \sigma(P_2) \cap \cdots \cap \sigma(P_n)$. Hence $(\sigma(R) \cap P_1) \cap \sigma(P_2) \cap \cdots \cap \sigma(P_n) = 0 \subseteq \sigma(P_1)$. Since the $\sigma(P_i)$ are the minimal prime ideals of $\sigma(R)$, we conclude that $\sigma(R) \cap P_1 \subseteq \sigma(P_1)$. Since R/P_1 is an integral domain, $\sigma(R) \cap P_1$ is a prime ideal of $\sigma(R)$. Therefore $\sigma(P_1) = \sigma(R) \cap P_1$. Similarly we can prove that $\sigma(P_i) \subseteq P_i$ for each $i = 2, \ldots, n$.
 (2)⇒(1). This follows from Corollary 1.3. □

Example 1.5. Let F be a field and let $F[x,y]$ be a polynomial ring on x, y over F. Consider the ring $R = F[x,y]/(xy)$. Then R is a reduced ring and the minimal prime ideals are $(x)/(xy)$ and $(y)/(xy)$. Let $\bar{x} = x + (xy)$ and let $\bar{y} = y + (xy)$. We can easily see that for any nonzero $f \in R\bar{x}$ and any nonzero $g \in R\bar{y}$ there exists a unique F-monomorphism σ of R such that $\sigma(\bar{x}) = f$ and $\sigma(\bar{y}) = g$. In this case, $R[z;\sigma]$ is reduced. Similarly there exists a unique F-monomorphism σ of R such that $\sigma(\bar{x}) = g$ and $\sigma(\bar{y}) = f$. In this case, $R[z;\sigma]$ is not reduced. Also we can see that every F-monomorphism of R is obtained by the above manner.

We close this section with the following question.

Question. Does Corollary 1.4 remain true for an arbitrary ring R with an endomorphism σ?

2. Conditions for $R[x;\sigma]$ to be Baer, quasi-Baer or p.q.-Baer

According to [3], a polynomial ring $R[x]$ is a right p.q.- Baer ring, if R is a right p.q.-Baer ring.

The following example shows that this is not true for a skew polynomial ring in general.

Example 2.1. Let $S = \prod_{i \in \mathbb{Z}} T_i$ with $T_i = \mathbb{Q}$. Consider the automorphism σ of S defined by $\sigma((a_i)_{i \in \mathbb{Z}}) = (a_{i+1})_{i \in \mathbb{Z}}$. Let $R = \mathbb{Q}1 + \oplus_{i \in \mathbb{Z}} T_i$. Then R is a reduced p.p.-ring and so is a p.q.-Baer ring. Clearly the restriction of σ to R is an automorphism of R. We denote it by α. Let e (resp. e') denote the element $(a_i)_{i \in \mathbb{Z}} \in S$ such that $a_0 = 1$ and $a_i = 0$ for all $i \neq 0$ (resp. $a_i = 0$ if $i < 0$ and $a_i = 1$ if $i \geq 0$). We can easily see that $r_{S[x;\sigma]}(eR[x;\alpha]) = e'S[x;\sigma]$. Then $r_{R[x;\alpha]}(eR[x;\alpha]) = e'S[x;\sigma] \cap R[x;\alpha]$. From this, we see that $r_{R[x;\alpha]}(eR[x;\alpha])$ can not be generated by any idempotent. Hence $R[x;\alpha]$ is not right p.q.-Baer.

Of course, for a reduced ring R, the following statements are equivalent:
(1) R is a right p.p.-ring.
(2) R is a left p.p.-ring.
(3) R is a right p.q.-Baer ring.
(4) R is a left p.q.-Baer ring.

A ring R is called a p.p.-ring if it is both a right and left p.p.-ring.

Theorem 2.2. *Let R be a reduced p.p.-ring with an automorphism σ of finite order. Then $R[x;\sigma]$ is a p.q.-Baer ring.*

Proof. Let k denote the order of σ. Let $f = a_n x^n + \cdots + a_1 x + a_0 \in R[x;\sigma]$. Suppose $g = b_m x^m + \cdots + b_1 x + b_0 \in r_{R[x;\sigma]}(fR[x;\sigma])$. Then, for any positive integer t, we have $fx^t g = a_n \sigma^{n+t}(b_m) x^{n+t+m} + \cdots + a_0 \sigma^t(b_0) x^t = 0$. Hence $a_n \sigma^{n+t}(b_m) = 0$ for any positive integer t. Let $S = \{a_n, \sigma(a_n), \ldots, \sigma^{k-1}(a_n)\}$. Then we can easily see that $b_m \in r_R(S)$. Since R is a reduced p.p.-ring, there exists a central idempotent e_n such that $r_R(S) = e_n R$. Since the set S is σ-stable, we see that $\sigma(e_n)R = \sigma(e_n R) = e_n R$. This implies that the idempotent $\sigma(e_n)$ is also the identity of the ring $e_n R$. So we obtain that $\sigma(e_n) = e_n$ and $b_m = e_n b_m$. Now, considering the coefficient of $x^{n+m+t-1}$ in $fx^t g = 0$, we obtain $a_n \sigma^{n+t}(b_{m-1}) + a_{n-1} \sigma^{n+t-1}(e_n b_m) = 0$. Hence $a_n \sigma^{n+t}(b_{m-1}) = -e_n a_{n-1} \sigma^{n+t-1}(b_m) \in eR$. Therefore we obtain that $a_n \sigma^{n+t}(b_{m-1}) = e_n a_n \sigma^{n+t}(b_{m-1}) = 0$. This implies that $b_{m-1} \in r_R(S) = e_n R$. Inductively, we obtain $b_j \in e_n R$ for all $j = 0, 1, \cdots, m$. Thus $a_n x^n R[x;\sigma]g = 0$, and hence $g \in r_{R[x;\sigma]}((f - a_n x^n)R[x;\sigma])$. By induction on the degree of f, we have a central idempotent $e' = e_{n-1} \cdots e_0 \in R$ such that $\sigma(e') = e'$ and $r_{R[x;\sigma]}((f - a_n x^n)R[x;\sigma]) = e'R[x;\sigma]$. Hence $g \in e_n R[x;\sigma] \cap e'R[x;\sigma] = e_n e'R[x;\sigma]$. Since $e_n e'R[x;\sigma] \subseteq r_{R[x;\sigma]}(fR[x;\sigma])$, this proves that $r_{R[x;\sigma]}(fR[x;\sigma]) = e_n e'R[x;\sigma]$. □

If R is a reduced ring, then R is Baer if and only if R is quasi-Baer. Using a similar method as in the proof of Theorem 2.2, we obtain the following theorem.

Theorem 2.3. *Let R be a reduced Baer ring with an automorphism σ of finite order. Then $R[x; \sigma]$ is a quasi-Baer ring.*

In [1], Armendariz proved that if R is a reduced ring, then R is a p.p.-(resp. Baer) ring if and only if the polynomial ring $R[x]$ is a p.p.- (resp. Baer) ring. We shall extend this result to reduced skew polynomial rings. To do it, we need two lemmas.

Lemma 2.4. *Let R be a ring with an endomorphism σ of R. Assume that $R[x; \sigma]$ is reduced. If $f \in R[x; \sigma]$ is an idempotent, then $f \in R$, that is, every idempotent of $R[x; \sigma]$ is an idempotent of R.*

Proof. Let $f = a_0 + a_1 x + \cdots + a_n x^n \in R[x; \sigma]$ be an idempotent. Then $0 = f - f^2 = f(1-f)$. By Proposition 1.2, $a_0(1 - a_0) = 0$ and $a_i^2 = 0$ for each i ($1 \leq i \leq n$), and hence we get $a_0 = a_0^2$ and $a_i = 0$ for each i ($1 \leq i \leq n$). Hence $f = a_0 \in R$. □

Lemma 2.5. *Let R be a ring with an endomorphism σ of R. Assume that $R[x; \sigma]$ is reduced. If $T \subseteq R[X; \sigma]$ and $S_f = \{a_0, a_1, \ldots, a_n\}$, where $f = a_0 + a_1 x + \cdots + a_n x^n \in T$, then $r_{R[x;\sigma]}(T) = r_R(S_T)[x; \sigma]$, where $S_T = \cup_{f \in T} S_f$.*

Proof. If $g = b_0 + b_1 x + \cdots + b_m x^m \in r_{R[x;\sigma]}(T)$, then $Tg = 0$, i.e., $fg = 0$ for all $f \in T$. By Proposition 1.2, $a_i b_j = 0$ for all i and j ($0 \leq i \leq m$, $0 \leq j \leq n$), which implies that $b_j \in r_R(S_T)$, and so $g \in r_R(S_T)[x; \sigma]$. Hence $r_{R[x;\sigma]}(T) \subseteq r_R(S_T)[x; \sigma]$. The other inclusion is obvious. □

Theorem 2.6. *Let R be a ring with an endomorphism σ of R. Assume that $R[x; \sigma]$ is reduced. Then $R[x; \sigma]$ is a p.p.-ring if and only if R is a p.p.-ring.*

Proof. ($\Rightarrow$). If $R[x; \sigma]$ is a p.p.-ring and $a \in R$, then $r_R(a) = R \cap r_{R[x;\sigma]}(a) = R \cap eR[x; \sigma]$ for some idempotent $e \in R[x; \sigma]$. By Lemma 2.4, $e \in R$, and so $r_R(a) = eR$. Hence R is a p.p.-ring.

($\Leftarrow$). Assume that R is a p.p.-ring. Note that for any finite subset T of R, $r_R(T) = eR$ for some idempotent $e \in R$. If $f \in R[x; \sigma]$, then by Lemma 2.5, $r_{R[x;\sigma]}(f) = r_R(S_f)[x; \sigma] = eR[x; \sigma]$ for some idempotent $e \in R$, because S_f is a finite subset of R and e is central. Hence $R[x; \sigma]$ is a p.p.-ring □

Similarly we can obtain the following.

Theorem 2.7. *Let R be a ring with an endomorphism σ of R. Assume that $R[x; \sigma]$ is reduced. Then $R[x; \sigma]$ is a Baer ring if and only if R is a Baer ring.*

Proof. ($\Rightarrow$). If $R[x;\sigma]$ is a Baer, then for any subset T of R, $r_{R[x;\sigma]}(T) = fR[x;\sigma]$ for some idempotent $f \in R[x;\sigma]$. By Lemma 2.4, $f \in R$, and then $r_R(T) = R \cap r_{R[x;\sigma]}(T) = R \cap fR[x;\sigma] = fR$. Hence R is a Baer ring.

($\Leftarrow$). Suppose that R is Baer and T is an arbitrary subset of $R[x;\sigma]$. Let $S_T = \cup_{f \in T} S_f$. Since R is Baer, $r_R(S_T) = eR$ for some idempotent $e \in R$. By Lemma 2.5, $r_{R[x;\sigma]}(T) = r_R(S_T)[x;\sigma] = eR[x;\sigma]$. Thus $R[x;\sigma]$ is a Baer ring. □

Note that Theorems 2.6 and 2.7 extend Armendariz's results [1, Theorems A and B].

Let R be a ring with an endomorphism σ. By Proposition 1.2, if $R[x;\sigma]$ is reduced, then σ must be a monomorphism. We close this paper with the following example which shows that even if R is an integral domain, if the endomorphism σ is not monic, then $R[x;\sigma]$ does not need to be a p.p.-ring.

Example 2.8. [cf 4, p.18] Let F be a field and $R = F[t]$ a polynomial ring over F with the endomorphism σ given by $\sigma(f(t)) = f(0)$ for all $f(t) \in R$. Then R is a principal ideal domain but the skew polynomial ring $R[x;\sigma]$ is not an integral domain because $xt = \sigma(t)x = 0$. We will show that the skew polynomial ring $R[x;\sigma]$ is neither right p.q.-Baer nor right p.p.

Consider a right ideal $xR[x;\sigma]$. Then $x\{f_0(t)+f_1(t)x+\cdots+f_n(t)x^n\} = f_0(0)x + f_1(0)x^2 + \cdots + f_n(0)x^{n+1}$ for all $f_0(t) + f_1(t)x + \cdots + f_n(t)x^n \in R[x;\sigma]$ and hence $xR[x;\sigma] = \{a_1x + a_2x^2 + \cdots + a_nx^n | n \in N \cup \{0\}, a_i \in F(i = 0,1,\ldots,n)\}$. Note that $R[x;\sigma]$ has only idempotents 0 and 1 by simple computation. Since $(a_1x + a_2x^2 + \cdots + a_nx^n)1 = (a_1x + a_2x^2 + \cdots + a_nx^n) \neq 0$ for some nonzero element $a_1x + a_2x^2 + \cdots + a_nx^n \in xR[x;\sigma]$, we get $1 \notin r_{R[x;\sigma]}(xR[x;\sigma])$ and so $r_{R[x;\sigma]}(xR[x;\sigma]) \neq R[x;\sigma]$. Also, since $(a_1x + a_2x^2 + \cdots + a_nx^n)t = 0$ for all $a_1x + a_2x^2 + \cdots + a_nx^n \in xR[x;\sigma]$, $t \in r_{R[x;\sigma]}(xR[x;\sigma])$ and hence $r_{R[x;\sigma]}(xR[x;\sigma]) \neq 0$. Thus $r_{R[x;\sigma]}(xR[x;\sigma])$ is not generated by an idempotent. Therefore $R[x;\sigma]$ is not a right p.q.-Baer ring and so neither a quasi-Baer ring nor a Baer ring. Similarly, we can verify that $R[x;\sigma]$ is not a right p.p.-ring.

Acknowledgements. The authors thank Professors Gary F. Birkenmeier and Jae Keol Park for their valuable suggestions and comments. Also the authors would like to thank the referee for kind comments.

References

[1] E. P. Armendariz, *A note on extensions of Baer and p.p.-rings*, J. Australian Math. Soc. **18** (1974), 470–473.

[2] G. F. Birkenmeier, J. Y. Kim and J. K. Park, *On extensions of quasi-Baer and principally quasi-Baer rings,* preprint.

[3] K. R. Goodearl and R. B. Warfield, Jr., *An Introduction to Noncommutative Noetherian Rings,* Cambridge University Press, Cambridge, 1989.

[4] J. C. McConnell and J. C. Robson, *Noncommutative Noetherian Rings,* Wiley, New York, 1987.

Juncheol Han
Department of Computational Mathematics
Kosin University
Pusan 606-701, Korea
e-mail: jchan@sdg.kosin.ac.kr

Yasuyuki Hirano
Department of Mathematics
Okayama University
Okayama 700-8530, Japan
e-mail: yhirano@math.okayama-u.ac.jp

Hongkee Kim
Department of Mathematics
Gyeongsang National University
Jinju 660-701, Korea
e-mail: hkkim@gshp.gsnu.ac.kr

Derived Equivalences and Tilting Theory

Dieter Happel

Dedicated to K. W. Roggenkamp on the occasion of his 60th birthday

Abstract

The aim of this article is to survey aspects of tilting theory. In the last 25 years the idea of tilting has emerged from rather special situations in terms of reflection functors [BGP] to a thorough investigation of derived categories of abelian categories admitting a tilting complex. It is impossible to survey here a full account of these developments. For some historic remarks see [H1] or [K]. Instead we will focus after a general introduction on one particular aspect. This is the theory of quasitilted algebras or equivalently that of hereditary abelian categories with tilting object as introduced in the work with Reiten and Smalø [HRS1].

1. Preliminaries

Let k be an algebraically closed field and let $\mathcal{A}$ be an abelian k-category satisfying the following finiteness conditions.

For each $i \in \mathbb{N}$ and $X, Y \in \mathcal{A}$ we have $\dim_k \mathrm{Ext}^i(X,Y) < \infty$ and there exists $n \in \mathbb{N}$ such that $\mathrm{Ext}^i(X,Y) = 0$ for all $i > n$.

These assumptions ensure that the theorem of Krull/Remak/Schmidt will hold in $\mathcal{A}$ and that $\mathcal{A}$ has finite global dimension. We denote by $D^b(\mathcal{A})$ the bounded derived category of $\mathcal{A}$. For $i \in \mathbb{Z}$ and $X^\bullet \in D^b(\mathcal{A})$ we denote by $X^\bullet[i]$ the i-th fold application of the shift functor in $D^b(\mathcal{A})$. We refer to [V], [Gr] and [Har] for a detailed account of the construction of $D^b(\mathcal{A})$.

We are mainly interested in the following two kinds of examples. Let Λ be a finite dimensional k-algebra of finite global dimension. We denote by $\mathrm{mod}\,\Lambda$ the category of finitely generated left Λ-modules. In this case we denote by $D^b(\Lambda)$ the bounded derived category of $\mathrm{mod}\,\Lambda$. For the second example let $\mathbb{X}$ be a smooth projective variety. We denote by $\mathrm{coh}\,\mathbb{X}$ the category of coherent sheaves on $\mathbb{X}$. Also in this case we denote by $D^b(\mathbb{X})$ the bounded derived category of $\mathrm{coh}\,\mathbb{X}$. We refer to the next section for further examples derived from noncommutative geometry.

An object $T^\bullet \in D^b(\mathcal{A})$ is called a *tilting complex* if the following two conditions are satisfied: (i) $\mathrm{Hom}_{D^b(\mathcal{A})}(T^\bullet, T^\bullet[i]) = 0$ for all $i \neq 0$ and (ii) $\mathrm{add}\, T^\bullet$ generates $D^b(\mathcal{A})$ as a triangulated category. Here we have denoted by $\mathrm{add}\, T^\bullet$ the full subcategory of $D^b(\mathcal{A})$ whose objects are direct sums of direct summands of $T^\bullet$.

The purpose of tilting theory is to form the endomorphism algebra Γ of a tilting complex $T^\bullet$ and then to compare $\mathcal{A}$ with $\mathrm{mod}\,\Gamma$. Usually $\mathcal{A}$ and $\mathrm{mod}\,\Gamma$ will be quite different on a first glance. But the philosophy is that in certain situations a lot of information is available on $\mathcal{A}$ and some information may then be deduced via the tilting procedure on $\mathrm{mod}\,\Gamma$. This reasoning will be explained in examples in the later sections of this article.

The following basic result goes back to Beilinson [B] and Bondal [Bon], but compare also [H2] and [Ri].

Theorem. *Let $\mathcal{A}$ be an abelian k-category and let $T^\bullet \in D^b(\mathcal{A})$ be a tilting complex. Let $\Gamma = \mathrm{End}_{D^b(\mathcal{A})} T^\bullet$ be its endomorphism algebra. Then there exists a derived equivalence $D^b(\mathcal{A}) \simeq D^b(\Gamma)$.*

In the next section we will explain in a special situation the construction of such a derived equivalence.

Before we consider some examples we will introduce the following results and notation which will be used throughout this article. For this let Λ be a finite dimensional k-algebra. If $\mathcal{C}$ is a full subcategory of $\mathrm{mod}\,\Lambda$ we denote by $\mathrm{add}\,\mathcal{C}$ the full subcategory whose objects are direct sums of direct summands of modules in $\mathcal{C}$. Let $X \in \mathrm{mod}\,\Lambda$, then the subcategory of epimorphic images of $\mathrm{add}\, X$ is denoted by $\mathrm{fac}\, X$. We denote by $\mathrm{ind}\,\Lambda$ the full subcategory of $\mathrm{mod}\,\Lambda$ containing one representative from each isomorphism class of indecomposable Λ-modules. For an indecomposable Λ-module X we denote by τX the *Auslander-Reiten translation DTr* and by $\tau^- X$ the inverse Auslander-Reiten translation TrD. The *Auslander-Reiten quiver* of Λ is denoted by $\overrightarrow{\Gamma}_\Lambda$. A *path* in $\mathrm{mod}\,\Lambda$ is a sequence $(X_0, \ldots, X_s)$ of (isomorphism classes of) indecomposable Λ-modules X_i, $0 \leq i \leq s$ such that $\mathrm{Hom}(X_{i-1}, X_i) \neq 0$ and $X_{i-1} \not\simeq X_i$ for all $1 \leq i \leq s$. We will say that $(X_0, \ldots, X_s)$ is a path from X_0 to X_s of *length* s, and we write $X \preceq X'$, or $X \preceq_\Lambda X'$ to indicate that a path from X to X' exists. We say that X is a *predecessor* of X' or X' is a *successor* of X. A path from X to X of length greater than or equal to one is called a *cycle*. An indecomposable module X is called *directing* if X does not lie on a cycle. If X is a Λ-module we denote by $\mathrm{pd}_\Lambda X$ (resp. $\mathrm{id}_\Lambda X$) the *projective* (resp. *injective*) dimension of X. Also we denote by $\mathrm{gl.dim}\,\Lambda$ the *global dimension* of Λ.

If Λ is a finite dimensional algebra, we may assume that Λ is basic. In this case Λ is isomorphic to the quotient algebra $k\vec{\Delta}/I$ of the path algebra $k\vec{\Delta}$ over k of a finite quiver $\vec{\Delta}$ by a two-sided admissible ideal I. We may assume that the vertices of $\vec{\Delta}$ are given by $\{1,\ldots,n\}$. The simple Λ-modules corresponding to these vertices are denoted by $S(1),\ldots,S(n)$. The indecomposable projective Λ-module whose simple top is $S(i)$ will be denoted by $P(i)$, the indecomposable injective Λ-module whose simple socle is $S(i)$ is denoted by $I(i)$. If $X \in \mathrm{mod}\,\Lambda$, then we denote by $\dim X = (\dim_k \mathrm{Hom}_\Lambda(P(i),X))_i$ for $1 \leq i \leq n$ the *dimension vector* of X. This yields the well-known isomorphism from the Grothendieck group $K_0(\Lambda)$ to $\mathbb{Z}^n$.

Given a full subcategory $\mathcal{C} \subset \mathrm{mod}\,\Lambda$ we say that $X \in \mathcal{C}$ is *Ext-projective* (resp. *Ext-injective*) provided $\mathrm{Ext}^1_\Lambda(X,\mathcal{C}) = 0$ (resp. $\mathrm{Ext}^1_\Lambda(\mathcal{C},X) = 0$).

A pair $(\mathcal{T},\mathcal{F})$ of full subcategories of $\mathrm{mod}\,\Lambda$ is called a *torsion pair* provided that $\mathrm{Hom}\,(\mathcal{T},\mathcal{F}) = 0$ and both $\mathcal{T}$ and $\mathcal{F}$ are maximal with respect to this property. Given a torsion pair $(\mathcal{T},\mathcal{F})$ we have for each $X \in \mathrm{mod}\,\Lambda$ a canonical exact sequence

$$0 \to t(X) \to X \to X/t(X) \to 0$$

with $t(X) \in \mathcal{T}$ and $X/t(X) \in \mathcal{F}$. We say that the torsion pair $(\mathcal{T},\mathcal{F})$ is *split* if each indecomposable Λ-module X is either torsion, so in $\mathcal{T}$, or torsionfree, so in $\mathcal{F}$. This is equivalent to $\mathrm{Ext}^1_\Lambda(\mathcal{F},\mathcal{T}) = 0$.

We denote the composition of morphisms $f: X \to Y$ and $g: Y \to Z$ in a given category $\mathcal{K}$ by fg.

The notation and terminology introduced here will be fixed throughout this article. For unexplained representation-theoretic terminology we refer to [R1] or [ARS]. These two sources will also contain the proofs of certain results from the general representation theory we will need in the next sections. Moreover they will also contain the references to the original articles we usually will not refer to.

We point out that most of the results are valid or have a simple analogue in case we start with an arbitrary Artin algebra. But for expository reasons we will restrict to the case described above.

We will now consider the following two examples of tilting complexes in the bounded derived category of a finite dimensional algebra and in the category of coherent sheaves on a smooth projective variety.

Let $\vec{\Delta} = (\Delta_0, \Delta_1)$ be a quiver without oriented cycle which admits a level function $\lambda: \Delta_0 \to \{0,\ldots,r\}$ for some $r \geq 0$. So λ is surjective and for each arrow $s(\alpha) \xrightarrow{\alpha} t(\alpha)$ in Δ_1 we have that $\lambda(t(\alpha)) = \lambda(s(\alpha)) + 1$.

For example:

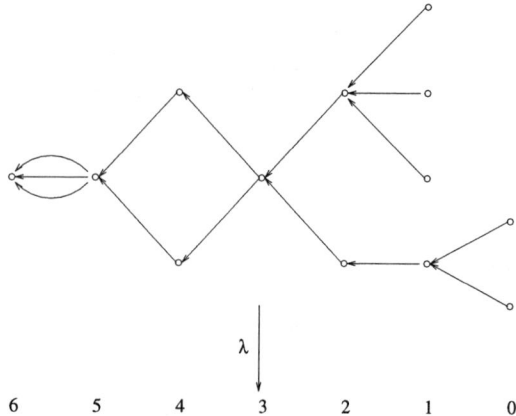

Let $\Lambda = k\overrightarrow{\Delta}$ be the path algebra of $\overrightarrow{\Delta}$. Then

$$T^\bullet = \bigoplus_{i=0}^{r} \bigoplus_{\lambda(x)=i} S(x)[i]$$

is a tilting complex in $D^b(\Lambda)$, where $S(x)$ is the simple module associated with the vertex x.

Then $\Lambda = \operatorname{End} T^\bullet$ is given by $k\overrightarrow{\Delta}/I$ where I is the two-sided ideal generated by all paths of length 2.

We point out that in general the endomorphism algebra of a tilting complex in $D^b(\Lambda)$ will not be a factor algebra of Λ.

The next example is due to [B]. Let $\mathbb{X}$ be the n dimensional projective space $\mathbb{P}^n(\mathbb{C})$ over the field of complex numbers $\mathbb{C}$.

Then

$$T = \bigoplus_{i=0}^{n} \mathcal{O}(i)$$

is a tilting object in $D^b(\mathbb{X})$, where $\mathcal{O}(i)$ is the structure sheaf twisted i times.

Then $\Lambda = \operatorname{End} T$ is given by $k\overrightarrow{\Delta}/I$, where I is a two-sided ideal and $\overrightarrow{\Delta}$ is the following quiver with $n+1$ vertices and $n+1$ arrows between two consecutive vertices. The generators of I are similar to those in the symmetric algebra.

2. Hereditary Categories

In this section we will recall the basic theory of tilting objects in hereditary categories and of quasitilted algebras. Most of the results are contained in [HRS1]. Note that some of the general constructions can be formulated for arbitrary abelian categories. Since here we are mainly interested in the application to quasitilted algebras we will restrict to the case of hereditary abelian categories.

Let $\mathcal{H}$ be a hereditary, abelian category (i.e., $\text{Ext}^2_{\mathcal{H}}(X,Y) = 0$ for all $X, Y \in \mathcal{H}$ and $\mathcal{H}$ is abelian). We will always assume that $\mathcal{H}$ is a locally-finite k-category (i.e., for all $X, Y \in \mathcal{H}$ we have that $\text{Hom}_{\mathcal{H}}(X,Y)$ and $\text{Ext}^1_{\mathcal{H}}(X,Y)$ are k-vector spaces which are finite dimensional over k and we have that the composition is bilinear over k). Note that these assumptions imply that the theorem of Krull/Remak/Schmidt holds in $\mathcal{H}$.

Given $X \in \mathcal{H}$ we denote by add X the full subcategory of $\mathcal{H}$ with objects the finite direct sums of direct summands of X. Further we denote by fac X the full subcategory of $\mathcal{H}$ with objects the epimorphic images of objects in add X and by $\mathcal{E}(X)$ the full subcategory of $\mathcal{H}$ with objects Y satisfying $\text{Ext}^1_{\mathcal{H}}(X,Y) = 0$.

A pair $(\mathcal{T}, \mathcal{F})$ of full subcategories of $\mathcal{H}$ is called a *torsion pair* provided that $\text{Hom}\,(\mathcal{T}, \mathcal{F}) = 0$ and both $\mathcal{T}$ and $\mathcal{F}$ are maximal with respect to this property and for each $X \in \text{mod}\,\Lambda$ we have an exact sequence

$$0 \to t(X) \to X \to X/t(X) \to 0$$

with $t(X) \in \mathcal{T}$ and $X/t(X) \in \mathcal{F}$. Note that in our specific examples of torsion pairs the existence of the exact sequence will follow from the assumption on $\mathcal{H}$ to be locally-finite. We say that the torsion pair $(\mathcal{T}, \mathcal{F})$ is split if each indecomposable $X \in \mathcal{H}$ is either torsion, so in $\mathcal{T}$, or torsion-free, so in $\mathcal{F}$. This is equivalent to $\text{Ext}^1_{\mathcal{H}}(\mathcal{F}, \mathcal{T}) = 0$.

Given a full subcategory $\mathcal{C} \subset \mathcal{H}$ we say that $X \in \mathcal{C}$ is Ext-projective (resp. Ext-injective) provided $\text{Ext}^1_{\mathcal{H}}(X, \mathcal{C}) = 0$ (resp. $\text{Ext}^1_{\mathcal{H}}(\mathcal{C}, X) = 0$).

We say that $T \in \mathcal{H}$ is a *tilting object* if fac $T = \mathcal{E}(T)$. So T is the direct sum of the Ext-projective objects in fac T. If $T \in \mathcal{H}$ is a tilting object, then $\text{End}_{\mathcal{H}} T$ is called a quasitilted algebra. We point out that this was not the original definition of a tilting object. This definition is sometimes easier to verify in concrete situations. It is shown in [H3] that this definition is equivalent to the old one from [HRS1]. Also we point out that a tilting complex which is contained in $\mathcal{H}$ is a tilting object and that a tilting object is a tilting complex. Moreover it is shown in [HRe2] that a hereditary abelian category $\mathcal{H}$ will admit a tilting object if and only if

$D^b(\mathcal{H})$ admits a tilting complex.

Given a tilting object $T \in \mathcal{H}$, we set $\mathcal{T} = \operatorname{fac} T = \mathcal{E}(T)$. This is a torsion class of a torsion pair $(\mathcal{T}, \mathcal{F})$ on $\mathcal{H}$, where

$$\mathcal{F} = \{X \in \mathcal{H} \mid \operatorname{Hom}_{\mathcal{H}}(T, X) = 0\}.$$

Indeed, $\mathcal{T}$ is clearly closed under extensions and factor objects and since $\mathcal{H}$ is locally-finite we have for each $X \in \mathcal{H}$ a short exact sequence $0 \to t(X) \to X \to X/t(X) \to 0$ where $t(X) \in \mathcal{T}$ is just the trace of T in X and $X/t(X) \in \mathcal{F}$. The torsion pair $(\mathcal{T}, \mathcal{F})$ will be called the torsion pair induced by T on $\mathcal{H}$.

Since $\mathcal{H}$ is locally-finite we also have for each $X \in \mathcal{H}$ a short exact sequence

$$0 \to X \to E \to T^r \to 0$$

for some $r \geq 0$ with $E \in \mathcal{T}$. In particular, $\mathcal{T}$ is a cogenerator for $\mathcal{H}$. So the torsion pair $(\mathcal{T}, \mathcal{F})$ is a tilting torsion pair in the sense of [HRS1].

The following fact shows that the case of hereditary abelian categories is very special.

Theorem. *Let $\mathcal{H}$ be a hereditary abelian category. Then each indecomposable object $X^\bullet \in D^b(\mathcal{H})$ is isomorphic to $X[i]$ for an indecomposable object $X \in \mathcal{H}$ and some $i \in \mathbb{Z}$.*

The main theorem can now be formulated. For a detailed proof we refer to [HRS1].

Theorem. *Let $\mathcal{H}$ be a hereditary abelian category with tilting object T and $\Lambda = \operatorname{End}_{\mathcal{H}} T$. Then there exists a triangle equivalence $F : D^b(\mathcal{H}) \to D^b(\Lambda)$ with the following property. If $(\mathcal{T}, \mathcal{F})$ is the torsion pair on $\mathcal{H}$ induced by T, then $(\mathcal{X}, \mathcal{Y})$ is a split torsion pair on $\operatorname{mod} \Lambda$, where $\mathcal{Y} = F(\mathcal{T})$ and $\mathcal{X} = F(\mathcal{F}[1])$.*

The functor F is constructed as follows. For $X \in \mathcal{H}$, choose an exact sequence $0 \to X \to E \to T^r \to 0$ with $E \in \mathcal{T}$ as before. Then $F(X)$ is the complex of Λ-modules

$$\cdots \to 0 \to \operatorname{Hom}(T, E) \to \operatorname{Hom}(T, T^r) \to 0 \to \cdots,$$

where $\operatorname{Hom}(T, E)$ is in degree zero.

The following figure gives a pictorial description of the tilting process described above:

Derived Equivalences

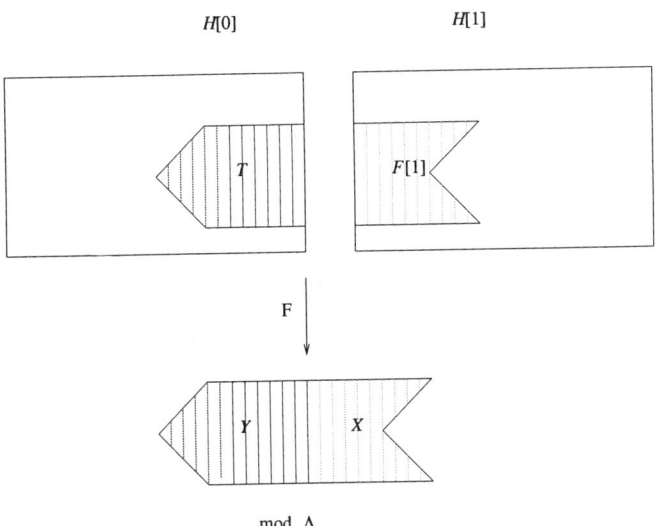

The torsion pair $(\mathcal{X}, \mathcal{Y})$ on mod Λ will be called the torsion pair induced by T on mod Λ. The splitting of $(\mathcal{X}, \mathcal{Y})$ is shown as follows. Let $X \in \mathcal{X}$ and $Y \in \mathcal{Y}$. Then $X = F(X'[1])$ for some $X' \in \mathcal{F}$ and $Y = F(Y')$ for some $Y' \in \mathcal{T}$. Then we have that $\operatorname{Ext}^1_\Lambda(Y, X) = \operatorname{Hom}_{D^b(\Lambda)}(Y, X[1]) \simeq \operatorname{Hom}_{D^b(\mathcal{H})}(Y', X'[2]) \simeq \operatorname{Ext}^2_\mathcal{H}(Y', X') = 0$.

We will now collect several consequences of the theorem above for quasitilted algebras as well as for hereditary abelian categories with tilting object. For a proof we refer to [HRS1].

Corollary 1. *Let Λ be a quasitilted algebra. Then the following hold:*
(1) $\operatorname{gl.dim} \Lambda \leq 2$.
(2) *If X is an indecomposable Λ-module, then $\operatorname{pd}_\Lambda X \leq 1$ or $\operatorname{id}_\Lambda X \leq 1$.*
(3) *Λ is directed (i.e., if $\Lambda = k\overrightarrow{\Delta}/I$, then $\overrightarrow{\Delta}$ has no oriented cycle).*
(4) *If X is an indecomposable Λ-module with $\operatorname{Ext}^1_\Lambda(X, X) = 0$, then $\operatorname{End}_\Lambda X \simeq k$.*
(5) *If X, Y are indecomposable Λ-modules which lie on a cycle in $\operatorname{ind} \Lambda$, then $\operatorname{Ext}^2_\Lambda(X, Y) = 0$.*

Next we observe that the Grothendieck groups $K_0(\mathcal{H})$ and $K_0(\Lambda)$ are both endowed with a bilinear form $\langle -, - \rangle$ defined by:

$$\langle X, Y \rangle = \sum_{i \geq 0} (-1)^i \dim_k \operatorname{Ext}^i(X, Y).$$

The following is another straightforward consequence of the theorem above.

Corollary 2. *Let $\mathcal{H}$ be a hereditary abelian category with tilting object T and $\Lambda = End_{\mathcal{H}}T$. Then the triangle equivalence $F : D^b(\mathcal{H}) \to D^b(\Lambda)$ induces an isometry*

$$f : (K_0(\mathcal{H}), \langle -, - \rangle_{\mathcal{H}}) \to (K_0(\Lambda), \langle -, - \rangle_\Lambda)$$

with $f(X) = Hom(T, X) - Ext^1(T, X)$. In particular, $\langle -, - \rangle_{\mathcal{H}}$ is non-degenerate and $K_0(\mathcal{H})$ is free abelian of finite rank.

We point out that the existence of such an isometry is one of the reasons for the name tilting due to S. Brenner and M. Butler.

We will now give several characterizations of quasitilted algebras. A finite dimensional k-algebra is called *almost hereditary* if it satisfies the homological properties (1) and (2) from Corollary 1. So a quasitilted algebra is almost hereditary. We will indicate that the converse holds as well.

If Λ is an almost hereditary algebra the following two subsets of ind Λ are very important.

Let $\mathcal{L}_\Lambda$ denote the subset of ind Λ given by $\mathcal{L}_\Lambda = \{X \in$ ind $\Lambda \mid$ for all Y with $Y \preceq X$ we have $pd_\Lambda Y \leq 1\}$, and let $\mathcal{R}_\Lambda$ denote the subset of ind Λ given by $\mathcal{R}_\Lambda = \{X \in$ ind $\Lambda \mid$ for all Y with $X \preceq Y$ we have $id_\Lambda Y \leq 1\}$. When there is no danger of confusion we simply write $\mathcal{L}$ for $\mathcal{L}_\Lambda$ and $\mathcal{R}$ for $\mathcal{R}_\Lambda$.

With this terminology we have the following characterization of almost hereditary algebras from [HRS1].

Theorem. *The following are equivalent for a finite dimensional algebra Λ where $\mathcal{R}$ and $\mathcal{L}$ are as before.*
 (a) *Λ is almost hereditary.*
 (b) *$\mathcal{R}$ contains all injective modules in ind Λ.*
 (b') *$\mathcal{L}$ contains all projective modules in ind Λ.*

We also point out that we have a trisection of ind Λ when Λ is almost hereditary, namely the three parts $\mathcal{L}\backslash\mathcal{R}$, $\mathcal{L}\cap\mathcal{R}$ and $\mathcal{R}\backslash\mathcal{L}$ which satisfy the following mapping properties: $Hom_\Lambda(\mathcal{R}\backslash\mathcal{L}, \mathcal{L}\cap\mathcal{R}) = Hom_\Lambda(\mathcal{R}\backslash\mathcal{L}, \mathcal{L}\backslash\mathcal{R}) = Hom_\Lambda(\mathcal{L}\cap\mathcal{R}, \mathcal{L}\setminus\mathcal{R}) = 0$ (compare [HRS1].) As a main open problem we state: Is $\mathcal{L}\cap\mathcal{R} \neq \emptyset$?

We will indicate now that an almost hereditary algebra is actually a quasitilted algebra. But before doing so we will recall from [HRS1] a quite

Derived Equivalences

general construction and refer to [HRS1] for some of its properties. This then will be applied to the situation above.

For the general construction let $\mathcal{A}$ be a locally-finite abelian category with a torsion pair $(\mathcal{X}, \mathcal{Y})$. We consider $\mathcal{B} = \Phi(\mathcal{A}; (\mathcal{X}, \mathcal{Y}))$ defined by

$$\mathcal{B} = \{X^\bullet \in D^b(\mathcal{A}) \mid H^0(X^\bullet) \in \mathcal{Y}, H^1(X^\bullet) \in \mathcal{X}, H^i(X^\bullet) = 0 \text{ for } i \neq 0, 1\},$$

where for a complex $X^\bullet = (X^i, d^i)_i$ the cohomology objects $H^i(X^\bullet)$ are defined by $H^i(X^\bullet) = \ker d^i / \operatorname{im} d^{i-1}$. Moreover we set $\mathcal{T} = \mathcal{Y} \subset \mathcal{B}$ and $\mathcal{F} = \mathcal{X}[-1] \subset \mathcal{B}$. It is shown in [HRS1] that $\mathcal{B}$ is the heart of a t-structure in the sense of [BBD], so $\mathcal{B}$ is an abelian category. In general $\mathcal{A}$ and $\mathcal{B}$ will not be derived equivalent. We refer to [HRS1] for conditions which ensure such a derived equivalence.

We now come back to our particular situation of almost hereditary algebras. For this let Λ be almost hereditary. Then $(\mathcal{X}, \mathcal{Y}) = (\operatorname{add}(\mathcal{R} \setminus \mathcal{L}), \operatorname{add} \mathcal{L})$ is a split torsion pair for $\mathcal{A} = \operatorname{mod} \Lambda$, where $\operatorname{add} \mathcal{L}$ is a generator, since ${}_\Lambda \Lambda \in \operatorname{add} \mathcal{L}$. Then it is rather easy to see that $\mathcal{H} = \Phi(\operatorname{mod} \Lambda; (\operatorname{add}(\mathcal{R} \setminus \mathcal{L}), \operatorname{add} \mathcal{L}))$ is a hereditary category and $T = {}_\Lambda\Lambda \in \operatorname{add} \mathcal{L} = \mathcal{T}$ is a tilting object in $\mathcal{H}$ with $\operatorname{End}_\mathcal{H} T = \Lambda$. In particular Λ is a quasitilted algebra.

3. Basic Examples and their Characterization

In this section we will recall the basic examples of hereditary abelian categories having a tilting object. The first two classes of examples are well known. We will include characterizations of these classes of examples. The first one follows from the torsion theoretic characterization of tilted algebras (compare [HRS1] and [HRe1]) and the second one is due to Lenzing [Le]. We will also include some further results on hereditary categories with tilting object.

First we consider the case of finite dimensional hereditary k-algebras over an algebraically closed field k. Let $H = k\overrightarrow{\Delta}$ be the path algebra of a finite quiver $\overrightarrow{\Delta}$ without oriented cycles. The underlying graph of $\overrightarrow{\Delta}$ is denoted by Δ. Note that all basic finite dimensional hereditary k-algebras are of this form. Then clearly $\operatorname{mod} H$ is a hereditary abelian category. It was shown in [HRS1] that in this situation the tilting objects are precisely the tilting modules in the sense of [HR].

It is well known that the representation type of $k\overrightarrow{\Delta}$ can be read off from Δ. For this we may assume that Δ is connected. Then $k\overrightarrow{\Delta}$ is representation-finite if and only if Δ is a Dynkin diagram, so of type $\mathbb{A}_n, \mathbb{D}_n, \mathbb{E}_6, \mathbb{E}_7$ or $\mathbb{E}_8$. Also $k\overrightarrow{\Delta}$ is representation-tame if and only if Δ is an affine diagram, so of type $\tilde{\mathbb{A}}_n, \tilde{\mathbb{D}}_n, \tilde{\mathbb{E}}_6, \tilde{\mathbb{E}}_7$ or $\tilde{\mathbb{E}}_8$. In all the other cases $k\overrightarrow{\Delta}$ is representation-wild.

The following two results, which were obtained jointly with I. Reiten, characterizes these module categories. For a proof we refer to [H3] and [HRe2]

Theorem. *Let $\mathcal{H}$ be a connected hereditary abelian category with tilting object T. If $\mathcal{H}$ contains a nonzero projective object P, then there exists a finite dimensional hereditary algebra H such that $\mathcal{H} \simeq \operatorname{mod} H$.*

Theorem. *Let $\mathcal{H}$ be a connected hereditary abelian category with tilting object T. If $\mathcal{H}$ contains a directing object, then there exists a finite dimensional hereditary algebra H such that $\mathcal{H}$ is derived equivalent to $\operatorname{mod} H$.*

The next class of examples deals with the categories of coherent sheaves over weighted projective curves in the sense of [GL1]. We recall the main ingredients of the definition but refer for details to the original work of [GL1].

Let $p = (p_1, \ldots, p_t)$ with $p_i > 0$ be a weight sequence and $\lambda = (1 = \lambda_3, \ldots, \lambda_t)$ with $\lambda_i \in k \setminus \{0\}$ be a set of distinct parameters. We always will assume that $t \geq 2$.

We consider the following k-algebra defined by generators and relations:

$$R = R(p, \lambda) = k[X_1, \ldots, X_t]/(X_i^{p_i} - X_2^{p_2} - \lambda_i X_1^{p_1}, 3 \leq i \leq t)$$

Let $\mathbb{X} = \mathbb{X}(p, \lambda)$ be the weighted projective curve associated to R (compare [GL1]). Then R is G-graded by the group $G = G(p, \lambda)$ defined by the quotient of the free abelian group with generators $\alpha_1, \ldots, \alpha_t$ satisfying the relations $p_1\alpha_1 = \cdots = p_t\alpha_t$. We set $\deg X_i = \alpha_i$.

The category of coherent sheaves $\operatorname{coh} \mathbb{X} = \operatorname{coh} \mathbb{X}(p, \lambda)$ is then defined as

$$\operatorname{coh} \mathbb{X} := \operatorname{mod}^G R / \operatorname{mod}^G_{fl} R,$$

where $\operatorname{mod}^G R$ is the category of finitely generated G-graded R-modules and $\operatorname{mod}^G_{fl} R$ is the full subcategory of G-graded R-modules of finite length.

Then $\operatorname{coh} \mathbb{X}$ is a hereditary abelian category and it follows from [GL1] that in this case the tilting objects are the tilting sheaves. It is also shown in [GL1] that there exists a tilting sheaf in $T \in \operatorname{coh} \mathbb{X}(p, \lambda)$ such that $\operatorname{End} T$ is a canonical algebra (compare [R2]). The canonical algebras $C(p, \lambda)$ are defined by the following quiver satisfying the relations

Derived Equivalences

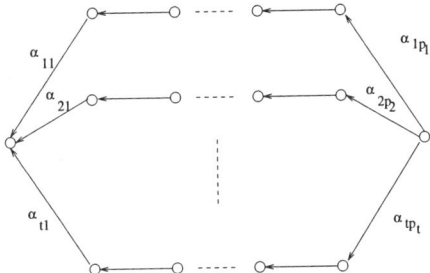

$$\alpha_{1p_1}\cdots\alpha_{11} - \alpha_{2p_2}\cdots\alpha_{21} - \lambda_i\alpha_{ip_i}\cdots\alpha_{i1} = 0 \text{ for } 3 \leq i \leq t.$$

Given the weight sequence $p = (p_1, \ldots, p_t)$ we define $\delta_p = (t-2) - \sum_{i=1}^{t} 1/p_i$. If $\delta_p < 0$, then it is well known that $\coh \mathbb{X}(p,\lambda)$ is derived equivalent to $\mod H$ for a finite dimensional hereditary k-algebra. So in order to distinguish the cases we will assume that $\delta_p \geq 0$; note that this implies that $t \geq 3$.

The category of coherent sheaves over a weighted projective curve admits the following categorical characterization [Le]. For this recall that an abelian category is called noetherian if each ascending chain of subobjects becomes stationary. In this case each object X has a maximal subobject.

Theorem. *Let $\mathcal{H}$ be a connected noetherian hereditary abelian category without nonzero projective objects containing a tilting object T. Then there is a weighted projective curve $\mathbb{X}$ such that $\mathcal{H} \simeq \coh\mathbb{X}$.*

Let us indicate one instance in the proof where the assumption on $\mathcal{H}$ being noetherian is used. For this let $\mathcal{H}_0 \subset \mathcal{H}$ be the full subcategory of objects of finite length. Since $\mathcal{H}$ is noetherian each $X \in \mathcal{H}$ has a simple quotient. So $\mathcal{H}_0 \neq \emptyset$. Since $\mathcal{H}$ has no nonzero projective object the Auslander-Reiten translation is an equivalence on $\mathcal{H}$. By a result of Gabriel [G] and using the fact that $\rk K_0(\mathcal{H}) < \infty$, we have that $\mathcal{H}_0 = \coprod_{i \in I} \mathcal{T}_i$, where $\mathcal{T}_i$ are stable tubes of rank n_i. Moreover, again using that $\rk K_0(\mathcal{H}) < \infty$, there are at most finitely many $\mathcal{T}_i$ where $n_i > 1$. Then one shows that $(\ind(\mathcal{H} \setminus \mathcal{H}_0), \ind \mathcal{H}_0)$ induces a split torsion pair on $\mathcal{H}$. The main step is then to identify a tilting object T with $\End_{\mathcal{H}} T = C(p,\lambda)$. This is done by induction on $\rk K_0(\mathcal{H})$ using perpendicular categories [GL2].

We note that $\coh \mathbb{X}$ trivially satisfies the assumptions of the theorem above.

We will now formulate some further results. The first one generalizes the previous result and is contained in [HRe3].

Theorem. *Let $\mathcal{H}$ be a connected hereditary abelian category without nonzero projective objects containing a tilting object T. If $\mathcal{H}$ contains a simple object, then there is a weighted projective curve $\mathbb{X}$ such that $\mathcal{H}$ is derived equivalent to $\operatorname{coh} \mathbb{X}$.*

We point out that there exist examples of hereditary abelian categories $\mathcal{H}$ containing a tilting object which neither contain simple objects nor directing objects. These examples are however derived equivalent to $\operatorname{coh} \mathbb{X}$ for a weighted projective curve $\mathbb{X} = \mathbb{X}(p, \lambda)$ of weight type $\delta_p = 0$. These are the so-called tubular types. For details we refer to [H3].

In the proof of the previous result the following step is important and seems to be also of independent interest. If E is an *exceptional* object in $\mathcal{H}$ (i.e., E is indecomposable and satisfies $\operatorname{Ext}^1(E, E) = 0$), then we denote by

$$E^\perp = \{X \in \mathcal{H} \mid \operatorname{Hom}(E, X) = \operatorname{Ext}^1(E, X) = 0\}$$

the perpendicular category of E. For properties of this category we refer to [GL2] and [HRe3].

Theorem. *Let $\mathcal{H}$ be a connected hereditary abelian category containing a tilting object T. If E is an exceptional torsion object of infinite length, then there is a finite dimensional hereditary algebra H such that $E^\perp \simeq \operatorname{mod} H$.*

This has the following direct consequence (compare also [H3] for further properties.) For this we need the definition of a one-point extension algebra. Let Σ be a finite dimensional algebra and let M be a Σ-module. The one-point extension $\Sigma[M]$ of Σ by M is then defined as the triangular matrix ring

$$\Sigma[M] = \begin{bmatrix} \Sigma & M \\ 0 & k \end{bmatrix}$$

with the obvious multiplication.

Corollary. *Any quasitilted algebra Λ is derived equivalent to a quasitilted algebra $\Gamma = H[M]$ which is the one-point algebra of a finite dimensional hereditary algebra H by an indecomposable H-module M.*

We refer to [HS] for further investigations of quasitilted one-point extension algebras.

The previous results on the structure of hereditary abelian categories containing a tilting object suggest the following conjecture.

Derived Equivalences

Conjecture. *Let $\mathcal{H}$ be a connected hereditary abelian category containing a tilting object T. Then there is either a weighted projective curve $\mathbb{X}$ such that $\mathcal{H}$ is derived equivalent to $\operatorname{coh}\mathbb{X}$ or a finite dimensional hereditary k-algebra H such that $\mathcal{H}$ is derived equivalent to $\operatorname{mod} H$.*

We point out that this conjecture has been verified in [Sk] in case there exists $T \in \mathcal{H}$ such that $\Lambda = \operatorname{End} T$ is of tame representation type.

In [H3] we determined all hereditary abelian categories which are derived equivalent to $\operatorname{coh}\mathbb{X}$ for a weighted projective curve $\mathbb{X}$ or to $\operatorname{mod} H$ for a finite-dimensional hereditary k-algebra H. This is done by classifying all split torsion pairs in these two cases.

4. Hochschild Cohomology

Let Λ be a finite dimensional algebra over an algebraically closed field k. We say that Λ is *piecewise hereditary* of type $\mathcal{H}$ if there exists a hereditary abelian category $\mathcal{H}$ such that $D^b(\Lambda)$ is triangle-equivalent to the bounded derived category $D^b(\mathcal{H})$ of complexes over $\mathcal{H}$. The type of Λ is of course only defined up to derived equivalence. This class of algebras has been previously studied in several articles (compare for example [H1], [HRS2], [HRe2] and the references in those). We will show here the computation of the Hochschild cohomology of piecewise hereditary algebras and will give some applications. For details we refer to [H4].

We denote by $H^i(\Lambda)$ the i-th *Hochschild cohomology space* (i.e., $H^i(\Lambda) = \operatorname{Ext}^i_{\Lambda^e}(\Lambda, \Lambda)$, where Λ^e denotes the enveloping algebra).

Theorem. *Let Λ be a connected piecewise hereditary algebra. Then it holds that $H^0(\Lambda) \simeq k$ and $H^i(\Lambda) = 0$ for $i \geq 3$.*

In the next two results we deal with the two main cases of hereditary categories mentioned above. We keep the notation introduced above. Moreover for an arrow α in $\overrightarrow{\Delta}$ we denote by $s(\alpha)$ the starting point and by $e(\alpha)$ the end point of α, where vertices are interpreted as idempotents in $k\overrightarrow{\Delta}$. Let α be an arrow in $\overrightarrow{\Delta}$, then let $\nu(\alpha) = \dim_k s(\alpha)(k\overrightarrow{\Delta})e(\alpha)$. Moreover let n be the number of vertices in $\overrightarrow{\Delta}$.

Theorem. *Let Λ be a connected piecewise hereditary algebra of type $\operatorname{mod} H$ for a basic connected finite dimensional hereditary k-algebra $H = k\overrightarrow{\Delta}$, where $\overrightarrow{\Delta}$ is a finite quiver without cycles. Then $H^0(\Lambda) = k$, $\dim_k H^1(\Lambda) = 1 - n + \sum_{\alpha \in \Delta_1} \nu(\alpha)$ and $H^i(\Lambda) = 0$ for $i \geq 2$.*

We recall that a finite-dimensional k-algebra Λ is *representation-directed* if every indecomposable Λ-module is directing or equivalently the Auslander-Reiten quiver of Λ does not contain an oriented cycle. Note that a representation-directed algebra is necessarily representation-finite, so admits up to isomorphism only finitely many indecomposable modules. There exist special representation-directed algebras. These are the simply connected algebras. For a definition we refer to [R1].

Corollary. *Let Λ be a connected piecewise hereditary algebra of type $\mathrm{mod}\, H$ for a basic connected finite dimensional hereditary k-algebra $H = k\vec{\Delta}$, where $\vec{\Delta}$ is a finite quiver without cycles. If Λ is representation-finite, then Λ is simply connected if and only if the underlying graph Δ of $\vec{\Delta}$ is a tree.*

Theorem. *Let Λ be piecewise hereditary of type $\mathrm{coh}\,\mathbb{X}(p,\lambda)$ with $d_p \geq 0$. Then $H^0(\Lambda) \simeq k$, $H^1(\Lambda) = 0$ and $\dim_k H^2(\Lambda) = t - 3$.*

The next corollary was previously shown in [M] by using completely different methods such as the transitivity of the braid group action on the set of complete exceptional sequences. We give the short proof from [H4].

Corollary. *Let Λ be piecewise hereditary of type $\mathrm{coh}\,\mathbb{X}(p,\lambda)$ with $d_p \geq 0$. If $t \geq 4$, then Λ is representation-infinite.*

Proof. If $t \geq 4$ we infer by the last theorem that $H^2(\Lambda) \neq 0$. If Λ is representation-finite, then Λ is representation-directed by [RS2]. But then it was shown in [H5] that $H^i(\Lambda) = 0$ for $i \geq 2$, a contradiction. $\square$

Note that there are examples of representation-finite piecewise hereditary algebras of type $\mathrm{coh}\,\mathbb{X}(p,\lambda)$ with $d_p \geq 0$. However it is an open question for which weight sequences there actually exists a representation-finite piecewise hereditary algebra of the corresponding type. The corresponding problem in the case that the piecewise hereditary algebra is of type $\mathrm{mod}\, H$ for a basic connected finite dimensional hereditary k-algebra H is also unsolved.

It is also an open question if there exists a connected piecewise hereditary algebra Λ with $H^1(\Lambda) \neq 0 \neq H^2(\Lambda)$. Note that there are no such examples if the conjecture formulated at the end of the last section holds.

Next we will show how our previous results can be used to obtain alternative proofs of assertions about the quasi length of exceptional vector bundles in $\mathrm{coh}\,\mathbb{X}$ (compare [LP1] and [LP2]). For this we have to recall some terminology. We will need the structure of the Auslander-Reiten quiver of coh $\mathbb{X}$. For this we will need some particular translation quivers. Let $\vec{\Sigma}$ be an arbitrary quiver with vertex set Σ_0. We denote by $\mathbb{Z}$ the set of integers. We then can define the translation quiver $\mathbb{Z}\vec{\Sigma}$ as follows: The vertex set is $\mathbb{Z} \times \Sigma_0$. There are an arrow $(n, x) \to (n, y)$ and an arrow $(n, y) \to (n+1, x)$ for each arrow $x \to y$ in $\vec{\Sigma}$. The translation τ is defined by $\tau(n, x) = (n-1, x)$. We denote by $\mathbb{A}_\infty$ the infinite quiver $\circ \to \circ \to \circ \to \circ \to \dots$. Let $r > 0$, then a translation quiver of the form $\mathcal{T}_r = \mathbb{Z}\mathbb{A}_\infty/\langle \tau^r \rangle$ is called a stable tube of rank r. Let I be some set. A translation quiver $\coprod_{i \in I} \mathcal{T}_i$ is called a tubular family parametrized by I, if each $\mathcal{T}_i$ is a stable tube of rank n_i. The tubular type is the function $\rho : I \to \mathbb{N}$ defined by $\rho(i) = n_i$ for $i \in I$. In most situations all but finitely many values of ρ will be 1. In this case we denote by $(n_1, \dots, n_t)$ the set of exceptional values, meaning that there exist precisely t stable tubes of the indicated rank $n_i \geq 1$ for $1 \leq i \leq t$ and all other tubes are of rank one. Let $\mathbb{X}(p, \lambda)$ be a weighted projective curve. We are interested in the case $\delta_p > 0$. Then the corresponding sheaf category is representation-wild. The Auslander-Reiten quiver consists of components of type $\mathbb{Z}\mathbb{A}_\infty$ and a tubular family parametrized by the projective line $\mathcal{P}^1(k)$ over k of tubular type p. The full subcategory of indecomposable objects which are not of finite length is denoted by $\mathcal{V}$. They form the components of type $\mathbb{Z}\mathbb{A}_\infty$. For further properties we refer to [GL1], [LP1] and [LP2]. Let $X \in \mathcal{V}$, so X lies in a component $\mathbb{Z}\mathbb{A}_\infty$. We define the *quasi length* $\mathrm{ql}(X)$ of the length of a sectional path from X to Y, where Y has an indecomposable middle term in the Auslander-Reiten sequence ending at Y. An indecomposable $X \in \mathcal{V}$ with $\mathrm{ql}(X) = 1$ is called *quasi-simple*. If $X \in \mathcal{V}$ is exceptional it is well known that $\mathrm{ql}(X) \leq n - 2$, if $n = \mathrm{rk}\, K_0(\mathbb{X})$. The following theorem was shown with an alternative proof in [LP1].

Theorem. *Let $\mathbb{X}(p, \lambda)$ be a weighted projective curve with $\delta_p > 0$ and $t \geq 4$. If $E \in \mathcal{V}$ is exceptional, then E is quasi-simple.*

Proof. Let $E \in \mathcal{V}$ be exceptional with $\mathrm{ql}(E) = r$. If $r > 1$, let F be exceptional quasi-simple in $\mathcal{V}$ with $E \to \cdots \to F$ a chain of irreducible surjective maps. Then $\mathrm{ql}(F) = 1$. Let

$$0 \to \tau F \to M \to F \to 0$$

be the Auslander-Reiten sequence ending at F. Then M is indecomposable and it is easy to see that $\mathrm{Ext}^1(M, M) = 0$. Now let $F^\perp \simeq \mathrm{mod}\, H$ for a finite dimensional hereditary k-algebra H. From [H5] we have an exact sequence

$$0 \to H^1(C(p, \lambda)) \to H^1(H) \to \mathrm{Ext}^1_H(M, M) \to H^2(C(p, \lambda)) \to 0.$$

Since $\mathrm{Ext}^1(M,M) = 0$ we infer that $H^2(C(p,\lambda)) = 0$, hence $t = 3$, which gives the required contradiction. So E is quasi-simple. □

Note that this result is just an example of how to use Hochschild cohomology for estimates of this kind. For further applications see [H6].

References

[ARS]   M. Auslander, I. Reiten and S. Smaløm, *Representation theory of artin algebras,* Cambridge Univ. Press, Cambridge, 1995.

[B]   A. A. Beilinson, *Coherent sheaves on $\mathbb{P}^n$ and problems of linear algebra,* Func. Anal. and Appl. **12** (1978), 214–216.

[BBD]   A. A. Beilinson, J. Bernstein and P. Deligne, *Faisceaux pervers,* Astérique no **100** (1982).

[BGP]   I. N. Bernstein, I. M. Gelfand and V. A. Ponomarev, *Coxeter functors and Gabriel's theorem,* Russian Math. Surveys **28** (1973), 17–32.

[Bon]   A. Bondal, *Representations of associative algebras and coherents sheaves,* Math. USSR Izv. **34** (1990), 23–42.

[G]   P. Gabriel, *Indecomposable representations II,* Symp. Math. Ist. Naz. **11** (1973), 81–104.

[Gr]   P. P. Grivel, *Catégories derivées et foncteurs derives,* in Algebraic D-modules, Perspectives in Mathematics **2**, Academic Press, New York, 1987.

[GL1]   W. Geigle and H. Lenzing, *A class of weighted projective curves arising in representation theory of finite dimensional algebras,* In: Singularities, representations of algebras, and vector bundles, Lecture Notes in Math. **1273**, Springer-Verlag, Heidelberg, New York (1987), 265–297.

[GL2]   _____, *Perpendicular categories with applications to representations and sheaves,* J. Algebra **144** (1991), 273–343.

[H1]   D. Happel, *Triangulated categories in the representation theory of finite dimensional algebras,* London Math. Soc. Lecture Note Series **119**, Cambridge, 1988.

[H2]   _____, *On the derived category of a finite dimensional algebra,* Comment. Math. Helvet. **62** (1987), 339–389.

[H3]   ——, *Quasitilted algebras,* Algebras and Modules I, CMS Conference Proceedings **23** (1996), 55–82.

[H4]   ——, *Hochschild cohomology of piecewise hereditary algebras,* Coll. Math. **78** (1998), 261–266.

[H5]   ——, *Hochschild cohomology of finite dimensional algebras,* Sémi- naire d'Algèbre Paul Dubreil et Marie-Paul Malliavin, Lecture Notes in Math. **1404**, Springer-Verlag, Heidelberg, New York (1989), 108–126.

[H6]   ——, *Applications of Hochschild cohomology to hereditary categories with tilting object,* in preparation.

[Har]   R. Hartshorne, *Residues and duality,* Lecture Notes in Math. **20**, Springer-Verlag, Heidelberg, New York, 1966.

[HRe1]   D. Happel and I. Reiten, *An introduction to quasitilted algebras* An. St. Univ. Ovidius Constantza, **4** (1996), 137–149.

[HRe2]   ——, *Directing objects in hereditary categories,* Proceedings Seattle Conference 1997, Cont. Math. **229** (1998), 169–180.

[HRe3]   ——, *On hereditary categories with tilting objects,* Math. Z., to appear.

[HR]   D. Happel and C. M. Ringel, *Tilted algebras,* Trans. Amer. Math. Soc. **274** (1982), 399–443.

[HS]   D. Happel and I. H. Slungård, *One-point extensions of hereditary algebras,* Proceedings ICRA VIII, CMS Conference Proceedings **24** (1996), 285–292.

[HRS1]   D. Happel, I. Reiten and S. O. Smalø, *Tilting in Abelian Categories and Quasitilted Algebras,* Memoirs **575**, Amer. Math. Soc., 1996.

[HRS2]   ——, *Piecewise hereditary algebras,* Arch. Math. **66** (1996), 182–186.

[K]   St. König, *Auslander-Reiten sequences and tilting theory,* An. St. Univ. Ovidius Constantza, **4** (1996), 136–169.

[Le]   H. Lenzing, *Hereditary noetherian categories with a tilting complex,* Proc. Amer. Math. Soc. **125** (1997), 1893–1901.

[LP1] H. Lenzing and J. A. de la Peña, *Wild canonical algebras*, Math. Z. **224** (1997), 403–425.

[LP2] _____, *Concealed canonical algebras and separating tubular family*, preprint.

[M] H. Meltzer, *Exceptional vector bundles, tilting sheaves and tilting complexes on weighted projective lines*, preprint.

[Ri] J. Rickard, *Derived equivalences as derived functors*, J. London Math. Soc. **43** (1991), 37–48.

[R1] C. M. Ringel, *Tame algebras and integral quadratic forms*, Lecture Notes in Math. **1099**, Springer-Verlag, Heidelberg, New York, 1984.

[R2] _____, *The canonical algebras*, In: Topics in Algebra, Banach Center Publications, **26** PWN, Warsaw 1990, 407–432.

[Sk] A. Skowroński, *Tame quasitilted algebras*, J. Agebra **203** (1998), 470–490.

[V] J. L. Verdier, *Catégories dérivées, état 0*, In: SGA 4 1/2, Lecture Notes in Math. **569**, Springer-Verlag, Heidelberg, New York (1977), 262–311.

Fakultät für Mathematik
Technische Universität Chemnitz
D-09107 Chemnitz
Germany
e-mail: `happel@mathematik.tu-chemnitz.de`

CS-Property of Direct Sums of Uniform Modules

J. Kado, Y. Kuratomi and K. Oshiro

Abstract

Let P be an R-module with a decomposition $P = M_1 \oplus \cdots \oplus M_n$, where each M_i is a CS-module. In [6], a necessary and sufficient condition for P to be a CS-module for the decomposition is given, by introducing generalized relative injectivity.

In this paper, we study the problem when a direct sum of uniform modules is CS for the given direct sum. In the case when each uniform module has a local endomorphism ring, this problem has been studied by several authors, e.g., Baba-Harada [2], Dung [3], Harada-Oshiro [5].

1. Preliminaries

Throughout this paper R is a ring with identity and modules are unitary right R-modules.

Let M be a module and N a submodule of M. We denote by $N \subseteq_e M$ (resp. $N <_\oplus M$) when N is an essential submodule (resp. direct summand) of M. For an element $m \in M$, $(0:m)$ denotes the annihilator right ideal of m.

An R-module M is said to be a *CS-module* if for any submodule A of M, there exists a direct summand A^* of M with $A \subseteq_e A^*$.

Let $\{M_i \mid i \in I\}$ be a family of modules and let $M = \oplus_I M_i$. M is said to be a *CS-module for* $M = \oplus_I M_i$, if for any submodule X of M, there exist $X^* \subseteq M$ and $\overline{M_i} <_\oplus M_i$ ($i \in I$) such that $X \subseteq_e X^*$ and $M = X^* \oplus (\oplus_I \overline{M_i})$, that is, M is CS and satisfies the internal exchange property in the direct sum $M = \oplus_I M_i$.

We note that, in general, a CS-module M is not a CS-module for a given decomposition of M. For example, a torsion free abelian group $G = \mathbb{Z} \oplus \cdots \oplus \mathbb{Z}$ (n-times, $n \geq 2$) is CS but not CS for $G = \mathbb{Z} \oplus \cdots \oplus \mathbb{Z}$, since G does not satisfy the internal exchange property in $G = \mathbb{Z} \oplus \cdots \oplus \mathbb{Z}$.

Let $M = M_1 \oplus M_2$ and let $\varphi : M_1 \to M_2$ be a homomorphism. Then $< M_1 \xrightarrow{\varphi} M_2 > = \{m_1 + \varphi(m_1) \mid m_1 \in M_1\}$ is a direct summand of

$M = M_1 \oplus M_2$; $M = < M_1 \xrightarrow{\varphi} M_2 > \oplus M_2$. Let $f : N \to M$ be a monomorphism. f is called a proper monomorphism if f is not an epimorphism. If there exists a proper monomorphism from N to M, we write $N \prec M$. If there does not exist a proper monomorphism from N to M, we write $N \not\prec M$.

Let N be a module and let $\{M_\alpha \mid \alpha \in \Lambda\}$ be a family of modules. For $\alpha_i \in \Lambda$, we write M_i as M_{α_i} for convenience.

We consider the following two chain conditions :

(A_2) For every choice of $n \in N$ and $m_i \in M_i$ for distinct $\alpha_i \in \Lambda$ ($i \in \mathbb{N}$) such that $(0 : m_i) \supseteq (0 : n)$, the ascending sequence $\bigcap_{i \geq k}(0 : m_i)$ ($k \in \mathbb{N}$) becomes stationary.

(A_2') For every choice of $n \in N$ and $m_i \in M_i$ for distinct $\alpha_i \in \Lambda$ ($i \in \mathbb{N}$) such that $(0 : m_i) \supseteq (0 : n)$ and $\bigcap_{i=1}^{\infty} Ker\varphi_i \subseteq_e nR$ for the canonical homomorphism $\varphi_i : nR \to m_i R$, the ascending sequence $\bigcap_{i \geq k}(0 : m_i)$ ($k \in \mathbb{N}$) becomes stationary.

The following result is well known (cf. [7, Theorem 1.7]).

Proposition 1.1. *The following conditions are equivalent*:
(1) $\oplus_\Lambda M_\alpha$ is N-injective.
(2) $\oplus_I M_i$ is N-injective for every countable subset $I \subseteq \Lambda$.
(3) M_α is N-injective for every $\alpha \in \Lambda$ and (A_2) holds.

By a quite similar proof, we can show the following.

Proposition 1.2. *The following conditions are equivalent*:
(1) $\oplus_\Lambda M_\alpha$ is essentially N-injective.
(2) $\oplus_I M_i$ is essentially N-injective for every countable subset $I \subseteq \Lambda$.
(3) M_α is essentially N-injective for every $\alpha \in \Lambda$ and (A_2') holds.

Definition 1.3. Let M and N be modules. M is said to be generalized N-injective, if for any submodule X of N and any homomorphism $\varphi : X \to M$, there exist decompositions $N = \overline{N} \oplus \overline{\overline{N}}$, $M = \overline{M} \oplus \overline{\overline{M}}$, a homomorphism $\overline{\varphi} : \overline{N} \to \overline{M}$ and a monomorphism $\psi : \overline{\overline{M}} \to \overline{\overline{N}}$ satisfying the following properties (*) and (**).

(*) $X \subseteq \overline{N} \oplus \psi(\overline{\overline{M}})$.

(**) For $x \in X$, we express x in $N = \overline{N} \oplus \overline{\overline{N}}$ as $x = \overline{x} + \overline{\overline{x}}$ where $\overline{x} \in \overline{N}$, $\overline{\overline{x}} \in \overline{\overline{N}}$. Then $\varphi(x) = \overline{\varphi}(\overline{x}) + \overline{\overline{\varphi}}(\overline{\overline{x}})$ where $\overline{\overline{\varphi}} = \psi^{-1}$.

Proposition 1.4. (1) *If M is N-injective, then clearly M is generalized N-injective.*

(2) [6] If M is generalized N-injective, then M is essentially N-injective.

Remark 1.5. N-injective $\Rightarrow$ generalized N-injective $\Rightarrow$ essentially N-injective.

The following extending lemma is due to Oshiro-Rizvi [8, Lemma 2.1].

Lemma 1.6. Let $\{M_\alpha \mid \alpha \in \Lambda\}$ be a family of CS-modules and let $P = \oplus_\Lambda M_\alpha$. We consider the index set Λ as a well ordered set: $\Lambda = \{0, 1, \cdots, \omega, \omega+1, \cdots\}$, and let X be a submodule of P. Then there exist submodules $T(\alpha) \subseteq_e T(\alpha)^* <_\oplus M_\alpha$, decompositions $M_\alpha = T(\alpha)^* \oplus N_\alpha$ and a submodule $\oplus_\Lambda X(\alpha) \subseteq_e X$ for which the following properties hold:
  (1) $X(0) = T(0) \subseteq_e T(0)^*$.
  (2) $X(k) \subseteq T(k) \oplus (\oplus_{i<k} N_i)$ for all $k \in \Lambda$.
  (3) $\sigma(X(k)) = T(k) \subseteq_e T(k)^*$, $X(k) \simeq \sigma(X(k))$ (by $\sigma|_{X(k)}$) for all $k \in \Lambda$, where σ is a projection $P = (\oplus_\Lambda T(\alpha)^*) \oplus (\oplus_\Lambda N_\alpha) \to \oplus_\Lambda T(\alpha)^*$.
  (4) $X \simeq \sigma(X)$ (by $\sigma|_{X(k)}$).

Let M be a module. Two decompositions $M = \oplus_A M_\alpha = \oplus_B N_\beta$ of M are said to be *equivalent* in case there is a bijection $\sigma : A \to B$ such that $M_\alpha \simeq N_{\sigma(\alpha)}$ ($\alpha \in A$).

A decomposition $M = \oplus_\Lambda M_\alpha$ is said to *complement direct summands* (*complement maximal direct summands*) in case for every (maximal) direct summand K of M there is a subset $\Lambda' \subseteq \Lambda$ with $M = (\oplus_{\Lambda'} M_\beta) \oplus K$.

Of course, a decomposition that complements direct summands complements maximal direct summands. We recall the following (cf. [1, 12.4]).

Proposition 1.7. Let $M = \oplus_A M_\alpha$ be an indecomposable decomposition. If the decomposition $M = \oplus_A M_\alpha$ complements maximal direct summands, then all indecomposable decompositions of M are equivalent.

Proposition 1.8. Let M be an R-module with an indecomposable decomposition $M = \oplus_\Lambda M_\alpha$. Then the following conditions are equivalent:
  (1) M is CS for $M = \oplus_\Lambda M_\alpha$.
  (2) M is a CS-module and $M = \oplus_\Lambda M_\alpha$ complements direct summands.

Proof. Evident. □

2. Main Theorem

We need the following proposition which will be useful for our results.

Proposition 2.1. Let $\{M_\alpha \mid \alpha \in \Lambda\}$ be a family of modules and let $P = \oplus_\Lambda M_\alpha$. If P is CS for $P = \oplus_\Lambda M_\alpha$, then $P' = \oplus_\Lambda \overline{M_\alpha}$ is CS for $P' = \oplus_\Lambda \overline{M_\alpha}$ for any decomposition $M_\alpha = \overline{M_\alpha} \oplus \overline{\overline{M_\alpha}}$ $(\alpha \in \Lambda)$.

Proof. Since P' is a direct summand of P, P' is CS. Let $X <_\oplus P'$. For $X \oplus (\oplus_\Lambda \overline{\overline{M_\alpha}}) <_\oplus P$, there exists a direct summand M'_α of M_α $(\alpha \in \Lambda)$ such that $P = X \oplus (\oplus_\Lambda \overline{\overline{M_\alpha}}) \oplus (\oplus_\Lambda M'_\alpha)$ by assumption. Let $m'_\alpha \in M'_\alpha$ and express m'_α in $M_\alpha = \overline{M_\alpha} \oplus \overline{\overline{M_\alpha}}$ as $m'_\alpha = \overline{m_\alpha} + \overline{\overline{m_\alpha}}$ $(\overline{m_\alpha} \in \overline{M_\alpha}, \overline{\overline{m_\alpha}} \in \overline{\overline{M_\alpha}})$.

Since $\overline{\overline{M_\alpha}} \cap M'_\alpha = 0$, $\overline{m_\alpha} = 0$ implies $\overline{\overline{m_\alpha}} = 0$. So there exists a homomorphism $f_\alpha : \overline{M'_\alpha} \to \overline{\overline{M_\alpha}}$ such that $M'_\alpha = <\overline{M'_\alpha} \xrightarrow{f_\alpha} \overline{\overline{M_\alpha}} >$. Thus $P = X \oplus (\oplus_\Lambda M'_\alpha) \oplus (\oplus_\Lambda \overline{\overline{M_\alpha}}) = X \oplus (\oplus_\Lambda < \overline{M'_\alpha} \xrightarrow{f_\alpha} \overline{\overline{M_\alpha}} >) \oplus (\oplus_\Lambda \overline{\overline{M_\alpha}}) = X \oplus (\oplus_\Lambda \overline{M'_\alpha}) \oplus (\oplus_\Lambda \overline{\overline{M_\alpha}})$. Then $P' = X \oplus (\oplus_\Lambda \overline{M'_\alpha})$. □

Theorem 2.2. [6] *Let $M_1, \cdots, M_n$ be CS-modules and put $P = M_1 \oplus \cdots \oplus M_n$. Then:*

(1) *When $n = 2$, P is CS for $P = M_1 \oplus M_2$ if and only if M_i is a generalized M_j-injective for $i \neq j$.*

(2) *When $n \geq 3$, P is CS for $P = M_1 \oplus \cdots \oplus M_n$ if and only if M_i is a generalized $\oplus_{j \neq i} M_j$-injective for any distinct i, and, if and only if $\oplus_{j \neq k} M_j$ is a generalized M_k-injective for any k.*

(3) *When each M_i is a uniform module, P is CS for $P = M_1 \oplus \cdots \oplus M_n$ if and only if M_i is a generalized M_j-injective for $i \neq j$.*

The main purpose of this paper is to show the following.

Theorem 2.3. *Let $\{M_\alpha \mid \alpha \in \Lambda\}$ be a family of uniform modules and let $P = \oplus_\Lambda M_\alpha$. Then the following conditions are equivalent:*

(1) *P is CS for $P = \oplus_\Lambda M_\alpha$.*

(2) (a) *M_α is generalized M_β-injective for all $\alpha \neq \beta \in \Lambda$.*

(b) *(A'_2) holds for all M_α and $\{M_\beta \mid \beta \neq \alpha, \beta \in \Lambda\}$.*

(c) *There does not exist an infinite sequence of proper monomorphisms $\{f_k : M_{i_k} \to M_{i_{k+1}} \mid k \in \mathbb{N}\}$ with all $i_k \in \Lambda$ distinct.*

Proof. (1)⇒(2). (a) holds by Theorem 2.2.

(b) By Theorem 2.2 and Proposition 1.4, $\oplus_{\beta \neq \alpha} M_\beta$ is essentially M_α-injective. Thus (A'_2) holds by Proposition 1.2.

(c) Assume that there exists an infinite sequence of proper monomorphisms $\{f_k : M_{i_k} \to M_{i_{k+1}} \mid k \in \mathbb{N}\}$. Now we put $P' = \oplus_{k=1}^\infty M_{i_k}$ and $X = \oplus_{k=1}^\infty < M_{i_k} \xrightarrow{f_k} M_{i_{k+1}} >$.

Since P' is CS for $P' = \oplus_{k=1}^\infty M_{i_k}$ by Proposition 2.1, there exist $X^* (_e \supseteq X)$ and $\overline{M_{i_k}} <_\oplus M_{i_k}$ $(k \in \mathbb{N})$ such that $P' = X^* \oplus (\oplus_{k=1}^\infty \overline{M_{i_k}})$. Assume

$0 \neq \overline{M_{i_l}} (= M_{i_l})$ for some $l \in \mathbb{N}$. Since $\overline{M_{i_{l+1}}}(= M_{i_{l+1}})$ is a uniform module, we see
$$< M_{i_l} \xrightarrow{f_l} M_{i_{l+1}} > \oplus M_{i_l} \subseteq_e M_{i_{l+1}} \oplus M_{i_l}.$$

However, we have $< M_{i_l} \xrightarrow{f_l} M_{i_{l+1}} > \oplus M_{i_l} <_\oplus P'$ because of $< M_{i_l} \xrightarrow{f_l} M_{i_{l+1}} > <_\oplus X^*$ and $M_{i_l} <_\oplus (\oplus_{k=1}^\infty \overline{M_{i_k}})$. Hence we have $< M_{i_l} \xrightarrow{f_l} M_{i_{l+1}} > \oplus M_{i_l} = M_{i_{l+1}} \oplus M_{i_l}$. This implies that f_l is an epimorphism, a contradiction. Thus $P' = X^*$.

In particular, $M_{i_1} \cap X \subseteq_e M_{i_1}$. Pick $0 \neq x_1 \in M_{i_1} \cap X$. Then there exist $n \in \mathbb{N}$ and $m_k \in M_{i_k}$ ($k = 1, \cdots, n$) such that $x_1 = m_1 + f_1(m_1) + m_2 + f_2(m_2) + \cdots + m_n + f_n(m_n)$. Since each f_k is a monomorphism, it follows that $x_1 = m_1 = 0$, a contradiction.

(2)$\Rightarrow$(1). Put $\Gamma = \{\lambda \in \Lambda \mid M_\lambda \not\prec M_\lambda\}$ and $\Omega = \Lambda - \Gamma = \{\lambda \in \Lambda \mid M_\lambda \prec M_\lambda\}$. For $\alpha, \beta \in \Omega$, we define $\alpha \sim \beta$ when $M_\alpha \prec M_\beta$ and $M_\beta \prec M_\alpha$. Clearly, $\sim$ is an equivalence relation. Put $\Omega_k = \{\alpha \in \Omega \mid \alpha \sim k\}$. By (c), we see that Ω_k is a finite subset of Ω, and, by Theorem 2.2, $Q_k = \oplus_{\Omega_k} M_\alpha$ is CS for $Q_k = \oplus_{\Omega_k} M_\alpha$. Let Δ be a representative subset of Ω for $\sim$, that is, $\Omega = \cup_{\alpha \in \Delta} \Omega_\alpha$ and $\Omega_\alpha \cap \Omega_\beta = \emptyset$ for $\alpha \neq \beta$. Put $I = \Gamma \cup \Delta$ and define V_λ for $\lambda \in I$ as

$$V_\lambda = \begin{cases} M_\lambda & \text{if } \lambda \in \Gamma \\ Q_\lambda = \oplus_{\Omega_\lambda} M_\alpha & \text{if } \lambda \in \Delta \end{cases}.$$

Then we can easily see that for any direct summands $V'_\alpha <_\oplus V_\alpha$ and $V'_\beta <_\oplus V_\beta$ ($\alpha \neq \beta \in I$), $V'_\alpha \prec V_\beta$ and $V'_\beta \prec V_\alpha$ do not occur. Now, by (c), we can take $k \in I$ such that $V'_k \not\prec V_t$ for all $t \in I$ ($t \neq k$) and any $0 \neq V'_k <_\oplus V_k$. Here we transfinitely define $I_1, I_2, \ldots, I_\omega, \ldots$ as follows: $I_1 = \{k \in I \mid V'_k \not\prec V_t \text{ for all } t \in I (t \neq k) \text{ and all } V'_k <_\oplus V_k\}$, $I_2 = \{k \in I - I_1 \mid V'_k \not\prec V_t \text{ for all } t \in I - I_1 (t \neq k) \text{ and all } V'_k <_\oplus V_k\}, \ldots, I_\omega = \{k \in I - \cup_{i<\omega} I_i \mid V'_k \not\prec V_t \text{ for all } t \in I - \cup_{i<\omega} I_i (t \neq k) \text{ and all } V'_k <_\oplus V_k\}, \ldots$, and so on. We have a partition of I; $I = I_1 \cup I_2 \cup \cdots \cup I_\omega \cup I_{\omega+1} \cup \cdots \cup I_\alpha \cup \cdots$ where $\{1, 2, \ldots, \omega, \omega+1, \ldots, \alpha, \ldots\}$ is a well ordered set. Then we see that $i < j$ ($i, j \in I$) implies $V'_i \not\prec V_j$ for any $V'_i <_\oplus V_i$.

Now, by transfinite induction on $\alpha \in I$, we shall prove that M is CS for $M = \oplus_I V_i$.

Let $\alpha \in I$ and put $J = \{i \in I \mid i < \alpha\}$. Assume that $S = \oplus_J V_i$ is CS for $S = \oplus_J V_i$. Let X be a submodule of $S^* = \oplus_{J \cup \{\alpha\}} V_i$. By Lemma 1.6, there exists a submodule $\oplus_{J \cup \{\alpha\}} X_i \subseteq_e X$ for which the following properties hold:

(1) $X_k \subseteq T(k)^* \oplus (\oplus_{i<k} N_i)$ for all $k \in J \cup \{\alpha\}$;
(2) $\sigma(X_k) = T(k) \subseteq_e T(k)^*$, $X_k \simeq \sigma(X_k)$ (by $\sigma|_{X_k}$) for all $k \in J \cup \{\alpha\}$, where σ is the projection : $S^* \to \oplus_{J \cup \{\alpha\}} T(i)^*$.

In the case of $X_\alpha = 0$, by assumption, there exist $Y^*{}_e \supseteq \oplus_J X_i$ and a direct summand $\overline{V_i}$ of V_i such that $S = Y^* \oplus (\oplus_J \overline{V_i})$. Put $\overline{V_\alpha} = V_\alpha$. Then we have
$$S^* = S \oplus V_\alpha = Y^* \oplus (\oplus_{J \cup \{\alpha\}} \overline{V_i}).$$
Let τ_1 and τ_2 be the projections: $S^* \to Y^*$, $S^* \to \oplus_{J \cup \{\alpha\}} \overline{V_i}$ respectively. The natural map
$$\phi : \tau_1(X) \to \tau_2(X) \text{ via } \phi : \tau_1(x) \mapsto \tau_2(x)$$
is a homomorphism with $\mathrm{Ker}\phi \subseteq_e \tau_1(X)$. By (b), there exists a homomorphism $\phi^* : Y^* \to \oplus_{J \cup \{\alpha\}} \overline{V_i}$ with $\phi^*|_{\tau_1(X)} = \phi$. Thus we obtain
$$X = <\tau_1(X) \xrightarrow{\phi} \tau_2(X)> \subseteq_e <Y^* \xrightarrow{\phi^*} \oplus_{J \cup \{\alpha\}} \overline{V_i}> <_\oplus S^*$$
as desired.

In the case of $X_\alpha \neq 0$, let π be the projection : $S^* \to \oplus_{J \cup \{\alpha\}} N_i$. Since $X_\alpha \simeq \sigma(X_\alpha)$, the natural map $f : \sigma(X_\alpha) \to \pi(X_\alpha)$ via $\sigma(x_\alpha) \to \pi(x_\alpha)$ is a homomorphism.

Let $\alpha \in \Gamma$. Then $V_\alpha = M_\alpha = T(\alpha)^*$ is a uniform module. If the homomorphism f above is not a monomorphism, as $T(\alpha)^*$ is a uniform module, $\mathrm{Ker} f \subseteq_e \sigma(X_\alpha)$. Thus there exists a homomorphism $f^* : T(\alpha)^* \to \oplus_{I \cup \{\alpha\}} N_k$ with $f^*|_{\sigma(X_\alpha)} = f$. Thus we see
$$X_\alpha = <\sigma(X_\alpha) \xrightarrow{f} \pi(X_\alpha)> \subseteq_e <T(\alpha)^* \xrightarrow{f^*} \oplus_{I \cup \{\alpha\}} N_k>.$$

On the other hand, if f above is a monomorphism, then $\pi(X_\alpha)$ is a uniform module. By assumption and Proposition 2.1, there exist $Y(\supseteq_e \pi(X_\alpha))$ and a direct summand $\overline{N_k}$ of N_k such that $\oplus_{k<\alpha} N_k = Y \oplus (\oplus_{k<\alpha} \overline{N_k})$. Since Y is a uniform module, there exists $k(<\alpha)$ such that $Y \simeq N'_k$ for a direct summand N'_k of N_k. Then N'_k is generalized $T(\alpha)^*$-injective. So there exist decompositions $T(\alpha)^* = \overline{T(\alpha)^*} \oplus \overline{\overline{T(\alpha)^*}}$ and $Y = \overline{Y} \oplus \overline{\overline{Y}}$, and there exist a homomorphism $\overline{f} : \overline{T(\alpha)^*} \to \overline{Y}$ and a monomorphism $g : \overline{\overline{Y}} \to \overline{\overline{T(\alpha)^*}}$. If $\overline{\overline{T(\alpha)^*}} = 0$, then $X_\alpha \subseteq_e <T(\alpha)^* \xrightarrow{\overline{f}} \oplus_{k<\alpha} N_k>$. In the case of $\overline{T(\alpha)^*} = 0$, the homomorphism g is an isomorphism since $N'_k \not< V_\alpha$. So we see $X_\alpha \subseteq_e <T(\alpha)^* \xrightarrow{g^{-1}} \oplus_{k<\alpha} N_k>$. Therefore we have a decomposition $M = Y \oplus <T(\alpha)^* \to \oplus_{k<\alpha} N_k> \oplus (\oplus_{k<\alpha} \overline{V_k})$ such that $\oplus_{J \cup \{\alpha\}} X_i \subseteq_e Y \oplus <T(\alpha)^* \to \oplus_{k<\alpha} N_k>$. By the same argument as in the case $X_\alpha = 0$, we see
$$X \subseteq_e <Y \oplus <T(\alpha)^* \to \oplus_{k<\alpha} N_k> \to \oplus_{k<\alpha} \overline{V_k}> <_\oplus S^*.$$

If $\alpha \in \Delta$, then there exists $n \in \mathbb{N}$ such that $V_\alpha = M_{\alpha_1} \oplus \cdots \oplus M_{\alpha_n}$. By the proof of Lemma 1.6 (see [8]), $N_\alpha = M_{t_1} \oplus \cdots \oplus M_{t_k}$ (where $t_i \in \{\alpha_1, \cdots, \alpha_n\}$). So we see

$$M_{t_{k+1}} \oplus \cdots \oplus M_{t_n} \stackrel{\phi}{\simeq} T(\alpha)^*.$$

Since M_{t_i} is a uniform module, $T_i = \phi(M_{t_i})$ is a uniform module. So we see $T_i \cap \sigma(X) \subseteq_e T_i$. We put $f_i = f|_{T_i \cap \sigma(X)}$. By the same argument as when V_α is uniform, there exists a homomorphism $f_i^* : T_i \to \oplus_{j<\alpha} N_j$ with $f_i^*|_{T_i \cap \sigma(X)} = f_i$. Thus we get

$$< T_i \cap \sigma(X) \stackrel{f_i}{\longrightarrow} \oplus_{j<\alpha} N_j > \subseteq_e < T_i \stackrel{f_i^*}{\longrightarrow} \oplus_{j<\alpha} N_j >.$$

Now we put $X'_\alpha = < \oplus_{i=t+1}^n (T_i \cap \sigma(X)) \stackrel{\sum f_i}{\longrightarrow} \oplus_{j<\alpha} N_j >$. Then we get

$$X'_\alpha = \oplus_{i=t+1}^n < T_i \cap \sigma(X) \stackrel{f_i}{\longrightarrow} \oplus_{j<\alpha} N_j >$$

$$\subseteq_e \oplus_{i=t+1}^n < T_i \stackrel{f_i^*}{\longrightarrow} \oplus_{j<\alpha} N_j >$$

$$= < T(\alpha)^* \stackrel{\sum f_i^*}{\longrightarrow} \oplus_{j<\alpha} N_j >.$$

Since $f|_{\oplus_{i=t+1}^n (T_i \cap \sigma(X))} = \sum_{i=t+1}^n f_i$, we see

$$X'_\alpha \subseteq_e < \sigma(X) \stackrel{f}{\longrightarrow} \pi(X) > = X_\alpha.$$

By the same argument as in the case $X_\alpha = 0$, we get

$$X \subseteq_e < Y \oplus < T(\alpha)^* \to \oplus_J V_i > \to \oplus_J \overline{V_i} \oplus V_\alpha >.$$

Consequently, M is CS for $M = \oplus_I V_\alpha$ by transfinite induction.

Finally, we show that M is CS for $M = \oplus_\Lambda M_\alpha$. Using the above argument for all $\alpha \in I$ with $X_\alpha \neq 0$ we obtain an essential submodule X'_α of X_α such that $X'_\alpha \subseteq_e < T(\alpha)^* \to \oplus_I N_i >$. Thus we get

$$\oplus_I X'_\alpha \subseteq_e \oplus_I < T(\alpha)^* \to \oplus_I N_k >, \quad \oplus_I X'_\alpha \subseteq_e X.$$

By the same arguments which we used to show that S^* is CS for $S^* = \oplus_{J \cup \{\alpha\}} V_i$, we can obtain a decomposition $M = X^* \oplus (\oplus_I N_k)$ with $X \subseteq_e X^*$. Since $N_k = M_{k_1} \oplus \cdots \oplus M_{k_n}$, there exists a subset Λ' of Λ such that $\oplus_I N_k = \oplus_{\Lambda'} M_\alpha$. Thus we get

$$M = X^* \oplus (\oplus_{\Lambda'} M_\alpha).$$

Accordingly M is CS for $M = \oplus_\Lambda M_\alpha$. $\square$

Definition 2.4. A module P is said to be a CS-module with the internal exchange property if P is CS for any direct sum decomposition of P.

Theorem 2.5. Let $\{M_\alpha \mid \alpha \in \Lambda\}$ be a family of uniform modules and let $P = \oplus_\Lambda M_\alpha$. Then the following conditions are equivalent:
   (1) P is a CS-module with the internal exchange property.
   (2) P is CS for $M = \oplus_\Lambda M_\alpha$.
   (3) (a) M_α is generalized M_β-injective for all $\alpha \neq \beta \in \Lambda$.
   (b) (A'_2) holds for all M_α and $\{M_\beta \mid \beta \neq \alpha, \beta \in \Lambda\}$.
   (c) There does not exist an infinite sequence of non-isomorphic monomorphisms $\{f_k : M_{i_k} \to M_{i_{k+1}} \mid k \in \mathbb{N}\}$ with all $i_k \in \Lambda$ distinct.

Proof. (2)⇔(3) follows from Theorem 2.3 and (1)⇒(2) is clear.
   (2)⇒(1). Let $P = \oplus_\Gamma N_\gamma$ be a decomposition of P. Then, for any direct summand N_γ ($\gamma \in \Gamma$), there exists a subset Λ_γ of Λ such that $P = N_\gamma \oplus (\oplus_{\Lambda_\gamma} M_\lambda)$. Thus we see

$$N_\gamma \simeq \oplus_{\Lambda - \Lambda_\gamma} M_\lambda.$$

So there exist uniform submodules K_λ of N_γ such that $N_\gamma = \oplus_{\Lambda - \Lambda_\gamma} K_\lambda$. Hence we have the indecomposable decomposition

$$P = \oplus_\Gamma (\oplus_{\Lambda - \Lambda_\gamma} K_\lambda).$$

By Proposition 1.8, $P = \oplus_\Lambda M_\alpha$ complements direct summands. Hence two decompositions $P = \oplus_\Lambda M_\alpha$ and $P = \oplus_\Gamma (\oplus_{\Lambda - \Lambda_\gamma} K_\lambda)$ are equivalent by Proposition 1.7. Since conditions (a), (b) and (c) are inherited by isomorphic image, P is CS for $P = \oplus_\Gamma (\oplus_{\Lambda - \Lambda_\gamma} K_\lambda)$. Therefore P is CS for $P = \oplus_\Gamma N_\gamma$. So P is a CS-module with the internal exchange property. □

A family of modules $\{M_\alpha \mid \alpha \in \Lambda\}$ is said to be *locally semi-transfinitely nilpotent* ($lsTn$), if for any subfamily M_{α_i} ($i \in \mathbb{N}$) with distinct α_i and any family of non-isomorphisms $f_i : M_{\alpha_i} \to M_{\alpha_{i+1}}$, and for every $x \in M_{\alpha_1}$, there exists $n \in \mathbb{N}$ (depending on x) such that $f_n \cdots f_2 f_1(x) = 0$. Let $\{X_\alpha \mid \alpha \in \Lambda\}$ be a family of submodules of M. Then $\sum_{\alpha \in \Lambda} X_\alpha$ is said to be a *local summand*, if $\sum_{\alpha \in \Lambda} X_\alpha$ is direct and $\sum_{\alpha \in F} X_\alpha$ is a direct summand of M for every finite subset $F \subseteq \Lambda$.

We need the following two lemmas (the first lemma is due to Oshiro).

Lemma 2.6. [9] *If every local summand of M is a direct summand, then M has an indecomposable decomposition.*

Lemma 2.7. *Assume that (a), (b) in Theorem 2.3 hold. Then:*

(1) Every uniform submodule of P can be extended to a direct summand of P.

(2) If X is an indecomposable summand of P, then $X \simeq M_\alpha$ for some $\alpha \in \Lambda$.

(3) If $\{X_1, X_2, \ldots, X_n\}$ is an independent set of uniform submodules of P and $\oplus_{i=1}^n X_i <_\oplus P$, then there exists a finite subset $F \subseteq \Lambda$ such that $P = (\oplus_{i=1}^n X_i) \oplus (\oplus_{\Lambda - F} M_\alpha)$.

Proof. (1) Let X be a non-zero uniform submodule of P, and take $0 \neq x \in X$. Then $xR \subseteq M_{\alpha_1} \oplus \cdots \oplus M_{\alpha_n}$ for some $F = \{\alpha_1, \cdots, \alpha_n\} \subseteq \Lambda$. As $T = M_{\alpha_1} \oplus \cdots \oplus M_{\alpha_n}$ is CS for $M_{\alpha_1} \oplus \cdots \oplus M_{\alpha_n}$, xR can be extended to a direct summand $(xR)^* <_\oplus T$ and $T = (xR)^* \oplus (\oplus_{F - \{\alpha_i\}} M_\alpha)$ for some α_i. Let ϕ_1, ϕ_2 be the projections: $P \to (xR)^*$ and $P \to \oplus_{\Lambda - \{\alpha_i\}} M_{\alpha_j}$, respectively. Then

$$X = < \phi_1(X) \xrightarrow{\varphi} \phi_2(X) > \text{ and } Ker\varphi \subseteq_e \phi_1(X),$$

where φ is the mapping given by $\phi_1(x) \mapsto \phi_2(x)$. Then φ can be extended to a mapping $\varphi^* : (xR)^* \to \oplus_{\Lambda - \{\alpha_i\}} M_j$ and

$$X \subseteq_e < (xR)^* \xrightarrow{\varphi^*} \oplus_{\Lambda - \{\alpha_i\}} M_j > <_\oplus P.$$

(2) Let X be an indecomposable summand of P. We can take a uniform submodule $(0 \neq) xR$ of X. Let $(xR)^*$ be a closed submodule of X (that is, $xR \subseteq_e (xR)^*$ and $(xR)^*$ has no proper essential extension in X). Then $(xR)^*$ is also a closed uniform submodule of P. So, by (1), $(xR)^* <_\oplus P$ and hence $(xR)^* = X$. Moreover, in view of the proof of (1), there exists $\alpha \in \Lambda$ such that $P = X \oplus (\oplus_{I - \{\alpha\}} M_\beta)$. Hence $X \simeq M_\alpha$.

(3) Put $X = X_1 \oplus \cdots \oplus X_n$, and take $0 \neq x_i \in X_i$. Then $x_1 R \oplus \cdots \oplus x_n R \subseteq_e X$. As above $x_1 R \oplus \cdots \oplus x_n R \subseteq \oplus_F M_\alpha$ for some finite subset $F \subseteq \Lambda$, and $\oplus_F M_\alpha = (x_1 R \oplus \cdots \oplus x_n R)^* \oplus (\oplus_{F - E} M_\alpha)$ for some direct summand $(x_1 R \oplus \cdots \oplus x_n R)^* \supseteq_e x_1 R \oplus \cdots \oplus x_n R$ and some $E \subseteq F$. Then, as above, there exists a homomorphism $\varphi^* : (x_1 R \oplus \cdots \oplus x_n R)^* \to \oplus_{\Lambda - F} M_\alpha$ such that $X \subseteq_e < (x_1 R \oplus \cdots \oplus x_n R)^* \xrightarrow{\varphi^*} \oplus_{\Lambda - F} M_\alpha >$; whence $X = < (x_1 R \oplus \cdots \oplus x_n R)^* \xrightarrow{\varphi^*} \oplus_{\Lambda - F} M_\alpha >$ and $P = X \oplus (\oplus_{\Lambda - F} M_\alpha)$. □

The following is essentially due to Harada [4].

Theorem 2.8. *Let $P = \oplus_I M_\alpha$, where each M_α has a local endomorphism ring. Then the following conditions are equivalent:*

(1) *P has the internal exchange property in the direct sum $P = \oplus_I M_\alpha$.*

(2) *P has the (finite) exchange property.*

(3) *$P = \oplus_I M_\alpha$ satisfies lsTn.*

(4) *Every local summand of P is a direct summand.*

Remark 2.9. (See the proof of [10, Proposition 1]) Let X be a uniform module. If $X \oplus X$ has an internal exchange property, then X has a local endomorphism ring.

Theorem 2.10. *Let $P = \oplus_I M_\alpha$, where each M_α is a uniform module. Assume that $P = \oplus_I M_\alpha$ satisfies the conditions (a) and (b) in Theorem 2.3. Then the following are equivalent:*
  (1) *(c) in Theorem 2.3 holds.*
  (2) *$P = \oplus_I M_\alpha$ satisfies lsTn.*
  (3) *Every local summand of P is a direct summand.*

Proof. (1)⇒(2). By Theorem 2.3 and Proposition 1.7, $P = \oplus_I M_\alpha$ complements direct summands. Hence, by [7, Theorem 2.26], $P = \oplus_I M_\alpha$ satisfies lsTn.

(2)⇒(1) is clear.

(1)⇒(3). By a quite similar proof of [7, (2)⇒(3) in Theorem 2.22], we can show that every local summand of P is a direct summand.

(3)⇒(1). Let $\{M_i \mid i \in \mathbb{N}\} \subseteq \{M_\alpha\}_I$ and $f_i : M_i \to M_{i+1}$. Assume that there exists a proper monomorphism for $i = 1, 2, \ldots$. By Theorem 2.8, we may assume that $End(M_i)$ is not a local ring. Put

$$T = <M_1 \xrightarrow{f_1} M_2> \oplus <M_2 \xrightarrow{f_2} M_3> \oplus \cdots.$$

T is a local summand. So, T is a direct summand of P; let $P = T \oplus X$.

If $X = 0$, then for $0 \neq x \in M_1$, there exist n and $x_i \in M_i$ ($i = 1, \ldots, n$) such that

$$x = x_1 + f_1(x_1) + x_2 + f_2(x_2) + \cdots + x_n + f_n(x_n).$$

Then we see $x_n = x_{n-1} = \cdots = x_1 = 0$, whence $x = 0$, a contradiction.

Hence $X \neq 0$. By Lemma 2.6, X is expressed as $X = \oplus_J X_\alpha$ where X_α is indecomposable. Take $\alpha \in J$. Then $X_\alpha \simeq M_i$ for some $i \in \mathbb{N}$ by Lemma 2.7(2). Then

$$P = <M_1 \xrightarrow{f_1} M_2> \oplus \cdots \oplus X_\alpha \oplus (\oplus_{J-\{\alpha\}} X_\beta).$$

Here, $X_\alpha \oplus <M_i \xrightarrow{f_i} M_{i+1}> \simeq M_i \oplus M_i$. So, by Remark 2.9, $End(M_i)$ is a local ring, a contradiction. Thus (1) holds. □

References

[1] F. W. Anderson and K. R. Fuller, *Rings and Categories of Modules,* Springer-Verlag, Heidelberg-New York, 1973.

[2] Y. Baba and M. Harada, *On almost M-projectives and almost M-injectives*, Tsukuba J. Math. **14** (1990), 53–69.

[3] N. V. Dung, *On indecomposable decomposition of CS-modules II*, J. Pure and Applied Algebra **119** (1997), 139–153.

[4] M. Harada, *Factor categories with applications to direct decomposition of modules*, Lecture Notes in Pure Appl. Math. **88**, Dekker, New York, 1983.

[5] M. Harada and K. Oshiro, *On extending property of direct sums of uniform modules*, Osaka J. Math. **18** (1981), 767–785.

[6] K. Hanada, Y. Kuratomi, and K. Oshiro, *On direct sums of extending modules and internal exchange property*, Preprint.

[7] S. H. Mohamed and B. J. Müller, *Continuous and Discrete Modules*, London Math. Soc., Lecture Notes **147**, Cambridge Univ. Press, Cambridge, 1990

[8] K. Oshiro and T. Rizvi, *The exchange property of quasi-continuous modules with the finite exchange property*, Osaka J. Math. **33** (1996), 217–234.

[9] K. Oshiro, *Lifting modules, extending modules and their applications to QF-rings*, Hokkaido Math. J. **13** (1984), 310–338.

[10] R. B. Warfield, *A Krull-Schmidt theorem for infinite sums of modules*, Proc. Amer. Math. Soc. **22** (1969), 460–465.

J. Kado
Department of Mathematics
Osaka City University
Osaka 558-8585, Japan

Y. Kuratomi and K. Oshiro
Department of Mathematics
Yamaguchi Universit
Yamaguchi 753-8512, Japan
yoshi@po.yb.cc.yamaguchi-u.ac.jp

Generalized Principally Injective Maximal Ideals

Jin Yong Kim, Nam Kyun Kim and Sang Bok Nam

Abstract

We investigate in this paper von Neumann regularity of rings whose maximal right ideals are GP-injective. Actually, it is proved that a ring R is strongly regular if and only if R is a 2-primal ring whose maximal right ideals are GP-injective. It is also proved that if R is a PI-ring whose maximal right ideals are GP-injective, then R is strongly π-regular.

Throughout this paper, R denotes an associative ring with identity, and all modules are unitary. $P(R)$ and $J(R)$ denote the prime radical and the Jacobson radical, respectively. A right R-module M is called *generalized right principally injective* (briefly *right GP-injective*) if, for any $0 \neq a \in R$, there exists a positive integer n such that $a^n \neq 0$ and any right R-homomorphism of $a^n R$ into M extends to one of R into M.

In particular, M is called *right principally injective* in case $n = 1$. As a well-known fact, the following are equivalent: (1) every right ideal of R is injective; (2) every maximal right ideal of R is injective; (3) R is semisimple Artinian. While every right ideal of R is principally injective if and only if R is von Neumann regular. But it is well known that if every maximal right ideal of R is principally injective, then R is a right SF-ring (i.e., every simple right R-module is flat). Von Neumann regularity of SF-rings has been studied by several authors in [4] and [15].

Rings whose maximal right ideals are GP-injective have been studied in [4, 5, 21, 22]. In these papers, the authors proved that a ring R is strongly regular if and only if R is a reduced ring whose maximal right ideals are GP-injective if and only if every left annihilator of R is two-sided and every maximal right ideals are GP-injective.

We also investigate in this paper von Neumann regularity of rings whose maximal right ideals are GP-injective. Actually we prove the following theorems which extend several known results:

Theorem. *A ring R is strongly regular if and only if R is a 2-primal ring whose maximal right ideals are GP-injective.*

Theorem. *A ring R is strongly regular if and only if R is a right (or left) quasi-duo ring whose maximal right ideals are GP-injective.*

We also prove the following theorem which characterizes rings whose proper right ideals are GP-injective in a 2-primal ring:

Theorem. *Assume that R is 2-primal. Then the following statements are equivalent:*

(1) *Every proper right ideal is GP-injective.*
(2) *Every proper principal right ideal is GP-injective.*
(3) *R is strongly regular.*
(4) *R is von Neumann regular.*

Finally, we prove the following:

Theorem. *If R is a PI-ring whose maximal right ideals are GP-injective, then R is strongly π-regular.*

For any nonempty subset S of R, $r(S)$ and $\ell(S)$ denote the right annihilator of S and the left annihilator of S, respectively.

Proposition 1. *Assume that R is a semiprime ring whose maximal right ideals are GP-injective. Then the center $C(R)$ of R is von Neumann regular.*

Proof. First we will show that $aR + r(a) = R$ for any nonzero $a \in C(R)$. Suppose not. Then there exists a maximal right ideal M of R containing $aR + r(a)$. Thus M is GP-injective. So there exists a positive integer n such that any R-homomorphism of $a^n R$ into M extends to one of R into M.

Let $i : a^n R \to M$ be an inclusion map. Since M is GP-injective, there exists $c \in M$ such that $a^n = i(a^n) = ca^n$, whence $((1-c)aR)^n = 0$. This implies that $(1-c)a = 0$ since R is semiprime. So $1 - c \in r(a) \subseteq M$, which is a contradiction. So $aR + r(a) = R$. Then $a = aba$ for some $b \in R$. By [8, Theorem 1.14], we have $C(R)$ is von Neumann regular. □

Recall that a ring R is called *2-primal* [1] (also called an *N*-ring in [10]) if its prime radical coincides with the set of all nilpotent elements of R. A ring R is called *right quasi-duo* [2] if every maximal right ideal of R is a two-sided ideal. Left quasi-duo rings are defined similarly. A ring R is *quasi-duo* if R is right and left quasi-duo.

Lemma 2. [11, Lemma 8] *If R is von Neumann regular, then every right R-module is GP-injective.*

The following theorem extends known results [21, Theorem 5.1 and Proposition 7].

Theorem 3. *The following statements are equivalent*:
(1) *R is strongly regular.*
(2) *R is a 2-primal ring whose maximal right ideals are GP-injective.*

Proof. By Lemma 2, (1) implies (2). For (2)⇒(1), assume that R is 2-primal. Then $\bar{R} = R/P(R)$ is a reduced ring. Let $\bar{0} \neq \bar{a} \in \bar{R}$. We first claim that $\bar{a}\bar{R} + r(\bar{a}) = \bar{R}$. Suppose not. Then there exists a maximal right ideal $\overline{M} = M/P(R)$ of $\bar{R}$ containing $\bar{a}\bar{R} + r(\bar{a})$. Note that M is also a maximal right ideal of R. Thus M is GP-injective.

So there exists a positive integer n such that any R-homomorphism of $a^n R$ into M extends to one of R into M. Let $i : a^n R \to M$ be an inclusion map. Since M is GP-injective, there exists $c \in M$ such that $a^n = i(a^n) = ca^n$. Thus $\bar{a}^n = \bar{c}\bar{a}^n$ in $\bar{R}$, whence $\bar{1} - \bar{c} \in l(\bar{a}^n) = l(\bar{a}) = r(\bar{a})$. Hence $\bar{1} \in \overline{M}$, which is a contradiction. Therefore $\bar{R}$ is strongly regular and so R is quasi-duo.

Now we will prove that $P(R) = 0$. It is enough to show that $P(R)$ is reduced. Suppose that there exists a nonzero element x of $P(R)$ such that $x^2 = 0$. Then $l(x) \neq R$, so there exists a maximal left ideal M of R containing $l(x)$. Since R is quasi-duo, M is itself a maximal right ideal of R. For, let K be a maximal right ideal of R such that $M \subseteq K$. Since R is right quasi-duo, K is also a left ideal of R and so $M = K$. Consider the inclusion map $i : xR \to M$.

Since M is GP-injective and there exists $y \in M$ such that $x = i(x) = yx$. Hence $1 - y \in l(x) \subseteq M$, so $1 \in M$. It is a contradiction. Therefore $P(R) = 0$, whence R is a strongly regular ring. □

Shin [17] proved that if every right (or left) annihilator is a two-sided ideal, then R is 2-primal.

Corollary 4. [21, Theorem 5.1] *The following statements are equivalent*:
(1) *R is strongly regular.*
(2) *R is a reduced ring whose maximal right ideals are GP-injective.*
(3) *Every maximal right ideal is GP-injective and $\ell(a)$ is two-sided for any $a \in R$.*

The condition "R is 2-primal" in Theorem 3 is not superfluous. For example, let R be the full matrix ring over a field. Clearly, every maximal right ideal is GP-injective. But we can easily check that R is neither 2-primal nor strongly regular.

Recall that a ring R is *weakly right duo* [18] if for any $a \in R$, there exists a positive integer n such that $a^n R$ is a two-sided ideal of R. Weakly left duo rings are defined similarly. Recently, Yu [19] proved the following.

Lemma 5. [19, Proposition 2.2] *Every weakly right (left) duo ring is right (left) quasi-duo.*

The following theorem also extends a result [22, Proposition 7].

Theorem 6. *The following statements are equivalent:*
  (1) R *is strongly regular.*
  (2) R *is a weakly right (or left) duo ring whose maximal right ideals are GP-injective.*
  (3) R *is a right (or left) quasi-duo ring whose maximal right ideals are GP-injective.*

Proof. By Lemma 2 and Lemma 5, (1)$\Rightarrow$(2)$\Rightarrow$(3). Assume that R is right quasi-duo. Then by [15, Proposition 4.4], $R/J(R)$ is reduced. By the same methods in the proof of Theorem 3, R is strongly regular. □

A ring R is called *abelian* if every idempotent element of R is central. A ring R is called π-*regular* [9] if for every $x \in R$, there exists a positive integer n, depending on x, such that $x^n \in x^n R x^n$. Ohori [12] proved that the set of all nilpotent elements of an abelian π-regular ring is a two-sided ideal. From this result, we can see that an abelian π-regular ring is quasi-duo. Hence we have the following.

Corollary 7. *The following statements are equivalent:*
  (1) R *is strongly regular.*
  (2) R *is an abelian π-regular ring whose maximal right ideals are GP-injective.*

Note that a ring whose principal right ideals are principally injective is von Neumann regular. However we have the following.

Lemma 8. *If every proper principal right ideal of R is GP-injective, then R is semiprime.*

Proof. Let I be a nonzero proper right ideal of R. Suppose that $I^2 = 0$. Then there exists a nonzero element a in I, but $(aR)^2 = 0$. Since aR is GP-injective, any R-homomorphism of aR into aR extends to one of R into aR. Consider the identity map $i : aR \to aR$. Since aR is GP-injective, there exists $c \in aR$ such that $a = i(a) = ca$. But $ca \in aRa \subseteq (aR)^2 = 0$, whence $a = 0$, which is a contradiction. Therefore R is semiprime. □

Theorem 9. *Assume that R is 2-primal. Then the following statements are equivalent:*
  (1) *Every proper right ideal of R is GP-injective.*

(2) *Every proper principal right ideal of R is GP-injective.*
(3) *R is strongly regular.*
(4) *R is von Neumann regular.*

Proof. Obviously, (1) implies (2) and (3) implies (4). Also by Lemma 2, (4) implies (1). Thus it remains to prove that (2) implies (3). By Lemma 8, R is semiprime and so R is reduced. Let $0 \neq a \in R$. Then aR is GP-injective, so there exists a positive integer n such that any R-homomorphism of $a^n R$ into aR extends to one of R into aR. Consider the inclusion map $i: a^n R \to aR$. Then $a^n = i(a^n) = ca^n$ for some $c \in aR$. Thus $1 - c \in l(a^n) = r(a^n) = r(a)$, whence $a = ac \in a^2 R$. This implies that R is strongly regular. □

A ring R is called *right (left) weakly regular* [14] if $I^2 = I$ for each right (left) ideal I of R; equivalently, $a \in aRaR$ ($a \in RaRa$) for every $a \in R$. R is *weakly regular* if it is both right and left weakly regular. Recently, we proved that if R is a 2-primal ring whose simple right R-modules are GP-injective, then R is a reduced weakly regular ring [11]. But we have the following.

Corollary 10. *A ring R is strongly regular if and only if R is a 2-primal ring whose cyclic right R-modules are GP-injective.*

A ring R is called *MERT* [20] if every maximal essential right ideal of R is two-sided. Here $Soc_r(R)$ denotes the right socle of R.

Theorem 11. *Assume that R is a MERT ring. Then the following statements are equivalent:*
(1) *R is von Neumann regular.*
(2) *Every maximal right ideal is GP-injective and $R/Soc_r(R)$ is semiprimitive.*

Proof. Obviously, (1) implies (2).
(2)⇒(1). Since R is MERT, $\overline{R} = R/Soc_r(R)$ is right quasi-duo. Then by [15, Proposition 4.4], $\overline{R}/J(\overline{R})$ is reduced and hence $\overline{R}$ is reduced. Let $\overline{0} \neq \bar{a} \in \overline{R}$. First we will show that $\bar{a}\overline{R} + r(\bar{a}) = \overline{R}$. Suppose not. Then there exists a maximal right ideal $\overline{M} = M/Soc_r(R)$ of $\overline{R}$ containing $\bar{a}\overline{R} + r(\bar{a})$. Thus M is GP-injective, so there exists a positive integer n such that any R-homomorphism of $a^n R$ into M extends to one of R into M. Let $i: a^n R \to M$ be an inclusion map. Since M is GP-injective, there exists $c \in M$ such that $a^n = i(a^n) = ca^n$. Thus $\bar{a}^n = \bar{c}\bar{a}^n$ in $\overline{R}$, whence $\bar{1} - \bar{c} \in l(\bar{a}^n) = l(\bar{a}) = r(\bar{a}) \subseteq \overline{M}$, which is a contradiction. Therefore $\overline{R}$ is strongly regular. Hence $J(R) \subseteq Soc_r(R)$ which implies $(J(R))^2 \subseteq Soc_r(R)J(R) = 0$.

Now we will prove that $J(R) = 0$. Suppose that $J(R) \neq 0$. Then there exists a nonzero element $a \in J(R)$. Now $Soc_r(R) \subseteq \ell(J(R)) \subseteq \ell(a)$. Since $\overline{R}$ is strongly regular, $\ell(a)/Soc_r(R)$ is a two-sided ideal of $\overline{R}$ and so $\ell(a)$ is a two-sided ideal of R. Then there exists a maximal right ideal K of R containing $l(a)$. Since K is GP-injective, the inclusion map $i : aR \to K$ can be extended to one of R into K. So there exists $b \in K$ such that $a = i(a) = ba$. Thus $1 - b \in l(a) \subseteq K$, whence $1 \in K$, which is a contradiction. Hence $J(R) = 0$ and so $Soc_r(R)$ is von Neumann regular. Therefore R is von Neumann regular. □

Finally, we turn our attention to a PI-ring whose maximal right ideals are GP-injective. Recall that a ring R is called *strongly π-regular* if for every $x \in R$, there exists a positive integer n, depending on x, such that $x^n \in x^{n+1}R$.

For PI-rings, the following result due to Fisher and Snider [7] is well known.

Lemma 12. [7, Theorem 2.3] *Assume that R is a PI-ring. Then the following statements are equivalent:*
(1) *R is strongly π-regular.*
(2) *Each prime factor ring of R is von Neumann regular.*

The following lemma is also well known.

Lemma 13. [16, Theorem 6.1.28] *Let R be a semiprime PI-ring. Then $I \cap C(R) \neq 0$ for every nonzero two-sided ideal I of R, where $C(R)$ denotes the center of R.*

Theorem 14. *Assume that R is a PI-ring whose maximal right ideals are GP-injective. Then R is strongly π-regular.*

Proof. First we will show that every prime factor ring of R is right weakly regular. If not, then there exists a prime ideal P such that $\overline{R} = R/P$ is not right weakly regular. This implies that $\overline{R}\bar{a}\overline{R} \neq \overline{R}$ for some $\bar{0} \neq \bar{a} \in \overline{R}$.

Then there exists maximal right ideal $\overline{M} = M/P$ of $\overline{R}$ such that $\overline{R}\bar{a}\overline{R} \subseteq \overline{M}$. Also since $\overline{R}$ is a semiprime PI-ring, by Lemma 13 there exists $\bar{0} \neq \bar{x} \in \overline{R}\bar{a}\overline{R} \cap C(\overline{R})$. Thus M is GP-injective, so there exists a positive integer n such that any R-homomorphism of $x^n R$ into M extends to one of R into M. Let $i : x^n R \to M$ be an inclusion map. Since M is GP-injective, there exists $y \in M$ such that $x^n = i(x^n) = yx^n$. Thus $\bar{x}^n = \bar{y}\bar{x}^n$ and so $(\bar{1} - \bar{y})\bar{x}^n = \bar{0}$. Since $\bar{x} \in C(\overline{R})$, $(\bar{1} - \bar{y})\overline{R}\bar{x}^n = \bar{0}$. If $\bar{x}^n = \bar{0}$, then $(\bar{x}\overline{R})^n = \bar{x}^n \overline{R} = \bar{0}$. Since $\overline{R}$ is semiprime, $\bar{x}\overline{R} = \bar{0}$ and so $\bar{x} = \bar{0}$, which is a contradiction. Hence $\bar{1} - \bar{y} = \bar{0}$ and so $\bar{1} \in \overline{M}$. It is also a contradiction. Therefore every prime factor ring of R is right weakly regular and so is von Neumann regular since R is a PI-ring. Hence by Lemma 12, R is strongly π-regular. □

Corollary 15. *Assume that R is a semiprime PI-ring. Then the following statements are equivalent*:

(1) *R is semisimple Artinian.*

(2) *R is a right Goldie ring whose maximal right ideals are GP-injective.*

Proof. The implication (1)⇒(2) is well known. Assume (2). Then it follows from Theorem 14 that R is strongly π-regular. Let I be an essential right ideal of R. By [3, Theorem 1.10], I contains a regular element $c \in R$. Since R is strongly π-regular, there exists a positive integer n such that $c^n = c^{n+1}d$ for some $d \in R$. So $c^n(1 - cd) = 0$ and hence $1 = cd \in I$. Therefore $I = R$ and so R is semisimple Artinian. □

Corollary 16. *Assume that R is a prime PI-ring. Then the following statements are equivalent*:

(1) *R is simple Artinian.*

(2) *R is von Neumann regular.*

(3) *Every maximal right ideal is GP-injective.*

(4) *R is strongly π-regular.*

Proof. Obviously, (1) implies (2) and (2) implies (3). Also by Theorem 14, we have (3) implies (4). Thus it remains to show that (4) implies (1). Since R is prime, it is enough to show that R is semisimple Artinian. If not, there exists a maximal essential right ideal M of R. By Posner's Theorem [13], R is right Goldie. Then by [3, Theorem 1.10], M contains a regular element c of R. Observe that c is a unit element of R, which is a contradiction. □

Acknowledgements. The first named author was supported in 1997 by the Academic Research Fund of Ministry of Education, Republic of Korea, Project No. BSRI-97-1432 and also partially by Kyung Hee University. The second named author wishes to acknowledge the financial support of the Korea Research Foundation made in the Program Year 1997.

References

[1] G. F. Birkenmeier, H. E. Heatherly and Enoch K. Lee, *Completely prime ideals and associated radicals*, Proc. Biennial Ohio State-Denison Conference 1992, edited by S. K. Jain and S. T. Rizvi, World Scientific, New Jersey (1993), 102–129.

[2] S. H. Brown, *Rings over which every simple module is rationally complete*, Canad. J. Math. **25** (1973), 693–701.

[3] A. W. Chatters and C. R. Hajarnavis, *Rings with Chain Conditions*, Pitman, Boston, 1980.

[4] J. Chen, *On von Neumann regular rings and SF-rings*, Math. Japonica **36**(6) (1991), 1123–1127.

[5] N. Ding and J. Chen, *Rings whose simple singular modules are YJ-injective*, Math. Japonica **40**(1) (1994), 191–195.

[6] ———, *On maximal essential right ideals of rings*, Acta Math. Sinica **38**(3) (1995), 303–309.

[7] J. W. Fisher and R. L. Snider, *On the von Neumann regularity of rings with prime factor rings*, Pacific J. Math. **54** (1974), 135–144.

[8] K. R. Goodearl, *Von Neumann Regular Rings*, Pitman, Boston, 1979.

[9] V. Gupta, *Weakly π-regular rings and group rings*, Math J. Okayama Univ. **19** (1977), 123–127.

[10] Y. Hirano, *Some studies on strongly π-regular rings*, Math. J. Okayama Univ. **20** (1978), 141–149.

[11] S. B. Nam, N. K. Kim and J. Y. Kim, *On simple GP-injective modules*, Comm. Algebra **23**(14) (1995), 5437–5444.

[12] M. Ohori, *On abelian π-regular rings*, Math. Japonica **37** (1985), 21–31.

[13] E. C. Posner, *Prime rings satisfying a polynomial identity*, Proc. Amer. Math. Soc. **11** (1960), 180–183.

[14] V. S. Ramamurthi, *Weakly regular rings*, Canad. Math. Bull. **13** (1973), 317–321.

[15] M. B. Rege, *On von Neumann regular rings and SF-rings*, Math. Japonica **31**(6) (1986), 927–936.

[16] L. H. Rowen, *Ring Theory II*, Academic Press, New York, 1988.

[17] G. Shin, *Prime ideals and sheaf representation of a pseudo symmetric ring*, Trans. Amer. Math. Soc. **84** (1973), 43–60.

[18] Xue Yao, *Weakly right duo rings*, Pure and Applied Math. Sciences **21** (1985), 19–24.

[19] H. P. Yu, *On quasi-duo rings*, Glasgow Math. J. **37** (1995), 21–31.

[20] R. Yue Chi Ming, *On regular rings and self-injective rings*, Mh. Math. **91** (1981), 153–166.

[21] _____, *On regular rings and Artinian rings (II)*, Riv. Math. Univ. Parma **11**(4) (1985), 101–109.

[22] _____, *On von Neumann regular rings, XII*, Tamkang J. Math. **14**(4) (1985), 67–75.

Jin Yong Kim and Nam Kyun Kim
Department of Mathematics
Kyung Hee University
Suwon 449-701, Korea
e-mail for N.K. Kim: `nkkim@yonsei.ac.kr`

Sang Bok Nam
Department of Mathematics
Kyungdong University
Kosung 219-830, Korea

The Module of Differentials of a Noncommutative Ring Extension

Hiroaki Komatsu

Abstract

We study elementary properties of modules of differentials of noncommutative ring extensions and give a commutativity condition for separable algebras.

The notion of the module of differentials of a commutative algebra has been used for a long time. In [S], Sweedler studied *right* derivations and defined the module of differentials of a noncommutative algebra as a *right* module. In commutative ring theory, 'derivations' usually mean 'right derivations' in the sense of [S]. Therefore his theory includes the commutative case. Recently, in [HK], Hongan and the author of this article took notice of *central* derivations and constructed the module of differentials of an algebra extension as a *bimodule*. This bimodule structure played an important role in the study of biderivations in [HK]. In the case of an algebra our module of differentials is isomorphic to Sweedler's module of differentials as a right module.

In [HK], the existence of an identity element of an algebra was not assumed. In this article, using the same method described in [K], we rebuild the theory on rings with an identity element. First, in §2, we give a simple definition of modules of differentials. In §3, we generalize some results in [HK] about relations between modules of differentials of related rings. Finally in §4, we study quasi-separable ring extensions and their commutativity.

Throughout this article, every ring has an identity element 1, every ring homomorphism preserves 1, and every module over a ring is unitary. For a ring R, R-**Mod**-R, R-**Mod**, and **Mod**-R represent the category of R-bimodules, the category of left R-modules, and the category of right R-modules, respectively. Given $M, N \in R$-**Mod**-R, the set of all R-bimodule homomorphisms from M to N is denoted by $\text{Hom}_{R\text{-}R}(M, N)$.

Let R and S be rings. For brevity we shall call R an S-*ring* if there exists a ring homomorphism $\alpha \colon S \to R$. If R is an S-ring, then all R-modules are viewed as S-modules via α, and in this view point, $S \cdot 1$ is a subring of R.

1. Central Derivations and One-Sided Derivations

Let R be an S-ring. Given an R-bimodule M, an S-bimodule homomorphism $d\colon R \to M$ is called an *S-derivation* of R to M if $d(xy) = xd(y) + d(x)y$ for all $x,\ y \in R$, and d is said to be *central* if $xd(y) = d(y)x$ for all $x,\ y \in R$. The set of all S-derivations and the set of all central S-derivations of R to M are denoted by $\mathrm{Der}_S(R, M)$ and $\mathrm{CDer}_S(R, M)$, respectively. For any $M,\ N \in R\text{-}\mathbf{Mod}\text{-}R$, if $f \in \mathrm{Hom}_{R\text{-}R}(M, N)$ and $d \in \mathrm{CDer}_S(R, M)$, then we have $fd \in \mathrm{CDer}_S(R, N)$. Hence $\mathrm{CDer}_S(R, -)$ is viewed as a subfunctor of $\mathrm{Hom}_{\mathbb{Z}}(R, -)\colon R\text{-}\mathbf{Mod}\text{-}R \to \mathbb{Z}\text{-}\mathbf{Mod}$.

Given a left R-module M, an S-homomorphism $d\colon R \to M$ is called a *left S-derivation* of R to M if $d(xy) = xd(y) + yd(x)$ for all $x,\ y \in R$. The set of all left S-derivations of R to M is denoted by $\mathrm{LDer}_S(R, M)$. If $d \in \mathrm{LDer}_S(R, M)$ then, for any $r,\ x,\ y \in R$, we have $d(xyr) = xyd(r) + rd(xy)$. On the other hand, we have $d(xyr) = xd(yr) + yrd(x) = xyd(r) + xrd(y) + yrd(x)$. Hence $rd(xy) = xrd(y) + yrd(x)$. This shows that the mapping $rd\colon R \to M$ defined by $(rd)(x) = rd(x)$ belongs to $\mathrm{LDer}_S(R, M)$. It is easy to see that $\mathrm{LDer}_S(R, M)$ becomes a left R-module by this multiplication and that $\mathrm{LDer}_S(R, -)$ is viewed as a subfunctor of $\mathrm{Hom}_{\mathbb{Z}}(R, -)\colon R\text{-}\mathbf{Mod} \to R\text{-}\mathbf{Mod}$.

Symmetrically, *right S-derivations* of R to a right R-module M are defined and the set of all right S-derivations of R to M is denoted by $\mathrm{RDer}_S(R, M)$. Then $\mathrm{RDer}_S(R, M)$ has a right R-module structure and $\mathrm{RDer}_S(R, -)$ is viewed as a subfunctor of $\mathrm{Hom}_{\mathbb{Z}}(R, -)\colon \mathbf{Mod}\text{-}R \to \mathbf{Mod}\text{-}R$.

2. Module of Differentials

Let R be an S-ring. Let J denote the kernel of the mapping $R \otimes_S R \ni x \otimes y \mapsto xy \in R$. The mapping $\tau\colon R \to J$ defined by $\tau(x) = x \otimes 1 - 1 \otimes x$ is an S-derivation. For any $\sum x_i \otimes y_i \in J$, we see that $\sum x_i \otimes y_i = \sum \tau(x_i) y_i = -\sum x_i \tau(y_i)$. Hence we have $J = R\tau(R) = \tau(R)R$. It is easy to see that, for any R-bimodule M, the additive mapping

$$\mathrm{Hom}_{R\text{-}R}(J, M) \to \mathrm{Der}_S(R, M) \qquad f \mapsto f\tau$$

is an isomorphism (cf. [B, III §10, Proposition 17]). Put $\Gamma = \{x\tau(y) - \tau(y)x \mid x,\ y \in R\}$. Since $x\bigl(y\tau(z) - \tau(z)y\bigr) + \bigl(x\tau(z) - \tau(z)x\bigr)y = xy\tau(z) - \tau(z)xy \in \Gamma$ for all $x,\ y,\ z \in R$, we have $R\Gamma = \Gamma R$. Put $\Omega_{R/S} = J/R\Gamma$ and $d_{R/S} = p\tau$, where $p\colon J \to \Omega_{R/S}$ is the canonical mapping. Then $d_{R/S}$ is a central S-derivation. We shall call $\Omega_{R/S}$ the *module of differentials* of R over S and $d_{R/S}$ the *canonical S-derivation* of R.

If R is commutative, then $R \otimes_S R$ is a commutative ring and we see that $\tau(x)\tau(y) = x\tau(y) - \tau(y)x$ for all $x, y \in R$, and hence we have $R\Gamma = J^2$ as an ideal of $R \otimes_S R$.

Theorem 1. *Let R be an S-ring. Then:*
(1) *$d_{R/S}$ is a central S-derivation and $\Omega_{R/S} = Rd_{R/S}(R) = d_{R/S}(R)R$;*
(2) *For each $M \in R$-**Mod**-R, the additive mapping*

$$\eta \colon \operatorname{Hom}_{R\text{-}R}(\Omega_{R/S}, M) \to \operatorname{CDer}_S(R, M)$$

defined by $\eta(f) = fd_{R/S}$ is a natural isomorphism.

Proof. (1) is clear by definition.

(2) It is trivial that η is natural. Since $\Omega_{R/S} = Rd_{R/S}(R)$, η is injective. Let d be an arbitrary element in $\operatorname{CDer}_S(R, M)$. We can define a left R-module homomorphism $g \colon R \otimes_S R \to M$ such that $g(x \otimes y) = -xd(y)$ for all $x, y \in R$. The restriction mapping $g' \colon J \to M$ of g is an R-bimodule homomorphism. Actually, for any $\sum x_i \otimes y_i \in J$, we see that $g'(\sum x_i \otimes y_i) = -\sum x_i d(y_i) = -\sum d(x_i y_i) + \sum d(x_i) y_i = \sum d(x_i) y_i$. Since $g'\tau = d$, we have $g'(x\tau(y) - \tau(y)x) = 0$ for all $x, y \in R$. Hence there exists $f \in \operatorname{Hom}_{R\text{-}R}(\Omega_{R/S}, M)$ such that $g' = fp$. Then $d = \eta(f)$. Thus η is surjective. □

In the language of category theory, this theorem asserts that the pair $(\Omega_{R/S}, d_{R/S})$ represents the functor $\operatorname{CDer}_S(R, -) \colon R\text{-}\mathbf{Mod}\text{-}R \to \mathbb{Z}\text{-}\mathbf{Mod}$. For any $M \in R\text{-}\mathbf{Mod}$, $\operatorname{Hom}_R(\Omega_{R/S}, M)$ has a left R-module structure

$$(rf)(\omega) = f(\omega r) \quad (r \in R,\ f \in \operatorname{Hom}_R(\Omega_{R/S}, M),\ \omega \in \Omega_{R/S}),$$

and for any $M \in \mathbf{Mod}\text{-}R$, $\operatorname{Hom}_R(\Omega_{R/S}, M)$ has a right R-module structure.

Theorem 2. *For an S-ring R, the pair $(\Omega_{R/S}, d_{R/S})$ represents functors*

$$\operatorname{LDer}_S(R, -) \colon R\text{-}\mathbf{Mod} \to R\text{-}\mathbf{Mod} \quad \text{and}$$
$$\operatorname{RDer}_S(R, -) \colon \mathbf{Mod}\text{-}R \to \mathbf{Mod}\text{-}R.$$

Proof. We show that, for each $M \in R\text{-}\mathbf{Mod}$, the mapping

$$\eta' \colon \operatorname{Hom}_R(\Omega_{R/S}, M) \to \operatorname{LDer}_S(R, M)$$

defined by $\eta'(f) = fd_{R/S}$ is a natural R-isomorphism. For any $f \in \operatorname{Hom}_R(\Omega_{R/S}, M)$ and $r, x \in R$, we see that

$$\eta'(rf)(x) = (rf)d_{R/S}(x) = f(d_{R/S}(x)r) = f(rd_{R/S}(x))$$
$$= r(fd_{R/S}(x)) = r\eta'(f)(x).$$

Hence η' is an R-module homomorphism. The rest of the proof is quite similar to Theorem 1(2).

Symmetrically we can get the assertion on $\operatorname{RDer}_S(R, -)$. □

3. Elementary Properties of Modules of Differentials

The next theorem improves [HK, Theorems 5.1 and 5.3].

Theorem 3. *Consider the following commutative diagram of ring homomorphisms:*

$$\begin{array}{ccc} S & \longrightarrow & R \\ \downarrow & & \downarrow \rho \\ S' & \longrightarrow & R' \end{array}$$

(1) *There exists a unique R-bimodule homomorphism $\nu : \Omega_{R/S} \to \Omega_{R'/S'}$ such that the following diagram commutes:*

$$\begin{array}{ccc} R & \xrightarrow{d_{R/S}} & \Omega_{R/S} \\ \downarrow \rho & & \downarrow \nu \\ R' & \xrightarrow{d_{R'/S'}} & \Omega_{R'/S'} \end{array}$$

(2) *If ρ is surjective, then ν is surjective and*

$$\operatorname{Ker}\nu = Rd_{R/S}\bigl(\rho^{-1}(S' \cdot 1)\bigr).$$

Proof. (1) It is easy to see that $d_{R'/S'}\rho \in \operatorname{CDer}_S(R, \Omega_{R'/S'})$. Hence, by Theorem 1(2), there exists a unique $\nu \in \operatorname{Hom}_{R\text{-}R}(\Omega_{R/S}, \Omega_{R'/S'})$ such that $\nu d_{R/S} = d_{R'/S'}\rho$.

(2) By Theorem 1(1), we can see that ν is surjective. Now we put $T = \rho^{-1}(S' \cdot 1)$. Since $\nu d_{R/S} = d_{R'/S'}\rho$, we have $Rd_{R/S}(T) \subseteq \operatorname{Ker}\nu$. Let $p\colon \Omega_{R/S} \to \Omega_{R/S}/Rd_{R/S}(T)$ be the canonical mapping. Since $I = \operatorname{Ker}\rho$ is an ideal, we see that

$$I\Omega_{R/S} = Id_{R/S}(R) \subseteq d_{R/S}(IR) + d_{R/S}(I)R \subseteq Rd_{R/S}(T),$$

and also we have $\Omega_{R/S}I \subseteq Rd_{R/S}(T)$. Hence $\Omega_{R/S}/Rd_{R/S}(T)$ is viewed as an R'-bimodule. Since $\operatorname{Ker}\rho \subseteq T \subseteq \operatorname{Ker} pd_{R/S}$, there exists an additive mapping $d\colon R' \to \Omega_{R/S}/Rd_{R/S}(T)$ such that $d\rho = pd_{R/S}$. We can see that d is a central S'-derivation. Hence, by Theorem 1(2), there exists $f \in \operatorname{Hom}_{R'\text{-}R'}(\Omega_{R'/S'}, \Omega_{R/S}/Rd_{R/S}(T))$ such that $d = fd_{R'/S'}$. Then we see that $f\nu d_{R/S} = fd_{R'/S'}\rho = d\rho = pd_{R/S}$. By Theorem 1(1), we have $f\nu = p$. Hence $\operatorname{Ker}\nu \subseteq \operatorname{Ker} p = Rd_{R/S}(T)$. We have thus shown that $\operatorname{Ker}\nu = Rd_{R/S}(T)$. □

Corollary 4. *Let S be a T-ring and R an S-ring. Then $\Omega_{R/S}$ is isomorphic to $\Omega_{R/T}/Rd_{R/T}(S \cdot 1)$ as R-bimodule.*

Proof. This is obtained by applying Theorem 3 to the following commutative diagram.

$$\begin{array}{ccc} T & \longrightarrow & R \\ \downarrow & & \| \\ S & \longrightarrow & R \end{array}$$

□

Corollary 5. *Let R be an S-ring and I an ideal of R. Put $\overline{R} = R/I$. Then $\Omega_{\overline{R}/S} \simeq \Omega_{R/S \cdot 1 + I} \simeq \Omega_{R/S}/Rd_{R/S}(I)$ as R-bimodules, and $Rd_{R/S}(I) = d_{R/S}(I) + I\Omega_{R/S} = d_{R/S}(I) + \Omega_{R/S}I$.*

Proof. Consider the following commutative diagram:

$$\begin{array}{ccc} S & \longrightarrow & R \\ \| & & \downarrow \pi \\ S & \longrightarrow & \overline{R} \end{array}$$

where $\pi \colon R \to \overline{R}$ is the canonical mapping. Since $\pi^{-1}(S \cdot 1) = S \cdot 1 + I$ and $d_{R/S}(S \cdot 1 + I) = d_{R/S}(I)$, we have $\Omega_{\overline{R}/S} \simeq \Omega_{R/S}/Rd_{R/S}(I)$ by Theorem 3. Moreover, by Corollary 4, we have $\Omega_{R/S \cdot 1 + I} \simeq \Omega_{R/S}/Rd_{R/S}(I)$. We see that $Rd_{R/S}(I) \subseteq d_{R/S}(RI) + d_{R/S}(R)I \subseteq d_{R/S}(I) + \Omega_{R/S}I$, and, by Theorem 1(1), we see that $\Omega_{R/S}I = d_{R/S}(R)RI = d_{R/S}(R)I \subseteq d_{R/S}(RI) + Rd_{R/S}(I) \subseteq Rd_{R/S}(I)$. Hence we have $Rd_{R/S}(I) = d_{R/S}(I) + \Omega_{R/S}I$; and similarly $d_{R/S}(I)R = d_{R/S}(I) + I\Omega_{R/S}$. Since $d_{R/S}$ is central, we get the assertion. □

The next theorem shows a limit of information given by modules of differentials. As usual, for elements x, y in a ring R, we write $[x, y] = xy - yx$. If A and B are additive subgroups of R, we define $[A, B]$ to be the additive subgroup generated by all $[a, b]$ where $a \in A$ and $b \in B$.

Theorem 6. *Let $\alpha \colon S \to R$ be a ring homomorphism and I the ideal of R generated by $[\operatorname{Ker} d_{R/S}, R]$. Put $\overline{R} = R/I$ and $\overline{S} = S/\alpha^{-1}(I)$.*

(1) $\overline{S}$ is commutative and $\overline{R}$ is an $\overline{S}$-algebra satisfying the identity $[[X, Y], Z] = 0$.

(2) $\Omega_{\overline{R}/\overline{S}} \simeq \Omega_{R/S}$ as R-bimodules.

Proof. (1) Since $S \cdot 1 \subseteq \operatorname{Ker} d_{R/S}$, $\overline{S}$ is commutative and $\overline{R}$ is an $\overline{S}$-algebra. Since $d_{R/S}$ is central, it is easy to see that $d_{R/S}(xy) = d_{R/S}(yx)$ for all $x, y \in R$, and so $[R, R] \subseteq \operatorname{Ker} d_{R/S}$. Hence $\overline{R}$ satisfies the identity $[[X, Y], Z] = 0$.

(2) By the definition of $\overline{S}$, $\Omega_{\overline{R}/\overline{S}}$ is isomorphic to $\Omega_{\overline{R}/S}$, and, by Corollary 5, this is isomorphic to $\Omega_{R/S}/Rd_{R/S}(I)$. Hence it suffices to show

that $d_{R/S}(I) = 0$. We showed in the proof of (1) that $[R, R] \subseteq \operatorname{Ker} d_{R/S}$. Since $I = R[\operatorname{Ker} d_{R/S}, R]$, we have $I \subseteq [R \operatorname{Ker} d_{R/S}, R] + [R, R] \operatorname{Ker} d_{R/S} \subseteq \operatorname{Ker} d_{R/S}$. □

4. Quasi-Separable Extensions

According to Nakai [N], an S-ring R is said to be *quasi-separable* if $\Omega_{R/S} = 0$. By [K, Theorem 2.4], every separable S-ring in the sense of Miyashita [M] is quasi-separable.

We study commutators of a quasi-separable S-ring R. We define $[R, R]_n$ as follows: let $[R, R]_0 = R$ and proceed inductively $[R, R]_n = [[R, R]_{n-1}, R]$. The ideal of R generated by $[R, R]_n$ is denoted by $D_n(R)$.

Theorem 7. *Let R be a quasi-separable S-ring and n a positive integer. If $[[R, R]_i, S] \subseteq D_{i+2}(R)$ for all $i = 0, 1, \ldots, n-2$, then $D_n(R)$ coincides with $D_1(R)$.*

Proof. Let $\alpha\colon S \to R$ be the ring homomorphism which defines the S-ring structure. Put $\overline{R} = R/D_n(R)$. Let $d\colon R \to \overline{R}$ be an inner derivation effected by an element in $[\overline{R}, \overline{R}]_{n-2}$. Then, by assumption, d is a central S-derivation. Hence we have $d = 0$ by Theorem 1(2). This means that $[R, R]_{n-1} \subseteq D_n(R)$, and hence $D_{n-1}(R) = D_n(R)$. Continuing this method, we have $D_1(R) = D_n(R)$. □

The following corollaries are immediate consequences of this theorem.

Corollary 8. *If R is a quasi-separable algebra over a commutative ring, then $D_n(R) = D_1(R)$ for all positive integers n.*

Corollary 9. *Let R be a quasi-separable algebra over a commutative ring. If $[R, R]_n = 0$ for some positive integer n, then R is commutative.*

Finally, we characterize the quasi-separability of finitely generated algebras by modules of differentials of free algebras.

Let $A = k\langle x_1, \ldots, x_n\rangle$ be the free algebra on the finite set $\{x_1, \ldots, x_n\}$ over a commutative ring k, and let $\dfrac{\partial}{\partial x_i}$ denote the k-derivation of A determined by $\dfrac{\partial x_i}{\partial x_i} = 1$ and $\dfrac{\partial x_j}{\partial x_i} = 0$ $(j \neq i)$. Then the mapping $d\colon A \to A^n$ defined by $d(f) = \left(\dfrac{\partial f}{\partial x_1}, \ldots, \dfrac{\partial f}{\partial x_n}\right)$ is a k-derivation. Let N be the submodule of A^n generated by $\{fd(g) - d(g)f \mid f, g \in A\}$ as an A-bimodule.

Theorem 10. Let $A = k\langle x_1, \ldots, x_n \rangle$ be the free algebra on the finite set $\{x_1, \ldots, x_n\}$ over a commutative ring k, and let I be an ideal of A. Under the above notation, A/I is a quasi-separable k-algebra if and only if $d(x_i)$ belongs to $d(I) + \sum_{j=1}^{n} I d(x_j) + N$ for all $i = 1, 2, \ldots, n$.

Proof. Put $\Omega = A^n/N$ and $\delta = \pi d$, where $\pi \colon A^n \to \Omega$ is the canonical mapping. Then, by [K, Example 1.3], there exists an A-bimodule isomorphism $\nu \colon \Omega_{A/k} \to \Omega$ such that $\delta = \nu d_{A/k}$. By Corollary 5, we have $\Omega_{(A/I)/k} \simeq \Omega/(\delta(I) + I\Omega)$. Hence A/I is a quasi-separable k-algebra if and only if $A^n = d(I) + I \cdot A^n + N = d(I) + \sum_{j=1}^{n} I d(x_j) + N$. □

References

[B] N. Bourbaki, *Algèbre I: Chapitres 1 à 3*, Hermann, Paris, 1970.

[HK] M. Hongan and H. Komatsu, *On the module of differentials of a noncommutative algebras and symmetric biderivations of a semiprime algebra*, Comm. Algebra **28** (2000), 669–692.

[K] H. Komatsu, *Quasi-separable extensions of noncommutative rings*, Comm. Algebra, to appear.

[M] Y. Miyashita, *Finite outer Galois theory of non-commutative rings*, J. Fac. Sci. Hokkaido Univ. Ser. I **19** (1966), 114–134.

[N] Y. Nakai, *On the theory of differentials in commutative rings*, J. Math. Soc. Japan **13** (1961), 63–84.

[S] M. E. Sweedler, *Right derivations and right differential operators*, Pacific J. Math. **86** (1980), 327–360.

Department of System Engineering
Okayama Prefectural University
Okayama 719-1197, Japan
e-mail: komatsu@cse.oka-pu.ac.jp

Dual Bimodules and Nakayama Permutations

Y. Kurata, K. Koike and K. Hashimoto

Abstract

Unifying results due to Azumaya, Hajarnavis and Norton, and Nicholson and Yousif, we shall give a characterization of the Nakayama permutation. Then the dual bimodule can be characterized by a Nakayama permutation. We shall show that any left PF ring also admits a Nakayama permutation.

A ring R is QF [11] if it is Artinian and admits a "Nakayama permutation" of its basic set $\{e_1, e_2, \ldots, e_n\}$ of primitive idempotents, i.e., a permutation σ on $\{1, 2, \ldots, n\}$ such that for each i, $1 \leq i \leq n$,

$$\mathrm{soc}(Re_i) \cong Re_{\sigma(i)}/J(R)e_{\sigma(i)} \quad \text{and} \quad \mathrm{soc}(e_{\sigma(i)}R) \cong e_iR/e_iJ(R).$$

In [3] Azumaya has introduced the notion of quasi-Frobenius two-sided R-S-modules and shown that if $_RR$ and S_S are both Artinian, then these are characterized by the existence of a Nakayama permutation. Hajarnavis and Norton [5] have shown that every dual ring admits a Nakayama permutation and recently Nicholson and Yousif [13] also have shown that every minfull ring admits a Nakayama permutation.

In this note, unifying these results, we shall give a characterization of the Nakayama permutation using min injectivity (Theorem 1.5) and then characterize the dual bimodules [8] using a Nakayama permutation and simple injectivity (Theorem 2.2). Finally we shall point out that if $_RQ_S$ is a dual bimodule, then so is $_{eRe}eQf_{fSf}$ for some idempotents e and f (Theorem 3.2) and that any left PF ring also admits a Nakayama permutation. Using this we shall give some criterion for a left PF ring to be a two-sided PF ring (Corollary 3.7).

Throughout this paper, R and S are rings with identity and $_RQ_S$ is an (R,S)-bimodule. The radicals of R and S are denoted by $J(R)$ and $J(S)$, and the socles of $_RQ$ and Q_S are denoted by $\mathrm{soc}(_RQ)$ and $\mathrm{soc}(Q_S)$, respectively. For a left R- or right S-module M, its Q-dual module will be denoted by M^*. We shall denote by $r_Q(*)$ and $\ell_R(*)$ the right annihilator in Q and the left annihilator in R, respectively.

1. Let R and S be rings with identity and $_RQ_S$ an (R,S)-bimodule. Then $_RQ$ is called min Q-injective (cf. [13, p.549]) if, for each simple submodule Q' of $_RQ$, every R-homomorphism from Q' to Q is given by a right multiplication of some element of S. The min Q-injectivity of Q_S is defined similarly. Slightly modifying [13, Theorem 1.14 and Proposition 2.2], we have the following Lemmas 1.1 and 1.2.

Lemma 1.1. *Let $_RQ_S$ be an (R,S)-bimodule and $_RQ$ min Q-injective. For u in Q, if Ru is simple, then so is uS. In particular,*
$$soc(_RQ) \leq soc(Q_S).$$

Proof. Suppose that Ru is simple. Then for any $us(\neq 0)$ in uS, the mapping $f : Ru \to Rus$ sending au to aus is an R-isomorphism. Hence Rus is simple and by assumption the mapping f^{-1} is given by t_R, the right multiplication of some t in S. Therefore, $u = f^{-1}(us) = (us)t$ and thus $uS = usS$, which means that uS is simple. □

Lemma 1.2. *Let $_RQ_S$ be an (R,S)-bimodule. Then the following conditions are equivalent:*

(1) $_RQ$ *is min Q-injective.*
(2) *If $_RM$ is simple, then M^* is either simple or 0.*
(3) *If A is a maximal left ideal of R, then $r_Q(A)$ is either simple or 0.*

In case R is semiperfect and $\{e_1, e_2, \ldots, e_n\}$ is a basic set of primitive idempotents for R, each one of (1) to (3) is also equivalent to

(4) $e_i soc(_RQ)$ *is either simple or 0 for $i = 1, 2, \ldots, n$.*

Proof. (1)⇒(2). Let $_RM$ be simple and $M^* \neq 0$. Then for any $f_1 \neq 0$, $f_2 \neq 0$ in M^*, by assumption $f_2 \circ f_1^{-1} = s_R$ for some s in S. Hence, for any x in M, $f_2(x) = (f_2 \circ f_1^{-1})(f_1(x)) = f_1(x)s = (f_1 s)(x)$. This means that $M^* = f_1 S$ and thus M^* is simple.

(2)⇒(3) is trivial.

(3)⇒(1). Let Q' be a simple submodule of $_RQ$ and $f : Q' \to Q$ an R-homomorphism. Let $Q' = Ru$ for some u in Q. Since $Q' \cong R/\ell_R(u)$, $\ell_R(u)$ is a maximal left ideal of R and $r_Q \ell_R(u)$ contains u. Hence $r_Q \ell_R(u)$ must be simple and is equal to uS. Let $\bar{f} \in (R/\ell_R(u))^*$ be the mapping induced by f. As $(R/\ell_R(u))^* \cong r_Q \ell_R(u) = uS$, $\bar{f}(1 + \ell_R(u)) = us$ for some s in S. Then for any au in Ru, $f(au) = \bar{f}(a + \ell_R(u)) = a \cdot \bar{f}(1 + \ell_R(u)) = (au)s$ and hence $_RQ$ is min Q-injective.

Suppose now that R is semiperfect. Then every simple left R-module is isomorphic to $Re_i/J(R)e_i$ for some i and $(Re_i/J(R)e_i)^* \cong e_i r_Q(J(R)) = e_i soc(_RQ)$. Hence (4) is equivalent to (2). □

Let $_RQ_S$ be an (R,S)-bimodule. We shall call $_RQ$ Kasch if Q contains an isomorphic copy of every simple left R-module, or equivalently, $M^* \neq 0$

for every simple left R-module M. In case R is semiperfect, $_RQ$ is Kasch if and only if, for any primitive idempotent e in R, $\mathrm{esoc}(_RQ) \neq 0$. Kasch right S-modules are defined similarly.

Lemma 1.3. *Let $_RQ_S$ be an (R,S)-bimodule. Suppose that R is semiperfect with a basic set $\{e_1, e_2, \ldots, e_n\}$ of primitive idempotents for R and that both $_RQ$ and Q_S are min Q-injective. If $_RQ$ is Kasch, then the following conditions are equivalent:*

(1) *There exist exactly n non-isomorphic simple right S-modules.*
(2) *Q_S is Kasch.*

Proof. Let $1 \leq i \leq n$. Since $_RQ$ is Kasch, there exists a u_i in Q such that $Re_i/J(R)e_i \cong Ru_i$. Then u_iS is simple by Lemma 1.1. Now $(Ru_i)^* \cong r_Q\ell_R(u_i) \ni u_i \neq 0$. Since $r_Q\ell_R(u_i)$ is simple by Lemma 1.2, $r_Q\ell_R(u_i) = u_iS$ and hence $(Ru_i)^* \cong u_iS$. Similarly, we have $(u_iS)^* \cong Ru_i$, since Q_S is min Q-injective.

Let uS, u in Q, be any simple right S-module contained in Q. Since Q_S is min Q-injective, Ru is simple by the right version of Lemma 1.1. There exists some j, $1 \leq j \leq n$, such that $Ru \cong Ru_j$. Hence $uS \cong u_jS$ and thus $u_1S, u_2S, \ldots, u_nS$ exhaust non-isomorphic simple right S-modules contained in Q. The equivalence of (1) and (2) are then clear. □

Lemma 1.4. *Let $_RQ_S$ be an (R,S)-bimodule. Suppose that R and S are semiperfect rings with basic sets consisting of the same number of primitive idempotents. If Q_S is min Q-injective, then the following conditions are equivalent:*

(1) *$_RQ$ is min Q-injective and $_RQ$ is Kasch.*
(2) *$_RQ$ is min Q-injective and Q_S is Kasch.*
(3) *For any basic sets $\{e_1, e_2, \ldots, e_n\}$ and $\{f_1, f_2, \ldots, f_n\}$ for R and S, respectively, there exists a permutation σ on $\{1, 2, \ldots, n\}$ such that for each i, $1 \leq i \leq n$,*

$$e_{\sigma(i)}\mathrm{soc}(_RQ) \cong f_iS/f_iJ(S).$$

Proof. The equivalence of (1) and (2) follows from Lemma 1.3 and its left version.

(1)⇒(3). Let $\{e_1, e_2, \ldots, e_n\}$ and $\{f_1, f_2, \ldots, f_n\}$ be basic sets for R and S, respectively. Then $f_1S/f_1J(S), f_2S/f_2J(S), \ldots, f_nS/f_nJ(S)$ is a non-redundant set of representatives of the simple right S-modules. Since $_RQ$ is Kasch, as is seen from the proof of Lemma 1.3, there exist $u_1, u_2, \ldots, u_n$ in Q such that $u_1S, u_2S, \ldots, u_nS$ exhaust non-isomorphic simple right S-modules. Therefore, for each i, $1 \leq i \leq n$, there exists a j, $1 \leq j \leq n$, such that $f_iS/f_iJ(S) \cong u_jS$. Let $j = \sigma(i)$. Then σ is a permutation on $\{1, 2, \ldots, n\}$ such that for each i, $1 \leq i \leq n$,

$$f_i S/f_i J(S) \cong u_{\sigma(i)} S \cong (R u_{\sigma(i)})^* \cong (Re_{\sigma(i)}/J(R)e_{\sigma(i)})^* \cong e_{\sigma(i)} \mathrm{soc}(_R Q).$$
(3)$\Rightarrow$(2) follows from Lemma 1.2. $\square$

Similarly, we have

Lemma 1.4′. *Let $_R Q_S$ be an (R, S)-bimodule. Suppose that R and S are semiperfect rings with basic sets consisting of the same number of primitive idempotents. If $_R Q$ is min Q-injective, then the following conditions are equivalent:*

(1′) Q_S *is min Q-injective and* Q_S *is Kasch.*
(2′) Q_S *is min Q-injective and* $_R Q$ *is Kasch.*
(3′) *For any basic sets $\{e_1, e_2, \ldots, e_n\}$ and $\{f_1, f_2, \ldots, f_n\}$ for R and S, respectively, there exists a permutation σ on $\{1, 2, \ldots, n\}$ such that for each i, $1 \leq i \leq n$,*
$$\mathrm{soc}(Q_S) f_i \cong Re_{\sigma(i)}/J(R) e_{\sigma(i)}.$$

Remark. Let $_R Q_S$ be an (R, S)-bimodule, R and S semiperfect rings with basic sets consisting of the same number of primitive idempotents and Q_S min Q-injective. For any basic sets $\{e_1, e_2, \ldots, e_n\}$ and $\{f_1, f_2, \ldots, f_n\}$ for R and S, respectively, assume (3) of Lemma 1.4. Then there exists a permutation σ on $\{1, 2, \ldots, n\}$ such that for each i, $1 \leq i \leq n$,
$$e_{\sigma(i)} \mathrm{soc}(_R Q) \cong f_i S/f_i J(S).$$

Hence we have

$$\mathrm{soc}(Q_S) f_i \cong (f_i S/f_i J(S))^* \cong (e_{\sigma(i)} \mathrm{soc}(_R Q))^* \cong (Re_{\sigma(i)}/J(R) e_{\sigma(i)})^{**}.$$

As is shown in the proof of Lemma 1.3, $Re_{\sigma(i)}/J(R) e_{\sigma(i)}$ is reflexive. Hence we have
$$\mathrm{soc}(Q_S) f_i \cong Re_{\sigma(i)}/J(R) e_{\sigma(i)}.$$
This means that the permutation σ in (3′) of Lemma 1.4′ is the same as σ in (3) of Lemma 1.4.

We can use these lemmas to prove that, for an (R, S)-bimodule $_R Q_S$, the Nakayama permutation can be characterized by the min Q-injectivity of Q in both sides. This may be seen as a unification of [3, Theorem 12], [5, Theorem 4.2] and [13, Corollary 3.10].

Theorem 1.5. *Let $_R Q_S$ be an (R, S)-bimodule. Suppose that R and S are semiperfect rings with basic sets consisting of the same number of primitive idempotents. Then the following conditions are equivalent:*

(1) $_RQ$ and Q_S are min Q-injective and $_RQ$ is Kasch.
(2) $_RQ$ and Q_S are min Q-injective and Q_S is Kasch.
(3) For any basic sets $\{e_1, e_2, \ldots, e_n\}$ and $\{f_1, f_2, \ldots, f_n\}$ of primitive idempotents for R and S, respectively, there exists a permutation σ on $\{1, 2, \ldots, n\}$ such that for each i, $1 \leq i \leq n$,

$$e_{\sigma(i)} soc(_RQ) \cong f_i S / f_i J(S) \quad \text{and} \quad soc(Q_S) f_i \cong Re_{\sigma(i)} / J(R) e_{\sigma(i)}.$$

Let $_RQ_S$ be an (R, S)-bimodule. Then we shall call $_RQ$ simple R-injective (cf. [14, p.980]) if every R-homomorphism with simple image from a left ideal of R to Q is given by a right multiplication by some element of Q, and we shall call $_RQ$ simple Q-injective if every R-homomorphism with simple image from a submodule of $_RQ$ to Q is given by a right multiplication by an element of S. Similarly we may define the simple S-injectivity and the simple Q-injectivity of Q_S. Injective modules are clearly simple injective and simple injective modules are also min injective, but the converses are not true in general (cf. [14, Examples 1 to 5]).

Using simple injectivity, we may give a sufficient condition for the existence of Nakayama permutations (cf. [14, Proposition 1]). We need the following lemma which is a converse of Lemma 1.1.

Lemma 1.6. (cf. [3, Proposition 3]) *Let $_RQ_S$ be an (R, S)-bimodule. If Q_S is simple Q-injective, then the following conditions are equivalent:*
(1) $soc(_RQ) \leq soc(Q_S)$.
(2) $_RQ$ is min Q-injective.

Proof. (1)$\Rightarrow$(2). Let A be a maximal left ideal in R and suppose that $r_Q(A) \neq 0$. Then, for any $u(\neq 0)$ in $r_Q(A)$, $\ell_R(u) = \ell_R r_Q(A) = A$. Hence Ru is simple and $u \in soc(_RQ) \leq soc(Q_S)$. Thus $r_Q(A)$ is semisimple. If $r_Q(A)$ is not simple, then there exist simple submodules Q_1, Q_2 in $r_Q(A)$ such that $Q_1 \oplus Q_2 \leq r_Q(A)$. By assumption, the projection map $p : Q_1 \oplus Q_2 \to Q_1$ is given by a_L for some a in R. As $aQ_2 = p(Q_2) = 0$, a is in $\ell_R(Q_2)$. But since A is maximal, $A = \ell_R r_Q(A) = \ell_R(Q_2)$ and hence $a \in A$. Therefore, $Q_1 = p(Q_1 \oplus Q_2) = aQ_1 \leq A \cdot r_Q(A) = 0$, a contradiction. Thus $r_Q(A)$ is simple. By Lemma 1.2, $_RQ$ is min Q-injective.
(2)$\Rightarrow$(1) follows from Lemma 1.1. □

Theorem 1.7. *Let $_RQ_S$ be an (R, S)-bimodule with R and S semiperfect. If $_RQ$ is faithful, simple Q-injective and $soc(_RQ)$ is essential in Q_S, then for any basic set $\{e_1, e_2, \ldots, e_n\}$ of primitive idempotents for R, there exist n primitive idempotents $f_1, f_2, \ldots, f_n$ in S and a permutation σ on $\{1, 2, \ldots, n\}$ such that for each i, $1 \leq i \leq n$,*

$$e_{\sigma(i)} soc(_RQ) \cong f_i S / f_i J(S) \quad \text{and} \quad soc(Q_S) f_i \cong Re_{\sigma(i)} / J(R) e_{\sigma(i)}.$$

Proof. Since $\mathrm{soc}(_RQ)$ is essential in Q_S, $\mathrm{soc}(Q_S) \leq \mathrm{soc}(_RQ)$ and hence by the right version of Lemma 1.6 Q_S is min Q-injective.

Let $1 \leq i \leq n$. Since $_RQ$ is faithful, $e_iQ \neq 0$. Hence $(Re_i/J(R)e_i)^* \cong e_i\mathrm{soc}(_RQ) = e_iQ \cap \mathrm{soc}(_RQ) \neq 0$, which means that $_RQ$ is Kasch.

As is seen from the proof of Lemma 1.3, there exist $u_1, u_2, \ldots, u_n$ in Q such that $u_1S, u_2S, \ldots, u_nS$ exhaust non-isomorphic simple right S-modules contained in Q. Then there exists a basic set $\{f_1, \ldots, f_n, \ldots, f_m\}$ of primitive idempotents for S such that $f_1, f_2, \ldots, f_n$ correspond to simple right S-modules contained in Q. Hence we can define a permutation σ on $\{1, 2, \ldots, n\}$ such that for each i, $1 \leq i \leq n$, $f_iS/f_iJ(S) \cong e_{\sigma(i)}\mathrm{soc}(_RQ)$, and moreover, as is seen from Remark, we have $\mathrm{soc}(Q_S)f_i \cong Re_{\sigma(i)}/J(R)e_{\sigma(i)}$. □

As the following example shows, the converse of Theorem 1.7 is not true in general.

Example 1.8. Camillo [4] has given an example of a commutative, local and semiprimary ring R which is principally injective, but not simple R-injective (cf. [14, Example 5]). Then R is min R-injective and is Kasch by [12, Theorem 2.4]. Hence R admits a Nakayama permutation by Theorem 1.5.

2. An (R, S)-bimodule $_RQ_S$ is called a left dual bimodule [8] if

$$l_R r_Q(A) = A \quad \text{and} \quad r_Q l_R(Q') = Q'$$

for every left ideal A of R and every submodule Q' of Q_S. A right dual bimodule is similarly defined and we shall call Q a dual bimodule if it is a left dual bimodule and is a right dual bimodule as well.

Now using Theorem 1.5 we shall point out that any dual bimodule admits a Nakayama permutation. Let $_RQ_S$ be a dual bimodule. Then R and S are semiperfect by [10, Corollary 1.6], and have basic sets consisting of the same number of primitive idempotents by [8, Theorem 2.1]. Moreover, by [8, Lemma 2.4], $_RQ$ is simple Q-injective and is Kasch. Similarly, Q_S is simple Q-injective and is Kasch. Hence by Theorem 1.5 Q admits a Nakayama permutation.

However, the converse of this is not true in general.

Example 2.1. As in [8, Example 4.2], let $Q' = Rp^{-1}/R$ and $\bar{R} = R/Rp$. Then $_{\bar{R}}Q'_R$ is a left dual bimodule but not a dual bimodule. Both $\bar{R}$ and R have just one simple module within isomorphisms and so trivially Q' admits a Nakayama permutation.

We now characterize the dual bimodule by means of a Nakayama permutation. The following theorem may be seen as a generalization of [13, Theorem 4.1].

Theorem 2.2. *An (R,S)-bimodule $_RQ_S$ is a dual bimodule if and only if the following conditions hold:*

(1) *R and S are semiperfect and have basic sets with the same number of primitive idempotents.*

(2) *$_RQ$ is simple R- and simple Q-injective and Q_S is simple S- and simple Q-injective.*

(3) *One of the following conditions holds:*

  (a) *$_RQ$ is Kasch.*

  (b) *Q_S is Kasch.*

  (c) *For any basic sets $\{e_1, e_2, \ldots, e_n\}$ and $\{f_1, f_2, \ldots, f_n\}$ of primitive idempotents for R and S, respectively, there exists a permutation σ on $\{1, 2, \ldots, n\}$ such that*

$$e_{\sigma(i)} soc(_RQ) \cong f_i S / f_i J(S).$$

  (d) *For any basic sets $\{e_1, e_2, \ldots, e_n\}$ and $\{f_1, f_2, \ldots, f_n\}$ of primitive idempotents for R and S, respectively, there exists a permutation σ on $\{1, 2, \ldots, n\}$ such that*

$$soc(Q_S) f_i \cong Re_{\sigma(i)} / J(R) e_{\sigma(i)}.$$

Proof. We have only to prove the "if" part and this follows from Lemmas 1.4 and 1.4', and [9, Theorem 3.3]. $\square$

In particular, we have

Corollary 2.3. *A ring R is a dual ring if and only if the following conditions hold:*

(1) *R is semiperfect.*

(2) *R is simple R-injective on both sides.*

(3) *One of the following conditions holds:*

  (a) *$_RR$ is Kasch.*

  (b) *R_R is Kasch.*

  (c) *For any basic set $\{e_1, e_2, \ldots, e_n\}$ of primitive idempotents for R, there exists a permutation σ on $\{1, 2, \ldots, n\}$ such that*

$$e_{\sigma(i)} soc(_RR) \cong e_i R / e_i J(R).$$

  (d) *For any basic set $\{e_1, e_2, \ldots, e_n\}$ of primitive idempotents for R, there exists a permutation σ on $\{1, 2, \ldots, n\}$ such that*

$$soc(R_R) e_i \cong Re_{\sigma(i)} / J(R) e_{\sigma(i)}.$$

3. Let $_RQ_S$ be a dual bimodule. Then R and S are semiperfect and have basic sets of primitive idempotents with the same number of elements.

Let $\{e_1, e_2, \ldots, e_n\}$ and $\{f_1, f_2, \ldots, f_n\}$ be any basic sets for R and S, respectively and σ a Nakayama permutation on $\{1, 2, \ldots, n\}$. For any non-empty subset I of $\{1, 2, \ldots, n\}$, let

$$e = \sum_{i \in I} e_{\sigma(i)} \text{ and } f = \sum_{i \in I} f_i.$$

With these notations, we have

Lemma 3.1. $\operatorname{soc}(_{eRe}eQf) = e \cdot \operatorname{soc}(_RQ) \cdot f = e \cdot \operatorname{soc}(Q_S) \cdot f = \operatorname{soc}(eQf_{fSf}).$

Proof. Since R and eRe are semiperfect and $er_Q(J(R))f \leq r_{eQf}(eJ(R)e)$, we have $e\operatorname{soc}(_RQ)f \leq \operatorname{soc}(_{eRe}eQf)$.

To prove the reverse inclusion, we may show that $J(R)euf = 0$ for any euf in $r_{eQf}(eJ(R)e)$ with u in Q. Suppose that $J(R)euf \neq 0$ for some euf. Then $J(R)euf_i \neq 0$ for some i in I. By [8, Proposition 1.8], $\operatorname{soc}(_RQ)f_i$ is essential in $_RQf_i$. Hence there exists a b in $J(R)$ such that $0 \neq beuf_i \in \operatorname{soc}(_RQ)f_i$. Since $Re_{\sigma(i)}/J(R)e_{\sigma(i)} \cong \operatorname{soc}(_RQ)f_i$, $e_{\sigma(i)}abeuf_i \neq 0$ for some a in R. This contradicts the fact that $e_{\sigma(i)}abe \in eJ(R)e$ and $euf_i \in r_{eQf}(eJ(R)e)$. Thus we have $e\operatorname{soc}(_RQ)f = \operatorname{soc}(_{eRe}eQf)$. By symmetry we have $e\operatorname{soc}(Q_S)f = \operatorname{soc}(eQf_{fSf})$. □

As an application of Theorem 2.2 we have

Theorem 3.2. *With the notation as above, $_{eRe}eQf_{fSf}$ is a dual bimodule.*

Proof. We may check conditions (1), (2), (3)(a) of Theorem 2.2 for the bimodule $_{eRe}eQf_{fSf}$.

(1) eRe and fSf are semiperfect and, as is easily seen, $\{e_{\sigma(i)}\}_{i \in I}$ and $\{f_i\}_{i \in I}$ are basic sets of primitive idempotents for eRe and fSf, respectively. Hence eRe and fSf have basic sets with the same number of primitive idempotents.

(2) Now we show that $_{eRe}eQf$ is simple eRe-injective. Let A be a left ideal of eRe and $\alpha : A \to eQf$ an eRe-homomorphism with simple image. Then $\bar{\alpha} : ReA \to ReQf$ sending $\sum_k r_k ea_k$ to $\sum_k r_k e \cdot \alpha(a_k)$ is a well-defined R-homomorphism, where $r_k \in R$ and $a_k \in A$. To see this, suppose that $\sum r_k ea_k = 0$ and $u = \sum r_k e\alpha(a_k) \neq 0$. Since $R(e_1 + e_2 + \cdots + e_n)R = R$, there exists i in $\{1, 2, \ldots, n\}$ such that $e_i Ru \neq 0$. Hence $e_i ru \neq 0$ for some r in R. If $i = \sigma(i')$ for some $i' \in I$, then $ee_i = e_i$ and $0 \neq e_i ru = \sum e_i rr_k e\alpha(a_k) = \alpha(\sum (ee_i rr_k e)a_k) = 0$, a contradiction. If $i \neq \sigma(i')$ for any $i' \in I$, then $ee_i = 0$ and $e_i rr_k e \in J(R)$. $\alpha(A)$ is a

simple eRe-module and is contained in $r_Q(J(R))$ by Lemma 3.1. Hence $0 \neq e_i r u = \sum(e_i r r_k e)\alpha(a_k) = 0$, a contradiction.

Since $\alpha(A)$ is simple, $\alpha(A) = eRe \cdot v$ for some v in Q. Hence $\bar{\alpha}(ReA) = Re \cdot \alpha(A) = Rev$ is finitely generated. Therefore by [8, Lemma 1.13] there exists a w in Q such that $\bar{\alpha} = w_R$, the right multiplication by w. Then for any a in A, $\alpha(a) = \alpha(ea) = \bar{\alpha}(ea) = eaw = aw = a \cdot ewf$. Thus we have $\alpha = (ewf)_R$, which means that eQf is simple eRe-injective.

Similarly we can show that eQf is simple eQf-injective. By symmetry, eQf is simple fSf-injective and simple eQf-injective.

(3) Finally we shall show that $_{eRe}eQf$ is Kasch. Since $\{e_{\sigma(i)}\}_{i \in I}$ is a basic set of primitive idempotents for eRe, any simple left eRe-module is isomorphic to some $eRee_{\sigma(i)}/J(eRe)e_{\sigma(i)}$. Hence we may show that $e_{\sigma(i)}\text{soc}(_{eRe}eQf) = e_{\sigma(i)}\text{soc}(Q_S)f \neq 0$. Since $\text{soc}(Q_S)f_i \cong Re_{\sigma(i)}/J(R)e_{\sigma(i)}$ and is a simple left R-module, $e_{\sigma(i)}\text{soc}(Q_S)f_i \neq 0$. As $f_i = ff_i$, it follows that $e_{\sigma(i)}\text{soc}(Q_S)f \neq 0$. □

Let $Q = R$ and assume that $e = \sum_{i \in I} e_{\sigma(i)} = \sum_{i \in I} e_i$, or equivalently, $\sigma(i) \in I$ for all i. Then we have

Corollary 3.3. *If R is a dual ring, then so is eRe.*

If R is a basic, dual ring with the identical Nakayama permutation, then by Corollary 3.3, $e'Re'$ is a dual ring for every nonzero idempotent e' in R. For QF-rings, this fact was noted by [6, Corollary 1.4].

Corollary 3.4. *If $_RQ_S$ defines a Morita duality, then so does $_{eRe}eQf_{fSf}$.*

Proof. Suppose that $_RQ_S$ defines a Morita duality. Then $_RQ_S$ is a dual bimodule. By Theorem 3.2, $_{eRe}eQf_{fSf}$ is also a dual bimodule and by [15, Lemma 4.9] $_{eRe}eRe$ and fSf_{fSf} are linearly compact. Therefore, by [16, Theorem 1.7] $_{eRe}eQf_{fSf}$ defines a Morita duality. □

Finally, as an application of Theorem 1.7, we shall point out that left PF rings also admit a Nakayama permutation. Recall that R is a left PF ring if $_RR$ is an injective cogenerator, or equivalently, R is semiperfect and $_RR$ is injective with essential socle.

Proposition 3.5. *Let R be a left PF ring with a basic set $\{e_1, e_2, \ldots, e_n\}$. Then there exists a permutation σ on $\{1, 2, \ldots, n\}$ such that for each i, $1 \leq i \leq n$,*
$$e_{\sigma(i)}\text{soc}(_RR) \cong e_iR/e_iJ(R) \quad \text{and} \quad \text{soc}(R_R)e_i \cong Re_{\sigma(i)}/J(R)e_{\sigma(i)}.$$

Proof. This follows from [7, Theorem 6] and Theorem 1.7. □

Proposition 3.6. *Let R be a left PF ring with a basic set $\{e_1, e_2, \ldots, e_n\}$. For any non-empty subset I of $\{1, 2, \ldots, n\}$, assume that $e = \sum_{i \in I} e_{\sigma(i)} = \sum_{i \in I} e_i$. Then eRe is also a left PF ring.*

Proof. We may show that $_{eRe}eRe$ is injective and has essential socle, since eRe is semiperfect. As $r_{Re}(eR) = 0$, the injectivity of $_{eRe}eRe$ follows from [15, Lemma 4.10]. Now let eae be any nonzero element of eRe. Since $_RR$ has essential socle, there exists some b in R such that $0 \neq beae \in \mathrm{soc}(_RR)$. Hence for some i in I, $0 \neq beae_i$. Since $\mathrm{soc}(_RR)e_i \cong Re_{\sigma(i)}/J(R)e_{\sigma(i)}$, $e_{\sigma(i)}cbeae_i \neq 0$ for some c in R. Hence $0 \neq ecbeae$ is in $\mathrm{esoc}(_RR)e$, which shows that $\mathrm{soc}(_{eRe}eRe)$ is essential in $_{eRe}eRe$. □

As a corollary of Proposition 3.5, we can give a criterion for a left PF ring to be a two-sided PF ring.

Corollary 3.7. *For a left PF ring R, the following conditions are equivalent:*

(1) *R is a two-sided PF ring.*
(2) *R is a dual ring and every cyclic left R-module is reflexive.*
(3) *R_R is simple R-injective and every cyclic left R-module is reflexive.*
(4) *Every cyclic left R-module is reflexive and every cyclic right R-module is torsionless.*

Proof. (1)⇒(2). R is clearly a dual ring and by [8, Theorem 3.2] every cyclic left R-module is reflexive.

(2)⇒(3). This follows from Corollary 2.3.

(3)⇒(4). Since R is left PF and R_R is simple R-injective, R is a dual ring by Corollary 2.3 and hence every cyclic right R-module is torsionless.

(4)⇒(1). R is a dual ring and R_R is injective by [8, Theorem 3.2]. Hence R is right PF. □

References

[1] F. W. Anderson and K. R. Fuller, *Rings and Categories of Modules*, Springer-Verlag, Heidelberg, New York, 1973.

[2] P. N. Ánh, *Characterization of two-sided PF-rings*, J. Algebra **141** (1991), 316–320.

[3] G. Azumaya, *A duality theory for injective modules*, Amer. J. Math. **81** (1959), 249–278.

[4] V. Camillo, *Commutative rings whose principal ideals are annihilators*, Portugaliae Math. **46** (1989), 33–37.

[5] C. R. Hajarnavis and N. C. Norton, *On dual rings and their modules*, J. Algebra **93** (1985), 253–266.

[6] M. Hoshino, *Strongly quasi-Frobenius rings*, preprint.

[7] T. Kato, *Self-injective rings*, Tohoku Math J. **19** (1967), 485–495.

[8] Y. Kurata and K. Hashimoto, *On dual-bimodules*, Tsukuba J. Math. **16** (1992), 85–105.

[9] _____, *Dual-bimodules and AB5* conditions*, Comm. Algebra **27** (1999), 4743–4751.

[10] M. Morimoto and T. Sumioka, *On dual pairs and simple-injective modules*, preprint.

[11] T. Nakayama, *On Frobeniusean algebras II*, Ann. Math. **42** (1941), 1–21.

[12] W. K. Nicholson and M. F. Yousif, *Principally injective rings*, J. Algebra **174** (1995), 77–93.

[13] _____, *Mininjective rings*, ibid. **187** (1997), 548–578.

[14] _____, *On perfect simple-injective rings*, Proc. Amer. Math. Soc. **125** (1997), 979–985.

[15] W. Xue, *Rings with Morita Duality*, Lecture Notes in Math. **1523**, Springer-Verlag, Heidelberg, New York, 1992.

[16] _____, *Characterizations of Morita duality*, Algebra Colloq. **2** (1995), 339–350.

[17] _____, *Characterizations of Morita duality via idempotents for semiperfect rings*, ibid. **5** (1998), 99–110.

Y. Kurata
Res. Inst. for Integrated Science
Kanagawa University
Hiratsuka, Japan

K. Hashimoto
System Development Laboratory
Hitachi Ltd.
Yokohama, Japan

K. Koike
Division of General Education
Oshima National College of Maritime Technology
Oshima-cho, Japan
e-mail: `koike@c.oshima-k.ac.jp`

The Coinduced Functor and Homological Properties of Hopf Modules

Tao Li and Zhixi Wang

Abstract

Let H be a commutative Hopf algebra over a field k and A a right H-comodule algebra. This paper is concerned with homological algebra for Hopf A-modules, especially with injective modules, and the transfer of homological properties of A to those of A^{coH}.

0. Introduction

Throughout this paper, let H be a commutative Hopf algebra over a field k with the antipode S, and A a right H-comodule algebra. Let $B := A^{coH} = \{a \in A \mid \rho_A(a) = a \otimes 1\}$ denote the subalgebra of coinvariants of A. An outline of the paper is as follows. In the first section we give basic definitions, conventions and notations, and summarize previous results. In the second section we introduce a bifunctor in the category of Hopf A-modules which acts like an internal Hom-functor, and use it to obtain its injective-preserving property. In the third section we introduce some properties of the coinduced functor $\mathrm{HOM}_B(A, -)$ in the case when H is cosemisimple.

1. Preliminaries

(1.1) By $\mathbb{M}^H$ we denote the category of right H-comodules. If M is in $\mathbb{M}^H$, M^{coH} denotes the subcomodule of coinvariants.

(1.2) A Hopf A-module M is a left A-module supplied with right H-comodule structure $\rho_M : M \longrightarrow M \otimes H$ such that $\rho_M(am) = \sum a_{(0)} m_{(0)} \otimes a_{(1)} m_{(1)}$, $m \in M$, $a \in A$. Let ${}_A\mathbb{M}^H$ denote the category of Hopf A-modules. Morphisms in this category are A-linear and H-colinear maps. If $M \in {}_A\mathbb{M}^H$, $E_A^H(M)$ denotes the injective hull of M in ${}_A\mathbb{M}^H$ and $E_B(M)$ denotes the injective hull of M in ${}_B\mathbb{M}$, the category of B-modules.

(1.3) We recall that a left (resp. right) integral of H is an element $\xi \in H^*$ such that $h^*\xi = h^*(1)\xi$ (resp. $\xi h^* = h^*(1)\xi$) for all $h^* \in H^*$. If H is

---

1991 *Mathematics Subject Classification.* 16E30, 16W30.
Project supported by the NSF of China and the NSF of Beijing City.

cosemisimple, there exists a left integral $\phi \in H^*$ such that $\phi(1) = 1$. It was proved in [3] that $\phi^2 = \phi$, and $\phi(h)1_H = \sum \phi(h_{(1)})h_{(2)} = \sum h_{(1)}\phi(h_{(2)})$. Thus from [3, Proposition 1.5] we get that the trace map $P_M : M \longrightarrow M^{coH}$ by $P_M(m) = \sum \phi(m_{(1)})m_{(0)}$ is a B-linear map. So we may define M_{coH} as the kernel of P_M. It is easy to see that $M = M^{coH} \oplus M_{coH}$.

2. The Functor HOM

In the sequel, H, A, B and S will be as above. Let $X, Y \in {}_A\mathbb{M}^H$. The H-comodule structure map of $X \otimes Y$ is $X \otimes Y \longrightarrow X \otimes Y \otimes H$, $x \otimes y \mapsto \sum x_{(0)} \otimes y_{(0)} \otimes x_{(1)}y_{(1)}$. Let us define the comodule $\text{HOM}_k(X, Y)$. For $f \in \text{Hom}_k(X, Y)$, we define $\rho(f) : X \longrightarrow Y \otimes H$ by

$$\rho(f)(x) = \sum f(x_{(0)})_{(0)} \otimes f(x_{(0)})_{(1)}S(x_{(1)}), \quad \text{for all} \quad x \in X.$$

Clearly $\rho(f) \in \text{Hom}_k(X, Y \otimes H)$.

Definition 2.1. Let $X, Y \in {}_A\mathbb{M}^H$. Then $f \in \text{Hom}_k(X, Y)$ is *rational* if $\rho(f) \in \text{Hom}_k(X, Y) \otimes H \subseteq \text{Hom}_k(X, Y \otimes H)$. We let $\text{HOM}_k(X, Y) = \{f \in \text{Hom}_k(X, Y) \mid f \text{ is rational}\}$.

We then have the following property:

$$\rho(f) = \sum f_{(0)} \otimes f_{(1)} \in \text{Hom}_k(X, Y) \otimes H$$

if and only if for all $x \in X$,

$$\sum f_{(0)}(x) \otimes f_{(1)} = \sum f(x_{(0)})_{(0)} \otimes f(x_{(0)})_{(1)}S(x_{(1)}).$$

Now we will prove that $\text{HOM}_k(X, Y)$ is again in $\mathbb{M}^H$.

Proposition 2.2. *Let X and Y be H-comodules. Then $\text{HOM}_k(X, Y)$ is the largest H-subcomodule of $\text{Hom}_k(X, Y)$.*

Proof. Clearly, every H-subcomodule of $\text{Hom}_k(X, Y)$ is in $\text{HOM}_k(X, Y)$. Now we only need to show that $\text{HOM}_k(X, Y)$ is an H-comodule.

We have to show that $\rho(\text{HOM}_k(X, Y)) \subseteq \text{HOM}_k(X, Y) \otimes H$. Using the method of [3, Lemma 2.1], we can easily get to this end. □

Let us now proceed to see that HOM is adjoint to the tensor product over k in $\mathbb{M}^H$.

Proposition 2.3. *Let X, Y and Z be H-comodules. Then there is a colinear isomorphism $\phi : \text{Hom}^H(X \otimes_k Y, Z) \longrightarrow \text{Hom}^H(X, \text{HOM}_k(Y, Z))$ given by $\phi(f)(x)(y) = f(x \otimes y)$.*

Proof. The standard "adjoint associativity" isomorphism gives an isomorphism $\mathrm{Hom}_k(X \otimes_k Y, Z) \longrightarrow \mathrm{Hom}_k(X, \mathrm{Hom}_k(Y, Z))$. We have to show that when restricted to $\mathrm{Hom}^H(X \otimes_k Y, Z)$, it yields an H-colinear homomorphism from X to $\mathrm{Hom}_k(Y, Z)$, which takes values in $\mathrm{HOM}_k(Y, Z)$.

For all $x \in X, f \in \mathrm{Hom}^H(X \otimes_k Y, Z), y \in Y$,

$$\begin{aligned}\rho(\phi(f)(x))(y) &= \sum \phi(f)(x)(y_{(0)})_{(0)} \otimes \phi(f)(x)(y_{(0)})_{(1)} S(y_{(1)}) \\ &= \sum f(x_{(0)} \otimes y_{(0)}) \otimes x_{(1)} y_{(1)} S(y_{(2)}) \\ &= \sum f(x_{(0)} \otimes y) \otimes x_{(1)} \\ &= \sum \phi(f)(x_{(0)})(y) \otimes x_{(1)}.\end{aligned}$$

Therefore, $\rho(\phi(f)(x)) \in \mathrm{Hom}_k(Y, Z) \otimes H$, and this implies that $\phi(f)(x) \in \mathrm{HOM}_k(Y, Z)$. Based on the above equality, we have $(\phi(f) \otimes I)\omega_X(x)(y) = \sum \phi(f)(x_{(0)})(y) \otimes x_{(1)} = \rho(\phi(f))(x)(y)$, where $\omega_X : X \longrightarrow X \otimes H$ is the comodule structure map of X. This implies that $\phi(f) : X \longrightarrow \mathrm{HOM}_k(Y, Z)$ is H-colinear. Thus $\mathrm{Hom}^H(X, \mathrm{HOM}_k(Y, Z))$ is the image of $\mathrm{Hom}^H(X \otimes_k Y, Z)$ under that isomorphism. □

We can now obtain the additional characterizations of "rational".

Proposition 2.4. *Let X and Y be H-comodules and assume $f \in \mathrm{Hom}_k(X, Y)$. Then the following are equivalent:*

(1) There are an H-comodule V, an element v of V and an H-comodule homomorphism $F : X \otimes_k V \longrightarrow Y$ such that $F(x \otimes v) = f(x)$ for all x in X.

(2) There are an H-comodule V, an element v of V and an H-comodule homomorphism $F' : X \longrightarrow \mathrm{HOM}_k(V, Y)$ such that $F'(x)(v) = f(x)$ for all x in X.

(3) f is rational.

Proof. Parts (1) and (2) are equivalent by (2.3): given F, take $F' = \phi(F)$, thus $F'(x)(v) = \phi(F)(x)(v) = F(x \otimes v) = f(x)$. Conversely, given F', take $F = \phi^{-1}(F')$, thus $F(x \otimes v) = \phi^{-1}(F')(x \otimes v) = F'(x)(v) = f(x)$.

To see that (1) implies (3), we note that for all $x \in X$,

$$\begin{aligned}\rho(f)(x) &= \sum f(x_{(0)})_{(0)} \otimes f(x_{(0)})_{(1)} S(x_{(1)}) \\ &= \sum F(x_{(0)} \otimes v)_{(0)} \otimes F(x_{(0)} \otimes v)_{(1)} S(x_{(1)}) \\ &= \sum F(x_{(0)} \otimes v_{(0)}) \otimes x_{(1)} v_{(1)} S(x_{(2)}).\end{aligned}$$

Then $\rho(f) \in \mathrm{Hom}_k(X, Y) \otimes H$, so (3) follows.

Given (3), we write $\rho(f) = \sum f_{(0)} \otimes f_{(1)}$ with $f_{(0)} \in \mathrm{HOM}_k(X, Y)$, since f is rational and $\mathrm{HOM}_k(X, Y)$ is an H-comodule. Let $V = k\{f_{(0)}\}$ be the span of $\{f_{(0)}\}$ and let $F : X \otimes_k V \longrightarrow Y$ be $F(x \otimes v) = v(x)$. Then

F is an H-comodule homomorphism and there clearly exists f in V such that $F(x \otimes f) = f(x)$ for all x in X. □

It is easy to obtain the functorial properties of HOM.

Corollary 2.5. *Let X, Y and Z be H-comodules, let $g \in \mathrm{HOM}_k(X, Y)$ and let $f \in \mathrm{HOM}_k(Y, Z)$. Then $fg \in \mathrm{HOM}_k(X, Z)$.*

Proof. For all $x \in X$, since

$$\sum g_{(0)}(x) \otimes g_{(1)} = \sum g(x_{(0)})_{(0)} \otimes g(x_{(0)})_{(1)} S(x_{(1)}),$$

it is easy to see that

$$\sum g_{(0)}(x)_{(0)} \otimes S(g_{(0)}(x)_{(1)}) g_{(1)} = \sum g(x_{(0)}) \otimes S(x_{(1)}).$$

Therefore,

$$\begin{aligned}
\rho(fg)(x) &= \sum (fg(x_{(0)}))_{(0)} \otimes (fg(x_{(0)}))_{(1)} S(x_{(1)}) \\
&= \sum f(g_{(0)}(x)_{(0)})_{(0)} \otimes f(g_{(0)}(x)_{(0)})_{(1)} S(g_{(0)}(x)_{(1)}) g_{(1)} \\
&= \sum f_{(0)} g_{(0)}(x) \otimes f_{(1)} g_{(1)}.
\end{aligned}$$

This implies that $\rho(fg) \in \mathrm{Hom}_k(X, Z) \otimes H$, i.e., $fg \in \mathrm{HOM}_k(X, Z)$. □

Corollary 2.5 shows that HOM is an internal bifunctor on $\mathbb{M}^H$. It is also clear that $\mathrm{HOM}_k(X, -)$ and $\mathrm{HOM}_k(-, Y)$ are left exact. We also have the following further consequence of (2.4).

Proposition 2.6. *Let E be an injective H-comodule. Then $\mathrm{HOM}_k(-, E)$ is an exact functor whose values are injective H-comodules.*

Proof. Since $\mathrm{HOM}_k(-, E)$ is left exact, we only need to show that $\mathrm{HOM}_k(-, E)$ is right exact. Let $g : X \longrightarrow Y$ be an H-comodule monomorphism and let $f \in \mathrm{HOM}_k(X, E)$. Using (2.4), we know that there is an H-comodule V, an element v of V, and an H-comodule homomorphism $F : X \otimes_k V \longrightarrow E$ such that $F(x \otimes v) = f(x)$. Since g is a monomorphism, $g \otimes I : X \otimes_k V \longrightarrow Y \otimes_k V$ is a monomorphism. So there is an H-comodule homomorphism $\overline{F} : Y \otimes_k V \longrightarrow E$ with $\overline{F} \circ (g \otimes I) = F$ since E is injective. Let $\overline{f}(y) = \overline{F}(y \otimes v)$ for all $y \in Y$. It is clear that $\overline{f} \in \mathrm{HOM}_k(Y, E)$. If $x \in X$, we have

$$\overline{f}(g(x)) = \overline{F}(g(x) \otimes v) = \overline{F} \circ (g \otimes I)(x \otimes v) = F(x \otimes v) = f(x).$$

So in $\mathrm{HOM}_k(Y, E) \longrightarrow \mathrm{HOM}_k(X, E)$, $\overline{f}$ restricts to f. It follows that $\mathrm{HOM}_k(-, E)$ is exact. Moreover, the adjoint isomorphism of (2.3) shows that $\mathrm{Hom}^H(-, \mathrm{HOM}_k(Y, E)) = \mathrm{Hom}^H(- \otimes_k Y, E)$ is the composite of the exact functors $- \otimes_k Y$ and $\mathrm{Hom}^H(-, E)$. So $\mathrm{HOM}_k(Y, E)$ is injective. □

Next we turn to the study of HOM for Hopf A-modules.

Definition 2.7. Let X and Y be Hopf A-modules. For given $f \in \mathrm{Hom}_A(X,Y)$ define $\rho(f) \in \mathrm{Hom}_A(X, Y \otimes H)$ by $\rho(f)(x) = \sum f(x_{(0)})_{(0)} \otimes f(x_{(0)})_{(1)} S(x_{(1)})$, for all $x \in X$. Then f in $\mathrm{Hom}_A(X,Y)$ is *rational* if $\rho(f) \in \mathrm{Hom}_A(X,Y) \otimes H$. We will denote that $\mathrm{HOM}_A(X,Y) = \{f \in \mathrm{Hom}_A(X,Y) \mid f \text{ is rational}\}$.

Lemma 2.8. *If X and Y are Hopf A-modules, so is $\mathrm{HOM}_A(X,Y)$.*

Proof. It is easy to see that $\mathrm{HOM}_A(X,Y)$ is an A-module and right H-comodule. We have to show that, for all $a \in A, \varphi \in \mathrm{HOM}_A(X,Y)$,

$$\rho(a\varphi) = \sum a_{(0)}\varphi_{(0)} \otimes a_{(1)}\varphi_{(1)}.$$

Indeed, using the method of [3, Lemma 2.2], we can get to this end. □

Analogous to (2.3) we establish the following adjointness property of HOM_A.

Proposition 2.9. *Let X, Y and Z be Hopf A-modules. There is a colinear isomorphism $\phi : \mathrm{Hom}_A^H(X \otimes_A Y, Z) \longrightarrow \mathrm{Hom}_A^H(X, \mathrm{HOM}_A(Y, Z))$ given by $\phi(f)(x)(y) = f(x \otimes y)$.*

Proposition 2.10. *Let M and N be Hopf A-modules with M finitely related.*
(1) $\varphi : \mathrm{HOM}_A(A, N) \longrightarrow N$ by $\varphi(f) = f(1)$ is an isomorphism of Hopf A-modules.
(2) $\mathrm{HOM}_A(M, N) = \mathrm{Hom}_A(M, N)$.

Proof. (1) φ is clearly an injective homomorphism of Hopf A-modules. If $n \in N$, define $f_n : A \longrightarrow N$ by $f_n(a) = na$. Since $f_n \in \mathrm{Hom}_A(A, N)$,

$$\begin{aligned} \rho(f_n)(a) &= \sum (na_{(0)})_{(0)} \otimes (na_{(0)})_{(1)} S(a_{(1)}) \\ &= \sum n_{(0)} a \otimes n_{(1)} \\ &= \sum f_{n_{(0)}}(a) \otimes n_{(1)} \end{aligned}$$

for all $a \in A$. Therefore, f_n is in $\mathrm{HOM}_A(A, N)$. Since $\varphi(f_n) = f_n(1) = n$, φ is onto.

(2) Since M is finitely related, by using [6, Lemma 3.83] we can obtain that $\mathrm{Hom}_A(M, N \otimes H) \cong \mathrm{Hom}_A(M, N) \otimes H$. Thus $\mathrm{HOM}_A(M, N) = \mathrm{Hom}_A(M, N)$. □

The method of the proof of (2.6) yields the following result about the functor $\mathrm{HOM}_A(-, I)$ when I is injective.

Proposition 2.11. *Assume that I is an injective Hopf A-module. Then $\mathrm{HOM}_A(-, I)$ is exact.*

3. The Coinduced Functor $\operatorname{HOM}_B(A,-)$

In this section, we use the functor HOM to induce modules from $B = A^{coH}$ to Hopf A-modules. Throughout, H is always a cosemisimple Hopf algebra.

Now let N be a B-module. With the trivial H action N becomes a Hopf B-module. We can thus consider the Hopf B-module $\operatorname{HOM}_B(A, N)$. If $a \in A$ and $f \in \operatorname{HOM}_B(A, N)$, let $(af)(b) = f(ab)$. Since $b \longrightarrow ab$ is in $\operatorname{HOM}_A(A, A)$, $af \in \operatorname{HOM}_B(A, N)$. This action makes $\operatorname{HOM}_B(A, N)$ a Hopf A-module.

In [3], the authors define the functor $\operatorname{HOM}_B(A, -) : {}_B\mathrm{M} \longrightarrow {}_A\mathrm{M}^H$ which is called the coinduced functor.

Theorem 3.1. *Let H be cosemisimple.*

(1) *If $X \longrightarrow Y$ is an essential monomorphism of B-modules, then $\operatorname{HOM}_B(A, X) \longrightarrow \operatorname{HOM}_B(A, Y)$ is an essential monomorphism of Hopf A-modules.*

(2) *Assume that X is a B-module and M is a Hopf A-submodule of $\operatorname{HOM}_B(A, X)$. Then if $M^{coH} = 0$, $M = 0$.*

Proof. At first, we prove (2): Assume $M^{coH} = 0$. So $\operatorname{Hom}_B(M^{coH}, X) = 0$. By [3, Theorem 2.3], $\operatorname{Hom}_A^H(M, \operatorname{HOM}_B(A, X)) = 0$. Since this later Hom contains the inclusion of M in $\operatorname{HOM}_B(A, X)$, we conclude $M = 0$.

Next, we prove (1): If M is a Hopf A-module of $\operatorname{HOM}_B(A, Y)$ and $M \neq 0$, then by (2), M^{coH} is a non-zero submodule of $\operatorname{HOM}_B(A, Y)^{coH}$. By [3, Theorem 2.3] this means $M(1)$ is a non-zero submodule of Y, and $M(1) \cap X$ is non-zero, where $M(1) = \{f(1) \mid f \in M\}$. Since $M(1) \cap X = (\operatorname{HOM}_B(A, X) \cap M)(1)$, M meets $\operatorname{HOM}_B(A, X)$ non-trivially. □

Definition 3.2. Let H be cosemisimple and let M be a Hopf A-module. Then we define $^*M = \{x \in M \mid P_M(ax) = 0 \text{ for all } a \in A\}$.

Lemma 3.3. *Let H be cosemisimple and let M be a Hopf A-module. Define $\phi_M : M \longrightarrow \operatorname{HOM}_B(A, M^{coH})$ by $\phi_M(m)(a) = P_M(am)$. Then:*

(1) *$^*M$ is the kernel of ϕ_M;*
(2) *if $f : M \longrightarrow M'$ is a Hopf A-homomorphism then $f(^*M) \subseteq {}^*(M')$;*
(3) *$^*(M/^*M) = 0$;*
(4) *if M is a Hopf A-submodule of N, then $^*N \cap M = {}^*M$;*
(5) *if $^*M = 0$, ϕ_M is an essential monomorphism.*

Proof. Assertion (1) is clear.

In (2), for all $m \in M$,

$$P_{M'}f(m) = \sum \phi(f(m)_{(1)})f(m)_{(0)} = \sum \phi(m_{(1)})f(m_{(0)}),$$

$$fP_M(m) = f(\sum \phi(m_{(1)})m_{(0)}) = \sum \phi(m_{(1)})f(m_{(0)}).$$

Therefore $P_{M'}f = fP_M$. So for all $a \in A, m \in {}^*M$,

$$P_{M'}(af(m)) = P_{M'}(f(am)) = fP_M(am) = 0.$$

In (3), we observe that $(M/{}^*M)^{coH} = M^{coH}/({}^*M)^{coH} = M^{coH}$ (since $({}^*M)^{coH} = 0$).

$$\phi_M : M \longrightarrow \mathrm{HOM}_B(A, M^{coH}) = \mathrm{HOM}_B(A, (M/{}^*M)^{coH}).$$

So $M/\ker\phi_M = M/{}^*M \longrightarrow \mathrm{HOM}_B(A, (M/{}^*M)^{coH})$ is a monomorphism. This implies that ${}^*(M/{}^*M) = 0$.

Assertion (4) is immediate from the definition and the fact that P_M restricts to P_N on N.

For (5), assume ${}^*M = 0$, so $\ker\phi_M = 0$, and this implies that $M = \phi_M(M)$. If X is a non-zero Hopf A-submodule of $\mathrm{HOM}_B(A, M^{coH})$, then by (3.1.2), $X^{coH} \neq 0$, and $M^{coH} = \mathrm{HOM}_B(A, M^{coH})^{coH}$ by [3, Theorem 2.3], so $X \cap M \neq 0$. □

We can use (3.3) to identify injective modules of the form $\mathrm{HOM}_B(A, I)$, where I is B-injective.

Theorem 3.4. *Let H be cosemisimple.*

(1) If E is an injective Hopf A-module with ${}^*E = 0$, then E^{coH} is B-injective and E is Hopf A-isomorphic to $\mathrm{HOM}_B(A, E^{coH})$.

(2) If M is any Hopf A-module with ${}^*M = 0$, then $E_A^H(M)$ is Hopf A-isomorphic to $\mathrm{HOM}_B(A, E_B(M^{coH}))$.

Proof. (1) Let $E' = E_B(E^{coH})$. Then $E^{coH} \longrightarrow E'$ is essential, so by (3.1.1) we have that the morphism $\mathrm{HOM}_B(A, E^{coH}) \longrightarrow \mathrm{HOM}_B(A, E')$ is essential. By (3.3.5) $E \longrightarrow \mathrm{HOM}_B(A, E^{coH})$ is also essential. As E is injective, we have isomorphisms $E \cong \mathrm{HOM}_B(A, E^{coH}) \cong \mathrm{HOM}_B(A, E')$. By [3, Theorem 2.3], $\mathrm{HOM}_B(A, E')^{coH} \cong E'$ is B-injective, so $E^{coH} = E'$ is B-injective.

(2) Let $E = E_B(M^{coH})$. By [3, Corollary 2.5], $\mathrm{HOM}_B(A, E)$ is injective, and we have essential monomorphisms $M \longrightarrow \mathrm{HOM}_B(A, M^{coH}) \longrightarrow \mathrm{HOM}_B(A, E)$ by (3.3.5) and (3.1.1). Thus $E_A^H(M)$ is Hopf A-isomorphic to $\mathrm{HOM}_B(A, E)$. □

Proposition 3.5. *Let H be cosemisimple, let I be an injective B-module, let M be a Hopf A-module with ${}^*M = 0$ and let $f : M \longrightarrow \mathrm{HOM}_B(A, I)$ be an essential Hopf A-monomorphism. Then $M^{coH} \longrightarrow \mathrm{HOM}_B(A, I)^{coH} = I$ is an essential monomorphism.*

Proof. By (3.3.5), ϕ_M is an essential monomorphism. From [3, Corollary 2.5] it follows that $\text{HOM}_B(A, I)$ is injective. Suppose Y is a B-submodule of I with $Y \cap M^{coH} = 0$. Then $\text{HOM}_B(A, M^{coH}) \cap \text{HOM}_B(A, Y) = 0$, so $\text{HOM}_B(A, Y) = 0$, and hence $0 = \text{HOM}_B(A, Y)^{coH} = Y$ by [3, Theorem 2.3]. □

Proposition 3.6. *Let H be cosemisimple and A flat over B. Then $(-)^{coH}$ carries Hopf A-injectives to B-injectives.*

Proof. Let E be an injective Hopf A-module, and X any B-module. Then we have $\text{Hom}_A^H(A \otimes_B X, E) = \text{Hom}_B(X, E^{coH})$, since $A \otimes_B -$ is the left adjoint of $(-)^{coH}$. So $\text{Hom}_B(-, E^{coH})$ is the composite of exact functors $A \otimes_B -$ and $\text{Hom}_A^H(-, E)$. □

References

[1] K. H. Ulbrich, *Smash products and comodules of linear maps*, Tsukuba J. Math. **14** (1990), 371–378.

[2] A. R. Magid, *Cohomology of rings with algebraic group action*, Adv. Math. **59** (1986), 124–151.

[3] B. Zhou, S. Canepeel and S. Raianu, *The coinduced functor for infinite dimensional Hopf algebras*, J. Pure and Applied Algebra **107** (1996), 141–151.

[4] T. Guédénon, *Algébre homologique dans la catégorie $Mod_{(R\#U(g))}$*, J. Algebra **197** (1997), 584–614.

[5] M. E. Sweedler, *Hopf Algebras*, Benjamin, New York, 1969.

[6] J. J. Rotman, *An Introduction to Homological Algebra*, Academic Press, New York, 1979.

Department of Mathematics
Capital Normal University
Beijing 100037, P. R. China
e-mail: `wangzhx@mail.cnu.edu.cn`

Hopf Algebra Coaction and Its Application to Group-Graded Rings

Gui-Long Liu

Abstract

Let H be a Hopf algebra, A a right H-comodule algebra and $A\#H^*$ a right smash product. Chen and Cai[5] give a Morita context connection between A^{coH} and $A\#H^{*rat}$ under the assumption that $\int_{H^*}^r = \int_{H^*}^l \neq 0$. In this paper we form a Morita context connection between A^{coH} and $A\#H^{*rat}$ under the condition $\int_{H^*}^l \neq 0$. We apply the results to G-graded algebras and give a Maschke type theorem for G-graded algebras.

0. Introduction

Let G be a group and kG the group algebra as usual. A G-graded algebra is just a kG-comodule algebra. If H is a Hopf algebra and A a right H-comodule algebra, then we can form a right smash product $A\#H^*$. Chen and Cai [5] give a Morita context connection between A^{coH} and $A\#H^{*rat}$ under the assumption that $\int_{H^*}^r = \int_{H^*}^l \neq 0$.

In Section 1 we form a Morita context connection between A^{coH} and $A\#H^{*rat}$ under the condition $\int_{H^*}^l \neq 0$ without assuming the unimodular condition $\int_{H^*}^l = \int_{H^*}^r$. Suppose that $0 \neq \lambda \in \int_{H^*}^l$ and there exists $c \in A$ such that $\lambda.c = 1$. Then $c\#\lambda$ is an idempotent element of $A\#H^*$, and $[e(A\#H^{*rat})e, e(A\#H^{*rat}), (A\#H^{*rat})e, A\#H^{*rat}]$ is also a Morita context via the multiplication of $A\#H^*$. We show that the two contexts are isomorphic.

In Section 2 we apply the results to G-graded algebras. Also we give a Maschke type theorem for G-graded algebras.

Throughout, we work over a fixed field k. Unless otherwise stated, all maps are k-linear, $\otimes$ means $\otimes_k$, we follow the notation in Sweedler's book [1], and H will always denote a Hopf algebra over k, with comultiplication Δ, counit ε and antipode S. When S is bijective $\overline{S}$ will denote its composition inverse. We shall use the sigma notation. We write $\Delta h = \Sigma h_1 \otimes h_2$, but we omit the summation index (h).

Let H be a Hopf algebra and A a right H-comodule algebra; then for all $a \in A$ we write $\rho(a) = \Sigma a_0 \otimes a_1 \in A \otimes H$. It is well known that we can form a right smash product $A\#H^*$ [5] which is the k-space $A \otimes H^*$ with multiplication defined by

$$(a\#f)(b\#g) = \Sigma ab_0 \#(f \leftharpoonup b_1)g$$

for all $a, b \in A$ and $f, g \in H^*$. Let H_R^{*rat} be the unique maximal rational submodule of the right regular H^*-module H^*; then $H_R^{*rat} = H_L^{*rat}$ by [4]. Denote them by H^{*rat}. Let $\int_{H^*}^l$ be the left integral space of H^*; then $\int_{H^*}^l \subseteq H^{*rat}$ and $\dim \int_{H^*}^l \leq 1$. If $\int_{H^*}^l \neq 0$, S is bijective.

1. A Morita Context Connection between A^{coH} and $A\#H^{*rat}$

Let A be a right H-comodule algebra, and the coinvariants $A^{coH} = \{a \in A \mid \rho(a) = a_0 \otimes a_1 = a \otimes 1\}$. We first give a Morita context connection between A^{coH} and $A\#H^{*rat}$ under the condition $\int_{H^*}^l \neq 0$ without assuming the unimodular condition.

Lemma 1. *Let $0 \neq t \in \int_{H^*}^l$. Then there exists a unique group-like element $h \in G(H)$ such that $tg = \langle g, h \rangle t$ for all $g \in H^*$.*

Proof. This is similar to the proof of [4]. □

Now let $0 \neq \lambda \in \int_{H^*}^l$ and $h \in G(H)$ fixed such that $\lambda g = \langle g, h \rangle \lambda$ for all $g \in H^*$. Then we have

Lemma 2. $S^*(\lambda) = h \rightharpoonup \lambda$.

Proof. Let $g \in H^*$. Then by Lemma 1 $(h \rightharpoonup \lambda)g = h \rightharpoonup [\lambda(S(h) \rightharpoonup g)] = h \rightharpoonup [\lambda \langle S(h) \rightharpoonup g, h \rangle] = (h \rightharpoonup \lambda)\langle g, 1 \rangle$. Thus $h \rightharpoonup \lambda \in \int_{H^*}^r$. Note that $\Sigma \langle h \rightharpoonup \lambda, l_1 \rangle l_2 = \langle h \rightharpoonup \lambda, l \rangle 1$ for all $l \in H$. Also $\langle S^*(\lambda), l \rangle \lambda = \langle \lambda, S(l) \rangle \lambda = (\lambda \leftharpoonup Sl)\lambda = \Sigma[\lambda(\lambda \leftharpoonup l_1)] \rightharpoonup Sl_2 = \Sigma[\lambda \langle \lambda \leftharpoonup l_1, h \rangle] \leftharpoonup Sl_2 = \Sigma \langle \lambda, l_1 h \rangle \lambda \leftharpoonup Sl_2 = \Sigma \langle h \rightharpoonup \lambda, l_1 \rangle \lambda \leftharpoonup Sl_2 = \Sigma \langle h \rightharpoonup \lambda, l \rangle \lambda$. This completes the proof. □

Define left and right $A\#H^{*rat}$-modules, respectively:

$$(A\#H^{*rat}) \otimes A \to A$$

by

$$(a\#f) \otimes b_1 \mapsto (a\#f) \cdot b = a(f.b) = \Sigma \langle f, b_1 \rangle a b_0$$

and

$$A \otimes (A\#H^{*rat}) \to A$$

by

$$a \otimes (b\#f) \mapsto \Sigma \langle f, \overline{S}(a_1 b_1) h \rangle a_0 b_0$$

for all $a, b \in A$, $f \in H^*$.

Lemma 3. Let H, A, t, h be as above. Then it follows that A is an $(A\#H^*, A\#H^*)$, $(A^{coH}, A\#H^*)$ and $(A\#H^*, A^{coH})$-bimodule.

Proof. We only show that A is a right $A\#H^*$-module. Let $a,b,c \in A$ and $f, g \in H^*$. Then

$$\begin{aligned}[a.(b\#f)].(c\#g) &= \Sigma\langle f, (a_1b_1)h\rangle (a_0b_0).(c\#g) \\ &= \Sigma\langle f, (a_2b_2)h\rangle\langle g, \overline{S}(a_1b_1c_1)h\rangle a_0b_0c_0 \\ &= \Sigma\langle f \leftharpoonup c_3, \overline{S}(a_2b_2c_2)h\rangle\langle g, \overline{S}(a_1b_1c_1)h\rangle a_0b_0c_0 \\ &= \Sigma\langle (f \leftharpoonup c_2)g, (a_1b_1c_1)h\rangle a_0b_0c_0 \\ &= \Sigma a.(bc_0\#(f \leftharpoonup c_1)g) \\ &= a.((b\#f)(c\#g)).\end{aligned}$$

The other conditions can be similarly checked. □

Now we define maps as follows:

$$[\,,\,] : A \otimes_{A^{coH}} A \to A\#H^{*rat} \text{ by } [a,b] = \Sigma ab_0\#\lambda \leftharpoonup b_1$$

and

$$(\,,\,) : A \otimes_{A\#H^{*rat}} A \to A^{coH} \text{ by } (a,b) = \lambda.(ab) = \Sigma\langle\lambda, a_1b_1\rangle a_0b_0$$

for all $a, b \in A$.

Theorem 1. Let H be a Hopf algebra and A a right H-comodule algebra. Let $0 \neq \lambda \in \int_{H^*}^l$, $h \in G(H)$, and let $(\,,\,)$, $[\,,\,]$ be as above. Then $[A^{coH}, {}_{A^{coH}}A_{A\#H^{*rat}}, {}_{A\#H^{*rat}}A_{A^{coH}}, A\#H^{*rat}]$ forms a Morita context.

Proof. The associativity $[a,b].c = a.(b,c)$ can be easily checked. By Lemma 2, the associativity $(a,b)c = a.[b,c]$ also holds for all $a, b, c \in A$. We only show that $[\,,\,]$ is an $A\#H^{*rat}$-bimodule map (discussing in $A\#H^*$). For $a, b, c, d \in A$ and $f, g \in H^{*rat}$, it follows that

$$\begin{aligned}(a\#f).[b,c](d\#g) &= (a\#f)(b\#\lambda)(cd\#g) \\ &= \Sigma(a\#f)(b\#\lambda)(1\#g \leftharpoonup \overline{S}(c_1d_1))(c_0d_0\#\varepsilon) \\ &= \Sigma(a\#f)(b\#\lambda)\langle g, \overline{S}(c_1d_1)h\rangle(c_0d_0\#\varepsilon) \\ &= \Sigma\langle g, \overline{S}(c_1d_1)h\rangle(ab_0\#(f \leftharpoonup b_1)\lambda)(c_0d_0\#\varepsilon) \\ &= \Sigma\langle f, b_1\rangle\langle g, \overline{S}(c_1d_1)h\rangle(ab_0\#\lambda)(c_0d_0\#\varepsilon) \\ &= [(a\#f).b, c.(d\#g)].\end{aligned}$$

The other conditions of Morita context can be checked similarly. □

Assume that there exists $c \in A$ such that $\lambda.c = \Sigma\langle\lambda, c_1\rangle c_0 = 1$. Then it follows that $e = c\#\lambda$ is a nonzero idempotent. Therefore we have that $[e(A\#H^{*rat})e, e(A\#H^{*rat}), (A\#H^{*rat})e, A\#H^{*rat}]$ forms a Morita context via the multiplication of $A\#H^{*rat}$. One can define a map between two Morita contexts $M = [R, V, W, S]$ and $M_0 = [R_0, V_0, W_0, S_0]$ to be a set of four homomorphisms $\tau = [\tau_R, \tau_V, \tau_W, \tau_S]$ where $\tau_R : R \to R_0$ and $\tau_S : S \to S_0$ are ring homomorphisms while $\tau_V : V \to V_0$ and $\tau_W : W \to W_0$ are module homomorphisms satisfying

$$\tau_V(rvs) = \tau_R(r)\tau_V(v)\tau_S(s)$$
$$\tau_W(swr) = \tau_S(s)\tau_W(w)\tau_R(r)$$
$$\tau_R((v,w)) = (\tau_R(v), \tau_W(w))$$

and

$$\tau_S([w,v]) = [\tau_W(w), \tau_V(v)].$$

We say that Morita contexts M and M_0 are isomorphic if τ_R, τ_S, τ_W and τ_V are bijective.

Theorem 2. *Let H, A, λ be as above. Assume that $\lambda.c = 1$ and $e = c\#\lambda$. Then the Morita context $[A^{coH}, A, A, A\#H^{*rat}]$ is isomorphic to $[e(a\#H^{*rat})e, e(A\#H^{*rat}), (A\#H^{*rat})e, A\#H^{*rat}]$ defined by the multiplication of $A\#H^{*rat}$.*

Proof. Put $\tau = [\tau_0, \tau_1, \tau_2, 1]$ as follows; $\tau_0 : A^{coH} \to eA^{coH}$ by $\tau_0(a) = ea$, $\tau_1 : A \to eA$ by $\tau_1(a) = ea$, $\tau_2 : A \to Ae = A\lambda$ by $\tau_2(a) = a\lambda$, and $1 : A\#H^{*rat} \to A\#H^{*rat}$ is the identity map. If $ea = 0$, then $0 = \lambda ea = \lambda a$. Thus $a = 0$ since $A\#H^*$ is a free right module over A. So τ_0 and τ_1 are bijective. It is easy to see that τ_0 and 1 are algebra isomorphisms. As for the other conditions, let $a \in A^{coH}, b \in A$ and $d\#f \in A\#H^{*rat}$. Then $\tau_2((d\#f).(ba)) = \tau_2(df.(ba)) = \Sigma db_0 a \langle f, b_1 \rangle \lambda = (d\#f)\tau_2(b)\tau_0(a)$. Similarly, we can prove that $\tau_1[(ab)(d\#f)] = \tau_0(a)\tau_1(b)(d\#f), \tau([a,b]) = [\tau_2(a), \tau_1(b)]$ and $\tau((a,b)) = (\tau_1(a), \tau_2(b))$. This completes the proof. □

Proposition 1. *If $\langle \lambda, 1 \rangle \neq 0$, then A is a coflat right H-comodule.*

Proof. Let $\langle \lambda, 1 \rangle \neq 0$. Then we can assume that $\langle \lambda, 1 \rangle = 1$. Note that $A \otimes H$ is a right H-comodule algebra via a structure map $id \otimes \Delta$ and $\rho : A \to A \otimes H$ a right H-comodule map. Now define $\psi : A \otimes H \to A$ by $\psi(a \otimes l) = \Sigma a_0 \langle S^*(\lambda), S(a_1) l \rangle$ for $a \in A$ and $l \in H$. Since $S^*(\lambda) \in \int_{H^*}^r$,

$$\rho\psi(a \otimes l) = \Sigma a_0 \langle S^*(\lambda), S(a_2) l \rangle \otimes a_1$$
$$= \Sigma a_0 \langle S^*(\lambda), S(a_3) l_1 \rangle \otimes a_1 S(a_2) l_2$$
$$= \Sigma a_0 \langle S^*(\lambda), S(a_1) l_1 \rangle \otimes l_2$$
$$= \psi(id \otimes \Delta)(a \otimes l).$$

This proves that ψ is a right H-comodule map and $\psi\rho(a) = \psi(\Sigma a_0 \otimes a_1) = \Sigma a_0 \langle S^*(\lambda), S(a_1)a_2 \rangle = \Sigma a_0 \varepsilon(a_1)\langle S^*(\lambda), 1\rangle = a\langle \lambda, 1\rangle = a$. Therefore $\psi\rho = id$. Thus A, as a right H-comodule, is a direct summand of the free right H-comodule $A \otimes H$. Therefore A is an injective coflat right H-comodule. □

2. Application to G-Graded Algebras

Let $G = \{1, g, \ldots\}$ be a group and A a G-graded algebra. Then A is a right kG-comodule algebra. Let $\{p_1, p_g \cdots\}$ be the dual basis of $(kG)^*$ and $p_1 \in \int_{H^*}^l$.

As an application of Theorems 1, 2 and Proposition 1 to G-graded algebras, we get the following corollary.

Corollary 1. *Let A be a G-graded algebra. Then $[A_1, A, A, A\#G]$ forms a Morita context (see [6]). The context is isomorphic to the Morita context $[A_1 p_1, p_1 A, A p_1, A\#G]$ defined by the multiplication of $A\#G$, and A is coflat as a right kG-comodule.*

Let H be a Hopf algebra and L a subHopf algebra of H. Then it follows that $HL^+ (L^+ = \ker \varepsilon \cap L)$ is a coideal of H and $\overline{H} = H/HL^+$ has a unique coalgebra structure. Define

$$H/L = \{f \in H^{*rat} \mid \langle f, hl\rangle = \varepsilon(l)\langle f, h\rangle \text{ for all } h \in H, l \in L\}.$$

Proposition 2. *Let H and L be as above. Then H/L is a subalgebra (without unit) of H^{*rat}.*

Proof. For $f, g \in H/L$, we have that $\langle fg, hl\rangle = \Sigma \langle f, h_1 l_1\rangle\langle g, h_2 l_2\rangle = \varepsilon(l)\langle fg, h\rangle$. Thus H/L is a subalgebra of H^{*rat}. This proves the proposition. □

By the definition of H^{*rat}, note that $H/L = \overline{H}^* \cap H^{*rat} \subseteq \overline{H}^{*rat}$.

Proposition 3. *H/L is a right H-module under $\leftharpoonup$.*

Proof. Let $f \in H/L$ and $h \in H$. By [1], it follows that $f \leftharpoonup h \in H^{*rat}$. We only need to show that $f \leftharpoonup h \in \overline{H}^*$. In fact, $\langle f \leftharpoonup h, gl\rangle = \langle f, hgl\rangle = \varepsilon(l)\langle f, hg\rangle = \varepsilon(l)\langle f \leftharpoonup h, g\rangle$ for all $g, h \in H$. □

By [3] and Proposition 3, we can form the smash product $A\#H/L$ which is a subalgebra of $A\#H^{*rat}$. As a G-graded algebra, $A\#H^{*rat}$ is just $A\#G$ and $A\#G = \bigoplus_{x \in G} A p_x$.

Let $H = kG$ and L be a subgroup of G. Denote H/kL by G/L, that is,
$$G/L = \{f \in (kG)^* \mid \langle f, xy \rangle = \langle f, x \rangle \text{ for all } x \in G \text{ and } y \in L\}.$$
Let $\{\sigma_i \mid i \in I\}$ be a representative of the left cosets of L in G. Then $G = \bigcup_{i \in I} \sigma_i L$. Put $\langle p_{\sigma_i L}, x \rangle = 1$ for $x \in \sigma_i L$ and $\langle p_{\sigma_i L}, x \rangle = 0$ for $x \notin \sigma_i L$.

For convenience we write $p_{\sigma_i L} = \sum_{x \in \sigma_i L} p_x$ (infinite form sum). We have the following properties for $A\#G/L$:

(1) $(a\#p_{\sigma_i L})(b\#p_{\sigma_j L}) = (a\# \sum_{x \in \sigma_i L} p_x)(b\# \sum_{y \in \sigma_j L} p_y) = \sum_{x \in \sigma_i L, y \in \sigma_j L} ab_{xy^{-1}}$
$= \sum_{x\sigma_j L = \sigma_i L} ab_x p_{\sigma_j L};$

(2) $\{p_{\sigma_i L} \mid i \in I\}$ is the set of orthogonal idempotents;

(3) $p_{\sigma_i L} a = \sum_{h \in \sigma_i L} p_h a = \sum_{g \in G, h \in \sigma_i L} a_{hg^{-1}} p_g = \sum_{x \in G, j \in I, x\sigma_j L = \sigma_i L} a_x p_{\sigma_j L};$

(4) $aP_{\sigma_i L} = \sum_{x \in G, x\sigma_i L = \sigma_j L} p_{\sigma_j L} a_x;$

(5) for each $a \in A^L (= \sum_{x \in L} A_x)$, $ap_{\sigma_i L} = \sum_{x \in G, j \in I, x\sigma_i L = \sigma_j L} p_{\sigma_j L} a_x$
$= \sum_{x \in L, x\sigma_i L = \sigma_j L} p_{\sigma_j L} a_x = p_{\sigma_i L} a.$

Lemma 4. *Let V be a right $A\#G$-module. Then V is a right $A\#G/L$-module. If, in addition, $V(A\#G) = V$, then $V(A\#G/L) = V$ and V is a unital right A-module (i.e., $v.1 = v$, $1 \in A$).*

Example. Let V be a G-graded A-module. Define $v.(ap_h) = (va)_{h^{-1}}$. Then V is a right $A\#G/L$-module and $V(A\#G/L) = V$.

Theorem 3. *(Maschke type) Let V be a right $A\#G/L$-module, W a right $A\#G/L$-submodule of V, $V(A\#G/L) = V$ and $W(A\#G/L) = W$. If W is a direct summand of V as a right A-module, then W is also a direct summand of V as a right $A\#G/L$-module.*

Proof. Assume an A-projection $\lambda : V \to W$ is given. Then we can define $\tilde{\lambda} : V \to W$ by $\tilde{\lambda}(v) = \sum_{i \in I} \lambda(v.(1\#p_{\sigma_i L}))(1\#p_{\sigma_i L})$ for $v \in V$. Observe that $V(A\#G/L) = V$. We write $v = \sum_{i \in I} v_i(a_i \# p_{\sigma_i L})$. Observe that $\lambda(v) = \sum_i \lambda[v_i(a_i\#p_{\sigma_i L})]$. Note $(1\#p_{\sigma_i L})$ is a finite sum. Thus $\tilde{\lambda}$ is well-defined.

Now we check that $\tilde{\lambda}$ is an $A\#G/L$-module map. In fact,
$$\tilde{\lambda}(v).(ap_{\sigma_i L} = \sum_{j \in I}(\lambda(vp_{\sigma_j L}).p_{\sigma_j L})(a\#p_{\sigma_i L})$$

$$= \sum_{j\in I, g\sigma_i L=\sigma_j L} \lambda(vp_{\sigma_j L})a_g p_{\sigma_i L} = \sum_{j\in I, g\sigma_i L=\sigma_j L} \lambda(vp_{\sigma_j L}a_g)p_{\sigma_i L}$$

$$= \sum_{j\in I, g\in G, g\sigma_i L=\sigma_j L} \lambda(va_g p_{\sigma_i L})p_{\sigma_i L} = \lambda(v(ap_{\sigma_i L}))p_{\sigma_i L} = \tilde{\lambda}(v.(a\#p_{\sigma_i L}))$$

for all $ap_{\sigma_i L} \in A\#G/L$. Let $v \in W$. We can write $v = \sum_i v_i(a_i\#p_{\sigma_i L})$ since $W = W(A\#G/L)$. Then $\tilde{\lambda}(v) = \sum_{i\in I} vp_{\sigma_i L} = \sum_i v_i(a_i p_{\sigma_i L}) = v$. That is, $\tilde{\lambda}|_W = id|_W$. This completes the proof. □

Note that $p_{1L} = p_L$ is an idempotent. So p_L induces a Morita context $[A^L, A, A, A\#G/L]$ defined by the multiplication of $A\#G/L$ and hence A is a left $A\#G$-module by $(a\#p_h).b = ab_h$. Similarly, A is a right $A\#G/L$-module via $b(a\#p_h) = (ba)_{h^{-1}}$. Observing that $p_L(A\#G/L)p_L \subseteq A^L p_L$, we have the following corollary.

Corollary 2. *Define* $(\,,\,) : A \otimes_{A\#G/L} A \to A^L$ *by* $(a, b) = p_L(ab) = \sum_{x\in L}(ab)_x$ *and* $[\,,\,] : A \otimes_{A^L} A \to A\#G/L$ *by* $[a, b] = ap_L b$. *Then* $[A^L, A, A, A\#G/L]$ *forms a Morita context and the context is isomorphic to the Morita context* $[p_L(A\#G/L)p_L, p_L(A\#G/L), (A\#G)p_L, A\#G/L]$ *defined by the multiplication of* $A\#G/L$.

We can use the context to discuss the relationship of the primeness, primitivity and simplicity between A^L and $A\#G/L$. But we omit them. We conclude the section by discussing the relation between smash products $A^L\#L$ and $A\#G$. We still use the idempotent $p_L = \sum_{x\in L} p_x$ of $A\#G$. Thus $[p_L(A\#G)p_L, p_L(A\#G), (A\#G)p_L, A\#G]$ is a Morita context via the multiplication of $A\#G$. Note that $p_L(A\#G)p_L = A^L\#L$, $p_L(A\#G) = (kL)^{*rat} \otimes A$ and $(A\#G)p_L = A \otimes (kL)^{*rat}$. Thus $[p_L(A\#G)p_L, p_L(A\#G), (A\#G)p_L, A\#G] = [A^L\#L, (kL)^{*rat} \otimes A, A \otimes (kL)^{*rat}, A\#G]$ is a Morita context and $[\,,\,], (\,,\,)$ are just ordinary multiplications of $A\#G$ such as

$$[ap_x, p_y b] = ap_x p_L p_y b = a\delta_{x,y} p_x b$$

and

$$(p_x a, bp_x) = p_x abp_y = (ab)_{xy^{-1}} p_y$$

for all $x, y \in L$. Here $\delta_{x,y}$ is the Kronecker delta.

Proposition 4. $(\,,\,)$ *is surjective.*

Proof. For $a\#p_y \in A^L\#L$, write $a = \sum_{x \in L} a_x$ (finite sum). Note that $a_{xy^{-1}}$ is almost always zero for $x \in L$, so $(\sum_{a_{xy^{-1}} \neq 0} p_x a, \ p_y) = \sum_x a_{xy^{-1}} p_y = a\#p_y$. This completes the proof. $\square$

References

[1] M. E. Sweedler, *Hopf Algebras,* Benjamin, New York, 1969.

[2] M. Beattie, *A generalization of smash product of a graded ring,* J. Pure and Applied Algebra **52** (1988), 219–226.

[3] ———, *Strongly inner actions, coactions and duality theorems,* Tsukuba J. Math. **16** (1992), 279–293.

[4] C. Cai and H. Chen, *Coactions, smash products and Hopf modules,* J. Algebra **167** (1994), 85–99.

[5] ———, *Hopf algebra coactions,* Comm. Algebra **22**(1) (1994), 253–267.

[6] D. Quinn, *Group-graded ring and duality,* Trans. Amer. Math. Soc. **292** (1985), 155–167.

Department of Mathematics and Computer Science
Beijing Language and Culture University
Beijing 100083, P. R. China
e-mail: `liuguilong@1115@263.net`

Non-Commutative Valuation Rings and Their Global Theories

Hidetoshi Marubayashi

Abstract

This is a survey of non-commutative valuation rings and their global theory, mainly, semi-hereditary orders and Prüfer orders, in which much progress has been made in the past fifteen years.

1. Non-Commutative Valuation Rings

Historically non-commutative valuation rings of division rings were first treated systematically in Schilling's book [S] and are nowadays called invariant valuation rings, though invariant valuation rings can be traced back to Hasse's work in [H]. Since then various attempts have been made to study the ideal theory of orders in finite dimensional algebras over fields and to describe the Brauer groups of fields by the use of "valuations", "places", "preplaces", "value functions" and "pseudoplaces".

In 1984, N. I. Dubrovin defined non-commutative valuation rings of simple Artinian rings with the notion of places in the category of simple Artinian rings and obtained significant results on non-commutative valuation rings (named Dubrovin valuation rings after him) which signify that these rings may be the correct definition of valuation rings of simple Artinian rings.

The aim of this section is to give a summary of his results and of some additional results obtained after him. We begin with the definitions of invariant valuation rings and their generalization. Let D be a division ring and let R be a subring of D. Consider the following two conditions:

(T) For every $d \in U(D)$, either $d \in R$ or $d^{-1} \in R$.
(I) For every $d \in U(D)$, $dRd^{-1} = R$.

Here we denote the *unit group* of a ring S by $U(S)$ and the *center* of S by $Z(S)$. If R satisfies the condition (T), then it is called a *total* valuation ring of D. If R satisfies the conditions (T) and (I), then it is called an *invariant* valuation ring of D. If R is invariant, then any one-sided ideal of R is two-sided and we can define the *value group* of R as follows: $\Gamma_R = U(D)/U(R)$, which is made into a totally ordered group by $d_1 U(R) \leq d_2 U(R)$ if and

only if $d_1 R \supseteq d_2 R$ for any $d_1, d_2 \in U(D)$. Then the natural mapping v: $U(D) \longrightarrow \Gamma_R$ given by $d \longmapsto dU(R)$ satisfies the following two conditions:
   (V1) $v(ab) = v(a) + v(b)$ (we use an additive notation for Γ_R).
   (V2) $v(a+b) \geq \min\{v(a), v(b)\}$ if $b \neq -a$.

In general, a surjective mapping v: $U(D) \longmapsto G$ which satisfies the conditons (V1) and (V2) is called a *valuation* on D, where G is a totally ordered group. This was an original definition given by Schilling [S]. If v is a valuation on D, then it is easily checked that $R = \{d \in U(D) \mid v(d) \geq 0\} \cup \{0\}$ is an invariant valuation ring of D and $\Gamma_R \simeq G$ naturally.

Let Q be a simple Artinian ring with $F = Z(Q)$ and let V be a valuation ring of F. An order R in Q is called an *extension* of V to Q provided $R \cap F = V$. The following is the classical result which is concerned with extension.

Theorem 1.1. [S] *Let D be a division ring with finite dimension over its center F and let V be a valuation of F. Suppose that V is either complete or Henselian. Then there exists an invariant valuation ring of D extending V to D.*

However not every valuation V of F can be extended to an invariant valuation ring of D (even to a total valuation ring of D). If D is the quaternion algebra over the rational field, then only the 2-adic valuation can be extended to an invariant valuation ring of D(see [W1]). Of course, the class of total valuation rings is much bigger than the class of invariant valuation rings (see [BG1], [BG2] and [M]).

As we have already seen, the invariant and total valuation rings have the following two problems:

(1) Not every valuation ring of F can be extended to an invariant (a total) valuation ring of D.

(2) They are not defined for a simple Artinian ring which is not a division ring.

In 1984, N. I. Dubrovin defined non-commutative valuation rings by using the concept of places as follows:

Let Q be a simple Artinian ring. We adjoin a new symbol ∞ to Q and define
$$q + \infty = \infty + q = \infty \text{ for any } q \in Q,$$
and
$$c \cdot \infty = \infty \cdot c = \infty \text{ for any } c \in U(Q).$$

Note that we do not define $\infty + \infty$ and $\infty \cdot q$ if $q \notin U(Q)$. We denote this set by (Q, ∞). Now let D be another simple Artinian ring. A mapping f of (Q, ∞) onto (D, ∞) is called a *place* of Q to D if

(i) $f(1) = 1$, $f(qr) = f(q)f(r)$ and $f(q + r) = f(q) + f(r)$ for any $q, r \in (Q, \infty)$ whenever the right-hand sides of (i) are defined, and for any $q \in Q$ with $f(q) = \infty$, there exist $r, s \in Q$ such that
(ii) $f(qr) \neq \infty, 0$ and $f(sq) \neq \infty, 0$ with $f(r) \neq \infty$ and $f(s) \neq \infty$.

The places are just a generalization of the ones in commutative rings (see [B] and [ZS]). However, Dubrovin has not only solved the problems (1) and (2) above but also obtained significant results which look like genuine non-commutative valuation rings as will be described in the following:

Proposition 1.2. [D1] *Let Q and D be simple Artinian rings and let f be a place of Q onto D. Then $R = \{q \in Q \mid f(q) \neq \infty\}$ is a subring of Q and $M = \{q \in R \mid f(q) = 0\}$ is an ideal of R such that*
(1) R/M is a simple Artinian ring, and
(2) for any $q \in Q \setminus R$ there exist $r, s \in R$ with $qr, sq \in R \setminus M$.
Conversely, if R is a subring of Q with ideal M satisfying (1) and (2), then the mapping defined by

$$f(a) = \begin{cases} [a + M], & \text{for } a \in R \\ \infty, & \text{for } a \in Q \setminus R. \end{cases}$$

is a place of Q onto $R \setminus M$.

A subring R of Q satisfying the conditions in Proposition 1.2 is called a *Dubrovin valuation ring* of Q. Let R be a Dubrovin valuation ring of a simple Artinian ring Q. Then Dubrovin first pointed out the following:
(a) $M = J(R)$, the Jacobson radical of R, and
(b) $qR \cap R \subseteq J(R)$ $(q \in Q) \Rightarrow q \in J(R)$.

Next, he proved the following by skillfully combining the facts (a) and (b) with the properties of the Jacobson radical, Nakayama's lemma etc.

Theorem 1.3. [D1] *Let R be a subring of a simple Artinian ring Q. The following are equivalent:*
(1) R is a Dubrovin valuation ring of Q.
(2) R is a local and semi-hereditary order in Q.
(3) R is a local and Bezout order in Q.

Here we explain some terminology used in Theorem 1.3. A ring S is called *local* if $S/J(S)$ is a simple Artinian ring. If any finitely generated one-sided ideal of S is projective, then S is called *semi-hereditary*. An order S in a simple Artinian ring is called *Bezout* provided any finitely generated one-sided ideal of S is principal.

Next we shall mention the localizations of prime ideals and the ideal theory of Dubrovin valuation rings. Let P be a prime ideal of a ring S. If

$\mathcal{C}(P) = \{c \in S \mid c + P \text{ is regular in } R/P\}$ is a regular Ore set of S, then we denote by S_P the localization of S at P. Concerning the localizations he obtained the following:

Theorem 1.4. [D1] *Let R be a Dubrovin valuation ring of a simple Artinian ring Q and let S be an overring of R. Then:*
  (1) $\tilde{R} = R/J(S)$ *is a Dubrovin valuation ring of* $\bar{S} = S/J(S)$.
  (2) S *is a Dubrovin valuation ring of Q, $J(S) \in Spec(R)$ and $S = R_{J(S)}$.*

Let $P \in Spec(R)$ and let R be a Dubrovin valuation ring. Then it is easy to see that R_P exists if and only if R/P is a Goldie ring (see [MMU, Theorem 14.5]) and Dubrovin gave an example of a total valuation ring with a prime ideal that is not completely prime(see [D4]). These suggest the following; a prime ideal P of a ring R is called *Goldie prime* if R/P is a Goldie ring. Note that if R is total, then Goldie primes are equivalent to completely primes. To classify all prime ideals of a Dubrovin valuation ring, we need some properties of ideals of Dubrovin valuation rings.

By using a sophisticated lemma (see [MMU Lemma 6.3]), he proved the following which showed one of the important properties of valuation rings.

Proposition 1.5. [D1] *Let R be a Dubrovin valuation ring of a simple Artinian ring. Then the set of all R-ideals is linearly ordered by inclusion.*

Let A be an R-ideal. Then $A^{-1} = \{q \in Q \mid AqA \subseteq A\}$, called the *inverse* of A and $A^* = A^{-1-1}$, an R-ideal containing A. Furthermore, set $O_r(A) = \{q \in Q \mid Aq \subset A\}$, the *right order* of A and $O_l(A) = \{q \in Q \mid qA \subset A\}$, the *left order* of A. He proved the following which was concerned with whether R-ideals are principal.

Proposition 1.6. [D1] *Let R be a Dubrovin valuation ring of Q, A be an R-ideal of Q and let $S = O_r(A)$.*
  (1) *The following are equivalent:*
    (i) A *is principal as a right S-ideal;*
    (ii) $A^{-1}A = S$;
    (iii) $A \supset AJ(S)$.
  (2) *If A is not a principal right S-ideal, then $A^{-1}A = J(S)$ and $J(S)$ is not a principal right S-ideal.*
  (3) *If $A \subset A^*$, then $A^* = cS$ and $A = cJ(S)$ for some regular element $c \in A^*$.*

Proposition 1.6 is used for getting the additional properties;

(c) For any family of Goldie primes P_i, $P_1 = \cup P_i$ and $P_2 = \cap P_i$ are both Goldie primes.

(d) Let I be any ideal, then $\cap I^n$ is also Goldie prime.

Let $P_1, P_2 \in Spec(R)$ with $P_1 \supset P_2$. We say that the prime pair $P_1 \supset P_2$ is a *prime segment* if there are no Goldie primes properly between P_1 and P_2. For any Goldie prime P_1, set $P_2 = \bigcup \{P_i \mid P_1 \supset P_i$ Goldie prime$\}$. If $P_1 \supset P_2$, then it is a prime segment by (c). If $P_1 = P_2$, then we say that P_1 is a *limit prime*. Therefore, either P_1 is a limit prime or there is a Goldie prime P_2 such that $P_1 \supset P_2$ is a prime segment. The following is a classification of prime segments.

Theorem 1.7. [BMO] *For a prime segment $P_1 \supset P_2$ of a Dubrovin valuation ring exactly one of the following possibilities occurs:*

(1) *The prime segment $P_1 \supset P_2$ is Archimedean, that is, for any $a \in P_1 \setminus P_2$, there is an ideal $I \subseteq P_1$ with $a \in I$ and $\cap I^n = P_2$;*

(2) *The prime segment $P_1 \supset P_2$ is simple, that is, there are no ideals properly between P_1 and P_2;*

(3) *The prime segment $P_1 \supset P_2$ is exceptional, that is, there exists non-Goldie prime C such that $P_1 \supset C \supset P_2$.*

In the case of (3), there are no ideals properly between P_1 and C and $P_2 = \cap C^n$.

In [M], Mathiak constructed a total valuation ring of rank one with a simple segment and, as we have pointed out before, Dubrovin gave a total valuation ring with non-Goldie primes [D4]. The idea of prime segments is used for a classification of primary ideals of Dubrovin valuation rings [BMU] and of indecomposable injective modules over total vauation rings [BT].

To obtain the further detailed results on Dubrovin valuation rings, we assume, in the remainder of this section, that Q is a simple Artinian ring with finite dimension over its center F. Then any order in Q is a P.I. ring. Combining these facts with Theorems 1.2 and 1.3, Dubrovin proved the following:

Theorem 1.8. [D2] *Let R be a Dubrovin valuation ring of Q. The following hold:*

(1) $V = F \cap R$ *is a valuation ring of F.*

(2) *For any $P \in Spec(R)$, $C(P)$ is a regular Ore set of R and R_P is a Dubrovin valuation ring of R.*

(3) *The set of all R-ideals is a commutative semigroup.*

(4) *There is a bijective mapping between $Spec(R)$ and $\mathcal{O}(R)$, the set of all overrings of R given by $P \longmapsto R_P$ and $S \longmapsto J(S)$, where $P \in Spec(R)$ and $S \in \mathcal{O}(R)$.*

The existence theorem has also been proved by Dubrovin, and later Brungs and Gräter gave an elementary proof of it. The conjugacy theorem is due to Wadsworth by using Henselization.

Theorem 1.9. *Let V be a valuation ring of F. (1) (The existence theorem) [D2] and [BG2] There is a Dubrovin valuation ring of Q which is an extension of V to Q.*

(2) (The conjugacy theorem) [W2] Any Dubrovin valuation rings of Q whose centers are V are conjugate.

Proposition 1.5 and Theorem 1.8(3) enable us to define the value groups as follows: Let $st(R) = \{q \in Q \mid qR = Rq\}$, a group and $U(R)$ be a normal subgroup of $st(R)$. The factor group $\Gamma_R = st(R)/U(R)$ is an abelian group by Theorem 1.8 and totally ordered in the following definition;
$$qU(R) \geq sU(R) \Leftrightarrow qR \subseteq sR,$$
where $q, s \in st(R)$. Γ_R is called a *value group* of R. It is easy to see that Γ_V is naturally embedded in Γ_R, where $V = R \cap F$. However, we can not freely handle Γ_R in order to get some good information on R as in the commutative case. In the case R is integral over V, Wadsworth obtained the following:

Theorem 1.10. [W2] *Let R be a Dubrovin valuation ring of Q with $V = F \cap R$. Then the following are equivalent:*
(1) Any element in R is integral over V.
(2) For any $q \in Q$, there exists $s \in st(R)$ with $RqR = sR = Rs$.

This theorem was extended by Gräter [G2] to the case R is a Bezout order in Q. By using the property (2) in Theorem 1.10, we can define a mapping $w : Q \longrightarrow \Gamma_R \cup \{\infty\}$, given by $w(q) = sU(R)$ and $w(0) = \infty$, where $q \in Q$ with $RqR = sR = Rs$ and the mapping w satisfies the following which look like valuations in commutative rings.
(V1) $w(q) = \infty \Longleftrightarrow q = 0$, $w(-1) = 0$.
(V2) $w(q+s) \geq min\{w(q), w(s)\}$.
(V3) $w(qs) \geq w(q) + w(s)$.
(V4) $Im w = \{w(q) \mid q \in Q\} = w(st(w))$, where $st(w) = \{q \in U(Q) \mid w(q^{-1}) = -w(q)\}$.

As we can see from the properties (V1) – (V4), in the case R is integral over V, we can use the value group of R. In fact, Wadsworth, Haile and Morandi have used the value group for further development of Dubrovin valuation rings and for characterizing Dubrovin valuation rings in crossed product algebras and tensor product algebras (see [HM],

[HMW] and [MW]). Invariant valuation rings of certain algebras of quantum type have been studied in [MV] and [VW]. These algebras are infinite dimensional over their centers. The study of valuation rings of algebras with infinite dimension over their centers is one of the fields which have scarcely been explored.

2. Prüfer Orders and Semi-hereditary Orders

Let D be a commutative domain with its quotient field F. The following are well known [G].

(e) *D is Prüfer, that is, any finitely generated ideal of D is invertible if and only if D_P is a valuation ring for any non-zero $P \in Spec(D)$.*

(f) *Let K be an algebraic field extension of F and let R be the integral closure of D in K. If D is Prüfer, then so is R.*

It is seen from (e) and (f) that Prüfer domains are a globalization of valuation rings and the class of Prüfer domains contains the class of infinite number theory. In commutative domains, an ideal is invertible if and only if it is projective. However, in non-commutative rings, an invertible ideal is, of course, projective and the converse is not necessarily true. This leads us to at least the following two orders; Prüfer orders and semi-hereditary orders, which are considered as global theories of Dubrovin valuation rings (see Theorem 1.3). Here an order in a simple Artinian ring Q is called *Prüfer* if any finitely generated one-sided ideal is a progenerator [AD].

In what follows, *Q is a simple Artinian ring with finite dimension over its center F; let R be an order in Q with $D = Z(R)$, that is, $F = Q(D)$, the quotient ring of D and $F \cdot R = Q$.* An order R in Q is called a *D-order* if any element in R is integral over D.

If R is a semi-hereditary D-order, then D is a Prüfer domain [MMU, Theorem 3.6]. The assumption that R is integral over D is not removed in order to obtain the Prüferness of D [G2]. Of course, if D is a Dedekind domain, then R is Prüfer iff it is Dedekind, that is, Noetherian Prüfer, and R is semi-hereditary iff it is hereditary.

The localizations are one of the most important methods to study the structure of orders. In [D3], Dubrovin obtained the following:

Theorem 2.1. [D3] *Let P be a non-zero prime ideal of a Prüfer D-order R. Then R_P exists and is a Dubrovin valuation ring.*

Since there are only finite maximal ideals in semi-local Prüfer orders, as a Corollary to Theorem 2.1, we have

Corollary 2.2. [D3] *A semi-local Prüfer D-order is Bezout.*

It is easily seen from Corollary 2.2 that if R is a Prüfer V-order, where V is a valuation ring, then R is Bezout. Hence Prüfer V-orders are much easier to study. In fact, we have plenty of nice properties on semi-local Bezout orders as follows: Let $R_1, \ldots, R_n$ be pairwise incomparable Dubrovin valuation rings of Q with $V_i = Z(Q_i)$, $1 \leq n$. Furthermore, let R_{ij} (resp. V_{ij}) be the least overring of R_i and R_j (resp. V_i and V_j). Then $\tilde{R}_i = R_i/J(R_{ij})$ and $\tilde{R}_j = R_j/J(R_{ij})$ are Dubrovin valuation rings of $\bar{R}_{ij} = R_{ij}/J(R_{ij})$ [MMU, Theorem 6.6]. In particular, $Z(\tilde{R}_i)$ and $Z(\tilde{R}_j)$ are valuation rings of $Z(\bar{R}_{ij})$. Under these conditions Morandi proved the following approximation theorem which is one of the most useful methods to study orders.

Theorem 2.3. [M1] *Suppose that $Z(\tilde{R}_i)$ and $Z(\tilde{R}_j)$ are independent in $Z(\bar{R}_{ij})$ for each pair $i \neq j$. If I_i are right ideals of R_i with $I_i R_{ij} = I_j R_{ij}$ and $q_i \in Q$ with $q_i - q_j \in I_i R_{ij}$, then there exists $x \in Q$ with $x \equiv q_i (\mathrm{mod}\, I_i)$ for all i.*

Set $R = R_1 \cap \cdots \cap R_n$, where R_i are Dubrovin valuation rings of Q. Gräter says that $R_1, \ldots, R_n$ have the *intersection property (IP)* if the map

$$\varphi : \mathcal{O}(R_1) \cup \cdots \cup \mathcal{O}(R_n) \longrightarrow \mathrm{Spec}(R)$$

defined by $\varphi(S) = J(S) \cap R$ is a well defined anti-inclusion-preserving isomorphism, where $S \in \mathcal{O}(R_i)$ for some i.

In [G1], he used the *IP* condition in order to get the following which is a generalization of Dubrovin valuation rings to semi-local Bezout rings.

Theorem 2.4. [G1] *(1) (The existence theorem) Let V be any valuation ring of F. Then there exists a Bezout V-order in Q.*

(2) (The conjugacy theorem) Any Bezout V-orders are conjugate.

Furthermore, Morandi's and Gräter's conditions are used for characterizing Bezout orders in Q as follows:

Theorem 2.5. [MMU] *Let $R_1, \ldots, R_n$ be pairwise incomparable Dubrovin valuation rings of Q and set $R = R_1 \cap \cdots \cap R_n$. The following are equivalent:*

(1) $R_1, \ldots, R_n$ have the IP.
(2) $Z(\tilde{R}_i)$ and $Z(\tilde{R}_j)$ are independent in $Z(\bar{R}_{ij})$ for each $i \neq j$.
(3) R is a semi-local Bezout order in Q.
(4) R is semi-local, and R_M exists and is a Dubrovin valuation ring for any maximal ideal M of R.
(5) R is semi-local, and R_P exists and is a Dubrovin valuation ring for any non-zero prime P of R.

A D-order R in Q is called *maximal* if $R \subseteq S$ and S is a D-order in Q, then $R = S$. It is easily seen that there always exists a maximal D-order containing a given D-order R by Zorn's lemma. If V is a discrete rank one valuation ring, then it is well known that a V-order is maximal if and only if it is Dedekind. However any semi-hereditary maximal V-order is not necessarily Prüfer if the rank of V is more than two (see [M2]). Furthermore, if V is a discrete rank one valuation ring, then $J(V)$ is principal. But, in general, $J(V)$ is either principal or idempotent. These are main obstructions to the study of semi-hereditary V-orders in the light of hereditary orders. The idealizer theory is one of the main techniques to study hereditary orders. Fortunately, the idealizers work well to study semi-hereditary orders. Let K be a semi-maximal right ideal of a ring S, that is, K is a finite intersection of maximal right ideals of S. The subring $\mathbf{I}_S(K) = \{s \in S \mid sK \subseteq K\}$ of S is called an *idealizer* of K in S. We use the following notation;

$\mathcal{E}(R) = \{A \mid A$ is an idempotent ideal of R which is finitely generated as a left ideal$\}$, $\mathcal{O}_V(R) = \{S \mid S$ is a V-order in Q containing $R\}$ and $(R : X)_l = \{q \in Q \mid qX \subseteq R\}$ for any subset X of Q.

Now we shall obtain some crucial properties of semi-hereditary V-orders in Q as follows:

(g) *If S is a semi-hereditary V-order in Q, then so is $R = \mathbf{I}_S(K)$.*

(h) *Let R be a semi-hereditary V-order. Then there is a bijection between $\mathcal{E}(R)$ and $\mathcal{O}_V(R)$ which is given by $A \longmapsto O_r(A)$ and $S \longmapsto (R : S)_l$, where $A \in \mathcal{E}(R)$ and $S \in \mathcal{O}_V(R)$.*

(i) *Any maximal ideal of a semi-hereditary V-order is either invertible or idempotent.*

In what follows, R is a semi-hereditary V-order in Q, where V is a valuation ring of R. To study the structure of overrings of a semi-hereditary V-order R, we need to study the structure of $\mathcal{E}(R)$ as it is seen from (h). The structure of $\mathcal{E}(R)$ is characterized in terms of cycles. Let $M_1 \ldots, M_n$ be distinct idempotent maximal ideals of a semi-hereditary V-order R $(n \geq 2)$ satisfying $O_r(M_i) = O_l(M_{i+1}) \supset R$ $(1 \leq i \leq n-1)$. We classify them into the following five types;

(i) $O_r(M_n) = O_l(M_1) \supset R$, a *first type of cycle*.
(ii) $O_r(M_n) = R = O_l(M_1)$, a *second type of cycle*.
(iii) $O_l(M_1) \supset R$, $O_r(M_n) \supset R$ and $O(M_n) \neq O_l(M_1)$, an *open cycle*.
(iv) $O_r(M_n) \supset R = O_l(M_1)$, a *right open cycle*.
(v) $O_l(M_1) \supset R = O_r(M_n)$, a *left open cycle*.

Note that an idempotent maximal ideal M is finitely generated as a left ideal if and only if $O_r(M) \supset R$. If a maximal ideal M is invertible, then it is considered as a (*trivial*) *first type of cycle*, because $O_r(M) = R = O_l(M)$ and M is finitely generated as a left and right ideal. If a maximal ideal M is not finitely generated as both a left and right ideal, then it is considered as a (*trivial*) *second type of cycle*, because $M = M^2$ and $O_r(M) = R = O_l(M)$. The type (i) only occurs in the case $J(V)$ is principal and any one of the types (ii), (iv) and (v) occurs in the case $J(V)$ is idempotent.

Let $D(R) = \{A \mid A$ is an invertible R-ideal and $J(V)^n \subseteq A \subseteq (R : J(V)^n R)_l$ for some $n > 0\}$. Then we have the following as in hereditary orders.

Proposition 2.6. [Ma] *Let R be a semi-hereditary V-order in Q.*

(1) *If $J(V)$ is principal, then $D(R)$ is a free abelian group generated by maximal invertible ideals and any maximal invertible ideal is the intersection of a first type of cycle. Furthermore, any maximal ideal belongs to a first type of cycle.*

(2) *If $J(V)$ is idempotent, then $D(R) = \{R\}$ and any maximal ideal belongs to a second type of cycle.*

By using the concept of cycles, any element in $\mathcal{E}(R)$ is characterized in the following:

Proposition 2.7. [Ma] *Let A be an ideal of a semi-hereditary V-order in Q. Then $A \in \mathcal{E}(R)$ if and only if*

(1) *any maximal ideal containing A is idempotent and finitely generated as a left ideal, and*

(2) $A = (M_{1m(1)} \cdots M_{11}) \cdots (M_{lm(l)} \cdots M_{l1})$, *where M_{ij} ($1 \leq i \leq l, 1 \leq j \leq m(i)$) are all maximal ideals containing A and $M_{i1}, \ldots, M_{im(i)}$ is either an open cycle or a right open cycle ($1 \leq i \leq l$).*

Assume $M_1, \ldots, M_n$ is a first (second) type of cycle of R with $\dim R/M_i = m_i$, the Goldie dimension of a right R-module R/M_i. We write $(M_1, \ldots, M_n) = (m_1, \ldots, m_n)$ to show implicitly the dimension of R/M_i as a right R-module and it is called the *form* of $M_1, \ldots, M_n$. Let $M_{11}, \ldots, M_{1n(1)}, \ldots, M_{k1}, \ldots, M_{kn(k)}$ be the set of all maximal ideals of R such that $M_{i1}, \ldots, M_{in(i)}$ is a first (second) type of cycle of R with $\dim R/M_{ij} = m_{ij}$ ($1 \leq i \leq k, 1 \leq j \leq n(i)$). We say that $M_{11}, \ldots, M_{1n(1)}, \ldots, M_{k1}, \ldots, M_{kn(k)}$ is the *series of maximal ideals* of R and $(M_{11}, \ldots, M_{1n(1)}, \ldots, M_{k1}, \ldots, M_{kn(k)}) = (m_{11}, \ldots, m_{1n(1)}, \ldots, m_{k1}, \ldots, m_{kn(k)})$ is the *form* of R. The following is used for characterizing the principalness of $J(R)$.

(j) *Let $M_1, \ldots, M_n$ be a first type of cyle in a semi-hereditary V-order R with its form $(M_1, \ldots, M_n) = (m_1, \ldots, m_n)$. Then:*

(1) $J = M_1 \cap \cdots \cap M_n$ is principal as a left and right ideal if and only if $m_1 = \cdots = m_n$;
(2) J^n is principal.

Theorem 2.8. [Ma] *Let R be a semi-hereditary V-order in Q with $M_{11}, \ldots, M_{1n(1)}, \ldots, M_{k1}, \ldots, M_{kn(k)}$ as the series of maximal ideals of R. Set $J_i(R) = M_{i1} \cap \cdots \cap M_{in(i)}$ ($1 \leq i \leq k$). Assume that $J(V)$ is principal. Then the following are equivalent:*
(1) $J(R)$ is principal.
(2) $J_i(R)$ is principal for each i.
(3) The form of R is $(M_{11}, \ldots, M_{1n(1)}, \ldots, M_{k1}, \ldots, M_{kn(k)}) = (m_1, \ldots, m_1, \ldots, m_k, \ldots, m_k)$.

The maximal ideals of a semi-hereditary V-order do not necessarily commute. We characterize the commutativity of maximal ideals by using the orders of ideals and an induction on overrings.

Proposition 2.9. [Ma] *Let M and N be distinct idempotent maximal ideals of a semi-hereditary V-order R in Q.*
(1) Assume that M is not finitely generated as a left and right ideal. Then $MN = M \cap N = NM$.
(2) Assume that M is finitely generated as a left ideal and is not finitely generated as a right ideal. Then $MN = M \cap N = NM$ if and only if $O_r(M) \neq O_l(N)$.
(3) Assume that M is finitely generated as a left and right ideal. Then $MN = M \cap N = NM$ if and only if $O_r(M) \neq O_l(N)$ and $O_r(N) \neq O_l(M)$.

We use the property (h), Propositions 2.7, 2.9 and Theorem 2.8 to obtain the following theorem, some parts of which were obtained in [K3] by different methods.

Theorem 2.10. [Ma] *Let R be a semi-hereditary V-order in Q with the maximal ideals series $M_{11}, \ldots, M_{1n(1)}, \ldots, M_{k1}, \ldots, M_{kn(k)}$.*
(1) Assume that $J(V)$ is principal. Then:
  (i) The number of V-overrings of R is $(2^{n(1)} - 1) \cdots (2^{n(k)} - 1)$.
  (ii) The number of semi-hereditary maximal V-orders containing R is $n(1) \cdots n(k)$, they are all conjugate and R is the intersection of all those semi-hereditary maximal V-orders containing R.
(2) Assume that $J(V)$ is idempotent. Then:
  (i) The number of V-overrings of R is $2^{n(1)+\cdots+n(k)-k}$.
  (ii) There is the unique semi-hereditary maximal V-order containing R.

The completion is the one of the important tools for studying Noetherian orders. However it does not work well for studying non-Noetherian orders so that we use the Henselization in the non-Noetherian case. Let V^h be the Henselization of a valuation ring V of F with its quotient field $F^h = F \otimes V^h$ and $Q^h = Q \otimes V^h = M_\mathcal{N}(D^h)$, the $\mathcal{N} \times \mathcal{N}$ matrix ring over D^h, where D^h is the division ring with $Z(D^h) = F^h$. There exists the unique invariant valuation ring Δ^h of D^h with $Z(\Delta^h) = V^h$. So the matrix size $\mathcal{N}$ of Q^h is unique up to V and Q. In [M3], Morandi conjectured: Is $R \otimes V^h$ semi-hereditary if R is a semi-hereditary V-order? Furthermore, he suspected that if R is a semi-hereditary *maximal* V-order, then so is $R \otimes V^h$. The following is the answer to these questions:

Theorem 2.11. (1) [MM], [K1] *If R is a semi-hereditary V-order, then so is $R \otimes V^h$.*

(2) [K3] *If R is a semi-hereditary maximal V-order, then $R \otimes V^h$ is a semi-hereditary maximal V^h-order.*

In [K1], Kauta proved that any semi-hereditary V^h-order S in Q^h is of the form; $S = (\Delta_{ij})$, the matrix ring whose (i,j) entries are Δ_{ij}, where Δ_{ij} are non-zero submodules of D^h with $\Delta^h = \Delta_{ii}$. Let M be a maximal ideal of R. Then R/M is isomorphic to a matrix ring over a division ring, which is called a *division part* of M. It is shown in [K1] and [MM] independently that $R/M \cong R^h/M^h$, where $M^h = M \otimes V^h$. We can use these properties in order to prove (3) and (4) in the following theorem:

Theorem 2.12. (Invariantness) [Ma] *Let R be a semi-hereditary V-order in Q and let S be a semi-hereditary V-order containing R.*

(1) *The number of first (second) type cycles of R is equal to the number of first (second) type cycles of S.*

(2) $D(R) \cong D(S)$.

(3) $\dim R/J(R) = \mathcal{N}$, *the matrix size of Q^h, that is, $\dim R/J(R)$ is unique up to V and Q.*

(4) *Let M be any maximal ideal of R. Then the division part of M is isomorphic to $\overline{\Delta}^h$, where $\overline{\Delta}^h = \Delta^h/J(\Delta^h)$, that is, the division part of any maximal ideal of any semi-hereditary V-order in Q is unique up to V and Q.*

For V a Henselian valuation ring, more detailed results on semi-hereditary V-orders are obtained in [M2], [K1] and [K2].

Let G be a finite group acting on a commutative ring R as automorphisms of R and let $R \star G$ be a (generalized) crossed product algebra of

G over R, that is, there is a factor set $f : G \times G \longrightarrow R^+ = R \setminus (0)$ and $t : G \longrightarrow Aut(R)$ satisfying; $\bar{g}\bar{h} = f(g,h)\overline{gh}$ and $r\bar{g} = \bar{g}r^{t(g)}$ for all $r \in R$ and $g \in G$, where $\bar{G} = \{\bar{g} \mid g \in G\}$ is a free basis of $R \star G$ as a right R-module. If $f(g,h) = 1$ for all $g, h \in G$, then $R \star G$ is called a *skew group ring*. Let A be any ideal of R and let $G(A) = \{g \in G \mid r - r^{t(g)} \in A \text{ for all } r \in R\}$, called the *inertial group* of M. In the case of skew group rings, we can characterize those skew group rings $R\star G$ which are semi-hereditary and Prüfer orders in a simple Artinian ring $Q(R\star G)$, the quotient ring of $R \star G$, as follows:

Theorem 2.13. [MY] *Let G be a finite group acting on a commutative ring R as automorphisms and let $R \star G$ be the skew group ring.*

(1) $R\star G$ is a semi-hereditary order in a simple Artinian ring $Q(R\star G)$ if and only if

(i) R is a semi-hereditary G-prime Goldie ring;

(ii) $G(P) = <1>$ for any minimal prime ideal P of R;

(iii) $G(M)$ contains no elements of order p for any maximal ideal M of R with $Char(R/M) = p > 0$.

(2) $R \star G$ is a Prüfer order in a simple Artinian ring $Q(R \star G)$ if and only if

(i) R is a semi-hereditary G-prime Goldie ring;

(ii) $G(M) = <1>$ for any maximal ideal M of R.

In [Ha] and [HM], Haile and Morandi studied Dubrovin valuation properties of generalized crossed product algebras. We finish by giving a few unsolved problems.

Problem 1. Find necessary and sufficient condition for the generalized crossed product algebras $R \star G$ to be semi-hereditary and Prüfer in terms of G and R.

Problem 2. Let P be a prime ideal of a fully bounded Prüfer order in a simple Artinian ring. Is R/P a Goldie ring? (If R is a Dubrovin valuation ring, then the answer is affirmative).

Problem 3. Does the approximation theorem hold in the case that the order is a semi-local Bezout order without the assumption that the order is a P.I. ring? We may assume that R_i are all fully bounded Dubrovin valuation rings.

We have given a rough survey of non-commutative valuation rings and their global theories. We refer the reader to the book [MMU] for more detailed results with proofs and for the references.

References

[AD]   J. H. Alajbegovic and N. I. Dubrovin, *Noncommutative Prüfer rings*, J. Algebra **135** (1990), 165–176.

[B]   N. Bourbaki, *Algebre Commutative, Chapters 1-7*, 2nd Printing, Springer-Verlag, Heidelberg, New York, 1989.

[BG1] H. H. Brungs and J. Gräter, *Valuation rings in finite dimensional division algebras*, J. Algebra **120** (1989), 90–99.

[BG2] _____, *Extensions of valuation rings in central simple algebras*, Trans. Amer. Math. Soc. **317** (1990), 287–302.

[BMO] H. H. Brungs, H. Marubayashi and E. Osmanagic, *A classification of prime segments in simple Artinian rings*, Proc. Amer. Math. Soc., to appear.

[BMU] H. H. Brungs, H. Marubayashi and A. Ueda, *A classification of primary ideals of Dubrovin valuation rings*, preprint.

[BT]   H. H. Brungs and G. Törner, *On the number of injective indecomposable modules*, preprint.

[D1]   N. I. Dubrovin, *Noncommutative valuation rings*, Trans. Moscow Math. Soc. **45** (1984), 273–287.

[D2]   _____, *Noncommutative valuation rings in simple finite dimensional algebras over a field*, Math. USSR Sbornik **51** (1985), 493–505.

[D3]   _____, *Noncommutative Prüfer rings*, ibid. **74** (1993), 1–8.

[D4]   _____, *The rational closure of group rings of left orderable groups*, ibid. **184** (1993), 3–48.

[G]   R. Gilmer, *Multiplicative Ideal Theory*, Queen's Papers in Pure and Applied Math. **90**, Queen's University, 1992.

[G1]   J. Gräter, *The Defektsatz for central simple algebras*, Trans. Amer. Math. Soc. **330** (1992), 823–843.

[G2]   _____, *Prime PI-rings in which finitely generated right ideals are principal*, Forum Math. **4** (1992), 447–463.

[H]   H. Hasse, *Über p-adische Schiefkörper und ihre Bedeutung für die Arithmetik hyperkomplexer Zahlsysteme,* Math. Ann. **104** (1931), 495–534.

[Ha] D. E. Haile, *Crossed-products orders over discrete valuation rings,* J. Algebra **105** (1987), 116–148.

[HM] D. E. Haile and P. J. Morandi, *On Dubrovin valuation rings in crossed product algebras,* Trans. Amer. Math. Soc. **338** (1993), 723–751.

[HMW] D. E. Haile, P. J. Morandi and A. R. Wadsworth, *Bezout orders and Henselization,* J. Algebra **173** (1995), 394–423.

[K1] J. Kauta, *Integral semihereditary orders, extremality and Henselization,* J. Algebra **189** (1997), 226–252.

[K2] _____, *Integral semihereditary orders inside a Bezout maximal order,* ibid. **189** (1997), 253–272.

[K3] _____, *On semihereditary maximal orders,* Bull. London Math. Soc. **30** (1998), 251–257.

[M]   K. Mathiak, *Bewertungen nichtkommutativer Körper,* J. Algebra **48** (1977), 217–235.

[M1] P. J. Morandi, *An approximation theorem for Dubrovin valuation rings,* Math. Z. **207** (1991), 71–82.

[M2] _____, *Maximal orders over valuation rings,* J. Algebra **152** (1992), 313–341.

[M3] _____, *Noncommutative Prüfer rings satisfying a polynomial identity,* ibid. **161** (1993), 324–341.

[Ma] H. Marubayashi, *On semi-hereditary orders integral over a commutative valuation ring,* preprint.

[MM] H. Marubayashi, H. Miyamoto, A. Ueda and Y. Zhao, *Semi-hereditary orders in a simple Artinian ring,* Comm. Algebra **22** (1994), 5209–5230.

[MMU] H. Marubayashi, H. Miyamoto and A. Ueda, *Non-Commutative Valuation Rings and Semi-hereditary Orders,* Kluwer Academic Publ., Dordrecht, 1997.

[MV] H. Moawad and F. Van Oystaeyen, *Discrete valuations extend to certain algebras of quantum type,* Comm. Algebra **24** (1996), 2551–2566.

[MW] P. J. Morandi and A. R. Wadsworth, *Integral Dubrovin valuation rings,* Trans. Amer. Math. Soc. **315** (1989), 623–640.

[MY] H. Marubayashi and Z. Yi, *Skew group rings which are semi-hereditary orders and Prüfer orders in simple Artinian rings,* Algebras and Representation Theory, to appear.

[PR] J. K. Park and K. W. Roggenkamp, *Bezout orders,* Forum Math. **7** (1995), 477–488.

[S] O. F. G. Schilling, *The Theory of Valuations,* Math. Surveys **4**, Amer. Math. Soc., Providence, 1950.

[VW] F. Van Oystaeyen and L. Willaert, *Valuations on extensions of Weyl skew fields,* J. Algebra **183** (1996), 359–364.

[W1] A. R. Wadsworth, *Dubrovin valuation rings, perspective in Ring Theory,* edited by F. Van Oystaeyen and L. LeBruyn, Kluwer Academic Publishers, Dordrecht, 1988, 359–374.

[W2] _____, *Dubrovin valuation rings and Henselization,* Math. Ann. **283** (1989), 301–328.

[ZS] O. Zariski and P. Samuel, *Commutative Algebra II,* Graduate Text in Math. **28**, Springer-Verlag, Heidelberg, New York, 1958.

Department of Mathematics
Naruto University of Education
Naruto 772, Japan
e-mail: `marubaya@naruto-u.ac.jp`

On the Maximal t-Corational Extensions of Modules

Shoji Morimoto

Abstract

The main aim of this paper is to relate the maximal t-corational extension with the t-colocalization and the t-projective cover of a module for a preradical t.

Courter [3] and Bican [1] studied some properties of corational extensions of modules. In this note, for a preradical t, we define t-corational extensions and maximal t-corational extensions of modules in the same way as they did. First, we determine the forms of t-corational extensions of a given module for an idempotent preradical t (Theorem 1.4). Next we investigate the maximal t-corational extensions of modules. In particular, if t is a cotorsion radical, then every module in $T(t)$ has the maximal t-corational extension uniquely up to isomorphism (Theorem 2.4). Moreover, if $B \xrightarrow{f} A \longrightarrow 0$ is the maximal t-corational extension of A, then B is t-projective (Theorem 2.8). Furthermore, we relate the maximal t-corational extension with the t-colocalization and the t-projective cover of a module for a preradical t. M. Sato proved that if t is a cotorsion radical, then $t(R) \otimes_R t(R) \otimes_R A \longrightarrow A$ is the t-colocalization of a module A [4]. Finally, we give a simple proof of this result (Theorem 3.1).

Throughout this note, R means a ring with identity and modules mean unitary left R-modules, unless otherwise stated. We denote the category of the modules by R-mod. For the notions and terminologies of preradicals and torsion theories, we refer to [5]. For each preradical t, we denote the t-torsion class (resp. t-torsionfree class) by $T(t)$ (resp. $F(t)$).

1. t-Corational Extensions of Modules

Let t be a preradical. We call an exact sequence $B \xrightarrow{f} A \longrightarrow 0$ of modules a *t-corational extension of A* if B is in $T(t)$ and every factor module of $Ker(f)$ is in $F(t)$.

As is easily seen, $B \xrightarrow{f} A \longrightarrow 0$ is a corational extension of A if and only if it is a t_B-corational extension of A, where t_B is a unique minimal idempotent preradical one of those preradical r for which $r(B) = B$.

Lemma 1.1. *Let t be a preradical. If $B \xrightarrow{f} A \longrightarrow 0$ is a t-corational extension of A, then $Ker(f)$ is small in B.*

Proof. Let C be a submodule of B with $B = C + Ker(f)$. Then $B/C = (C + Ker(f))/C \cong Ker(f)/(Ker(f) \cap C)$. Since B is in $T(t)$, B/C is in $T(t)$. Also since $B \xrightarrow{f} A \longrightarrow 0$ is a t-corational extension of A, $Ker(f)/(Ker(f) \cap C)$ is in $F(t)$. Thus $B/C = 0$; that is, $Ker(f)$ is small in B. □

Let t be an idempotent preradical. We call t a cotorsion radical if $t(M/N) = (t(M) + N)/N$ for all modules M and its submodules N.

Dually, we have

Proposition 1.2. *Let t be a cotorsion radical and $0 \longrightarrow K \longrightarrow B \xrightarrow{f} A \longrightarrow 0$ an exact sequence of modules such that K is small in B. Then A is in $T(t)$ if and only if B is in $T(t)$.*

Proof. Suppose that A is in $T(t)$. Since t is a cotorsion radical, $t(B) \xrightarrow{t(f)} A \longrightarrow 0$ is exact. Thus we have $B = t(B) + Ker(f)$. Hence $B = t(B)$ by assumption. □

Proposition 1.3. *Let t be a cotorsion radical and $B \xrightarrow{f} A \longrightarrow 0$ an exact sequence of modules. Then the following conditions are equivalent:*

(1) $B \xrightarrow{f} A \longrightarrow 0$ is a t-corational extension of A.
(2) A is in $T(t)$, $Ker(f)$ is in $F(t)$ and is small in B.
(3) B is in $T(t)$ and $Ker(f)$ is in $F(t)$.

Now, for an idempotent preradical t, we can determine the form of a t-corational extension of each module.

Theorem 1.4. *Let t be an idempotent preradical and let $0 \longrightarrow K \longrightarrow P \xrightarrow{g} A \longrightarrow 0$ be an exact sequence of modules with P projective. Then any t-corational extension of A is of the form $P/K' \xrightarrow{g^*} A \longrightarrow 0$ for some submodule K' of K that contains $t(K)$, where g^* is an induced homomorphism by g.*

Proof. Let $B \xrightarrow{f} A \longrightarrow 0$ be a t-corational extension of A. Since P is projective, there exists an epimorphism $h : P \longrightarrow B$ such that $f \circ h = g$. Thus we obtain the following diagram

$$\begin{array}{ccccccccc} 0 & \longrightarrow & K/Ker(h) & \longrightarrow & P/Ker(h) & \xrightarrow{\bar{g}} & A & \longrightarrow & 0 \\ & & {\scriptstyle k}\downarrow {\scriptstyle \cong} & & {\scriptstyle h^*}\downarrow {\scriptstyle \cong} & & \| & & \\ 0 & \longrightarrow & Ker(f) & \longrightarrow & B & \xrightarrow{f} & A & \longrightarrow & 0 \end{array}$$

where h^* is an isomorphism induced by h, $\bar{g}$ is an epimorphism induced by g and k is a restriction of h^* to $K/Ker(h)$. Since h^* and k are isomorphisms, $P/Ker(h)$ is in $T(t)$ and $K/Ker(h)$ is in $F(t)$. Moreover every homomorphic image of $K/Ker(h)$ is in $F(t)$. Hence $P/Ker(h) \xrightarrow{\bar{g}} A \longrightarrow 0$ is a t-corational extension of A. Since $f \circ h = g$, $Ker(h) \subseteq Ker(g) = K$. Also since $K/Ker(h) \supseteq (t(K) + Ker(h))/Ker(h) \cong t(K)/(t(K) \cap Ker(h))$, $t(K) \subseteq Ker(h)$. □

2. The Maximal t-Corational Extensions.

Let t be a preradical. An exact sequence $B \xrightarrow{f} A \longrightarrow 0$ of modules is called a *maximal t-corational extension of A* if $B \xrightarrow{f} A \longrightarrow 0$ is a t-corational extension of A and for any t-corational extension $B' \xrightarrow{f'} A \longrightarrow 0$ of A, there exists a homomorphism $\varphi : B \longrightarrow B'$ such that $f' \circ \varphi = f$.

Lemma 2.1. *Let both $B \xrightarrow{f} A \longrightarrow 0$ and $B' \xrightarrow{f'} A \longrightarrow 0$ be t-corational extensions of A. If there exists a homomorphism $\varphi : B \longrightarrow B'$ such that $f' \circ \varphi = f$, then such φ is an epimorphism and is unique.*

Proof. By assumption, $B' = Im(\varphi) + Ker(f')$. Since $Ker(f')$ is small in B', $B' = Im(\varphi)$. If there exists $\varphi' : B \longrightarrow B'$ such that $f' \circ \varphi' = f$, then $Im(\varphi - \varphi') \subseteq Ker(f')$. Thus $\varphi - \varphi' \in Hom_R(B, Ker(f'))$. Since B is in $T(t)$ and $Ker(f')$ is in $F(t)$, $\varphi = \varphi'$. □

Lemma 2.2. *Let $B \xrightarrow{f} A \longrightarrow 0$ be a maximal t-corational extension of A and let $B' \xrightarrow{f'} A \longrightarrow 0$ be a t-corational extension of A. If there exists a homomorphism $\varphi : B' \longrightarrow B$ such that $f \circ \varphi = f'$, then φ is an isomorphism.*

Proof. Since $B \xrightarrow{f} A \longrightarrow 0$ is a maximal t-corational extension of A, there exists an epimorphism $\psi : B \longrightarrow B'$ such that $f' \circ \psi = f$. Also, by assumption, $f \circ \varphi = f'$ and so $f' \circ \psi \circ \varphi = f'$. Thus $\psi \circ \varphi - 1_{B'} \in Hom_R(B', Ker(f')) = 0$. Hence $\psi \circ \varphi = 1_{B'}$; namely, ψ is a monomorphism. By Lemma 2.1, ψ is an isomorphism. □

Corollary 2.3. *Let t be a preradical. If a module A has a maximal t-corational extension, then it is unique up to isomorphism.*

Let t be a preradical. A module Q is called *t-projective* if the functor $Hom_R(Q, -)$ preserves the exactness for all exact sequences of modules $0 \longrightarrow A \longrightarrow B \longrightarrow C \longrightarrow 0$ with $A \in F(t)$.

Theorem 2.4. *Let t be a cotorsion radical. If a module A is in $T(t)$, then A has the maximal t-corational extension.*

Proof. Let $0 \longrightarrow K \longrightarrow P \xrightarrow{f} A \longrightarrow 0$ be an exact sequence with P projective. Then it induces an exact sequence $0 \longrightarrow K/t(K) \longrightarrow P/t(K) \xrightarrow{\bar{f}} A \longrightarrow 0$. By Theorem 1.4, any t-corational extension of A is of the form $P/K' \xrightarrow{f^*} A \longrightarrow 0$, where f^* is induced by f and $t(K) \subseteq K' \subseteq K$. By [3, Theorem 2.7], we obtain the following commutative diagram

$$\begin{array}{ccc} P/t(K) & \xrightarrow{\bar{f}} & A \\ {\scriptstyle \pi}\downarrow & \| & \\ P/K' & \xrightarrow{f^*} & A \end{array}$$

, where π is the canonical map.

Since t is a cotorsion radical, $t(P/t(K)) \xrightarrow{t(\bar{f})} A \longrightarrow 0$ is exact. Also since $Ker(t(\bar{f})) \subseteq Ker(\bar{f}) = K/t(K)$, $Ker(t(\bar{f}))$ is in $F(t)$. Thus $t(P/t(K)) \xrightarrow{t(\bar{f})} A \longrightarrow 0$ is a t-corational extension of A. Since π is an epimorphism and P/K' is in $T(t)$, $t(P/t(K)) \xrightarrow{t(\pi)} P/K' \longrightarrow 0$ is exact. Hence we have the following commutative diagram

$$\begin{array}{ccc} t(P/t(K)) & \xrightarrow{t(\bar{f})} & A \\ {\scriptstyle t(\pi)}\downarrow & \| & \\ P/K' & \xrightarrow{f^*} & A \end{array}.$$

Hence, $t(P/t(K)) \xrightarrow{t(\bar{f})} A \longrightarrow 0$ is the maximal t-corational extension of A. □

Lemma 2.5. *Let t be a cotorsion radical. If $B \xrightarrow{f} A \longrightarrow 0$ is the maximal t-corational extension of A and $B' \xrightarrow{f'} A \longrightarrow 0$ is an exact sequence for which $Ker(f')$ is in $F(t)$, then there exists a homomorphism $\varphi : B \longrightarrow B'$ such that $f' \circ \varphi = f$.*

Proof. Since t is a cotorsion radical, $t(B') \xrightarrow{t(f')} A \longrightarrow 0$ is a t-corational extension of A. Thus there exists an epimorphism $\psi : B \longrightarrow t(B')$ such that $t(f') \circ \psi = f$, and so $f' \circ i \circ \psi = f$, where $i : t(B') \longrightarrow B'$ is the inclusion map. Thus $i \circ \psi$ is a desired homomorphism. □

Lemma 2.6. *Let t be a cotorsion radical, $B \xrightarrow{f} A \longrightarrow 0$ the maximal t-corational extension of A and $B' \xrightarrow{f'} A \longrightarrow 0$ an exact sequence for which $Ker(f')$ is in $F(t)$. If there exists a homomorphism $g : B' \longrightarrow B$ such that $f \circ g = f'$, then B is isomorphic to a direct summand of B'.*

Proof. By assumption and Lemma 2.5, we obtain the following commutative diagram

$$\begin{array}{ccc} B & \xrightarrow{f} & A \\ \varphi \downarrow \uparrow g & & \| \\ B' & \xrightarrow{f'} & A \end{array}$$

where $\varphi : B \longrightarrow B'$ and $g : B' \longrightarrow B$ such that $f' \circ \varphi = f$ and $f \circ g = f'$. Thus $f \circ g \circ \varphi = f$; namely, $g \circ \varphi - 1_B \in Hom_R(B, Ker(f)) = 0$. Hence B is isomorphic to a direct summand of B'. □

Theorem 2.7. *Let t be a cotorsion radical. If $B \xrightarrow{f} A \longrightarrow 0$ is the maximal t-corational extension of A, then B is t-projective.*

Proof. Let $0 \longrightarrow K \longrightarrow P \xrightarrow{g} A \longrightarrow 0$ be an exact sequence with P projective. Since $B \xrightarrow{f} A \longrightarrow 0$ is the maximal t-corational extension of A and A is in $T(t)$, $t(P/t(K)) \cong_\varphi B$ by Theorem 2.4. Also $P/t(K) \xrightarrow{\bar{g}} A \longrightarrow 0$ is exact and $Ker(\bar{g}) = K/t(K)$ is in $F(t)$. Since $P/t(K)$ is t-projective, we obtain the following commutative diagram

$$\begin{array}{ccc} t(P/t(K)) & \longrightarrow & A \\ \downarrow \uparrow & & \| \\ P/t(K) & \longrightarrow & A \end{array}.$$

By Lemma 2.6, $t(P/t(K))$ is a direct summand of $P/t(K)$. Since $P/t(K)$ is t-projective, $t(P/t(K))$ is t-projective. Hence B is t-projective. □

Lemma 2.8. *Let t be an idempotent preradical and $\varphi \in Hom_R(B, A)$. If B is in $T(t)$ and $Coker(\varphi)$ is in $F(t)$, then $\varphi(B) = t(A)$.*

Proof. Since $B = t(B)$, $\varphi(B) \subseteq t(A)$. Also since $t(A/\varphi(B)) = 0$, $t(t(A)/\varphi(B)) = 0$. Thus $t(A) = \varphi(B)$. □

Let t be a preradical. We shall say $B \xrightarrow{f} A$ a t-colocalization of A if B is t-projective and is in $T(t)$ and both $Ker(f)$ and $Coker(f)$ are in $F(t)$. Also we call an exact sequence $Q \xrightarrow{g} A \longrightarrow 0$ a t-projective cover of A if Q is t-projective and $Ker(g)$ is in $F(t)$ and is small in Q.

Theorem 2.9. *Let t be a cotorsion radical and $B \xrightarrow{f} A \longrightarrow 0$ an exact sequence. Then the following conditions are equivalent:*

(1) $B \xrightarrow{f} A \longrightarrow 0$ *is the maximal t-corational extension of A.*
(2) $B \xrightarrow{f} A \longrightarrow 0$ *is a t-colocalization of A.*
(3) $B \xrightarrow{f} A \longrightarrow 0$ *is a t-projective cover of A and A is in $T(t)$.*
(4) $B \xrightarrow{f} A \longrightarrow 0$ *is a t-corational extension of A and B is t-projective.*
(5) A *is in $T(t)$, B is t-projective, $Ker(f)$ is in $F(t)$ and is small in B.*

Proof. (1)⇒(2). Let $B \xrightarrow{f} A \longrightarrow 0$ be the maximal t-corational extension of A. By Theorem 2.7, B is t-projective. Also by Lemma 1.1, $Ker(f)$ is small in B. Thus B is in $T(t)$ from Proposition 1.2. Hence $B \xrightarrow{f} A \longrightarrow 0$ is a t-colocalization of A.

(2)⇒(3) follows from Proposition 1.3. The implications (3)⇒(4)⇒(5) are clear.

(5)⇒(1). By Proposition 1.3, $B \xrightarrow{f} A \longrightarrow 0$ is the t-corational extension of A. Since B is t-projective, $B \xrightarrow{f} A \longrightarrow 0$ is the maximal t-corational extension of A. □

Corollary 2.10. *Let t be a cotorsion radical. If A is in $T(t)$, then there exists a t-projective cover and t-colocalization of A uniquely up to isomorphism.*

3. Applications

Theorem 3.1. *Let t be a cotorsion radical and A a module. Then $t(R) \otimes_R t(R) \otimes_R A \xrightarrow{f} t(A)$ is the maximal t-corational extension of $t(A)$, where $f(x \otimes y \otimes a) = xya$ for $x, y \in t(R)$ and $a \in A$.*

Proof. Since $t(A) = t(R)A$, $t(R) \otimes_R t(R) \otimes_R A \xrightarrow{f} t(A) \longrightarrow 0$ is exact. Also since $t(R)$ is an idempotent two-sided ideal of R, $t(R)(t(R) \otimes_R t(R) \otimes_R A) = t(R) \otimes_R t(R) \otimes_R A$; that is, $t(R) \otimes_R t(R) \otimes_R A$ is in $T(t)$. Also if $\sum x_i \otimes y_i \otimes a_i \in Ker(f)$, then $r(\sum x_i \otimes y_i \otimes a_i) = \sum ((rx_i) \otimes y_i \otimes a_i) = \sum (r \otimes (x_i y_i) \otimes a_i) = \sum (r \otimes x_i \otimes (y_i a_i))$ for all $r \in t(R)$. Since $t(R)$ is an idempotent ideal of R, we can write $r = \sum_{j=1}^{n} f_j f_{j'}$. Thus $r(\sum x_i \otimes y_i \otimes a_i) = 0$; namely, $Ker(f)$ is in $F(t)$. Hence $t(R) \otimes_R t(R) \otimes_R A \xrightarrow{f} t(A) \longrightarrow 0$ is a t-corational extension of $t(A)$. Let $B \longrightarrow t(A) \longrightarrow 0$ be a t-corational extension of $t(A)$. By Theorem 1.4, $B \cong P/K'$ for some projective module P and a submodule K' of P. Consider a diagram

$$\begin{array}{ccccc} t(R) \otimes_R t(R) \otimes_R A & \xrightarrow{f} & t(A) & \longrightarrow & 0 \\ & & \| & & \\ P/K' & \xrightarrow{g^\star} & t(A) & \longrightarrow & 0 \end{array}$$

We can define homomorphisms $\varphi : t(R) \otimes_R t(R) \otimes_R A \longrightarrow t(R) \otimes_R t(A)$ and $\psi : t(R) \otimes_R t(A) \longrightarrow P/K'$ by $\varphi(x \otimes y \otimes a) = x \otimes ya$ for $x, y \in t(R)$ and $a \in A$ and $\psi(x \otimes a') = xp + K'$ for $x \in t(R)$, $p \in P$ and $a' \in t(A)$, where $g^\star \circ \pi(p) = a'$ and π is the canonical map P to P/K'. Thus $g^\star \circ \psi \circ \varphi = f$. Therefore the above diagram commutes; that is, $t(R) \otimes_R t(R) \otimes_R A \xrightarrow{f} t(A) \longrightarrow 0$ is the maximal t-corational extension of $t(A)$. $\square$

In particular, if we take $t(A)$ as a module A, then $t(R) \otimes_R t(R) \otimes_R t(A) \xrightarrow{f'} t(A)$ is the maximal t-corational extension of $t(A)$.

Corollary 3.2. *Let t be a cotorsion radical. Then $t(R) \otimes_R t(R) \otimes_R A \cong t(R) \otimes_R t(R) \otimes_R t(A)$ for all modules A. In particular, $t(R) \otimes_R t(R) \cong t(R) \otimes_R t(R) \otimes_R t(R)$.*

Corollary 3.3. [4, Theorem 1] *Let t be a cotorsion radical and $t(R) \otimes_R t(R) \otimes_R A \xrightarrow{f} t(A)$ the maximal t-corational extension of $t(A)$. Then $i \circ f : t(R) \otimes_R t(R) \otimes_R A \longrightarrow A$ is a colocalization of A, where $i : t(A) \longrightarrow A$ is the inclusion map.*

References

[1] L. Bican, *Corational extensions and pseudo-projective modules*, Acta. Math. Acad. Sci. Hungaricae, Tomus **28**(1-2) (1976), 5–11.

[2] P. E. Bland, *Divisible and codivisible modules*, Math. Scand. **34** (1974), 153–161.

[3] R. C. Courter, *The maximal co-rational extension by a module*, Canad. J. Math. **18** (1966), 953–962.

[4] M. Sato, *The concrete description of the colocalization*, Proc. Japan Acad. **52** (1976), 501–504.

[5] B. Stenström, *Rings of Quotients*, Springer-Verlag, Heidelberg, New York, 1975.

Hagi Koen Gakuin
Hagi, Yamaguchi 758-0047, Japan
e-mail: shojim@ymg.urban.ne.jp

On Values of Cyclotomic Polynomials

K. Motose

Abstract

This short survey article consists of some results appearing in [7]. In Section 1, fundamental properties of cyclotomic polynomials and their applications to important theorems in algebra will be introduced, while in Section 2, a cipher using values of cyclotomic polynomials will be discussed.

1. Introduction

Cyclotomic polynomials are defined as

$$\Phi_n(x) = \prod_{(k,n)=1} (x - \zeta_n^k)$$

where $\zeta_n = \cos\left(\dfrac{2\pi}{n}\right) + \sqrt{-1}\sin\left(\dfrac{2\pi}{n}\right)$ and the product is extended over natural numbers k which are relatively prime to n with $1 \leq k \leq n$.

The *Möbius' function* μ is defined as

$$\mu(n) = \begin{cases} 1 & \text{if } n = 1, \\ 0 & \text{if } p^2 \mid n \text{ for a prime } p, \\ (-1)^r & \text{if } n = p_1 p_2 \ldots p_r \text{ where } p_k \text{ are all distinct primes.} \end{cases}$$

We state some fundamental properties of cyclotomic polynomials.

(1) $x^n - 1 = \prod_{d\mid n} \Phi_d(x)$, where d runs through positive divisors of n.
(2) $\Phi_n(x) \in \mathbb{Z}[x]$ and the leading coefficient of $\Phi_n(x)$ is 1.
(3) $\Phi_n(x) = \prod_{d\mid n} (x^d - 1)^{\mu(\frac{n}{d})}$.
(4) $\Phi_n(x)$ is irreducible in $\mathbb{Q}[x]$.

(1′) This formula is equivalent to the definition of cyclotomic polynomials.

(2′) It is easy to see from the above that $\Phi_n(a)$ is an integer for an integer a.

(3′) This formula is useful for calculations of cyclotomic polynomials.

(4′) This is essential for Gauss' theorem which gives necessary and sufficient conditions that regular polygons can be constructed by using only ruler and compass. Galois groups of cyclotomic fields are determined from this.

We provide an approximation of the values of cyclotomic polynomials.

Theorem 1. $\Phi_n(x)$ *are strictly increasing functions for* $x \geq 2$ *and*
$$a^{\varphi(n)+1} > \Phi_n(a) > a^{\varphi(n)-1} \quad \text{for } n, a \geq 2,$$
where $\varphi(n)$ *is the degree of* $\Phi_n(x)$, *which is the number of positive integers* $k < n$ *with* $(k, n) = 1$.

Fermat's Little Theorem. *If* p *is a prime and* a *is a positive integer with* $p \nmid a$, *then* $a^{p-1} \equiv 1 \bmod p$.

If p is a prime and a is a positive integer with $p \nmid a$, then the least positive integer s such that $a^s \equiv 1 \bmod p$ is called the order of a modulo p. We denote the order of a modulo p by $|a|_p$. It is easy to show that $|a|_p$ is a divisor of m if $a^m \equiv 1 \bmod p$.

Let q be a prime divisor of a Mersenne number $2^p - 1$ where p is prime. Then $p = |2|_q$ since p is prime. Thus p is a divisor of $q - 1$ and $q > p$. This shows that there exist infinitely many prime numbers. In fact, starting from 2 we have 3, 7, 127, ...

In this argument, the fact that $p = |2|_q$ is most important. We can generalize this to the next theorem which is easy to prove, but powerful for us because we can find explicitly a prime p with $|a|_p = n$ for a given natural number $a, n \geq 2$.

Theorem 2. *If* $p \mid \Phi_n(a)$, *then* $n = p^e |a|_p$.

The next theorem has been known for more than one hundred years and has been rediscovered by many mathematicians, but it is not so popular today. It follows from the above theorem and an approximation cited before.

Theorem 3. [Bang] *If* $n \geq 3, a \geq 2$ *and* $(n, a) \neq (6, 2)$, *then there exists a prime* p *with* $n = |a|_p$.

Cyclotomic polynomials provide some important theorems on algebra (see [1, 3, 7, 8]). The next theorem is an extension of Lucas' or Pepin's test and was proved in [7, III, Theorem 3].

Theorem 4. (1) $p > 3$ *is prime if and only if there exists an integer* c *such that* $\left(\frac{c}{p}\right) = -1$ *and* $\Phi_{p-1}(c) \equiv 0 \bmod p$.

(2) $p > 3$ is prime if and only if there exists an integer $c > 1$ such that $(c^3 - c, p) = 1, \gamma = c + \sqrt{c^2 - 1}, \left(\frac{2c+2}{p}\right) = \left(\frac{c^2-1}{p}\right) = -1$ and $\Phi_{p+1}(\gamma) \equiv 0 \bmod p\mathcal{O}_\gamma$ where $\mathcal{O}_\gamma$ is the ring of algebraic integers in $\mathbf{Q}(\gamma)$.

2. Cipher

Pseudo primes are useful for a cipher.

Definition. A composite number n is an a-pseudo prime if and only if $a^{n-1} \equiv 1 \bmod n$.

Roughly speaking, every a-pseudo prime is a product of divisors of cyclotomic numbers $\Phi_n(a)$'s for some n, and conversely. The next is a special case, but useful for a cipher.

Theorem 5. *If d is a divisor of $\Phi_n(a)$ and $(d, n) = 1$, then it follows that $a^{d-1} \equiv 1 \bmod d$.*

We shall present a cipher using cyclotomic polynomials. This cipher is similar to the RSA cipher. The RSA cipher needs two big primes, this cipher needs only the order n of cyclotomic polynomials $\Phi_n(x)$. In this cipher, $\boldsymbol{n}$ and $\boldsymbol{e}$ are public keys.

We shall represent by $\boldsymbol{Z}^s$ the direct sum of the ring $\boldsymbol{Z}$ of integers. Let $\mathbf{1}, \boldsymbol{a}, \boldsymbol{b}, \boldsymbol{n}$ be the elements of $\boldsymbol{Z}^s$ such that $\mathbf{1} = (1, 1, \ldots, 1)$ is the identity of $\boldsymbol{Z}^s$,

$$\boldsymbol{a} = (a_1, a_2, \ldots, a_s), \boldsymbol{b} = (b_1, b_2, \ldots, b_s), \text{ and}$$
$$\boldsymbol{n} = (n_1, n_2, \ldots, n_s).$$

We use the following notation:
(1) $\boldsymbol{a} \equiv \boldsymbol{b} \bmod \boldsymbol{n}$ if and only if $a_k \equiv b_k \bmod n_k$ for every k;
(2) $(\boldsymbol{a}, \boldsymbol{b}) = \mathbf{1}$ if and only if $(a_k, b_k) = 1$ for every k;
(3) $\boldsymbol{a}^{\boldsymbol{n}} = (a_1^{n_1}, a_2^{n_2}, \ldots, a_s^{n_s})$.

We can construct a cipher as follows: For a plain text $\boldsymbol{a}$, let $\boldsymbol{k} = (k_1, k_2, \ldots, k_s)$ and $\boldsymbol{\ell} = (\ell_1, \ell_2, \ldots, \ell_s)$ be vectors such that k_i and ℓ_i are divisors of $\Phi_{s_i}(a_i)$ with $(k_i, s_i) = 1$ and $\Phi_{t_i}(a_i)$ with $(\ell_i, t_i) = 1$ where $(s_i, t_i) = 1$, respectively. We set $\boldsymbol{n} = \boldsymbol{k}\boldsymbol{\ell}$ and $\boldsymbol{m} = (\boldsymbol{k} - \mathbf{1})(\boldsymbol{\ell} - \mathbf{1})$. Then we can see
$$\boldsymbol{a}^{\boldsymbol{m}} \equiv \mathbf{1} \bmod \boldsymbol{n}$$
because $(\boldsymbol{k}, \boldsymbol{\ell}) = \mathbf{1}$.

We choose an enciphering key e with $(e, m) = 1$ and calculate the deciphering key d with $ed \equiv 1 \bmod m$. Then the sender A enciphers a with $b \equiv a^e \bmod n$ and A sends b to the receiver B. Then B deciphers b using relationship
$$b^d \equiv a^{ed} \equiv a \bmod n.$$

References

[1] E. Artin, *The orders of the linear groups*, Comm. Pure Appl. Math. **8** (1955), 355–365.

[2] A. S. Bang, *Taltheoretiske Undersøgelser*, Tidsskrift for Math. **5** (1886), 70–80 and 130–137.

[3] L. E. Dickson, *History of the Theory of Numbers, Vol. 1*, Chelsea, 1971.

[4] M. Morimoto and Y. Kida, *Factorization of cyclotomic numbers*, Sophia Kokyuroku in Math. **26**, 1987 (in Japanese).

[5] M. Morimoto, Y. Kida and M. Saito, *Factorization of cyclotomic numbers*, Sophia II, Kokyuroku in Math. **29**, 1989 (in Japanese).

[6] M. Morimoto, Y. Kida and M. Kobayashi, *Factorization of cyclotomic numbers III*, Sophia Kokyuroku in Math. **35**, 1992 (in Japanese).

[7] K. Motose, *On values of cyclotomic polynomials, I $\sim$ IV*, Math. J. Okayama Univ. I: **35** (1993) 35–40, II: **37** (1995) 27–36, III: **38** (1996) 115–122, IV: Bull. Fac. Sci. Tech. Hirosaki Univ. **1** (1998), 1–7.

[8] J. H. M. Wedderburn, *A theorem on finite algebras*, Trans. Amer. Math. Soc. **6** (1905), 349–352.

Department of Mathematical System Science
Faculty of Science and Technology
Hirosaki University
Hirosaki 036, Japan
e-mail: skm@cc.hirosaki-u.ac.jp

Generalized Jordan Derivations

Atsushi Nakajima

Abstract

We define a notion of generalized Jordan (resp. Lie) derivations and give some elementary properties of generalized Jordan (resp. Lie) derivations. These categorical results correspond to the results of generalized derivations in [N]. Moreover, we extend Herstein's result of Jordan derivations on a prime ring to generalized Jordan derivations.

0. Introduction

Let A/k be an algebra over a commutative ring k. In [B], Brešar defined a notion of a *generalized derivation* $f : A \to A$ as follows: f is a k-linear map such that $f(xy) = f(x)y + xd(y)$ for a fixed derivation $d : A \to A$. It was already known as a derivation of higher order by Ribenboim [R]. In [N], the author defines generalized derivations without using derivations and gives some categorical properties of it. On the other hand, Jordan derivations in a ring and the relation between Jordan derivations and derivations were discussed in several papers including [Cu], [H1] and [H2], etc.

In this note, we first introduce a notion of generalized Jordan (resp. Lie) derivations in §1 and show a relation between derivations and corresponding homomorphisms. In §2, we show some categorical properties of the set of all generalized Jordan (resp. Lie) derivations $gJ\mathrm{Der}(A, M)$ (resp. $gLie\mathrm{Der}(A, M)$) from A to an A/k-bimodule M. We treat the common properties of $gJ\mathrm{Der}(A, M)$ and $gLie\mathrm{Der}(A, M)$ in this section. In the final §3, we show that some type of results of Jordan derivations are easily extended to generalized Jordan derivations using our method. Especially, we show that a generalized Jordan derivation is a generalized derivation on a 2-torsion free semiprime ring.

1991 *Mathematics Subject Classification*: 16W25.

Key words and phrases. Derivation, Jordan derivation, Lie derivation, exact sequence, functor.

This research was partially supported by Grant-in-Aid for Scientific Research (C)(2)(No.10640025), The Ministry of Education, Science Sport and Culture, Japan.

1. Preliminaries

In this section, we define a generalized notion of derivations, Jordan derivations and Lie derivations.

Let A/k be an algebra over a commutative ring k and M an A/k-bimodule, that is, M is a left and right A-module such that, for any $a, b \in A$, $r \in k$ and $m \in M$

$$a(mb) = (am)b, \quad r(am) = a(rm) \text{ and } rm = mr.$$

Let $f : A \to M$ be a k-linear map and ω an element of M. (f, ω) is called a *generalized derivation* if

$$f(ab) = f(a)b + af(b) + a\omega b \quad \text{for any } a, b \in A. \tag{1.1}$$

This is defined in [N]. (f, ω) is called a *generalized Jordan derivation* if

$$f(a^2) = f(a)a + af(a) + a\omega a \quad \text{for any } a \in A, \tag{1.2}$$

and (f, ω) is called a *generalized Lie derivation* if

$$f([a, b]) = [f(a), b] + [a, f(b)] + a\omega b - b\omega a \quad \text{for any } a, b \in A, \tag{1.3}$$

where $[a, b] = ab - ba$. If $\omega = 0$, then these are derivations, Jordan derivation and Lie derivations in the usual sense. Generalized Jordan or Lie derivations (f, ω) and (g, τ) are equal if $f = g$ and $\omega = \tau$. For a generalized Jordan derivation (f, ω), calculating $f((a+b)^2)$, we have

$$f(a \circ b) = f(a) \circ b + a \circ f(b) + a\omega b + b\omega a \quad \text{for any } a, b \in A, \tag{1.4}$$

where $a \circ b = ab + ba$. Conversely, if M is 2-*torsion free*, that is, $2m = 0$ implies $m = 0$ ($m \in M$), then (1.2) is induced from (1.4).

Let $g : A \to B$ be a k-linear map from A to a k-algebra B. g is called a *Jordan homomorphism* (cf., [H]) if

$$g(a^2) = g(a)g(a) \quad \text{for any } a \in A.$$

Then by $g((a+b)^2) = g(a+b)g(a+b)$, we see

$$g(a \circ b) = g(a) \circ g(b) \quad \text{for any } a, b \in A,$$

and of course if A is 2-torsion free, then $g(a \circ b) = g(a) \circ g(b)$ implies $g(a^2) = g(a)g(a)$. g is called a *Lie homomorphism* if

$$g([a, b]) = [g(a), g(b)] \quad \text{for any } a, b \in A.$$

Next, we construct a new type of algebra $A \ltimes_\omega M$ as follows. Let ω be a fixed element of M and let $A \ltimes_\omega M$ be the direct product of k-modules A and M. Define a multiplication on $A \ltimes_\omega M$ by

$$(a, m)(b, n) = (ab, an + mb + a\omega b) \quad \text{for any} \quad (a, m), (b, n) \in A \ltimes_\omega M.$$

Then it is easy to see that the multiplication is associative, and the distributive law holds. Thus we have

Lemma 1.1. *Let $A \ltimes_\omega M$ be a k-algebra as defined above. If A has an identity element 1, then $(1, -\omega)$ is an identity element of $A \ltimes_\omega M$. If a is invertible in A, then (a, m) is invertible in $A \ltimes_\omega M$ with inverse $(a^{-1}, -a^{-1}\omega - a^{-1}ma^{-1} - \omega a^{-1})$.*

We call this new type of algebra $A \ltimes_\omega M$ a *trivial extension* of A by M associated to ω. If $\omega = 0$, then $A \ltimes_0 M = A \ltimes M$ is a usual trivial extension.

The following lemma gives a relation of generalized (Jordan or Lie) derivations and corresponding homomorphisms.

Lemma 1.2. *For a k-linear map $f : A \to M$, we define $F : A \ni a \mapsto (a, f(a)) \in A \ltimes_\omega M$. Then*

(1) (f, ω) is a generalized derivation if and only if F is a ring homomorphism.

(2) (f, ω) is a generalized Jordan (resp. Lie) derivation if and only if F is a Jordan (resp. Lie) homomorphism.

Proof. (1) f is a generalized derivation if and only if $f(ab) = f(a)b + af(b) + a\omega b$. So by the definition of the multiplication in $A \ltimes_\omega M$, the result is clear. (2) is also easily seen. □

This lemma shows that the properties of Jordan (resp. Lie) homomorphisms give the properties of generalized Jordan (resp. Lie) derivations. These method were used in [Cr], [H1] and [H2] etc.

2. Elementary Properties of Generalized Jordan and Lie Derivations

In this section, we show some elementary properties of generalized Jordan and Lie derivations which are similar to the results of generalized derivations in [N, §2].

Let (f, ω) be a generalized Jordan derivation from A to M and let ω_ℓ be a left multiplication $\omega_\ell : A \ni a \mapsto \omega a \in M$. Then by

$$(f + \omega_\ell)(a)a + a(f + \omega_\ell)(a) = f(a^2) + \omega a^2 = (f + \omega_\ell)(a^2),$$

$f + \omega_\ell$ is a Jordan derivation. Conversely, if $f : A \to M$ is a Jordan derivation, then for any $\omega \in M$, $(f + \omega_\ell, -\omega)$ is a generalized Jordan derivation. These results hold for a right multiplication $\omega_r : A \ni a \mapsto a\omega \in M$. So we have

Lemma 2.1. *Let $f : A \to M$ be a k-linear map. Then the following hold.*

(1) If (f, ω) is a generalized Jordan derivation, then $f + \omega_\ell$ and $f + \omega_r$ are Jordan derivations.

(2) If f is a Jordan derivation, then $(f + \omega_\ell, -\omega)$ and $(f + \omega_r, -\omega)$ are generalized Jordan derivations for any $\omega \in M$.

This lemma is essential in our theory. First, we give a categorical application of Lemma 2.1.

Let $gJ\mathrm{Der}(A, M)$ (resp. $J\mathrm{Der}(A, M)$) be the set of all generalized Jordan derivations (resp. Jordan derivations). If $(f, \omega), (g, \tau) \in gJ\mathrm{Der}(A, M)$, then $(f+g, \omega+\tau), r(f, \omega) = (rf, r\omega) \in gJ\mathrm{Der}(A, M)$ $(r \in k)$. This means that $gJ\mathrm{Der}(A, M)$ and $J\mathrm{Der}(A, M)$ are k-modules. Then by Lemma 2.1, the following is easily seen.

Theorem 2.2. *The following sequence of k-modules is split exact.*

$$0 \to M \xrightarrow{\psi_M} gJ\mathrm{Der}(A, M) \xrightarrow{\varphi_M} J\mathrm{Der}(A, M) \to 0, \qquad (2.1)$$

where $\psi_M(\omega) = (\omega_\ell, -\omega)$ and $\varphi_M((f, \omega)) = f + \omega_\ell$ $(\omega \in M)$.

Since our definition of generalized Jordan derivations (1.2) is left and right symmetric, so the sequence (2.1) is also split exact by the maps $M \ni \omega \mapsto (\omega_r, -\omega) \in gJ\mathrm{Der}(A, M)$ and $gJ\mathrm{Der}(A, M) \ni (f, \omega) \mapsto f + \omega_r \in J\mathrm{Der}(A, M)$. Moreover the exact sequence (2.1) gives a functorial relation between $gJ\mathrm{Der}(A, -)$ and $J\mathrm{Der}(A, -)$ as follows. Let $\alpha : M \to N$ be a homomorphism of A/k-bimodule. Then α induces a k-module map

$$\alpha_* : gJ\mathrm{Der}(A, M) \ni (f, \omega) \mapsto (\alpha f, \alpha(\omega)) \in gJ\mathrm{Der}(A, N),$$

and $gJ\mathrm{Der}(A, -)$ is a functor from the category of A/k-bimodules to the category of k-modules. Moreover, the diagram below commutes.

$$\begin{array}{ccc} gJ\mathrm{Der}(A, M) & \xrightarrow{\alpha_*} & gJ\mathrm{Der}(A, N) \\ \Phi_M \downarrow & & \downarrow \Phi_N \\ J\mathrm{Der}_k(A, M) \oplus M & \xrightarrow{[\bar{\alpha}_*]} & J\mathrm{Der}_k(A, N) \oplus N, \end{array}$$

where $(\bar{\alpha}_*)(f,\omega) = (\alpha f, \alpha(\omega))$, $\Phi_X((f,x)) = (f + x_\ell, x)$ ($x \in X$, $X = M, N$). So we have a natural transformation of functors

$$\Phi : gJ\mathrm{Der}(A, -) \to J\mathrm{Der}(A, -) \oplus F, \qquad (2.2)$$

where F is the forgetful functor from the category of A/k-bimodules to the category of k-modules. Since Φ_M is an isomorphism for any A/k-bimodule M by Theorem 2.3, we get

Corollary 2.3. *The functors $gJ\mathrm{Der}(A, -)$ and $J\mathrm{Der}(A, -) \oplus F$ from the category of A/k-bimodules to the category of k-modules are naturally equivalent.*

As is shown in [N], the set of all generalized derivations $g\mathrm{Der}(A)$ from A to A has a Lie algebra structure and we have the following split exact sequence of Lie algebras:

$$0 \to A \xrightarrow{\psi_A} g\mathrm{Der}(A) \xrightarrow{\varphi_A} \mathrm{Der}(A) \to 0.$$

So, we consider the Lie algebra structure in $gJ\mathrm{Der}(A) = gJ\mathrm{Der}(A, A)$. Let J be a Jordan derivation. Then by (1.2) and (1.4), we see

$$\begin{aligned} 2J(aba) &= J(a \circ (ab + ba)) - J(a^2 \circ b) \\ &= 2(J(a)ba + aJ(b)a + abJ(a)). \end{aligned}$$

If 2 is a non-zero divisor in A, then for any $(f,\omega), (g,\tau) \in gJ\mathrm{Der}(A)$, we have

$$\begin{aligned} [f,g](a^2) &= f(g(a)a + ag(a) + a\tau a) - g(f(a)a + af(a) + a\omega a) \\ &= [f,g](a)a + a[f,g](a) + a(f(\tau) - g(\omega))a. \end{aligned}$$

Therefore $([f,g], f(\tau) - g(\omega))$ is a generalized Jordan derivation. We define a bracket operation on $gJ\mathrm{Der}(A)$, that is,

$$[(f,\omega), (g,\tau)] = ([f,g], f(\tau) - g(\omega)).$$

Then $gJ\mathrm{Der}(A)$ and $J\mathrm{Der}(A)$ have a Lie algebra structure and the corresponding sequence to (2.1) is

$$0 \to A \xrightarrow{\psi_A} gJ\mathrm{Der}(A) \xrightarrow{\varphi_A} J\mathrm{Der}(A) \to 0. \qquad (2.3)$$

In this case, ψ_A is a Lie algebra map and for φ_A, we have

$$\begin{aligned} &\varphi_A([(f,\omega),(g,\tau)])(x) - [\varphi_A((f,\omega)), \varphi_A((g,\tau))](x) \\ &= \{f(\tau)x + \tau f(x) + \tau(-\omega)x - f(\tau x)\} \\ &\quad - \{g(\omega)x + \omega g(x) + \omega(-\tau)x - g(\omega x)\}. \end{aligned}$$

Since there exists a generalized Jordan derivation which is not a generalized derivation (cf. [Cu]), φ_A is not a Lie algebra map in general. Thus (2.3) is not an exact sequence as in Lie algebras.

For generalized Lie derivations, we obtain similar results to generalized Jordan derivations. We give the corresponding results without proof.

Lemma 2.4. *Let $f : A \to M$ be a k-linear map. Then the following hold.*

(1) If (f, ω) is a generalized Lie derivation, then $f + \omega_\ell$ and $f + \omega_r$ are Lie derivations.

(2) If f is a Lie derivation, then $(f + \omega_\ell, -\omega)$ and $(f + \omega_r, -\omega)$ are generalized Lie derivations for any $\omega \in M$.

We denote $gLieDer(A, M)$ (resp. $LieDer(A, M)$) the set of all generalized Lie derivations (resp. Lie derivations). Then $gLieDer(A, M)$ and $LieDer(A, M)$ are k-modules with the usual addition and k-multiplication. Then by Lemma 2.5, the following is easily seen.

Theorem 2.5. *The following sequence of k-modules is split exact, too.*

$$0 \to M \xrightarrow{\psi_M} gLieDer(A, M) \xrightarrow{\varphi_M} LieDer(A, M) \to 0, \quad (2.4)$$

where $\psi_M(\omega) = (\omega_\ell, -\omega)$ and $\varphi_M((f, \omega)) = f + \omega_\ell$ ($\omega \in M$).

Corollary 2.6. *The functors $gLieDer(A, -)$ and $LieDer(A, -) \oplus F$ from the category of A/k-bimodules to the category of k-modules are naturally equivalent.*

In this case, $LieDer(A) = LieDer(A, A)$ is a Lie algebra by the bracket operation. But in the calculation of $[f, g]([a, b])$ for any $(f, \omega), (g, \tau) \in LieDer(A)$, we can not calculate the part

$$f(a\tau b - b\tau b) + g(a\omega b - b\omega a).$$

Thus we do not know if $gLieDer(A)$ has a Lie algebra structure or not.

3. Other Results of Generalized Jordan Derivations

In this section, as applications of Lemma 2.1, we give some results of generalized Jordan derivations. First, we show that the results of Jordan derivations by J. M. Cusack [Cu] are easily extended to generalized Jordan derivations. Our method is very simple. In the following, we assume that A is 2-torsion free.

An ideal P in A is called a *prime ideal* if $aAb \subseteq P$ implies $a \in P$ or $b \in P$. If 0 is a prime ideal in A, then A is called a *prime ring*. If the intersection of all prime ideals of A is zero, then A is said to be a *semiprime ring*.

For a generalized Jordan derivation $(f, \omega) : A \to M$ and a Jordan derivation d, we set

$$a^b = f(ab) - f(a)b - af(b) - a\omega b,$$
$$d(a, b) = d(ab) - d(a)b - ad(b).$$

Under these notations, we have

Lemma 3.1. (cf. [Cu, Lemma 2]) *Let (f, ω) be a generalized Jordan derivation from A to M. Then for any $a, b \in A$, the following hold.*
(1) $a^b[a, b] = [a, b]a^b = 0$.
(2) $[[a, b], x, a^b] = 0$ *for any* $x \in A$.

Proof. By [Cu, Lemma 2], it was shown that if d is a Jordan derivation, then

$$d(a,b)[a,b] = [a,b]d(a,b) = 0 \quad \text{and} \quad [[a,b], x, d(a,b)] = 0 \quad \text{for any } x \in A.$$

Thus if (f, ω) is a generalized Jordan derivation, then by Lemma 2.1(1), $f + \omega_\ell$ is a Jordan derivation. And by $(f + \omega_\ell)(a, b) = a^b$, (1) and (2) are clear. □

Using this method, we get the following generalization of [Cu, Theorem 4].

Theorem 3.2. *Let (f, x) be a generalized Jordan derivation from A to A $(x \in A)$. Then for any $a, b \in A$, $a^b = f(ab) - f(a)b - af(b) - a\omega b$ is contained in any prime ideal P of A.*

Proof. For Jordan derivation d, $d(a, b) \in P$ for any prime ideal in A by [Cu, Theorem 4]. Then since $f + \omega_\ell$ is a Jordan derivation, we have

$$(f + \omega_\ell)(ab) - a(f + \omega_\ell)(b) - (f + \omega_\ell)(a)b = (f + \omega_\ell)(a, b) = a^b \in P. \square$$

Corollary 3.3. *If A is a semiprime ring, then a generalized Jordan derivation on A is a generalized derivation.*

Theorem 3.4. *Let $(f, a), (g, b)$ be generalized derivations on A. If there exists $c \in A$ such that (fg, c) is a generalized Jordan derivation, then $fg(A)$ is contained in each minimal prime ideal in A.*

Proof. Let P be a minimal prime ideal in A. Since $f + a_\ell$ is a derivation, we see by T. Creedon [Cr], $f(P) \subseteq P$. Thus (f,a) induces a generalized derivation $f_p : A/P \ni x+P \mapsto f(x)+P \in A/P$. Now, by our assumption, since (fg,c) is a generalized Jordan derivation, then by Theorem 3.2, $(fg)_p(xy) - (fg)_p(x)y - x(fg)_p(y) - xcy \in P$ and thus $((fg)_p, c) = (f_p g_p, c)$ is a generalized derivation on a prime ring A/P. Therefore by Hvala [Hv, Corollary 1(1)], $f_p = 0$ or $g_p = 0$, which shows $fg(A) \subseteq P$. □

Corollary 3.5. *Let (f,a), (g,b) be generalized derivations on a semi-prime ring A. If there exists $c \in A$ such that (fg,c) is a generalized derivation, then $fg = 0$.*

A generalization of Herstein's Theorem for Jordan derivation is obtained in Corollary 3.3, but we have a simpler proof as follows. If (f, ω) is a generalized Jordan derivation on a semiprime ring A, then by Lemma 2.1(1), $f + \omega_\ell$ is a Jordan derivation, and so $f + \omega_\ell$ is a derivation by [Cu, Corollary 5]. Thus $(f + \omega_\ell - \omega_\ell, \omega) = (f, \omega)$ is a generalized derivation. Under these viewpoints, we will able to generalize many results of derivations and Jordan (resp. Lie) derivations to the generalized derivations and generalized Jordan (resp. Lie) derivations.

References

[B] M. Brešar, *On the distance of the composition of two derivations to the generalized derivations*, Glasgow Math. J. **33** (1991), 89–93.

[BV] M. Brešar and J. Vukman, *Jordan derivations on prime rings*, Bull. Austral. Math. Soc. **27** (1988), 321–322.

[Cr] T. Creedon, *Products of derivations*, Proc. Edinburgh Math. Soc. **41** (1998), 407–410.

[Cu] J. M. Cusack, *Jordan derivations on rings*, Proc. Amer. Math. Soc. **53** (1975), 321–324.

[H1] I. N. Herstein, *Jordan derivations of prime rings*, Proc. Amer. Math. Soc. **8** (1957), 1104–1110.

[H2] ———, *Topics in Ring Theory*, Univ. Chicago Press, Chicago, 1969.

[Hv] B. Hvala, *Generalized derivations in rings*, Comm. Algebra **26** (1998), 1147–1166.

[N] A. Nakajima, *On categorical properties of generalized derivations,* Scientiae Mathematicae **2** (1999), 345–352.

[R] P. Ribenboim, *Higher order derivations of modules,* Portugaliae Mathematica **39** (1980), 381–397.

Department of Mathematical Science
Faculty of Environmental Science and Technology
Okayama University
Okayama 700-8530, Japan
e-mail: `nakajima@math.ems.okayama-u.ac.jp`

On Quasi-Frobenius Rings

W. K. Nicholson and M. F. Yousif

Abstract

There are three outstanding conjectures about quasi-Frobenius rings: The Faith conjecture that every left perfect, right selfinjective ring is quasi-Frobenius; The FGF-conjecture that every ring for which each finitely generated right module embeds in a free module is quasi-Frobenius; and The Faith-Menal conjecture that every right noetherian ring in which all right ideals are annihilators is quasi-Frobenius. In this paper we survey recent work on these conjectures and provide some new results on the subject.

The work of Nakayama in the 1940s on duality between right and left ideals ([Nak1], [Nak2]) identified an important class of rings. A ring R is called *quasi-Frobenius* (*QF-ring*) if R satisfies any of the following equivalent conditions:

(1) R is left (or right) artinian and if $\{e_1, e_2, \cdots, e_n\}$ is a basic set of primitive idempotents of R, then there exists a (Nakayama) permutation σ of $\{1, 2, \cdots, n\}$ such that $soc(Re_k) \cong Re_{\sigma k}/Je_{\sigma k}$ and $soc(e_{\sigma k}R) \cong e_k R/e_k J$.

(2) R is left (or right) perfect and left and right selfinjective.

(3) Every left (or right) R-module embeds in a free module.

(4) R is left (or right) noetherian and every one-sided ideal of R is an annihilator.

The equivalence between (1) and (4) is essentially due to Nakayama [Nak1], (2) is due to Osofsky [O] and (3) is due to Faith and Walker [FW].

There are numerous other equivalent conditions that a ring is QF; we have chosen these because of their relevance to three conjectures of Faith which we will refer to as follows:

The Faith conjecture: Every left (or right) perfect, right selfinjective ring is QF.

The FGF-conjecture: Every right FGF-ring is QF. (A ring is called a right *FGF-ring* if every finitely generated right module embeds in a free module.)

The Faith-Menal conjecture: Every strongly right Johns ring is QF. (A ring R is called *right Johns* if R is right noetherian and every right ideal is an annihilator; and R is called strongly right Johns if the matrix ring $M_n(R)$ is right Johns for all $n \geq 1$.)

In this paper we review some recent work on these conjectures, and we also provide some new results on the subject.

If R satisfies any of these conditions, then every right ideal T in R is an annihilator. Indeed, this is a hypothesis in the third case; in the first case, R_R is a cogenerator so $(R/T)_R \hookrightarrow R^I$; and in the second case $(R/T)_R$ embeds in a free module. Furthermore, each of the properties in the conjectures is a Morita invariant, so R satisfies the first of the following equivalent conditions (see Theorem 2.2 below).

(1) *For every $n \geq 1$, every principal right ideal of $M_n(R)$ is an annihilator.*

(2) *If $_RK$ is a finitely generated submodule of a free module $_RF$, every R-map $\gamma : K \to {_RR}$ can be extended to $F \to R$.*

(3) *Every finitely presented right R-module is torsionless (embeds in R^I for some set I).*

A ring R is called left *FP-injective* if it satisfies any, and hence all, of these conditions. Thus FP-injectivity is a central theme in these conjectures.

Consequently, we study FP-injectivity in Sections 2 and 3, and obtain the following property: A right Kasch, right FP-injective ring is left FP-injective (Theorem 3.1). This result leads us to the following definition: A ring R will be called an FP-ring if R is semiperfect, right FP-injective and has an essential right socle. We give several characterizations of these rings in Section 3, and show that they are a Morita invariant class and retain many of the basic properties of quasi-Frobenius rings. Surprisingly, these FP-rings turn out to be left-right symmetric, unlike the pseudo-Frobenius rings (the selfinjective case). As an application we show that a left perfect, right FP-injective ring is quasi-Frobenius if and only if the second socle is finitely generated as a right module, extending most of the results in the selfinjective case.

In Section 4 we investigate a class of rings related to both the FGF-conjecture and the Faith-Menal conjecture. A ring R is called a right C2-ring if any right ideal that is isomorphic to a direct summand of R is itself a summand. We show that the FGF-conjecture is true if and only if every right FGF-ring is a right C2-ring. This result unifies and extends several theorems on the FGF-conjecture, and implies that every right FP-injective, right FGF-ring is quasi-Frobenius. In Section 5 we turn to the

Faith-Menal conjecture. We show that every right Johns, right C2-ring is right artinian, and that every strongly right Johns, right C2-ring is quasi-Frobenius. We conclude with two theorems (Theorems 5.8 and 5.9) proving the equivalence of several well known classes of rings which have been studied separately by several authors.

Throughout this paper all rings considered are associative with unity and all R-modules are unital. We write $J = J(R)$ for the Jacobson radical of R and $M_n(R)$ for the ring of $n \times n$ matrices over R. If M_R is a right R-module, we write $Z(M)$, $soc(M)$ and $M^* = hom_R(M, R)$ respectively, for the singular submodule, the socle and the dual of M. For a ring R, we write $soc(R_R) = S_r$, $soc(_RR) = S_l$, $Z(R_R) = Z_r$ and $Z(_RR) = Z_l$. The notations $N \subseteq^{ess} M$ and $N \subseteq^{max} M$ mean that N is an essential, (respectively maximal) submodule of M. Right annihilators will be denoted as $r(Y) = r_X(Y) = \{x \in X \mid yx = 0 \text{ for all } y \in Y\}$, with a similar definition of left annihilators $l_X(Y) = l(Y)$. Multiplication maps $x \mapsto ax$ and $x \mapsto xa$ will be donoted $a\cdot$ and $\cdot a$, respectively.

1. Preliminaries

We say that a ring R is right *mininjective* [NY2] if it has the following property: If K is any simple right ideal of R, every R-map $\gamma : K \to R_R$ extends to R, that is $\gamma = c\cdot$ for some $c \in R$. This is a large class of rings, including every right selfinjective ring, every semiprime ring and every ring with $S_r = 0$.

Lemma 1.1. *The following are equivalent for a ring R:*
 (1) *R is right mininjective.*
 (2) *M^* is simple or zero for every simple right R-module M.*
 (3) *$l(T)$ is simple or zero for every maximal right ideal T.*
 (4) *$lr(k) = Rk$ whenever kR is a simple right ideal of R.*
 (5) *If kR is simple and $r(k) \subseteq r(a)$, $k, a \in R$, then $Ra \subseteq Rk$.*

Proof. (1)⇒(2). Let M_R be simple. If $M^* = 0$ there is nothing to prove. Otherwise, let $0 \neq \delta \in M^*$ so that $\delta : M \to \delta(M)$ is an isomorphism. Given $\gamma \in M^*$ we have $\delta(M) \xrightarrow{\delta^{-1}} M \xrightarrow{\gamma} R$ so $\gamma \circ \delta^{-1} = a\cdot$ for some $a \in R$ by (1). It follows that $\gamma = a\delta \in R\delta$, proving (2).

(2)⇒(3). This is because $l(T) \cong (R/T)^*$ for every right ideal T.

(3)⇒(4). If $kR \subseteq R$ is simple, then $T = r(k)$ is maximal and $Rk \subseteq lr(k) = l(T)$.

(4)⇒(5). This is because $r(k) \subseteq r(a)$ implies that $a \in lr(k)$.

(5)⇒(1). Let $\gamma : kR \to R$ be R-linear where kR is a simple right ideal. If $a = \gamma(k)$ then $r(k) \subseteq r(a)$ so let $a = ck$, $c \in R$, by (5). Then $\gamma = c\cdot$, proving (1). □

The following basic properties of right mininjective rings will be used frequently.

Lemma 1.2. *Let R be a right mininjective ring.*
(1) *If kR is simple, $k \in R$, then Rk is also simple.*
(2) *If $kR \cong mR$ are simple, then $Rk \cong Rm$; in fact $Rk = (Rm)u$ for $u \in R$.*
(3) $S_r \subseteq S_l$.
(4) *Let e and f be local idempotents in R. If eR and fR contain isomorphic simple right ideals, then $eR \cong fR$.*

Proof. (1) If $0 \neq ak \in Rk$, define $\gamma = a\cdot : kR \to akR$. Then γ is an isomorphism so, as R is right mininjective, let $\gamma^{-1} = c\cdot$, $c \in R$. Thus $k = \gamma^{-1}(ak) = cak \in Rak$, and (1) follows.
(2) If $\sigma : kR \to mR$ is an isomorphism, write $\sigma(k) = mu$, $u \in R$. Clearly $muR = mR$ is simple and $\mathbf{r}(mu) = \mathbf{r}[\sigma(k)] = \mathbf{r}(k)$. Hence $Rmu = Rk$ by Lemma 1.1.
(3) This follows from (1).
(4) Let $\alpha : K \to fR$ be monic where $K \subseteq eR$ is a simple right ideal. By hypothesis $\alpha = a\cdot$ for some $a \in R$, and we may assume that $a \in fRe$. Hence $a\cdot : eR \to fR$ is R-linear. We have $S_r \subseteq S_l$ by (3) so $0 \neq \alpha(K) = aK \subseteq aS_r \subseteq aS_l$. This shows that $a \notin J$, so $aeR = aR \not\subseteq fJ$. Hence $a\cdot$ is onto fR because f is local. But then $a\cdot$ is one-to-one because fR is projective and eR is indecomposable. □

A ring R is called right *Kasch* if $M^* \neq 0$ for every simple right R-module M, equivalently if $\mathbf{l}(T) \neq 0$ for every $T \subseteq^{max} R_R$. Hence the proof of Lemma 1.1 shows that R is right mininjective and right Kasch if and only if M^* is simple for every simple right R-module M, if and only if $\mathbf{l}(T)$ is simple for all $T \subseteq^{max} R_R$.

For convenience, call a ring R right *minfull* if it is a semiperfect, right mininjective ring in which $soc(eR) \neq 0$ for each local idempotent $e \in R$. This class of rings contains every left perfect, right mininjective ring, and retains many of the properties of pseudo-Frobenius rings.

Theorem 1.3. *Let R be a right minfull ring. Then:*
(1) *R is right and left Kasch.*
(2) *$soc(eR)$ is homogeneous for each local $e^2 = e \in R$.*
(3) *$S_r e$ is a simple left ideal for each local $e^2 = e \in R$. Moreover, if $e_1, \ldots, e_n$ are basic, orthogonal, local idempotents, there exist $k_1, \ldots, k_n$ in R and a (Nakayama) permutation σ of $\{1, \ldots, n\}$ such that the following hold for each $i = 1, \ldots, n$:*

(4) $k_iR \subseteq e_iR$ and $Rk_i \subseteq Re_{\sigma i}$.
(5) $k_iR \cong e_{\sigma i}R/e_{\sigma i}J$ and $Rk_i \cong Re_i/Je_i$.
(6) $Rk_i = S_r e_{\sigma i}$.
(7) $\{k_1R, \ldots, k_nR\}$ and $\{Rk_1, \ldots, Rk_n\}$ are complete sets of distinct representatives of the simple right and left R-modules, respectively.
(8) The following conditions are equivalent:
  (a) $S_r = S_l$.
  (b) $\mathrm{lr}(K) = K$ for every simple left ideal K with $K \subseteq Re$ for some local $e^2 = e \in R$.
  (c) $\mathrm{soc}(Re) = S_r e$ for every local $e^2 = e \in R$.
  (d) $\mathrm{soc}(Re)$ is simple for every local $e^2 = e \in R$.

Proof. For each $i = 1, \ldots, n$ fix a simple right ideal $K_i \subseteq e_iR$. As R is semiperfect, choose $\sigma i \in \{1, \ldots, n\}$ such that $K_i \cong e_{\sigma i}R/e_{\sigma i}J$. This map σ is a permutation of $\{1, \ldots, n\}$ because $\sigma i = \sigma j$ implies that $K_i \cong K_j$, whence $e_iR \cong e_jR$ (by Lemma 1.2) and finally $i = j$ (because the e_i are basic). If $\gamma : e_{\sigma i}R/e_{\sigma i}J \to K_i$ is an isomorphism, write $k_i = \gamma(e_{\sigma i} + e_{\sigma i}J)$. Then $k_iR = K_i \cong e_{\sigma i}R/e_{\sigma i}J$ and $k_i \in e_iRe_{\sigma i}$ proving (4) and half of (5). Because $k_i \in S_r \subseteq S_l$ (using Lemma 1.2), we obtain $\mathbf{l}(k_i) \supseteq Je_i \oplus R(1-e_i)$. But $R/(Je_i \oplus R(1-e_i)) \cong Re_i/Je_i$ is simple, so it follows that $\mathbf{l}(k_i) = Je_i \oplus R(1-e_i)$, and hence that $Rk_i \cong Re_i/Je_i$. This proves (5). Now observe that $S_r = \mathbf{l}(J)$ because R/J is semisimple, so $S_r e_{\sigma i} = \mathbf{l}(J)e_{\sigma i} \cong (e_{\sigma i}R/e_{\sigma i}J)^*$ as is easily verified. Hence $S_r e_{\sigma i}$ is simple by Lemma 1.1 because $0 \neq k_i = k_i e_{\sigma i} \in S_r e_{\sigma i}$. This proves (6), and (3) follows because every local idempotent is part of a basic set.

Since R is semiperfect it has exactly n isomorphism classes of simple right (or left) modules, represented by e_jR/e_jJ where $1 \leq j \leq n$ and Re_i/Je_i where $1 \leq i \leq n$. Hence (5) implies both (7) and (1). To prove (2), let $K \subseteq e_iR$ be simple. Then $K \cong k_jR$ for some j by (5), so $j = i$ by Lemma 1.2. Hence $\mathrm{soc}(e_iR)$ is homogeneous, and (2) follows. Finally we prove (8).

(a)$\Rightarrow$(b). Let $K \subseteq Re$ be a simple left ideal. We have $KJ = 0$ by (a) so $\mathrm{r}(K) \supseteq J + (1-e)R (= M)$. But $M = eJ \oplus (1-e)R$ so $R/M \cong eR/eJ$ is simple, and it follows that $\mathrm{r}(K) = M$. Hence $K \subseteq \mathrm{lr}(K) = \mathbf{l}(M) = \mathbf{l}(J) \cap Re = S_r \cap Re = S_r e$, so $K = \mathrm{lr}(K)$ by (3).

(b)$\Rightarrow$(c). By (3) let $K \subseteq Re$ be simple, so that $\mathrm{r}(K) \supseteq (1-e)R$. Because e is local, $J + (1-e)R$ is the unique maximal right ideal containing $(1-e)R$, and so $\mathrm{r}(K) \subseteq J + (1-e)R$. But then (b) gives $K = \mathrm{lr}(K) \supseteq \mathbf{l}(J) \cap Re = S_r e$. Hence $K = S_r e$ by (3), and (c) follows.

(c)$\Rightarrow$(d). This follows by (3).

(d)$\Rightarrow$(a). Given (d) we have $\mathrm{soc}(Re) = S_r e$ for each local $e^2 = e$ by (3). Let $1 = e_1 + \cdots + e_m$ where the e_i are orthogonal local idempotents.

Then $S_l = \oplus_{i=1}^m soc(Re_i) = \oplus_{i=1}^m S_r e_i \subseteq S_r$, and (a) follows from Lemma 1.2. □

Theorem 1.4. *Every right artinian, right and left mininjective ring is quasi-Frobenius*

Proof. We have $S_r = S_l \ (= S)$ by Lemma 1.2 so, by [AF, Corollary 31.8], it remains to show that $soc(eR) = eS$ and $soc(Re) = Se$ are both simple for all local idempotents $e \in R$. The ideal S is essential in both R_R and $_R R$ because R is semiprimary. In particular $soc(eR) = eS$ and $soc(Re) = Se$ are nonzero for every local idempotent e. Hence R is right and left minfull, so eS and Se are both simple by (3) of Theorem 1.3. □

2. FP-Injective Rings

A module Q_R is called *FP-injective* (or *absolutely pure*) if every R-map $K \to Q$ from a finitely generated submodule K of a free right R-module F extends to F. Our interest is in the *right FP-injective rings*, that is the rings R for which R_R is FP-injective. Examples include regular and right selfinjective rings. These right FP-injective rings are closely associated to the larger class of right *principally injective* rings (right *P-injective* rings), that is the rings R satisfying the following equivalent conditions [NY1]:
 (1) Every R-linear map $aR \to R$, $a \in R$, extends to R.
 (2) $\mathrm{lr}(a) = Ra$ for all $a \in R$.
 (3) If $\mathrm{r}(a) \subseteq \mathrm{r}(b)$ where $a, b \in R$, then $b \in Ra$.
More generally, a ring R is called right *n-injective* if every R-linear map $\Sigma_{i=1}^n a_i R \to R$, $a_i \in R$, extends to R. We will use the following result several times.

Proposition 2.1. *If $M_n(R)$ is right P-injective for some $n \geq 1$ then R is right n-injective.*

Proof. Suppose $\gamma : \Sigma_{i=1}^n a_i R \to R$ is R-linear, and let $A = [a_1 \ a_2 \ \cdots \ a_n]$ denote the matrix with first row entries as shown and all other entries zero. If $B = [\gamma a_1 \ \gamma a_2 \ \cdots \ \gamma a_n]$ then $\mathrm{r}_{M_n(R)}(A) \subseteq \mathrm{r}_{M_n(R)}(B)$, so $B = CA$ for some $C \in M_n(R)$ by hypothesis. If $C = [c_{ij}]$ we have $\gamma = c_{11}\cdot$, as required. □

Our next result gives several matrix characterizations of the right FP-injective rings which will be used frequently. We write $M_{m \times n}(R)$ for the set of all $m \times n$ matrices over R, and we use the notation R^n (respectively R_n) for the set of all $1 \times n$ (respectively $n \times 1$) matrices over R.

Theorem 2.2. *The following are equivalent for a ring R:*
(1) R *is right FP-injective.*
(2) *If $\bar{b} \in R^n$ and $A \in M_n(R)$ satisfy $\mathbf{r}_{R_n}(A) \subseteq \mathbf{r}_{R_n}(\bar{b})$, then $\bar{b} = \bar{x}A$ for some $\bar{x} \in R^n$.*
(3) *If $\bar{b} \in R^n$ and $A \in M_{m \times n}(R)$ satisfy $\mathbf{r}_{R_n}(A) \subseteq \mathbf{r}_{R_n}(\bar{b})$, then $\bar{b} = \bar{x}A$ for some $\bar{x} \in R^m$.*
(4) *If $\bar{a}_1, \bar{a}_2, \ldots, \bar{a}_m$ and $\bar{b}$ in R^n satisfy $\cap_i \mathbf{r}_{R_n}(\bar{a}_i) \subseteq \mathbf{r}_{R_n}(\bar{b})$, then $\bar{b} \in \Sigma_i R \bar{a}_i$.*
(5) *If $n \geq 1$ and $_R K \subseteq R^n$ is finitely generated, then $K = \mathbf{1}_{R^n}(\mathcal{X})$ for some set $\mathcal{X} \subseteq M_n(R)$.*
(6) $M_n(R)$ *is a right P-injective ring for each $n \geq 1$.*

Proof. (1)$\Rightarrow$(2). Given $\mathbf{r}_{R_n}(A) \subseteq \mathbf{r}_{R_n}(\bar{b})$, let $\underline{c}_j$ denote column j of A. Then $AR_n = \Sigma_j \underline{c}_j R$ is a finitely generated R-submodule of R_n, and our hypothesis shows that $\alpha : AR_n \to R_R$ is well defined by $\alpha(A\underline{r}) = \bar{b}\underline{r}$. By (1), α extends to $R_n \to R$, so $\alpha = \bar{x} \cdot$ for some $\bar{x} \in R^n$. If we write $\bar{b} = [b_1 \cdots b_n] \in R^n$, we get $b_j = \bar{b}\underline{e}_j = \alpha(A\underline{e}_j) = \alpha(\underline{c}_j) = \bar{x}\underline{c}_j$ for each j, where $\underline{e}_j$ is the canonical basis vector in R_n. Hence, $\bar{b} = [\bar{x}\underline{c}_1 \cdots \bar{x}\underline{c}_n] = \bar{x}[\underline{c}_1 \cdots \underline{c}_n] = \bar{x}A$, as required.

(2)$\Rightarrow$(3). Let $\mathbf{r}_{R_n}(A) \subseteq \mathbf{r}_{R_n}(\bar{b})$ where A is $m \times n$. We are done if $m = n$. If $m < n$, let $A' = \begin{bmatrix} A \\ 0 \end{bmatrix}$ be $n \times n$. Then $\mathbf{r}_{R_n}(A') = \mathbf{r}_{R_n}(A) \subseteq \mathbf{r}_{R_n}(\bar{b})$, so (3) gives $\bar{b} = \bar{y}A'$ for some $\bar{y}$ in R^n. If we write $\bar{y} = [\bar{x}\ \bar{z}] \in R^n$, $\bar{x} \in R^m$, then $\bar{b} = \bar{x}A$. On the other hand, if $m > n$ we let $A' = [A\ 0]$ be $m \times m$. Then

$$\mathbf{r}_{R_m}(A') = \left\{ \begin{bmatrix} \underline{r} \\ \underline{s} \end{bmatrix} \mid \underline{r} \in R_n,\ \underline{s} \in R_{m-n},\ A\underline{r} = 0 \right\} = \begin{bmatrix} \mathbf{r}_{R_n}(A) \\ R_{m-n} \end{bmatrix}$$

$$\subseteq \begin{bmatrix} \mathbf{r}_{R_n}(\bar{b}) \\ R_{m-n} \end{bmatrix} = \mathbf{r}_{R_m}[\bar{b}\ 0].$$

By (3) there exists $\bar{x} \in R^m$ such that $[\bar{b}\ 0] = \bar{x}A' = [\bar{x}A\ 0]$. Thus $\bar{b} = \bar{x}A$.

(3)$\Rightarrow$(1). Let $K = \Sigma_{j=1}^m \underline{c}_j R \subseteq R_n$ and let $\gamma : K \to R_R$ be R-linear. We must extend γ to R_n. Write $\gamma(\underline{c}_j) = b_j \in R$, write $\bar{b} = [b_1 \cdots b_m] \in R^m$, and write $A = [\underline{c}_1 \cdots \underline{c}_m] \in M_{n \times m}(R)$. Then $\gamma(A\underline{r}) = \bar{b}\underline{r}$ for all $\underline{r} \in R_m$, so $\mathbf{r}_{R_m}(A) \subseteq \mathbf{r}_{R_m}(\bar{b})$. Thus (3) gives $\bar{x} \in R^m$ such that $\bar{b} = \bar{x}A$. Hence $b_j = \bar{x}\underline{c}_j$ for all j, that is $\gamma(\underline{c}_j) = \bar{x}\underline{c}_j$ for each j. Hence $\gamma = \bar{x} \cdot$, as required.

(3)$\Rightarrow$(4). Given the situation in (4), let $A \in M_{m \times n}(R)$ be the matrix with $\bar{a}_i$ as row i for each i. Then $\mathbf{r}_{R_n}(A) = \cap_{i=1}^m \mathbf{r}_{R_n}(\bar{a}_i) \subseteq \mathbf{r}_{R_n}(\bar{b})$ by hypothesis, so $\bar{b} = \bar{x}A$ for some $\bar{x} \in R^m$ by (3). If $\bar{x} = [x_1\ x_2 \cdots x_m]$ then $\bar{b} = \Sigma_i x_i \bar{a}_i \in \Sigma_i R \bar{a}_i$ as required.

(4)$\Rightarrow$(5). Let $K = R\bar{a}_1 + \cdots + R\bar{a}_m \subseteq R^n$ as in (5). We show that $K = \mathbf{1}_{R^n}(\mathcal{X})$ where $\mathcal{X} = \mathbf{r}_{M_n(R)}(K)$. Clearly $K \subseteq \mathbf{1}_{R^n}(\mathcal{X})$. If $\bar{b} \in \mathbf{1}_{R^n}(\mathcal{X})$

then $r_{M_n(R)}(K) = \mathcal{X} \subseteq r_{M_n(R)}(\bar{b})$. But this implies that $r_{R_n}(K) \subseteq r_{R_n}(\bar{b})$, and so $\bar{b} \in K$ by (4).

(5)$\Rightarrow$(6). If $A \in M_n(R)$, write $K = \sum_{i=1}^{n} R\bar{a}_i \subseteq R^n$ where $\bar{a}_i$ denotes row i of A. By (5), $K = 1_{R^n}(\mathcal{X})$, $\mathcal{X} \subseteq M_n(R)$, so $M_n(R)\,A = \begin{bmatrix} K \\ \vdots \\ K \end{bmatrix} =$
$\begin{bmatrix} 1_{R^n}(\mathcal{X}) \\ \vdots \\ 1_{R^n}(\mathcal{X}) \end{bmatrix} = 1_{M_n(R)}(\mathcal{X})$, and (6) follows.

(6)$\Rightarrow$(2). Suppose $r_{R_n}(A) \subseteq r_{R_n}(\bar{b})$ as in (2). If B is the matrix with every row equal to $\bar{b}$, then $r_{R_n}(A) \subseteq r_{R_n}(B)$. Hence $r_{M_n(R)}(A) \subseteq r_{M_n(R)}(B)$ so (6) gives $B = XA$ for some $X \in M_n(R)$. Hence $\bar{b} = \bar{x}A$ where $\bar{x}$ denotes row 1 of X. □

If S is a ring and $_SV_S$ is a bimodule, the additive group $T(S,V) = S \oplus V$ becomes a ring using the multiplication $(s,v)(s_1,v_1) = (ss_1, sv_1 + vs_1)$, called the *trivial extension* of S by V.

Example 2.3. The trivial extension $R = T(\mathbb{Z}, \mathbb{Q}/\mathbb{Z})$ is a commutative FP-injective ring R for which R/J is not regular, in contrast to the self-injective case.

Proof. A ring S is called a left *IF-ring* if every injective left S-module is flat. Colby [C] shows that these rings are all right P-injective (in fact $\aleph_0$-injective by [C, Theorem 1]). Clearly the left IF-rings are a Morita invariant class. Hence if S is a left IF-ring, the same is true of $M_n(S)$, whence $M_n(S)$ is right P-injective, proving that S is (right) FP-injective (by Theorem 2.2). Colby shows [C, Example 1] that R is a commutative IF-ring, and so is FP-injective. Moreover, $R/J \cong \mathbb{Z}$ is not regular. □

Corollary 2.4. *Being right FP-injective is a Morita invariant.*

This follows from Theorem 2.2 and the first item in the following lemma. The lemma contains some properties of right principally injective rings that will be needed later. A module M is said to satisfy the *C2-condition*, and M is called a *C2-module*, if every submodule that is isomorphic to a direct summand of M is itself a direct summand.

Lemma 2.5. *Let R be a right P-injective ring. Then:*
(1) *If $e^2 = e \in R$ satisfies $ReR = R$, then eRe is also right P-injective.*
(2) $J = Z_r$.
(3) R_R *satisfies the C2-condition.*

Proof. (1) Write $S = eRe$ and assume that $\mathbf{r}_S(a) \subseteq \mathbf{r}_S(b)$ where $a, b \in S$; we must show that $b \in Sa$. It suffices to show that $b \in Ra$; by hypothesis we show that $\mathbf{r}_R(a) \subseteq \mathbf{r}_R(b)$. Suppose $ar = 0$, $r \in R$. Given $x \in R$ we have $erxe \in \mathbf{r}_S(a)$ so $brxe = 0$. Thus $0 = brReR = brR$, so $br = 0$ as required.

(2) If $a \in Z_r$, then $\mathbf{r}(1-a) = 0$ because $\mathbf{r}(a) \cap \mathbf{r}(1-a) = 0$. Hence $R = \mathbf{lr}(1-a) = R(1-a)$, and it follows that $Z_r \subseteq J$. Conversely, if $a \in J$, suppose that $bR \cap \mathbf{r}(a) = 0$, $b \in R$. Then $\mathbf{r}(ab) \subseteq \mathbf{r}(b)$, so $b \in Rab$ by P-injectivity, say $b = rab$. This implies that $b = 0$ because $a \in J$, and it follows that $J \subseteq Z_r$.

(3) Suppose $a \in R$ and $\sigma : aR \to eR$ is an isomorphism where $e^2 = e$. Write $\sigma(a) = ed$ and $\sigma^{-1}(e) = ac$. Then $edc = \sigma(ac) = e$, so $f = ced$ is an idempotent. We have $af = aced = \sigma^{-1}(ed) = a$, so $Ra \subseteq Rf$. On the other hand, $\mathbf{r}(a) \subseteq \mathbf{r}(f)$ because $f = c\sigma(a)$, so $Rf \subseteq Ra$ by P-injectivity. Hence $Ra = Rf$, proving (3). □

3. FP-Rings

We begin with a symmetry property of FP-injectivity. Observe first that "right Kasch" is a Morita invariant (it is equivalent to every simple right module being embedded in a projective right module).

Theorem 3.1. *If R is right Kasch and right FP-injective, then R is left FP-injective.*

Proof. Write $S = M_n(R)$. Corollary 2.4 shows that S is right FP-injective, and S is right Kasch because "right Kasch" is a Morita invariant. By Theorem 2.2 it is enough to show that S is left P-injective (for any $n \geq 1$). To this end, let $y \in \mathbf{rl}(xS)$ where $x, y \in S$; we must show that $y \in xS$. If not, let T be a maximal right S-submodule of $xS + yS$ with $xS \subseteq T$. By the Kasch condition let $\sigma : (xS + yS)/T \to S$ be an embedding. Then define $\gamma : xS + yS \to S$ by $\gamma(s) = \sigma(s + T)$. Since S is right FP-injective, $\gamma = c \cdot$ for some $c \in S$. Thus $cx = \gamma(x) = 0$, so $c \in \mathbf{l}(xS)$. Since $y \in \mathbf{rl}(xS)$, it follows that $cy = 0$, and this is the desired contradiction because $cy = \gamma(y) \neq 0$. □

The converse to Theorem 3.1 is false: Any non-artinian regular ring is a left and right FP-injective which is neither left nor right Kasch (indeed, a regular ring that is left or right Kasch is semisimple). The following example is due to Colby.

Example 3.2. [C, Example 2] Let R be an algebra with basis $\{1, e_0, e_1, e_2, \cdots, x_1, x_2, \cdots\}$ over a field F such that 1 is the identity of R and, for all $i, j,$:

$$e_i e_j = \delta_{ij} e_j, \quad x_i e_j = \delta_{i,j+1} x_i, \quad e_i x_j = \delta_{ij} x_j \quad \text{and} \quad x_i x_j = 0.$$

Then R is left FP-injective (see the discussion in Example 2.3) but his discussion shows that it is not right P-injective.

There are a large number of theorems which show that some injectivity condition on a ring R implies that R is pseudo-Frobenius. Many of these results assume that R is semiperfect but often all that is needed is that R is *semilocal* (R/J is semisimple). Consequently, we digress briefly to present four lemmas about semilocal rings, of interest in their own right.

Lemma 3.3. *Assume that R is semilocal and right and left mininjective. Then R is right Kasch if and only if R is left Kasch.*

Indeed, given $k_i \in R$, $\{Rk_1, \ldots, Rk_n\}$ is a complete system of representatives of the simple left modules if and only if $\{k_1 R, \ldots, k_n R\}$ is a complete system of representatives of the simple right modules.

Proof. Suppose that $\{Rk_1, \ldots, Rk_n\}$, $k_i \in R$, is a complete system of representatives of the simple left modules. By Lemma 1.2, each $k_i R$ is simple, and $k_i R \cong k_j R$ implies $Rk_i \cong Rk_j$. Hence $i = j$ and so the set $\{k_1 R, \ldots, k_n R\}$ consists of pairwise nonisomorphic simple right ideals. Since R has exactly n isomorphism classes of simple right modules (being semilocal), it follows that $\{k_1 R, \ldots, k_n R\}$ is a complete system of representatives of the simple right modules. The proof of the converse is similar. □

The following is a useful formulation of the Ikeda-Nakayama argument [IN] which encompasses many special situations arising in injectivity proofs.

Lemma 3.4. *Let A and B be right ideals of R such that every R-map $A + B \to R$ extends to R. Then*

$$l(A \cap B) = l(A) + l(B).$$

In particular this holds if $A + B = R$, in which case $l(A \cap B) = l(A) \oplus l(B)$.

Proof. If $x \in l(A \cap B)$, then $\gamma : A + B \to R$ is well defined by $\gamma(a+b) = xa$, so $\gamma = c \cdot$ for some $c \in R$ by hypothesis. Hence for all $a \in A$ and $b \in B$ we have

$$ca = \gamma(a+0) = xa \quad \text{and} \quad cb = \gamma(0+b) = 0.$$

Thus $x - c \in l(A)$ and $c \in l(B)$, so $x = (x - c) + c \in l(A) + l(B)$. This proves that $l(A \cap B) \subseteq l(A) + l(B)$; the other inclusion always holds. □

Lemma 3.5. *If R is semilocal and right mininjective, then $S_r = 0$ or $S_r = Rk_1 \oplus \cdots \oplus Rk_m$ where both Rk_i and k_iR are simple for each i.*

Proof. If $S_r = 0$ there is nothing to prove. Otherwise, since R is semilocal $S_r = \mathbf{l}(J)$ and $J = T_1 \cap \cdots \cap T_n$ where each T_i is a maximal right ideal. Assume n is minimal so no T_i contains any intersection of the others. Then Lemma 3.4 gives $S_r = \mathbf{l}(J) = \mathbf{l}(T_1) + \cdots + \mathbf{l}(T_n)$. Since $S_r \neq 0$ we may assume that $\mathbf{l}(T_i) \neq 0$ for each i, so choose $0 \neq k_i \in \mathbf{l}(T_i)$. Then $T_i \subseteq \mathbf{r}(k_i) \neq R$, so $T_i = \mathbf{r}(k_i)$ and $k_iR \cong R/T_i$ is simple. But then $\mathbf{l}(T_i) = \mathbf{lr}(k_i) = Rk_i$ is simple by Lemma 1.2. The result follows. □

Lemma 3.6. *Let M_R be a finite dimensional module.*

(1) If M has the C2-condition, then monomorphisms in $\mathrm{end}(M)$ are isomorphisms.

(2) In this case, $\mathrm{end}(M)$ is semilocal.

In particular, every right finite dimensional right P-injective ring is semilocal.

Proof. If $\sigma : M \to M$ is monic, then $\sigma(M)$ is a direct summand of M by the C2-condition, say $M = \sigma(M) \oplus K$. If $K \neq 0$ then

$$dim(M) \geq dim[\sigma(M)] + dim(K) > dim[\sigma(M)] = dim(M),$$

a contradiction. Hence $K = 0$, and so σ is an isomorphism. This proves (1), and then (2) follows from a result of Camps and Dicks [CD] because M is finite dimensional. The last assertion follows from Lemma 2.5 (every right P-injective ring satisfies the right C2-condition). □

We now return to the FP-injective rings. With the above results in hand, the symmetry property in Theorem 3.1 leads to the following theorem.

Theorem 3.7. *The following conditions are equivalent for a ring R:*

(1) R is semilocal, right FP-injective and right Kasch.

(2) R is semilocal, left FP-injective and left Kasch.

(3) R is semilocal, right FP-injective and $J = \mathbf{r}\{k_1, \ldots, k_n\}$ where $\{k_1, \ldots, k_n\} \subseteq R$.

(4) R is semilocal, left FP-injective and $J = \mathbf{l}\{m_1, \ldots, m_n\}$ where $\{m_1, \ldots, m_n\} \subseteq R$.

(5) R is right finite dimensional, right FP-injective and right Kasch.

(6) R is left finite dimensional, left FP-injective and left Kasch.

(7) R is left finite dimensional, right FP-injective and right Kasch.

(8) R is right finite dimensional, left FP-injective and left Kasch.

Moreover the elements k_i and m_i in (3) and (4) can be chosen so that each of Rk_i, k_iR, Rm_i and m_iR is simple. Finally, if R satisfies any of these conditions we have:
  (a) $Z_r = J = Z_l$.
  (b) $S_r = S_l$ is essential in both $_RR$ and R_R.

Proof. (1)⇔(2). If R satisfies (1), it is left FP-injective by Theorem 3.1. Thus R is right and left mininjective, right Kasch and semilocal, so it is left Kasch by Lemma 3.3. Hence (1)⇒(2); the converse is analogous.

(1)⇒(6). By the equivalence of (1) and (2), we must show that R is left finite dimensional. As before, (1) and (2) show that R is left and right P-injective, so $S_r = S_l$ by Lemma 1.2. Hence by Lemma 3.5, it suffices to show that $S_l \subseteq^{ess} {_RR}$. If $0 \neq a \in R$ let $\mathbf{r}(a) \subseteq T \subseteq^{max} R_R$, so that $Ra = \mathbf{lr}(a) \supseteq \mathbf{l}(T)$. But $\mathbf{l}(T) \neq 0$ by the right Kasch hypothesis, so it is simple by Lemma 1.1.

(6)⇒(1). We have (6)⇒(2) by Lemma 3.6, and we already proved that (2)⇒(1).

(1)⇔(3). Assume (1). We have $J = \mathbf{r}(S_r)$ because R is right Kasch. By Lemma 3.5, $S_r = Rk_1 + \cdots + Rk_n$, $k_i \in R$, Rk_i simple. Hence $J = \mathbf{r}(S_r) = \mathbf{r}\{k_1, \cdots, k_n\}$. This proves (3). Conversely, if K is a simple right R-module, we must show that K embeds in R_R. As R/J is semisimple, K embeds in R/J as a right R/J-module, and hence as a right R-module. But (3) implies that R/J is R-embedded in $(R_R)^n$. Thus K embeds in $(R_R)^n$, and hence in R_R.

(1)⇔(7). We have (1)⇒(7) because we have already proved (2)⇒(6). Conversely, R is left P-injective by Theorem 3.1, and hence semilocal by Lemma 3.6.

(2)⇔(4), (1)⇔(5) and (2)⇔(8). These are analogs of (1)⇔(3) and (2)⇔(6) and (1)⇔(7) respectively.

Finally, R is left and right P-injective so (a) follows from Lemma 2.5 and $S_r = S_l$ by Lemma 1.2. But $S_r = S_l$ is essential on both sides by the proof of (1)⇒(6). □

Since "right Kasch" and "semilocal" are both Morita invariants, the rings identified in Theorem 3.7 form a Morita invariant class by Corollary 2.4. However, we do not know the answer to the following question:

Question. *Are the rings in Theorem 3.7 all semiperfect?*

A ring is called right *pseudo-Frobenius* (right *PF-ring*) if it satisfies any of the following equivalent conditions (due to Azumaya, Kato, Osofsky and Utumi).
  (1) *Every faithful right R-module is a generator.*
  (2) *R_R is an injective cogenerator.*

(3) R is right selfinjective and right Kasch.

(4) R is a right selfinjective ring with finitely generated, essential right socle.

(5) R is a semiperfect, right selfinjective ring with essential right socle.

Thus every right PF-ring satisfies the conditions in Theorem 3.7. Since all PF-rings are semiperfect, and since this fact is used frequently in the literature, we restrict our attention to the semiperfect rings which satisfy the conditions in Theorem 3.7.

Definition. A ring R is called an *FP-ring* if it is semiperfect and satisfies any of the equivalent conditions in Theorem 3.7.

Since "semiperfect" and "right Kasch" are Morita invariant properties, we have

Theorem 3.8. *The class of FP-rings forms a Morita invariant class.*

There are examples [DM] of right PF-rings which are not left PF-rings. The following theorem shows again how the right-left symmetry is restored in the class of FP-rings. The theorem also emphasizes how the FP-rings symmetrize condition (5) in the above characterization of PF-rings.

Theorem 3.9. *The following conditions are equivalent for a ring R:*
  (1) *R is an FP-ring.*
  (2) *R is semiperfect, right FP-injective and $S_r \subseteq^{ess} R_R$.*
  (3) *R is semiperfect, left FP-injective and $S_l \subseteq^{ess} {}_RR$.*

Proof. (1)$\Rightarrow$(2) by Theorem 3.7, and (2)$\Rightarrow$(1) by Theorem 1.3. □

The next theorem shows that the FP-rings enjoy many of the properties of QF-rings. Recall that if M is a module, submodules $soc_1(M) \subseteq soc_2(M) \subseteq \cdots$ are defined by setting $soc_1(M) = soc(M)$ and, if $soc_n(M)$ has been specified, by $soc_{n+1}(M)/soc_n(M) = soc[M/soc_n(M)]$. Recall also that M is called *finitely embedded* (or *finitely cogenerated*) if $soc(M)$ is finitely generated and essential in M. A ring R is called *right* (*left*) *finitely embedded* if R_R (respectively ${}_RR$) is a finitely embedded module. Finitely embedded modules are finite dimensional, but the converse is not true (for example $\mathbb{Z}$).

Theorem 3.10. *The following hold in any FP-ring R:*
  (1) *$soc(eR)$ and $soc(Re)$ are simple for all local idempotents $e \in R$.*
  (2) *R is left and right finitely embedded.*

(3) If $\{e_1,\ldots,e_n\}$ is a basic set of local idempotents in R, there exist $k_1,\ldots,k_n$ in R and a (Nakayama) permutation σ of $\{1,\ldots,n\}$ such that the following hold for all $i = 1,\ldots,n$:

(a) $Rk_i = soc(Re_{\sigma i}) \cong \dfrac{Re_i}{Je_i}$ and $k_i R = soc(e_i R) \cong \dfrac{e_{\sigma i} R}{e_{\sigma i} J}$.

(b) $\{k_1 R,\ldots,k_n R\}$ and $\{Rk_1,\ldots,Rk_n\}$ are sets of distinct representatives of the simple right and left R-modules, respectively.

(4) $soc_n(R_R) = soc_n({}_R R) = \mathbf{l}(J^n) = \mathbf{r}(J^n)$ for all $n \geq 1$.

Proof. (1) and (3) follow from Theorem 1.3, and (2) is by Theorem 3.7.

(4) We have $S_r = S_l = S$ by Theorem 3.7, and $S = \mathbf{l}(J) = \mathbf{r}(J)$ because R/J is semisimple. Suppose $soc_k(R_R) = soc_k({}_R R) = \mathbf{l}(J^k) = \mathbf{r}(J^k)$ for some $k \geq 1$. We have $soc_{k+1}(R_R) \subseteq \mathbf{l}(J^{k+1})$ because $soc_{k+1}(R_R)/soc_k(R)$ is right R-semisimple. On the other hand, if $aJ^{k+1} = 0$, then $aJ \subseteq soc_k(R)$ so $[aR + soc_k(R)]/soc_k(R)$ is right R-semisimple (because R/J is semisimple). Hence $aR \subseteq soc_{k+1}(R_R)$ and we have $soc_{k+1}(R_R) = \mathbf{l}(J^{k+1})$. Similarly, $soc_{k+1}({}_R R) = \mathbf{r}(J^{k+1})$. Finally, $\mathbf{l}(J^{k+1}) = \mathbf{r}(J^{k+1})$ follows easily from $\mathbf{l}(J^k) = \mathbf{r}(J^k)$. □

Lemma 3.11. *Suppose that R is a semiperfect ring in which $S_l \subseteq^{ess} R_R$. Then:*

(1) $\mathbf{rl}(T)$ *is essential in a summand of R_R for every right ideal T of R.*

(2) R *is left Kasch.*

Proof. (1). Since R is semiperfect, write $\mathbf{l}(T) = R(1-e) \oplus B$ where $e^2 = e$ and $B = \mathbf{l}(T) \cap Re \subseteq J$. We claim that $\mathbf{rl}(T) \subseteq^{ess} eR$. Since $\mathbf{rl}(T) = eR \cap \mathbf{r}(B)$, it suffices to show that $\mathbf{r}(B) \subseteq^{ess} R_R$. But $B \subseteq J$ so $\mathbf{r}(B) \supseteq \mathbf{r}(J) = S_l$ and the hypothesis applies.

(2). Let L be a maximal left ideal of R. As before, choose $e^2 = e \in R$ such that $(1-e) \in L$ and $Re \cap L \subseteq J$. Then $\mathbf{r}(Re \cap L) \supseteq \mathbf{r}(J) \supseteq S_l$, so $\mathbf{r}(Re \cap L)$ is essential in R_R by hypothesis. In particular $0 \neq eR \cap \mathbf{r}(Re \cap L) = \mathbf{r}[R(1-e) \oplus (Re \cap L)] = \mathbf{r}(L)$. □

A module M satisfies the *C1-condition* if every submodule of M is essential in a summand of M. The module M is called *continuous (finitely continuous)* if it satisfies the C1 and C2 conditions (respectively C1 for finitely generated submodules, and C2). If only C1 is required, M is called a *CS-module*, and M is a *min-CS-module* if C1 is required only for simple submodules. A ring R is called *right continuous (right finitely continuous, a right CS-ring, a right min-CS-ring)* if R_R has the corresponding property.

Proposition 3.12. *Every FP-ring is left and right finitely continuous.*

On Quasi-Frobenius Rings

Proof. Let L be a finitely generated left ideal of R, and let T be a finitely generated right ideal. Then $\mathbf{lr}(L) = L$ and $\mathbf{rl}(T) = T$ by (5) of Theorem 2.2, so L and T are essential in direct summands of R by Lemma 3.11. Since R is left and right P-injective, it satisfies the left and right C2-conditions by Lemma 2.5. □

Our next characterization of FP-rings requires showing that rings of a certain type are all semiperfect. In contrast to Lemma 3.6 on semilocal endomorphism rings, we have

Lemma 3.13. *Let M_R be a module and suppose $M = U_1 \oplus U_2 \oplus \cdots \oplus U_n$ where each U_i is uniform. If monomorphisms $M \to M$ are epic then $end(M)$ is semiperfect.*

Proof. It is enough to prove that $end(U_i)$ is local for each i. Given $\alpha \in end(U_i)$ we have $ker(\alpha) \cap ker(1 - \alpha) = 0$ in U_i so either α or $1 - \alpha$ is monic. But a routine argument shows that monomorphisms in $end(U_i)$ are epic, and the lemma follows. □

The proof of the following proposition is an adaptation of an argument in [GY, Theorem 2.1].

Proposition 3.14. *Let R be a left Kasch ring in which $\mathbf{r}(M)$ is essential in a summand of R_R for every maximal left ideal M of R. Then R is semiperfect.*

Proof. If $M \subseteq_R^{max} R$, we show that the simple left R-module R/M has a projective cover. Since R is left Kasch, let $Ma = 0$ where $0 \neq a \in R$. Then $M = \mathbf{l}(a)$ so, by hypothesis, let $\mathbf{rl}(a) = \mathbf{r}(M) \subseteq^{ess} eR$ where $e^2 = e \in R$. Define $\theta : Re \to Ra$ by $x\theta = xa$. Then θ is epic because $a \in eR$ and Ra is simple. Hence we are done if $ker\theta = Re \cap \mathbf{l}(a)$ is small in Re, equivalently if $Re \cap \mathbf{l}(a)$ is the only maximal submodule of Re (so e is local). So suppose that $N \subseteq^{max} Re$; we must show that $N \subseteq Re \cap \mathbf{l}(a)$, that is $N \subseteq \mathbf{l}(a)$. If not, then $N + \mathbf{l}(a) = R$ so $0 = \mathbf{r}(N) \cap \mathbf{rl}(a)$, whence $0 = [\mathbf{r}(N) \cap eR] \cap \mathbf{rl}(a)$. This gives $\mathbf{r}(N) \cap eR = 0$ by the choice of e. But Re/N is simple so let $\sigma : Re/N \to {}_RR$ be an embedding. If $b = (e + N)\sigma$ then $0 \neq b \in eR \cap \mathbf{r}(N)$, a contradiction. □

Note that the proof of Proposition 3.14 shows that, in a left Kasch ring, if $M \subseteq_R^{max} R$ and $\mathbf{r}(M) \subseteq^{ess} eR$ where $e^2 = e$, then e is a local idempotent.

Theorem 3.15. *The following conditions are equivalent for a ring R:*

(1) R is an FP-ring.
(2) R is right Kasch, right FP-injective and left min-CS.
(3) R is left Kasch, left FP-injective and right min-CS.
(4) R is right FP-injective, right finitely embedded and right min-CS.
(5) R is left FP-injective, left finitely embedded and left min-CS.

Proof. By the symmetry of (1), we prove only (1)⇔(3) and (1)⇔(4). Proposition 3.12 gives (1)⇒(3) and (1)⇒(4).

(3)⇒(1). Given (3), R is semiperfect by Proposition 3.14 and Lemma 1.1. Hence R is an FP-ring by Theorem 3.7.

(4)⇒(1). It suffices by Theorem 3.9 to show that R is semiperfect. Since R is right finite dimensional, write $S_r = K_1 \oplus \cdots \oplus K_n$ where each K_i is a simple right ideal. By the min-CS condition, let $K_i \subseteq^{ess} e_i R$ for each i where $e_i^2 = e_i \in R$. Since the K_i are uniform it follows by induction that $E = e_1 R \oplus \cdots \oplus e_n R$ is a direct sum. But R is right P-injective, and so satisfies the right C2-condition by Lemma 2.5. Hence, by [MM, Proposition 2.2], R satisfies the right C3-condition: If right ideals S and T are summands of R_R with $S \cap T = 0$, then $S \oplus T$ is also a summand. It follows that E is a direct summand of R_R. But $E \subseteq^{ess} R_R$ because $S_r \subseteq^{ess} R_R$, so $E = R$. Since monomorphsms $R_R \to R_R$ are epic (by Lemma 3.6), R is semiperfect by Lemma 3.13. □

The proof of the last theorem of this section requires the following characterization of right FP-injective modules due to S. Jain [J]. The original proof is homological so we include a short, elementary proof for completeness. A module M is said to be *finitely presented* if $M \cong F/K$ where F is free and both F and K are finitely generated, and M is called *torsionless* if it can be embedded in a direct product of copies of the ring.

Lemma 3.16. *A ring R is right FP-injective if and only if every finitely presented left module is torsionless.*

Proof. Assume that R is right FP-injective, and suppose $_RK$ is a finitely generated submodule of R^n, say $K = R\bar{a}_1 + \cdots + R\bar{a}_m$. We must embed $R^n/K \hookrightarrow (_RR)^I$ for some set I; equivalently, if $\bar{b} \in R^n, \bar{b} \notin K$, we must find $\gamma : R^n \to {_RR}$ such that $K\gamma = 0$ and $\bar{b}\gamma \neq 0$. Suppose no such γ exists so that $K\gamma = 0$ implies $\bar{b}\gamma = 0$. Since each such γ has the form $\gamma = \cdot \underline{c}$, $\underline{c} \in R_n$, this means $\cap_{i=1}^m \mathbf{r}_{R_n}(\bar{a}_i) = \mathbf{r}_{R_n}(K) \subseteq \mathbf{r}_{R_n}(\bar{b})$. But then $\bar{b} \in K$ by (4) of Theorem 2.2, contrary to hypothesis.

Conversely, assume the condition holds, and suppose $\cap_{i=1}^m \mathbf{r}_{R_n}(\bar{a}_i) \subseteq \mathbf{r}_{R_n}(\bar{b})$ as in (4) of Theorem 2.2. Write $K = \Sigma_{i=1}^m R\bar{a}_i$. If $\bar{b} \notin K$, our hypothesis gives $\gamma : R^n \to {_RR}$ such that $K\gamma = 0$ but $\bar{b}\gamma \neq 0$. But $\gamma = \cdot\underline{c}$

for some $\underline{c} \in R_n$, and so $\underline{c} \in \mathbf{r}_{R_n}(K) = \cap_{i=1}^{m}\mathbf{r}_{R_n}(\bar{a}_i) \subseteq \mathbf{r}_{R_n}(\bar{b})$. Hence $0 = \bar{b}\underline{c} = \bar{b}\gamma$, a contradiction. $\square$

In preparation for our last theorem we give a necessary and sufficient condition for the Jacobson radical of an FP-ring to be finitely generated.

Proposition 3.17. *Let R be an FP-ring. Then $R/\mathrm{soc}(R)$ is finitely embedded as a right (left) R-module if and only if J is finitely generated as a left (right) R-module.*

Proof. Write $S = S_r = S_l$. Since S_R is finitely generated by Theorem 3.7, R/S is right torsionless by Lemma 3.16 (as R is left FP-injective). If R/S is right finitely embedded, it follows from [AF, Propositions 10.2 and 10.7] that there exists an embedding $\sigma : R/S \to (R^n)_R$ for some $n \geq 1$. If $\sigma(1+S) = (a_1, \ldots, a_n)$, $a_i \in R$, then $S = \mathbf{r}\{a_1, \ldots, a_n\}$. Because R is right FP-injective, we have $\mathbf{l}(S) = \mathbf{lr}\{a_1, \ldots, a_n\} = \Sigma_{i=1}^{n} Ra_i$ by (5) of Theorem 2.2. But $J = \mathbf{l}(S)$ because R is left Kasch (being an FP-ring), so $J = \Sigma_{i=1}^{n} Ra_i$ as required.

Conversely, if $_RJ$ is finitely generated, then $S = \mathbf{r}(J) = \mathbf{r}\{a_1, \ldots, a_n\}$ with $a_i \in R$. It follows that $R/S \hookrightarrow (R^n)_R$. But R is right finitely embedded by Theorem 3.10, hence $(R^n)_R$ is finitely embedded, whence R/S is right finitely embedded. $\square$

The Faith conjecture asserts that every left perfect, left (or right) self-injective ring is quasi-Frobenius. This conjecture remains open even for a semiprimary, local, right selfinjective ring with $J^3 = 0$. Motivated by the work of Armendariz and Park [AP] and that of Ara and Park [ArP], it was shown by Clark and Huynh [CH] that, if R is a left and right perfect, right selfinjective ring, then R is quasi-Frobenius if and only if $\mathrm{soc}_2(R)$ is finitely generated as a right R-module. On the other hand, in [NY4] it was shown that, if R is a left perfect, right selfinjective ring, then R is quasi-Frobenius if and only if $\mathrm{soc}_2(R)$ is countably generated as a left R-module. In the next theorem we obtain a version of this result for right FP-injective rings.

Theorem 3.18. *Let R be a left perfect, right FP-injective ring. Then:*
(1) R is quasi-Frobenius if and only if $\mathrm{soc}_2(R)$ is finitely generated as a right R-module.
(2) R is quasi-Frobenius if and only if $R/\mathrm{soc}(R)$ is finitely embedded as a left R-module.
(3) If R is also right perfect, then R is quasi-Frobenius if and only if $\mathrm{soc}_2(R)$ is finitely generated as a left R-module.

Proof. Note that R is an FP-ring by Theorem 3.9 because $S_r \subseteq^{ess} R_R$ in a left perfect ring.

(1) R is right semi-artinian (being left perfect) so $R/soc(R)$ has an essential right socle. But then $R/soc(R)$ is finitely embedded as a right R-module because $soc_2(R)$ is right finitely generated. Hence $_RJ$ is finitely generated by Proposition 3.17, and so R is left artinian by [O, Lemma 11]. Hence R is quasi-Frobenius by Theorem 1.4.

(2) Proposition 3.17 shows that J_R is finitely generated, so R is right artinian by [O, Lemma 11]. But R is left and right mininjective, so it is quasi-Frobenius by Theorem 1.4.

(3) In this case R is right perfect, hence left semi-artinian, and so $R/soc(R)$ has an essential left socle. Thus $R/soc(R)$ is left finitely embedded by hypothesis, and (2) applies. □

4. The C2-Condition

If every right R-module embeds in a free module, it is a theorem of Faith and Walker [FW] that R is quasi-Frobenius; the FGF-conjecture asserts that the hypothesis can be weakened to finitely generated right modules. Right FGF-rings have been studied by many authors; see Faith [F2] for a detailed history of the problem. More recently, Gómez Pardo and Guil Asensio have carried out a fundamental study of the conjecture, and a thorough discussion of the recent work on the problem can be found in [GA2]. Here are four important results on this conjecture:

(1) Every left Kasch, right FGF-ring is R is QF. (Kato [K].)

(2) Every right selfinjective, right FGF-ring is QF. (Björk [B2]; also Osofsky [O].)

(3) Every right perfect, right FGF-ring is QF. (Rutter [R].)

(4) Every right CS, right FGF-ring is QF. (Gómez Pardo and Guil Asensio [GA].)

In this section we unify and extend the first three of the above results in a single theorem.

A ring R is called a right *C2-ring* if R_R has the C2-condition, and R is called a *strongly right C2-ring* if $M_n(R)$ is a right C2-ring for every $n \geq 1$. In Theorem 4.13 below we will show that if R is a strongly right C2-ring and every 2-generated right R-module embeds in a free module then R is quasi-Frobenius. This result implies that the FGF-conjecture is true in the right FP-injective case (Corollary 4.14). Moreover, the theorem is used to reformulate the conjecture by showing that it suffices to prove that every right FGF-ring is a right C2-ring.

Examples. (1) Every semiregular ring with $Z_r = J$ is a right C2-ring (see Proposition 4.12).

(2) If R is a right C2-ring with no infinite sets of orthogonal idempotents, then every monomorphism $R_R \to R_R$ is epic. If $\mathbf{r}(a) = 0$, $a \in R$, consider $R \supseteq aR \supseteq a^2R \supseteq a^3R \supseteq \cdots$. Since $a^k R \cong R$, we have $a^k R = e_k R$ for some $e_k^2 = e_k$. Hence $a^k R = a^{k+1} R$ for some k by hypothesis, whence $R = aR$ because $\mathbf{r}(a) = 0$.

(3) If 0 and 1 are the only idempotents in R then R is a right C2-ring if and only if every monomorphism $R_R \to R_R$ is epic. Indeed, if monomorphisms are epic and $aR \cong P$ where P is a summand of R, then either $P = 0$ (so $aR = 0$ is a summand) or $P = R$. In the second case, if $\sigma : R \to aR$ is an isomorphism let $\sigma(1) = ab$. Then b is a unit by hypothesis, whence a is a unit and $aR = R$. It follows that R is a right C2-ring. The converse follows from (2).

(4) Every right P-injective ring is a right C2-ring by Lemma 2.5. The converse is false: If V is a two-dimensional vector space over a field F, the trivial extension $R = T(F, V) = F \oplus V$ is a commutative, local, artinian ring with $J^2 = 0$ and $J = Z_r$, and so is a C2-ring by (1). However R is not P-injective. Indeed, if $V = vF \oplus wF$, let $\theta : V \to V$ be a linear transformation with $\theta(v) = w$. Then $(0, x) \mapsto (0, \theta(x))$ is an R-linear map from $(0, v)R \to R$ which does not extend to $R \to R$ because $w \notin vF$.

The next proposition provides another class of right C2-rings.

Proposition 4.1. *Let R be any ring. Then:*
$$R \text{ is left Kasch} \Rightarrow R \text{ is a right C2-ring} \Rightarrow Z_r \subseteq J.$$

Proof. For convenience we write $K \mid M$ to mean that K is a direct summand of the module M. Suppose R is left Kasch. If aR is isomorphic to a summand of R, $a \in R$, it suffices to show that $Ra \mid R$ (then $aR \mid R$ too). Since aR is projective, let $\mathbf{r}(a) = (1 - e)R$, $e^2 = e$. Then $a = ae$ so $Ra \subseteq Re$, and we claim that $Ra = Re$. If not let $Ra \subseteq M \subseteq^{max} Re$. By the Kasch hypothesis let $\sigma : Re/M \to {}_R R$ be monic and write $c = (e + M)\sigma$. Then $ec = c$ and (since $ae = a \in M$) $c \in \mathbf{r}(a) = (1 - e)R$. It follows that $c = ec = 0$ and hence that $e \in M$ since σ is monic, a contradiction. So $Ra = Re$, as required.

Now assume that R is a right C2-ring and let $a \in Z_r$. As $\mathbf{r}(a) \cap \mathbf{r}(1 - a) = 0$ it follows that $\mathbf{r}(1 - a) = 0$, whence $(1 - a)R \cong R$. By hypothesis $(1 - a)R \mid R$, whence $R(1 - a) \mid R$, say $R(1 - a) = Rg$, $g^2 = g$. It follows that $1 - g \in \mathbf{r}(1 - a) = 0$, so $R(1 - a) = R$. Since $a \in Z_r$ was arbitrary, this means that $Z_r \subseteq J$. □

Both converses in Proposition 4.1 are false even for commutative rings. If F is a field, then $F \times F \times F \times \cdots$ is a C2-ring (it is regular), but it is not

Kasch (regular right Kasch rings are artinian). The ring $\{\frac{m}{n} \mid m, n \in \mathbb{Z}, n$ odd$\}$ has $Z_r = 0 \subseteq J$ but it is not a C2-ring (it is a domain that is not a division ring).

Example 4.2. There exists a left C2-ring R that is not a right C2-ring, and hence not left Kasch. Faith and Menal [FM1] give an example of a right noetherian ring R in which every right ideal is an annihilator, but which is not right artinian. Thus R is a left P-injective ring, hence left C2. But R is not right C2 because it would then be right artinian by Theorem 5.8 below.

Example 4.3. The trivial extension $R = T(\mathbb{Z}, \mathbb{Z}_{2^\infty})$ is a commutative CS-ring with $Z_r = J \neq 0$ which does not satisfy the C2-condition. In fact, R has simple essential socle.

We begin by deriving some basic characterizations of the right C2-rings.

Proposition 4.4. *The following conditions are equivalent for a ring R:*
  (1) R is a right C2-ring.
  (2) Every R-isomorphism $aR \to eR$, $a \in R$, $e^2 = e \in R$, extends to $R_R \to R_R$.
  (3) If $\mathbf{r}(a) = \mathbf{r}(e)$, $a \in R$, $e^2 = e \in R$, then $e \in Ra$.
  (4) If $\mathbf{r}(a) = \mathbf{r}(e)$, $a \in R$, $e^2 = e \in R$, then $Re = Ra$.
  (5) If $Ra \subseteq Re \subseteq \mathbf{lr}(a), a \in R$, $e^2 = e \in R$, then $Re = Ra$.
  (6) If aR is projective, $a \in R$, then aR is a direct summand of R_R.

Proof. $(6) \Rightarrow (1) \Rightarrow (2) \Rightarrow (3) \Rightarrow (4) \Rightarrow (5)$ are routine computations. Assume that (5) holds. If aR is projective then $\mathbf{r}(a)$ is a direct summand of R, say $\mathbf{r}(a) = \mathbf{r}(e)$ for $e^2 = e$. Thus $a = ae$, so $Ra \subseteq Re$. But $e \in \mathbf{lr}(a)$ (because $\mathbf{r}(a) \subseteq \mathbf{r}(e)$) so we have $Ra \subseteq Re \subseteq \mathbf{lr}(a)$. Thus $Ra = Re$ by (5), so Ra is a direct summand of R, whence aR is a summand, proving (6). □

Condition (3) of Proposition 4.4 gives

Corollary 4.5. *The direct product $\Pi_i R_i$ of rings R_i is a right C2-ring if and only if each R_i is a right C2-ring.*

Corollary 4.6. *The following conditions are equivalent for a local ring R:*
  (1) R is a right C2-ring.
  (2) Every monomorphism $R_R \to R_R$ is epic.
  (3) $J = \{a \in R \mid \mathbf{r}(a) \neq 0\}$.
In particular, any local ring with nil radical is a right and left C2-ring.

Proof. We have already observed that (1)⇔(2) for any ring in which 0 and 1 are the only idempotents. Given (2), it is clear that $J \subseteq \{a \in R \mid \mathbf{r}(a) \neq 0\}$; this is equality in a local ring. Hence (2)⇒(3). Finally, if (3) holds, suppose $\mathbf{r}(a) = \mathbf{r}(e)$, $a \in R$, $e^2 = e \in R$. By Proposition 4.4 we must show that $e \in Ra$. This is clear if $e = 0$. If $e = 1$ then $\mathbf{r}(a) = 0$ so $a \notin J$ by (3). Hence $Ra = R$ because R is local, and so $e \in Ra$ as required. Thus (3)⇒(1). Finally the last statement follows from (3) because R is local. □

The upper triangular 2×2 matrix ring R over a field is a left and right artinian ring, so every monomorphism $R_R \to R_R$ is epic, but R is not a right C2-ring.

Corollary 4.7. *If R is a right C2-ring, so is fRf for any $f^2 = f \in R$ such that $RfR = R$.*

Proof. Write $S = fRf$ and suppose that $\mathbf{r}_S(a) = \mathbf{r}_S(e)$, $a \in S$, $e^2 = e \in S$. We must show that $e \in Sa$. It suffices to show that $e \in Ra$ so (by hypothesis) we show that $\mathbf{r}_R(a) = \mathbf{r}_R(e)$. If $r \in \mathbf{r}_R(a)$ then, for all $x \in R$, $a(frxf) = arxf = 0$ so $frxf \in \mathbf{r}_S(a) = \mathbf{r}_S(e)$. Thus $erxf = 0$ for all $x \in R$ so, as $RfR = R$, $er = 0$. Thus $\mathbf{r}_R(a) \subseteq \mathbf{r}_R(e)$; the other inclusion is proved in the same way. □

Corollary 4.7 is half of the proof that "right C2-ring" is a Morita invariant. We will characterize when this is true in Corollary 4.11 below.

Proposition 4.8. *The following conditions are equivalent for a module M_R with $E = \text{end}(M_R)$.*
   (1) M_R *satisfies the C2-condition.*
   (2) *If $\sigma : N \to P$ is an R-isomorphism where $N \subseteq M$ and P is a direct summand of M, then σ extends to some $\beta \in E$.*
   (3) *If $\alpha : P \to M$ is R-monic where P is a direct summand of M, there exists $\beta \in E$ with $\beta \circ \alpha = \iota$, where $\iota : P \to M$ is the inclusion.*
   (4) *If $\alpha : P \to M$ is R-monic where P is a direct summand of M, and if $\pi^2 = \pi \in E$ satisfies $\pi(M) = P$, there exists $\beta \in E$ with $\pi \circ \beta \circ \alpha = 1_P$.*

Proof. (1)⇒(2). If σ is as in (2), let $M = N \oplus N'$ by (1). Then $(n + n') \mapsto \sigma(n)$ extends σ.

(2)⇒(3). If α is as in (3) then $\sigma : \alpha(P) \to P$ is an R-isomorphism if we define $\sigma[\alpha(p)] = p$ for all $p \in P$. By (2) let $\beta \in E$ extend σ. Then $\beta \circ \alpha = \iota$.

(3)⇒(4). If α is as in (4), let $\beta \circ \alpha = \iota$ by (3) where $\beta \in E$. Then $\pi \circ \beta \circ \alpha = 1_P$.

(4)⇒(1). Suppose a submodule $N \subseteq M$ is isomorphic to P where P is a direct summand of M, say $\alpha : P \to N$ is an R-isomorphism. We must show that N is a direct summand of M. If $\pi^2 = \pi \in E$ satisfies $\pi(M) = P$, (4) provides $\beta \in E$ such that $\pi \circ \beta \circ \alpha = 1_P$. Define $\theta = \alpha \circ \pi \circ \beta \in E$. Then $\theta^2 = \theta$ and $\theta(M) \subseteq N$, so we are done if we can show that $N \subseteq \theta(M)$. But $\theta \circ \alpha = \alpha$ so $N = \alpha(P) = \theta[\alpha(P)] \subseteq \theta(M)$, as required. □

It is easy to verify that direct summands of a C2-module are again C2-modules. But the direct sum of C2-modules need not be a C2-module. If $R = \begin{bmatrix} F & F \\ 0 & F \end{bmatrix}$, $A = \begin{bmatrix} F & F \\ 0 & 0 \end{bmatrix}$ and $B = \begin{bmatrix} 0 & 0 \\ 0 & F \end{bmatrix}$ where F is a field, then R_R is not a C2-module, but $R = A \oplus B$ and both A_R and B_R are C2-modules (B_R is simple, and A_R has exactly one proper submodule $J \not\cong A$).

Theorem 4.9. *Let M_R be a module and write $E = end(M_R)$. Then:*

(1) If E is a right C2-ring then M_R is a C2-module.

(2) The converse in (1) holds if $ker(\alpha)$ is generated by M whenever $\alpha \in E$ is such that $\mathbf{r}_E(\alpha)$ is a direct summand of E_E.

Proof. (1). Let $\alpha : P \to M$ be R-monic where P is a direct summand of M, let $\pi^2 = \pi \in E$ satisfy $\pi(M) = P$, and write $ker(\pi) = Q$. Hence $M = P \oplus Q$ and we extend α to $\bar{\alpha} \in E$ by defining $\bar{\alpha}(p+q) = \alpha(p)$. Since α is monic, $ker(\bar{\alpha}) = Q = ker(\pi)$. It follows that

$$\mathbf{r}_E(\bar{\alpha}) = \{\lambda \in E \mid \lambda(M) \subseteq Q\} = \mathbf{r}_E(\pi).$$

Since E is a right C2-ring, Proposition 4.4 gives $\pi \in E\bar{\alpha}$, say $\pi = \beta \circ \bar{\alpha}$ with $\beta \in E$. Then $\pi \circ \beta \circ \alpha = 1_P$, and so M_R has the C2-property by Proposition 4.8.

(2) Let $\mathbf{r}_E(\alpha) = \mathbf{r}_E(\pi)$ where α and $\pi^2 = \pi$ are in E. By Proposition 4.4, we must show that $\pi \in E\alpha$.

Claim. $ker(\alpha) = ker(\pi)$.

Proof. $1 - \pi \in \mathbf{r}_E(\pi) = \mathbf{r}_E(\alpha)$, so $\alpha = \alpha \circ \pi$, whence $ker(\pi) \subseteq ker(\alpha)$. On the other hand, our hypothesis gives $ker(\alpha) = \Sigma\{\theta(M) \mid \theta \in E, \theta(M) \subseteq ker(\alpha)\}$. Since $\theta(M) \subseteq ker(\alpha)$ implies $\theta \in \mathbf{r}_E(\alpha) = \mathbf{r}_E(\pi)$, it follows that $\theta(M) \subseteq ker(\pi)$. Hence $ker(\alpha) \subseteq ker(\pi)$, proving the Claim.

Now write $\pi(M) = P$ and $ker(\pi) = Q$. Then $P \cap ker(\alpha) = 0$ by the Claim so $\alpha_{|P}$ is monic. Since M has the C2-condition, Proposition 4.8 provides $\beta \in E$ such that $\beta \circ (\alpha_{|P}) = \iota$ where $\iota : P \to M$ is the inclusion. We claim that $\beta \circ \alpha = \pi$, which proves (2). If $q \in Q$, then $(\beta \circ \alpha)(q) = 0 = \pi(q)$ by the Claim; if $p \in P$, then $(\beta \circ \alpha)(p) = p = \pi(p)$. As $M = P \oplus Q$ this shows that $\pi = \beta\alpha \in E\alpha$, as required. □

Since a free module generates all of its submodules, we obtain

Corollary 4.10. *If M_R is free, then M is a C2-module if and only if $\mathrm{end}(M_R)$ is a right C2-ring. In particular R^n is a right C2-module if and only if $M_n(R)$ is a right C2-ring.*

Question. Is "right C2-ring" a Morita invariant?

By Corollary 4.7, the answer is "yes" if we can show that $M_2(R)$ is a right C2-ring whenever R is a right C2-ring. Hence Corollary 4.10 gives

Corollary 4.11. *The following conditions are equivalent:*
  (1) *"Right C2-ring" is a Morita invariant.*
  (2) *If R is a right C2-ring then $(R \oplus R)_R$ is a C2-module.*

Recall that a ring R is called a strongly right C2-ring if $M_n(R)$ is a right C2-ring for every $n \geq 1$. This class of rings includes every Morita invariant class of right C2-rings. Examples are given in the next Proposition, and in Corollary 4.14 below.

Proposition 4.12. *If R is semiregular with $J = Z_r$ then R is a strongly right C2-ring.*

Proof. By Corollary 4.10 it is enough to show that R^n has the right C2-condition. Let $A \cong B$ where A and B are submodules of R^n and B is a direct summand. Then A is finitely generated (and projective) so, since R is semiregular, there is a decomposition $R^n = P \oplus Q$ where $P \subseteq A$ and $A \cap Q$ is small in R^n. Thus $A = P \oplus (A \cap Q)$ where $A \cap Q \subseteq \mathrm{rad}(R^n) = J^n = Z_r^n$ by hypothesis. This means that $A \cap Q$ is both projective and singular, and so $A \cap Q = 0$. Thus A is a summand of R^n, as required. □

Theorem 4.13. *Suppose that R is a strongly right C2-ring and every 2-generated right R-module embeds in a free module. Then R is quasi-Frobenius.*

Proof. Let $a \in E(R_R)$, where $E(M)$ denotes the injective hull of a module M. Since R is right FGF, let $\sigma : R + aR \to (R^n)_R$ be monic. Since R is a right C2-ring and $\sigma(R) \cong R$, it follows that $\sigma(R)$ is a summand of R^n, and hence of $\sigma(R+aR)$. But $\sigma(R) \subseteq^{ess} \sigma(R+aR)$ because $R \subseteq^{ess} R+aR$. This implies that $a \in R$, so $R = E(R_R)$ and hence R is right selfinjective. Since R is right Kasch, R is right finitely embedded by a result of Osofsky [O], and so every cyclic right R-module has a finitely generated, essential socle. Thus R is right artinian, and so is quasi-Frobenius. □

Corollary 4.14. *Suppose a ring R has the property that every 2-generated right R-module embeds in a free module. Then R is quasi-Frobenius if it has any of the following properties.*

(1) *R is semiregular with $J = Z_r$.*
(2) *R is left Kasch.*
(3) *R is semiperfect with $soc(Re) \neq 0$ for every local idempotent $e \in R$. (In particular if R is right perfect.)*
(4) *R is right FP-injective.*

Proof. In each case we show that R is a strongly right C2-ring and apply Theorem 4.13.

(1) R is a strongly right C2-ring by Proposition 4.12.

(2) Since "left Kasch" is Morita invariant, this follows from Proposition 4.1.

(3) $\mathrm{rl}(T) = T$ for every right ideal T because R/T embeds in a free module. In particular R is left P-injective, and so is left Kasch by Theorem 1.3. Hence (3) follows from (2).

(4) This follows from Corollary 2.4 and Lemma 2.5. □

Since being an FGF-ring is a Morita invariant, Theorem 4.13 and Corollary 4.10 immediately give the following simplification of what is required to prove the FGF-conjecture.

Theorem 4.15. *The following statements are equivalent:*
(1) *Every right FGF-ring is a right C2-ring.*
(2) *Every right FGF-ring is quasi-Frobenius.*

5. The Faith-Menal Conjecture

Recall that a ring R is right Johns if it is right noetherian and every right ideal is an annihilator, and that R is strongly right Johns if $M_n(R)$ is right Johns for every $n \geq 1$.

Strongly right Johns rings have been characterized by Faith and Menal [FM2] as the right noetherian left FP-injective rings, and the Faith-Menal conjecture asserts that these rings are quasi-Frobenius. In this regard we show that right Johns, right mininjective rings are quasi-Frobenius. In fact, this follows from a more general result (Theorem 5.3): A right minsymmetric ring is right artinian if and only if it is right noetherian with essential right socle. This can be viewed as a one-sided version of the result of Ginn and Moss [GM] that any (two-sided) noetherian ring with essential socle is artinian. As another consequence of Theorem 5.3, we can show that a ring is quasi-Frobenius if and only if it is a left and right

On Quasi-Frobenius Rings

mininjective, right Goldie ring with essential right socle, a result known earlier only in the right artinian case.

In Proposition 4.1 we showed that every left Kasch ring is a right C2-ring, and every C2-ring satisfies $Z_r \subseteq J$. The next result gives some conditions under which all three conditions are equivalent. The following rings are involved: A ring R is called right *minsymmetric* if, whenever kR is a simple right ideal of R, then Rk is also simple. This is a large class of rings, including all commutative rings, all semiprime rings and all right mininjective rings (by Lemma 1.2). Note that $S_r \subseteq S_l$ in any right minsymmetric ring.

Lemma 5.1. *Suppose R is a right finitely embedded, right minsymmetric ring. Then the following conditions are equivalent:*
(1) *R is left Kasch.*
(2) *R is a right C2-ring.*
(3) *$Z_r \subseteq J$.*
In this case R is semilocal and $\mathbf{l}(S_r) = \mathbf{l}(S_l) = J$.

Proof. (1)⇒(2)⇒(3) hold in any ring by Proposition 4.1.

(3)⇒(1). Assume that $Z_r \subseteq J$, and suppose that $\mathbf{r}(M) = 0$ where M is a maximal left ideal of R. We must deduce a contradiction.

Claim. If $k_1 R, k_2 R, \ldots, k_n R$ are simple, $k_i \in R$, then $(1-m)k_i = 0$ for some $m \in M$.

Proof. We use induction on n. If $n = 1$ then $M \not\subseteq \mathbf{l}(k_1)$ because $k_1 \neq \mathbf{r}(M)$. But $\mathbf{l}(k_1)$ is maximal because Rk_1 is simple (by hypothesis), so $M + \mathbf{l}(k_1) = R$ as required. If $n \geq 2$ let $m_1 \in M$ satisfy $(1-m_1)k_i = 0$ for each $i = 1, 2, \cdots, n-1$. If $(1-m_1)k_n = 0$ we are done. Otherwise $(1-m_1)k_n R$ is simple so (by the case $n = 1$) let $(1-m_2)(1-m_1)k_n = 0$ where $m_2 \in M$. Then $(1-m_2)(1-m_1)k_i = 0$ for each i, and the Claim follows with $m = m_2 + m_1 - m_2 m_1$.

As R is right finite dimensional, let $S_r = k_1 R + \cdots + k_n R$ where each $k_i R$ is simple, and choose m as in the Claim. Then $(1-m)S_r = 0$ so $S_r \subseteq \mathbf{r}(1-m)$. As $S_r \subseteq^{ess} R_R$ this gives $1 - m \in Z_r$. Hence (3) gives $1 - m \in J \subseteq M$, a contradiction. This proves (1).

Finally, we have $\mathbf{l}(S_l) = J$ by the left Kasch condition, so $J \subseteq \mathbf{l}(S_r)$ because $S_r \subseteq S_l$ (by hypothesis). Let $a \in \mathbf{l}(S_r)$. Then $S_r \subseteq \mathbf{r}(a)$, so $S_r \subseteq^{ess} R_R$ gives $a \in Z_r \subseteq J$ by (3). Hence $J = \mathbf{l}(S_r)$. It follows that $R/J \hookrightarrow \oplus_{i=1}^n Rk_i$, proving that R is semilocal. □

Lemma 5.2. *Suppose that R is a right finitely embedded, right minsymmetric ring with ACC on right annihilators. Then R is a semiprimary ring with $J = Z_r$.*

Proof. By the ACC on right annihilators, Z_r is nilpotent, and so $Z_r \subseteq J$. Hence R is semilocal by Lemma 5.1. But $S_r \subseteq S_l$ because R is right minsymmetric, so $JS_r = 0$. Since $S_r \subseteq^{ess} R_R$ it follows that $J \subseteq Z_r$, so $J = Z_r$ and R is semiprimary. This completes the proof. $\square$

A result of Ginn and Moss [GM, Theorem] asserts that a two-sided noetherian ring with essential right socle is right and left artinian. In the next result we obtain a one-sided version of this theorem in the presence of the right minsymmetric condition.

Theorem 5.3. *Every right noetherian, right minsymmetric ring with essential right socle is right artinian.*

Proof. Since a right noetherian, semiprimary ring is right artinian by Hopkins's theorem, the result is an immediate consequence of Lemma 5.2. $\square$

Note that the Faith-Menal counterexample in [FM1] shows that the minsymmetric hypothesis cannot be removed from Theorem 5.3; the ring $\mathbb{Z}$ of integers shows that the essential socle hypothesis cannot be removed.

The next theorem extends the result (in Theorem 1.4) that every right artinian, left and right mininjective ring is quasi-Frobenius. We need the following lemma proved in [CY, Lemma 6].

Lemma 5.4. *Let R be a semiprimary ring with ACC on left annihilators, in which $S_r = S_l$ is finite dimensional as a right R-module. Then R is right artinian.*

Proof. We use induction on the index of nilpotency n of J. If $n = 1$ then $J = 0$ and R is semisimple artinian. So assume that $n \geq 2$. As R is semilocal, we have $S_r = \mathbf{l}(J)$ and $S_l = \mathbf{r}(J)$, and we write $\bar{R} = R/A$ where $A = \mathbf{l}(J) = S_r = S_l = \mathbf{r}(J)$. Since $A_R = S_r$ is artinian it suffices to show that $\bar{R}_R = \bar{R}_{\bar{R}}$ is artinian.

Since R has the ACC on left annihilators, and since $A = \mathbf{r}(J)$ is a right annihilator, the ring $\bar{R}$ inherits the ACC on left annihilators, and hence has the DCC on right annihilators. Moreover $J(\bar{R}) = (J + A)/A = \bar{J}$, so $\bar{R}/\bar{J} \cong R/(J + A)$ is semisimple and $\bar{J}^{n-1} \subseteq (J^{n-1} + A)/A = 0$ because $J^{n-1} \subseteq \mathbf{l}(J) = A$. Hence $\bar{R}$ is semiprimary and $\bar{J}$ has index of nilpotency at most $n - 1$, so it suffices to show that $soc(_{\bar{R}}\bar{R}) = soc(\bar{R}_{\bar{R}})$ is finite dimensional as a right $\bar{R}$-module.

If $\bar{x} \in soc(_{\bar{R}}\bar{R}) = \mathbf{r}_{\bar{R}}(\bar{J})$, we have $\bar{J}\bar{x} = \bar{0}$ so $Jx \subseteq A = \mathbf{l}(J)$, whence $JxJ = 0$. Thus $xJ \subseteq \mathbf{r}(J) = A$, so $\bar{x}\bar{J} = \bar{0}$ and $\bar{x} \in \mathbf{l}_{\bar{R}}(\bar{J}) = soc(\bar{R}_{\bar{R}})$. Thus $soc(_{\bar{R}}\bar{R}) \subseteq soc(\bar{R}_{\bar{R}})$, and the other inclusion is similarly proved. Finally, since R has the DCC on right annihilators, we have $\mathbf{r}(J) = \mathbf{r}\{b_1, \cdots, b_m\}$ where $\{b_1, \cdots, b_m\} \subseteq J$. Then $\theta : \bar{R} \to R^m$ given by $\theta(r + A) = (b_1 r, \cdots, b_m r)$ is a well defined monomorphism of right R-modules. Moreover, $\theta[soc(\bar{R}_{\bar{R}})] = \theta[soc(\bar{R}_R)] \subseteq soc(R_R^m) = S_r^m$, so $soc(\bar{R}_{\bar{R}})$ is right finite dimensional by hypothesis. This completes the proof. □

Theorem 5.5. *The following are equivalent for a ring R:*
  (1) *R is right and left mininjective, right noetherian, and $S_r \subseteq^{ess} R_R$.*
  (2) *R is right and left mininjective, right finitely embedded with ACC on right annihilators.*
  (3) *R is quasi-Frobenius.*

Proof. (1)$\Rightarrow$(2) and (3)$\Rightarrow$(1) are clear. Given (2), R is semiprimary by Lemma 5.2. If we can show that R is left artinian, (3) follows from Theorem 1.4. We have $S_l = S_r$ by Lemma 1.2, and it is left finite dimensional because $soc(Re) = S_l e = S_r e$ is simple for each local $e = e^2$ by Theorem 1.3. Hence R is left artinian by Lemma 5.4. □

A ring R is called a right *CF-ring* if every cyclic right R-module embeds in a free module, and R is called a right *CEP-ring* if every cyclic right R-module C can be essentially embedded in a projective module P, that is if an embedding $\sigma : C \to P$ exists with $\sigma(C) \subseteq^{ess} P$. In Theorems 5.8 and 5.9 we show that right Johns rings, right CF-rings and right CEP-rings are closely related. We need the following major results of Gómez Pardo and Guil Asensio in [GA1] and [GA].

Theorem 5.6. (1) *Every right Kasch, right CS-ring is right finitely embedded.*
  (2) *Every right CEP-ring is right artinian.*

The following lemma is essentially due to Johns [Johns], see also [FM1] and [GA2].

Lemma 5.7. *Let R be a right Johns ring. Then:*
  (1) *J is nilpotent.*
  (2) *$\mathbf{r}(J) = \mathbf{l}(J) = S_r$.*
  (3) *$\mathbf{r}(S_r) = \mathbf{l}(S_r) = J$.*
  (4) *$S_r \subseteq^{ess} R_R$ and $S_r \subseteq^{ess} {}_R R$.*
  (5) *$J = Z_r = Z_l$.*

Proof. (1) We have $\mathbf{l}(J) \subseteq \mathbf{l}(J^2) \subseteq \cdots$ so $\mathbf{l}(J^k) = \mathbf{l}(J^{k+1})$ for some k. Hence $J^k = \mathbf{rl}(J^k) = \mathbf{rl}(J^{k+1}) = J^{k+1}$, so $J^k = 0$ by Nakayama's Lemma.

(2) If $x \neq 0$ in R, then either $xJ = 0$ or (by (1)) $xJ^k \neq 0$ and $xJ^{k+1} = 0$ for some k. Either way, $xR \cap \mathbf{1}(J) \neq 0$, proving that $\mathbf{1}(J) \subseteq^{ess} R_R$. It follows that $\mathbf{11}(J) \subseteq Z_r$. But Z_r is nilpotent (as R is right noetherian) so $\mathbf{11}(J) \subseteq J$. If we write $\mathbf{1}^1(J) = \mathbf{1}(J)$ and define $\mathbf{1}^{k+1}(J) = \mathbf{1}(\mathbf{1}^k(J))$ for each k, it follows that $\mathbf{1}(J) \subseteq \mathbf{1}^3(J) \subseteq \mathbf{1}^5(J) \subseteq \cdots$, whence $\mathbf{1}^k(J) = \mathbf{1}^{k+2}(J)$ for some k. But then $\mathbf{rl}^k(J) = \mathbf{rl}^{k+2}(J)$, so $\mathbf{1}^{k-1}(J) = \mathbf{1}^{k+1}(J)$. Continuing in this way, we get $J = \mathbf{1}^2(J)$, and finally $\mathbf{1}(J) = \mathbf{r}(J)$.

Clearly $S_r \subseteq \mathbf{1}(J)$. If $T \subseteq^{ess} R_R$, then $\mathbf{1}(T) \subseteq Z_r \subseteq J$, so $\mathbf{r}(J) \subseteq \mathbf{rl}(T) = T$. Hence $\mathbf{r}(J) \subseteq \cap \{T \mid T \subseteq^{ess} R_R\} = S_r$, proving (2).

(3) Since $S_r = \mathbf{1}(J)$ by (2), we have $\mathbf{r}(S_r) = \mathbf{rl}(J) = J$. Next, we proved in (2) that $\mathbf{1}(J) \subseteq^{ess} R_R$ and $\mathbf{1}^2(J) = J$, whence $J \subseteq Z_r$. Since $Z_r S_r = 0$ it follows that $J \subseteq \mathbf{1}(S_r)$. Furthermore, $\mathbf{r}(J) \subseteq S_r$ from (2), whence $J \subseteq \mathbf{1}(S_r) \subseteq \mathbf{lr}(J) = \mathbf{11}(J) = J$. This proves (3).

(4) We have $S_r = \mathbf{1}(J) \subseteq^{ess} R_R$ by (2) and its proof. To see that $\mathbf{1}(J) \subseteq^{ess}_R R$, suppose $\mathbf{1}(J) \cap Rb = 0$. Then $\mathbf{1}(bJ) \subseteq \mathbf{1}(b)$ so $\mathbf{rl}(b) \subseteq \mathbf{rl}(bJ) = bJ$. It follows that $b \in bJ$, whence $b = 0$.

(5). $J = Z_l$ because R is left P-injective, and $J = Z_r$ by the proofs of (2) and (3). □

Theorem 5.8. *The following are equivalent for a ring R:*
  (1) R is a right CEP-ring.
  (2) R is a right Johns, left Kasch ring.
  (3) R is a right Johns, right C2-ring.
  (4) R is right artinian and every right ideal is an annihilator.
  (5) R is both a right CS-ring and a right CF-ring.
  (6) R is a semiperfect, right CF-ring with $S_l \subseteq^{ess}_R R$.
  (7) R is a semilocal, right CF-ring with $S_l \subseteq^{ess} R_R$.

Proof. (1)⇒(2). R is right artinian by Theorem 5.6, and $\mathbf{rl}(T) = T$ for every right ideal T by hypothesis. As R is semiprimary and left mininjective, it is left Kasch by Theorem 1.3.

(2)⇒(3). This follows from Proposition 4.1.

(3)⇒(4). It suffices to show that R is semiprimary, since it is then right artinian by Hopkins's Theorem. But Lemma 3.6 shows that R is semilocal, and J is nilpotent by Lemma 5.7.

(4)→(5). As R is semiprimary and left mininjective, we have $S_r = S_l$ (by Theorem 1.3) and $S_l \subseteq^{ess} R_R$. If T is a right ideal of R, then $T = \mathbf{rl}(T)$ by hypothesis, and $\mathbf{rl}(T) \subseteq^{ess} eR$ for some $e^2 = e \in R$ by Lemma 3.11. Hence R is a right CS-ring. If C is a cyclic right R-module, then C is torsionless (because $T = \mathbf{rl}(T)$ for each right ideal T) and C is finitely embedded because R is right artinian. Hence C embeds in a free module by [AF, Propositions 10.2 and 10.7]. Thus R is a right CF-ring.

(5)⇒(6). R is right finitely embedded by Theorem 5.6 because it is right Kasch (being a right CF-ring). Then the right CF-condition shows that every cyclic right R-module is finitely embedded. This implies that R is right artinian.

(6)⇒(7). We must show that $S_l \subseteq^{ess} R_R$. Since right CF-rings are left P-injective, R is a left Kasch ring by Theorem 1.3. Let $0 \neq a \in R$, and suppose that $\mathbf{l}(a) \subseteq L \subseteq^{max} {}_R R$. Then $\mathbf{r}(L) \subseteq \mathbf{rl}(a) = aR$, and we are done by Lemma 1.1 because (since R is left Kasch) $\mathbf{r}(L)$ is a simple right ideal of R.

(7)⇒(1). We have $\mathbf{rl}(T) = T$ for each right ideal T because R is a right CF-ring. Hence R is left P-injective, and so $S_l \subseteq S_r$ by Lemma 1.2. Thus $S_l = S_r$ because $S_l \subseteq^{ess} R_R$. Write $S = S_r = S_l$. Since R is semilocal, S_R is finitely generated by Lemma 3.5, whence R_R is finitely embedded. Thus every cyclic right R-module is finitely embedded (because R is a right CF-ring), so R is right artinian. Moreover, if $e^2 = e$ is local, $eS = soc(eR)$ is simple by Theorem 1.3, and so $soc(eR)$ is simple and essential in eR. Suppose that C is a cyclic right R-module. Then C embeds in R^n for some n. Since R is semiperfect, let $\sigma : C \to \oplus_{i=1}^m e_i R$ be monic where each e_i is a local idempotent. We may assume that m is minimal. Thus $\sigma(C) \cap e_i R \neq 0$ for each i and so contains $soc(e_i R)$. Hence $\oplus_{i=1}^m [\sigma(C) \cap e_i R] \subseteq^{ess} R_R$, and so $\sigma(C) \subseteq^{ess} R_R$. This proves that R is a right CEP-ring. □

Björk [B1] has an example of a left and right artinian, left CEP-ring which is not quasi-Frobenius. However, we have the following theorem.

Theorem 5.9. *The following are equivalent:*

(1) R *is quasi-Frobenius.*

(2) R *is a strongly right Johns, right C2-ring.*

(3) R *is strongly right Johns and* $S_r \subseteq S_l$.

(4) $M_2(R)$ *is right Johns and* $S_r \subseteq S_l$.

(5) R *is a right Johns, right mininjective ring.*

(6) R *is a semilocal, right mininjective right CF-ring.*

(7) R *is a right CS-ring and every 2-generated right R-module embeds in a free module.*

(8) *Every 2-generated right R-module essentially embeds in a projective module.*

Proof. (1)⇔(7)⇔(8). If either (7) or (8) is satisfied, then R is right artinian by (5) and (1) respectively of Theorem 5.8. Hence R is quasi-Frobenius by Corollary 4.14.

(1)⇒(2). This is obvious.

(2)⇒(3). R is right artinian by Theorem 5.8, and hence $S_r = S_l$ by Theorem 1.3.

(3)⇒(4). This is obvious.

(4)⇒(5). Since $M_2(R)$ is right Johns, it is not difficult to see that R is also right Johns. Moreover, $M_2(R)$ is left P-injective, so R is left 2-injective by Proposition 2.1. In particular R is left mininjective, so $S_l \subseteq S_r$ by Lemma 1.2 and $S_l = S_r$ by hypothesis. To show that R is right mininjective, let kR be a simple right ideal of R; by Lemma 1.1 we must show that $\mathbf{lr}(k) = Rk$. We have $\mathbf{l}(J) = S_r$ and $J = \mathbf{r}(S_r)$ by Lemma 5.7. Hence $Rk \subseteq \mathbf{lr}(k) \subseteq \mathbf{lr}(S_r) = S_r = S_l$, so it suffices to show that $Rk \subseteq^{ess} \mathbf{lr}(k)$. To this end, suppose that $Rt \cap Rk = 0$ where $t \in \mathbf{lr}(k)$. As R is left 2-injective, Lemma 3.4 gives $R = \mathbf{r}[Rt \cap Rk] = \mathbf{r}(t) + \mathbf{r}(k)$. But $\mathbf{r}(k) \subseteq \mathbf{r}(t)$ because $t \in \mathbf{lr}(k)$, so $R = \mathbf{r}(t)$. This means $t = 0$, as required.

(5)⇒(6). R is right minsymmetric by Lemma 1.2, and $S_r \subseteq^{ess} R_R$ by Lemma 5.7. Hence R is right artinian by Theorem 5.3. By Theorem 5.8 (4), R is a right CF-ring.

(6)⇒(1). R is left mininjective (it is a right CF-ring), so it suffices to show that R is right finitely embedded (then R is right artinian because it is a right CF-ring, and so is quasi-Frobenius by Theorem 1.4). We have $S_r = S_l (= S)$ by Lemma 1.2, and S_R is finitely generated by Lemma 3.5 because R is left mininjective. So it remains to show that $S \subseteq^{ess} R_R$. Let $0 \neq a \in R$, and suppose that $\mathbf{l}(a) \subseteq L \subseteq_R^{max} R$. Then $\mathbf{r}(L) \subseteq \mathbf{rl}(a) = aR$. Since R is left Kasch (by Lemma 3.3) and left mininjective, it follows from Lemma 1.1 that $\mathbf{r}(L)$ is a simple right ideal. Thus $aR \cap S \neq 0$. □

Acknowledgements. The work of the first author was supported by NSERC Grant A8075, and the work of the second author was supported by a Visiting Scholar Award from the University of Calgary and by the Ohio State University. The work was presented at the Third Korea-China-Japan International Ring Theory Symposium in June-July, 1999, and the authors wish to thank the organizers for their hospitality.

References

[AF] F. W. Anderson and K. R. Fuller, *Rings and Categories of Modules,* Second Edition. Springer-Verlag, Heidelberg, New York, 1991.

[ArP] P. Ara and J. K. Park, *On continuous semiprimary rings,* Comm. Algebra **19** (1991), 1945–1957.

[AP] E. P. Armendariz and J. K. Park, *Self-injective rings with restricted chain conditions,* Arch. Math. **58** (1992), 24–33.

[B1] J.-E. Björk, *Rings satisfying certain chain conditions,* J. Reine Angew. Math. **245** (1970), 63–73.

[B2] _____, *Radical properties of perfect modules*, J. Reine Angew. Math. **253** (1972), 78–86.

[CY] V. Camillo and M. F. Yousif, *Continuous rings with ACC on annihilators*, Canad. Math. Bull. **34** (1991), 462–464.

[CD] R. Camps and W. Dicks, *On semilocal rings*, Israel J. Math. **81** (1993), 203–211.

[CH] J. Clark and D. V. Huynh, *A note on perfect self-injective rings*, Quart. J. Math. Oxford **45** (1994), 13–17.

[C] R. R. Colby, *Rings which have flat injective modules*, J. Algebra **35** (1975), 239–252.

[DM] F. Dischinger and W. Müller, *Left PF is not right PF*, Comm. Algebra **14** (1986), 1223–1227.

[DHSW] N. V. Dung, D. V. Huynh, P. F. Smith and R. Wisbauer, *Extending Modules*, Longman, Harlow, 1994.

[F1] C. Faith, *Rings with ascending chain condition on annihilators*, Nagoya Math. J. **27** (1966), 179–191.

[F2] _____, *Embedding modules in projectives. A report on a problem*, Lecture Notes in Math. **951**. Springer-Verlag, Heidelberg, New York (1982), 21–40.

[F3] _____, *Finitely embedded commutative rings*, Proc. Amer. Math. Soc. **112** (1991), 657–659.

[FM1] C. Faith and P. Menal, *A counter example to a conjecture of Johns*, Proc. Amer. Math. Soc. **116** (1992), 21–26.

[FM2] _____, *The structure of Johns rings*, ibid. **120** (1994), 1071–1081.

[FW] C. Faith and E. A. Walker, *Direct-sum representations of injective modules*, J. Algebra **5** (1967), 203–221.

[GM] S. M. Ginn and P. B. Moss, *Finitely embedded modules over noetherian rings*, Bull. Amer. Math. Soc. **81** (1975), 709-710.

[GA] J. L. Gómez Pardo and P. A. Guil Asensio, *Essential embeddings of cyclic modules in projectives*, Trans. Amer. Math. Soc. **349** (1997), 4343–4353.

[GA1] _____, *Rings with finite essential socle*, Proc. Amer. Math. Soc. **125** (1997), 971–977.

[GA2] _____, *Torsionless modules and rings with finite essential socle*, Preprint.

[GY] J. L. Gómez Pardo and M. F. Yousif, *Semiperfect min-CS rings*, Glasgow Math. J. **41** (1999), 231–238.

[IN] M. Ikeda and T. Nakayama, *On some characteristic properties of quasi-Frobenius and regular rings*, Proc. Amer. Math. Soc. **5** (1954), 15–19.

[J] S. Jain, *Flat and FP-injectivity*, Proc. Amer. Math. Soc. **41** (1973), 437–442.

[Johns] B. Johns, *Annihilator conditions in noetherian rings*, J. Algebra **49** (1977), 222–224.

[Kasch] F. Kasch, *Modules and Rings*, London Math. Soc. Monograph **17**, Academic Press, New York, 1982.

[K] T. Kato, *Torsionless modules*, Tôhoku Math. J. **20** (1968), 234–243.

[MM] S. H. Mohamed and B. J. Müller, *Continuous and Discrete Modules*, Cambridge Univ. Press, Cambridge, 1990.

[M] P. Menal, *On the endomorphism ring of a free module*, Publ. Mat. **27** (1983), 141–154.

[Nak1] T. Nakayama, *On Frobeniusean algebras I, II*, Ann. Math. **40** (1939), 611–633, and ibid. **42** (1941), 1–21.

[Nak2] _____, *On Frobeniusean algebras III*, Japan J. Math. **18** (1942), 49–65.

[Nak3] _____, *Supplementary remarks on Frobeniusean algebras II*, Osaka Math. J. **2** (1950), 7–12.

[N] W. K. Nicholson, *Semiregular modules and rings*, Canad. J. Math. **28** (1976), 1105–1120.

[NY1] W. K. Nicholson and M. F. Yousif, *Principally injective rings*, J. Algebra **174** (1995), 77–93.

[NY2] _____, *Mininjective rings*, ibid. **187** (1997), 548–578.

[NY3] _____, *On perfect simple-injective rings*, Proc. Amer. Math. Soc. **125** (1997), 979–985.

[NY4] _____, *Annihilators and the CS-condition*, Glasgow Math J. **40** (1998), 213–222.

[NY5] _____, *On a theorem of Camillo*, Comm. Algebra **23** (1995), 5309–5314.

[NY6] _____, *FP-rings*, Preprint.

[NY7] _____, *Weakly continuous and C2-rings*, Preprint.

[NY8] _____, *On finitely embedded rings*, Preprint.

[O] B. Osofsky, *A generalization of quasi-Frobenius rings*, J. Algebra **4** (1966), 373–387.

[RS] J. Rada and M. Saorín, *On semiregular rings whose finitely generated modules embed in free modules*, Canad. Math. Bull. **40** (1997), 221–230.

[R] E. A. Rutter, *Two characterizations of quasi-Frobenius rings*, Pacific J. Math. **30** (1969), 777–784.

[R1] _____, *Rings with the principal extension property*, Comm. Algebra **3** (1975), 203–212.

[U] Y. Utumi, *On continuous and self injective rings*, Trans. Amer. Math. Soc. **118** (1965), 158–173.

[Y] M. F. Yousif, *On continuous rings*, J. Algebra **191** (1997), 495–509.

W. K. Nicholson
Department of Mathematics
University of Calgary
Calgary, Alberta T2N 1N4, Canada
e-mail: wknichol@ucalgary.ca

M. F. Yousif
Department of Mathematics
Ohio State University
Lima, OH 45804-3576, USA
e-mail: yousif.1@osu.edu

Theories of Harada in Artinian Rings and Applications to Classical Artinian Rings

Kiyoichi Oshiro

Abstract

In the early 1980s, Harada introduced extending and lifting properties for modules and, simultaneously, considered two new classes of artinian rings which contain QF-rings and Nakayama rings. The main purpose of this note is to survey his work and discuss its development and influence on the theory of rings and modules.

1. Introduction

Throughout our discussion, rings R are associative with identity and R-modules are unital. M_R (resp. $_RM$) is used to stress that M is a right (resp. left) R-module. For an R-module M, $J(M)$ and $S(M)$ denote its Jacobson radical and socle, respectively. For X, $Y \subseteq R$, we put $l_X(Y) = \{x \in X \mid xY = 0\}$ and $r_X(Y) = \{x \in X \mid Yx = 0\}$. Let A and B be submodules of the R-module M with $B \subseteq A$. If B is an essential submodule of A, we write $B \subseteq_e A$, and, dually, if A/B is small in M/B, we say B is a *co-essential submodule* of A (in M) and write $B \subseteq_c A$. M is called an *extending module* (or CS-module) if for any submodule X of M, there is a decomposition $M = X^* \oplus X^{**}$ with $X \subseteq_e X^*$. M is called a *lifting module* (or an LI-module) if for any submodule X of M, there is a decomposition $M = X^* \oplus X^{**}$ with $X^* \subseteq_c X$. In the early days of ring theory, the extending property explicitly appears in Utumi [48] and the lifting property appears in Bass [6] as:

Theorem 1.1. *A regular ring R is right continuous if and only if the right R-module R_R is extending.*

Theorem 1.2. *A ring R is semiperfect if and only if R_R is an LI-module.*

Generalizing the concept of continuous rings, continuous and quasi-continuous modules were introduced by Jeremy [25] in 1974. These generalizations lay mainly dormant until Harada's work on lifting and extending properties. Harada is well known for completing the final version of the Krull-Remak-Schmidt-Azumaya Theorem, which we now recall. (Here a module X is called *completely indecomposable* if $\text{End}(X)$ is a local ring).

Theorem 1.3. *The following conditions are equivalent for an R-module M with completely indecomposable decompositions $M = \oplus_I M_\alpha = \oplus_J N_\beta$:*

(1) For any subset $K \subseteq I$, there exists $L \subseteq J$ such that $M = (\oplus_K M_\alpha) \oplus (\oplus_L N_\beta)$.

(2) $\{M_\alpha\}_I$ satisfies condition LsTn: For any subfamily $\{M_{\alpha_i}\}_{i \in \mathbb{N}}$ with distinct α_i and any family of non-isomorphisms $f_i: M_{\alpha_i} \to M_{\alpha_{i+1}}$, for any $x \in M_{\alpha_1}$ we have $f_n \ldots f_2 f_1(x) = 0$ for some n (depending on x).

(3) M satisfies condition LSS: If $A = \oplus_T A_\alpha$ is a submodule of M such that $\oplus_F A_\alpha$ is a direct summand of M for any finite subset F of T, then A is a direct summand of M.

(4) M satisfies the (finite) exchange property.

Remark. More on Theorem 1.3 appears in Harada-Kanbara [20], Harada-Ishii [21], Yamagata [54] and Zimmermann-Huisgen and Zimmermann [56].

We note that the LsTn, LSS, and exchange properties in Theorem 1.3 play important roles in the study of extending and lifting properties. Harada first studied very local extending and lifting properties for modules with completely indecomposable decompositions. Specifically, if F is a family of submodules of the module M, then M is said to satisfy the extending (resp. lifting) property for F if, for any A in F there is a decomposition $M = A^* \oplus A^{**}$ with $A \subseteq_e A^*$ (resp. $A^* \subseteq_c A$). Harada's work (see [18]) first considered the extending property for simple submodules and the lifting property for maximal submodules of modules with completely indecomposable decompositions. Some of his work appears in Harada and Oshiro [22] and Oshiro [38, 39], where the extending property for uniform submodules and lifting property of hollow submodules are studied and generalized. At the same time, Müller and Rizvi became interested in our work and studied (quasi-)continuous modules and (quasi-)discrete modules from both their own viewpoint and using our methods. Their paper [37] (and Rizvi's thesis [46]) motivated much interest in the extending and lifting properties and in the last 20 years this field has developed at an incredible rate, as evidenced by the texts of Harada [18], Mohamed and Müller [33], and Dung, Huynh, Smith, and Wisbauer [9].

2. Fuller's Theorem

We frequently use the following well-known theorem due to Fuller [12].

Theorem 2.1. *Let e be a primitive idempotent of the right artinian ring R. Then eR_R is injective if and only if there exists a primitive idempotent f in R such that (eR, Rf) is an i-pair, i.e., $S(eR_R) \simeq fR/fJ$, $S(Rf) \simeq Re/Je$ where $J = J(R)$. In this case, $_R Rf$ is also injective.*

This theorem is very useful but its original proof was quite complicated. For our purposes, it was necessary to extend it as follows. The proof uses the concept of simple N-injectivity due to Harada [19].

Theorem 2.2. [4] *Let R be a semiprimary ring.*

(1) *For a primitive idempotent e in R, eR_R is injective if and only if there exists a primitive idempotent f in R such that (a) (eR, Rf) is an i-pair and (b) $\mathrm{r}_{Rf}\mathrm{l}_{eR}(X) = X$ for any $X_{fRf} \subseteq Rf_{fRf}$.*

(2) *For primitive idempotents e, f in R such that (eR, Rf) is an i-pair, the following are equivalent:*

    (a) $_{eRe}eR$ *is artinian;*
    (b) Rf_{fRf} *is artinian;*
    (c) eR_R *is Σ-injective;*
    (d) $_RRf$ *is Σ-injective;*
    (e) eR_R, $_RRf$ *are injective.*

More on Fuller's theorem appears in [3], [23], [35], [52], and [53].

3. Harada Rings

As a prelude to his work on extending and lifting properties, Harada considered two new types of artinian rings, which we now survey. An R-module M is called *small* if it is small in its injective hull; otherwise M is called *non-small*. Dually, M is called a *cosmall* module if, for any projective module P and any epimorphism $f\colon P \to M$, $\mathrm{Ker}\, f$ is an essential submodule of P; otherwise M is called *non-cosmall*.

In [15]–[17], Harada studied the following two conditions:

$(*)$ Every non-small right R-module contains a non-zero injective submodule.

$(*)^*$ Every non-cosmall right R-module contains a non-zero projective summand.

Harada characterized these conditions as follows, using the descending and ascending Loewy chains of M, $\{J_i(M)\}$ and $\{S_i(M)\}$ respectively.

Theorem 3.1. [15, 16] *A right artinian ring R satisfies $(*)$ if and only if, for any primitive idempotent e in R such that eR_R is non-small, there exists $t \geq 0$ for which (a) $eR/S_k(eR)$ is injective for $0 \leq k \leq t$ and (b) $eR/S_{t+1}(eR)$ is a small module.*

Theorem 3.2. [15, 17] *A semiperfect ring R satisfies $(*)^*$ if and only if, for a complete set $\{e_i\} \cup \{f_j\}$ of orthogonal primitive idempotents of R such that each $e_i R_R$ is non-small and each $f_j R_R$ is small,*

(a) each e_iR_R is injective,

(b) for each e_iR, there exists $t_i \geq 0$ such that $J_t(e_iR)$ is projective for $0 \leq t \leq t_i$ and $J_{t_i+1}(e_iR)$ is a singular module,

(c) for each f_jR, there exists e_i such that $f_jR \subsetneq e_iR$, where '$X \subsetneq Y$' means X is isomorphic to a submodule of Y.

Remarks. (1) When R is a left or right artinian ring, the condition "$J_{t_i+1}(e_iR)$ is singular" is superfluous [40].

(2) QF-rings satisfy the conditions in the theorems above and are also left-right symmetric. However, $(*), (*)^*$ are not left-right symmetric [40].

Conditions $(*), (*)^*$ can be characterized by more familiar ones:

Theorem 3.3. [40] *The following are equivalent for a ring R:*

(1) *R is a right artinian ring with $(*)$.*

(2) *Every injective right R-module is a lifting module.*

(3) *R is a right perfect ring and the family of all injective right R-modules is closed under taking small covers.*

(4) *Every right R-module can be expressed as a direct sum of an injective module and a small module.*

In this case R is a QF-3 ring, i.e., the injective hulls of R_R and $_RR$ are projective.

Theorem 3.4. [40] *The following are equivalent for a ring R:*

(1) *Every projective right R-module is an extending module.*

(2) *R satisfies $(*)^*$ and ACC on right annihilators.*

(3) *Essential extensions of projective right R-modules are projective.*

(4) *Every right R-module can be expressed as a direct sum of a projective module and a singular module.*

A ring satisfying the conditions of Theorem 3.3 (resp. 3.4) is called a right *H-ring* (resp. *co-H-ring*). In [43], we proved the unexpected result:

Theorem 3.5. *R is a left H-ring if and only if R is a right co-H-ring.*

Remark. To show that a right co-H-ring is a left H-ring, the right artinianness of R is used to show that R is left artinian. Our original complicated proof first showed that R is semiprimary and then left artinian. However, this proof was simplified using Theorem 2.2.

Condition (3) in Theorem 3.4 shows that the left H-ring property is Morita invariant. So, in discussing H-rings, we may restrict our attention to basic H-rings. We now look at the internal structure of H-rings.

Let R be a basic left H-ring with a complete set E of orthogonal primitive idempotents. Then E can be arranged as $E = \{e_{11}, e_{12}, \ldots, e_{1n(1)},$ $e_{21}, \ldots, e_{2n(2)}, \ldots, e_{m1}, \ldots, e_{mn(m)}\}$ for which
  (1) each $e_{i1}R_R$ is injective,
  (2) $J(e_{i,k-1}, R_R) \simeq e_{ik}R_R$ for all i,k,
  (3) $e_{ik}R_R \not\simeq e_{jt}R_R$ if $i \neq j$.

For example, if $m = 1$, $n(1) = 2$, with $e = e_{11}$, $f = e_{12}$, we can represent R as $R = \begin{pmatrix} (e,e) & (f,e) \\ (e,f) & (f,f) \end{pmatrix}$ where $(h,k) = \mathrm{Hom}_R(hR, kR)$. Let θ be an isomorphism $fR \to J(eR)$. Then the mapping

$$\tau = \begin{pmatrix} \tau_{11} & \tau_{12} \\ \tau_{21} & \tau_{22} \end{pmatrix} : \begin{pmatrix} (e,e) & (e,e) \\ (e,e)^* & (e,e) \end{pmatrix} \to \begin{pmatrix} (e,e) & (f,e) \\ (e,f) & (f,f) \end{pmatrix}$$

where $\tau((\alpha_{ij})) = \begin{pmatrix} \alpha_{11} & \alpha_{12}\theta \\ \theta^{-1}\alpha_{21} & \theta^{-1}\alpha_{22}\theta \end{pmatrix}$ is a ring epimorphism. Since eR_R is injective, (eR, Rg) is an i-pair for some g in $\{e, f\}$. If $g = f$, Ker $\tau = 0$ and, if $g = e$, Ker $\tau = \begin{pmatrix} 0 & S((e,e)) \\ 0 & S((e,e)) \end{pmatrix}$. Thus R is represented as

$$R = \begin{pmatrix} Q & Q \\ J & Q \end{pmatrix} \text{ or } \begin{pmatrix} Q & Q/S \\ J & Q/S \end{pmatrix} \text{ where } Q = (e,e),\ J = J(Q),\ S = S(Q).$$

In both cases, Q is a QF-ring. Thus R can be constructed from a QF-ring. If $m = 2$, $n(1) = 1$, $n(2) = 2$, R can be represented as one of

$$\begin{pmatrix} Q & Q & \overline{Q} \\ J & Q & Q \\ J & J & Q \end{pmatrix}, \begin{pmatrix} Q & \overline{Q} & \overline{Q} \\ J & Q & \overline{Q} \\ J & J & \overline{Q} \end{pmatrix}, \begin{pmatrix} Q & A & A \\ B & T & T \\ B & K & T \end{pmatrix}, \begin{pmatrix} Q & A & \overline{A} \\ B & T & T \\ B & K & T \end{pmatrix}$$

where $Q = (e,e)$, $T = (f_1, f_1)$, $A = (f_1, e)$, $B = (e, f_1)$, $J = J(Q)$, $K = J(T)$, $\overline{Q} = Q/S(Q)$, $\overline{A} = A/S(A)$. Moreover Q is a QF-ring in the first two cases and $\begin{pmatrix} Q & A \\ B & T \end{pmatrix}$ is a QF-ring in the second two cases, so, in each case, R can be constructed from a QF-ring.

Similarly, we can construct every right co-H-ring from QF-rings. To see this, let R be a basic indecomposable QF-ring. Let ρ be a Nakayama permutation on $\{e_1, \ldots, e_n\}$, a complete set of orthogonal primitive idempotents in R, i.e., $S(e_iR) \simeq e_{\rho(i)}R/J(e_{\rho(i)}R)$ for each i. Express R as

$$R = \begin{pmatrix} Q_1 & A_{12} & \cdots & A_{1n} \\ A_{21} & Q_2 & \cdots & A_{2n} \\ & & \cdots & \\ A_{n1} & \cdots & & Q_n \end{pmatrix} \text{ where } Q_i = e_iRe_i,\ A_{ij} = e_iRe_j.$$

Define the $k(i) \times k(i)$ matrix P_i and the $k(i) \times k(j)$ matrix P_{ij} by

$$P_i = \begin{pmatrix} Q_i & \cdots & Q_i \\ & \ddots & \vdots \\ J(Q_i) & & Q_i \end{pmatrix} \text{ and } P_{ij} = \begin{pmatrix} A_{ij} & \cdots & A_{ij} \\ & \cdots & \\ A_{ij} & \cdots & A_{ij} \end{pmatrix}.$$

Let $\overline{P_{i\rho(i)}}$, for $1 < i \neq \rho(i)$, and $\overline{P_i}$, for $1 < i = \rho(i)$, denote the factor rings of $P_{i\rho(i)}$ and P_i respectively by the ideals

$$\begin{pmatrix} \overset{i}{\vphantom{|}} \\ S(A_{i\rho(i)}) \\ 0 \end{pmatrix} \text{ and } \begin{pmatrix} \overset{i}{\vphantom{|}} \\ S(Q_i) \\ 0 \end{pmatrix} \text{ respectively.}$$

We put $P = \begin{pmatrix} P_1 & P_{12} & \cdots & P_{1n} \\ P_{21} & P_2 & \cdots & P_{2n} \\ & & \cdots & \\ P_{n1} & P_{n2} & \cdots & P_n \end{pmatrix}.$

Replacing P_i by $\overline{P_i}$ if $i = \rho(i)$ and $P_{i\rho(i)}$, by $\overline{P_{i\rho(i)}}$ if $i \neq \rho(i)$, gives the ring

$$\overline{P} = \begin{pmatrix} P_1 & P_{12} & \cdot & \cdots & \cdot & \overline{P_{1\rho(1)}} & \cdot & \cdot & P_{1n} \\ \cdot & \cdot & \cdot & & & \cdot\cdot & & & \cdot\cdot \\ P_{n1} & P_{n2} & \cdot\cdot & \overline{P_{n\rho(n)}} & \cdot & \cdot & \cdot & \cdots & P_{nn} \end{pmatrix}$$

which is canonically isomorphic to the factor ring of P by

$$\begin{pmatrix} 0 & \cdots & \cdot & & \cdot & \cdots & \cdot & S(P_{1\rho(1)}) & \cdot & \cdot & 0 \\ \cdot & \cdot & & & & & \cdot\cdot & & & & \\ 0 & \cdots & \cdot\cdot & S(P_{n\rho(n)}) & & \cdot & \cdot & & & \cdot\cdot & 0 \end{pmatrix}.$$

We call P a *J-extension ring* of R and $\overline{P}$ a *J/S-extension ring* of R.

Now we can state the following fundamental structure theorem.

Theorem 3.6. [43] *Every right co-H-ring can be expressed as a suitable J-extension ring or J/S-extension ring of a QF-ring; In particular right co-H-rings and hence left H-rings are left and right artinian rings.*

Next we shall state some applications of left H-rings.

A right R-module M is said to be *non-cosingular* (as the dual notion of non-singular) if $\{x \in M : xR \text{ is a small module}\} = 0$.

By conditions (4) in Theorems 3.3 and 3.4, we get:

Theorem 3.7. [40] *If R is a left H-ring, then every non-singular right R-module is projective and every non-cosingular left R-module is injective.*

Theorem 3.8. [40, 13] *If R a left non-singular ring, the following conditions are equivalent:*
   (1) *R is a left H-ring.*
   (2) *R is a right H-ring.*
   (3) *Every non-singular right R-module is projective.*
   (4) *R is Morita equivalent to a finite direct sum of upper triangular matrix rings over division rings.*

For more on left H-rings, see [5], [7], [8], [9], [26], [27], [47], [49], and [50].

4. Nakayama Rings

In this section, we discuss more exciting applications of H-rings.

A left and right artinian ring R is said to be a *Nakayama ring* if R is a right and left serial ring, that is, for every primitive idempotent e in R, the submodules of eR_R and submodules of $_RRe$ form a chain by inclusion. Now, we again look at the following equivalent dual conditions:
   (#) Every injective *left* R-module is a lifting module.
   (#)$^\#$ Every projective *right* R-module is an extending module.
Further, we shall consider the following conditions:
   (a) Every injective *left* R-module is a projective module.
   (a)* Every projective *right* R-module is an injective module.
   (b) Every extending *left* R-module is a lifting module.
   (b)* Every lifting *right* R-module is an extending module.
As stated in section 3, R is a left H-ring $\Leftrightarrow$ R is a right co-H-ring $\Leftrightarrow$ (#) $\Leftrightarrow$ (#)$^\#$. Also, R is a QF-ring $\Leftrightarrow$ (a) $\Leftrightarrow$ (a)*. We also have:

Theorem 4.1. [41] *R is Nakayama ring if and only if R satisfies (b), and if and only if R is a right perfect ring with (b)*.*

Thus, all the conditions (a), (a)*, (b), right perfect ring with (b)* are left-right symmetric, since QF-rings and Nakayama rings are so. However, interestingly, as noted above, (#) and (#)$^\#$ are not left-right symmetric.

As a corollary of this theorem, we can obtain the following.

Theorem 4.2. (see [9]) *For a given ring R, the following are equivalent:*
   (1) *R is a Nakayama ring with $J(R)^2 = 0$,*
   (2) *Every left R-module is lifting,*
   (3) *Every right R-module is extending.*

Hence the following are also equivalent:
(2)* Every right R-module is lifting.
(3)* Every left R-module is extending.

By Theorem 4.1, we clearly see that Nakayama rings are left and right H-rings. This produces the following analogue of Theorem 3.7:

Theorem 4.3. [43] *Every basic indecomposable Nakayama ring can be represented as a suitable J-extension ring or J/S-extension ring of a basic indecomposable QF Nakayama ring.*

Since QF right serial rings are Nakayama, Theorem 4.1 gives:

Theorem 4.4. *A ring R is a Nakayama ring if and only if R is a right serial right QF-3 ring.*

For the history of Nakayama rings, see Faith's text [11].

5. Skew Matrix Rings

We observed above that all Nakayama rings can be constructed from QF-Nakayama rings. Now we look at QF-Nakayama rings in more detail, first introducing Nakayama automorphisms and skew matrix rings.

Let R be a basic indecomposable QF-ring and $\{e_1, \cdots, e_n\}$ be a complete set of orthogonal primitive idempotents of R. Let $\rho : e_i \mapsto f_i$ be a Nakayama permutation of $\{e_1, \cdots, e_n\}$. Then a ring automorphism τ of R is said to be a *Nakayama automorphism* if $\tau(e_i) = f_i$ for $i = 1, \cdots, n$.

Now, let us state the definition of a skew matrix ring.

Let Q be a ring, $c \in Q$ and $\sigma \in \mathrm{End}(Q)$ be such that $\sigma(c) = c$ and $\sigma(q)c = cq$ for all $q \in Q$. Let $R = M_n(Q)$ be the set of $n \times n$ matrices over Q. We define a multiplication in R with respect to σ and c as follows:

For $(x_{ik}), (y_{ik})$ in R, we set $(z_{ik}) = (x_{ik})(y_{ik})$ where (z_{ik}) is defined by:

(1) If $i \leq k$, $z_{ik} = \sum_{j<i} x_{ij}\sigma(y_{ik})c + \sum_{i \leq j \leq k} x_{ij}y_{jk} + \sum_{k<j} x_{ij}y_{jk}c.$

(2) If $k < i$, $z_{ik} = \sum_{j \leq k} x_{ij}\sigma(y_{ik}) + \sum_{k<j<i} x_{ij}\sigma(y_{jk})c + \sum_{i \leq j} x_{ij}y_{jk}.$

Denoting the matrix whose (i,j)-entry is a but other entries are 0 by $\langle a \rangle_{ij}$ this gives $\langle a \rangle_{ij} \langle b \rangle_{jk} = \begin{cases} \langle a\sigma(b) \rangle_{ik} & (j \leq k < i) \\ \langle a\sigma(b)c \rangle_{ik} & (k < j < i \text{ or } j < i \leq k) \\ \langle ab \rangle_{ik} & (i = j \text{ or } k < i < j \text{ or } i < j \leq k) \\ \langle abc \rangle_{ik} & (i \leq k < j) \end{cases}.$

Then the multiplication is associative and, with addition as usual, R is a ring called *the skew matrix ring of degree n over Q*, and denoted by

$$R = \begin{pmatrix} Q & \cdots & Q \\ & \cdots & \\ Q & \cdots & Q \end{pmatrix}_{\sigma, c, n}.$$

For example, when $n = 2$, the multiplication is

$$\begin{pmatrix} x_1 & x_2 \\ x_3 & x_4 \end{pmatrix} \begin{pmatrix} y_1 & y_2 \\ y_3 & y_4 \end{pmatrix} = \begin{pmatrix} x_1 y_1 + x_2 y_3 c & x_1 y_2 + x_2 y_4 \\ x_3 \sigma(y_1) + x_4 y_3 & x_3 \sigma(y_2) c + x_4 y_4 \end{pmatrix}.$$

Theorem 5.1. [42] *The mapping $\tau : R \to R$ given by*

$$\begin{pmatrix} x_{11} & x_{12} & \cdots & x_{1n} \\ x_{21} & x_{22} & \cdots & x_{2n} \\ \cdots & \cdots & & \\ x_{n1} & x_{n2} & \cdots & x_{nn} \end{pmatrix} \mapsto \begin{pmatrix} x_{nn} & x_{n1} & \cdots & x_{n,n-1} \\ \sigma(x_{1n}) & \sigma(x_{11}) & \cdots & \sigma(x_{1,n-1}) \\ \cdots & \cdots & & \\ \sigma(x_{n-1,n}) & \sigma(x_{n-1,1}) & \cdots & \sigma(x_{n-1,n-1}) \end{pmatrix}$$

is a ring homomorphism. Thus $\tau \in \text{Aut}(R)$ if $\sigma \in \text{Aut}(Q)$. Furthermore, $\tau(e_n) = e_1, \tau(e_1) = e_2, \cdots, \tau(e_{n-1}) = e_n$.

Setting $e_i = \langle 1 \rangle_{ii}$ for each i, $E = \{e_1, \cdots, e_n\}$ is a set of orthogonal idempotents with $1 = e_1 + \cdots + e_n$, and, if Q is local, each e_i is primitive. Let W_i be the submodule of $e_i R$ of elements where the (i,i)-th entry belongs to Qc. Then, for $i > 1$, define homomorphisms $\phi_i : e_i R \to W_{i-1}$ by mapping the ith row $x_1 \cdots x_{i-1} x_i \cdots x_n$ to the $(i-1)$th row $x_1 \cdots x_{i-1} c x_i \cdots x_n$ and $\phi_1 : e_1 R \to W_n$ by mapping the first row $x_1 \cdots x_{i-1} x_i \cdots x_n$ to the last row $\sigma(x_1) \cdots \sigma(x_{n-1}) \sigma(x_n)$. The following theorems hold:

Theorem 5.2. *If Q is a local QF-ring, $\sigma \in \text{Aut}(Q)$ and $c \in J(Q)$, then:*
(1) R is a basic indecomposable QF-ring.
(2) The Nakayama permutation of E is $e_i \mapsto e_{i-1}$ (with $e_1 \mapsto e_n$).
(3) τ^{-1} of Theorem 5.1 is a Nakayama automorphism of $\{e_1, \ldots, e_n\}$.
(4) For any idempotent e in R, eRe is represented as a skew matrix ring with respect to σ, c and so is a basic indecomposable QF-ring.

Theorem 5.3. *If Q is a local Nakayama ring, $\sigma \in \text{Aut}(Q)$ and $c \in J(Q)$, then R is a basic indecomposable QF-Nakayama ring with a Nakayama automorphism.*

Theorem 5.4. [45, 24, 55] *If R is a basic indecomposable QF-ring, then eRe is a QF-ring for any idempotent $e \in R$ if and only if R can be represented as a skew matrix ring over a local QF-ring or the Nakayama permutation of R is trivial.*

6. QF-Nakayama Rings

For R-modules X_R and Y_R, we put $(X,Y) = \operatorname{Hom}_R(X,Y)$. Now, let R be a basic indecomposable Nakayama ring with a complete set $\{e_1, \ldots, e_n\}$ of orthogonal primitive idempotents. Put $J = J(R)$.

We may rearrange $\{e_1 R, \ldots, e_n R\}$ so that $e_i R$ is a projective cover of $e_{i-1}J$ for $i \geq 2$ and $e_1 R$ is a projective cover of $e_n J$. In this case, $e_n R, \ldots, e_1 R$ is called a *Kupisch series* for R and there are epimorphisms:

$$\theta_{i,i+1} : e_{i+1}R \to e_i J, \text{ for } i < n \text{ and } \theta_{n1} : e_1 R \to e_n J \text{ where } n \neq 1.$$

We put $\theta_{ij} = \theta_{i,i+1}\theta_{i+1,i+2}\cdots\theta_{j-1,j}$ for $i < j$ and $\theta_{ii} = $ the identity map of $e_i R$. Then θ_{ij} induces an isomorphism $\overline{\theta_{ij}} : e_j R/S_{j-i}(e_i R) \cong e_i J^{j-i}$.

If $\alpha \in (e_t R, e_j R/S_k(e_j R))$ and $\eta : e_j R \to e_j R/S_k(e_j R)$ is the natural map, $\alpha = \eta\beta$ for some $\beta \in (e_t R, e_j R)$. If $(e_t R, S_k(e_j R)) = 0$, β is unique and denoted by $[\alpha]$. The next two properties are important in calculations:

(A) If $i < j$, $\gamma \in (e_t R, e_i J^{j-i})$ and $(e_t R, S_{j-i}(e_j R)) = 0$, then $\theta_{ij}[\overline{\theta_{ij}}^{-1}\gamma] = \gamma$.

(B) If $(e_t R, S_{j-1}(e_j R)) = 0$ and $\theta_{ij}\alpha = 0$ for $\alpha \in (e_t R, e_j R)$, then $\alpha = 0$.

Now the following holds:

Theorem 6.1. *Let R be a basic indecomposable QF-Nakayama ring. If $e_n R, \ldots, e_1 R$ is a Kupisch series, then there exists s such that $e_i \mapsto e_j$ where $j \equiv i + s - 1 \pmod{n}$ is a Nakayama permutation.*

For this s, we consider the following 5 cases:

(I) $s = n$, (II) $s > n - s$, $s \neq n$, (III) $n = sq$,
(IV) $n = sq + r$, $0 < r < s$, (V) $s = 1$.

Case (I): $s = n$. Here R has the following condition: (KN): if $\{e_n R, e_{n-1}R, \ldots, e_1 R\}$ is a Kupisch series, then the map $e_i \mapsto e_{i-1}$ (with $e_1 \mapsto e_n$) is a Nakayama permutation.

For this case, we have the following:

Theorem 6.2. *If R is a basic indecomposable Nakayama ring with (KN), there exist a local Nakayama ring Q, $c \in J(Q)$, $\sigma \in \operatorname{Aut}(Q)$ satisfying*

$$\sigma(c) = c, \ \sigma(q)c = cq \text{ for all } q \in Q, \text{ and } J(Q) = cQ$$

and R can be represented as a skew matrix ring over Q with respect to σ, c.

Case (II): $s > n-s$, $s \neq n$. Put $l = n-s+1$, $w = l-1$, $t = n-w+1$, and represent R as $R = \begin{pmatrix} R_{11} & R_{12} \\ R_{21} & R_{22} \end{pmatrix}$ where

$$R_{11} = \begin{pmatrix} A_{11} & \cdots & A_{1w} \\ & \cdots & \\ A_{1w} & \cdots & A_{ww} \end{pmatrix}, \quad R_{12} = \begin{pmatrix} A_{1l} & \cdots & A_{1n} \\ & \cdots & \\ A_{wl} & \cdots & A_{wn} \end{pmatrix},$$

$$R_{21} = \begin{pmatrix} A_{l1} & \cdots & A_{lw} \\ & \cdots & \\ A_{t1} & & \\ \vdots & \ddots & \\ A_{n1} & \cdots & A_{nw} \end{pmatrix}, \quad R_{22} = \begin{pmatrix} A_{ll} & \cdots & A_{ln} \\ & \cdots & \\ A_{nl} & \cdots & A_{nn} \end{pmatrix}.$$

Now put $Z = \begin{pmatrix} A_{11} & A_{1l} \\ A_{t1} & A_{ll} \end{pmatrix}$ and define multiplication in Z by:

$$(x_{ij})(y_{ij}) = \begin{pmatrix} x_{11}y_{11} + x_{1l}\theta_{lt}x_{t1} & x_{11}y_{1l} + x_{1l}y_{ll} \\ x_{t1}y_{11} + [\overline{\theta_{lt}}^{-1}x_{ll}\theta_{l,l+1}]\theta_{l+1,t}y_{t1} & \theta_{lt}x_{t1}y_{1l} + x_{ll}y_{ll} \end{pmatrix}.$$

$[\overline{\theta_{lt}}^{-1}x_{ll}\theta_{l,l+1}]$ is well-defined since $(e_{l+1}R, S_{t-l}(e_tR)) = 0$. Using (A) and (B), we can verify that multiplication is associative and, with the usual addition, Z is a ring. Moreover, if f_1, f_2 are the usual matrix units e_{11}, e_{22} then $\{f_1, f_2\}$ is a set of orthogonal primitive idempotents and Z is a basic indecomposable QF-Nakayama ring with condition (KN). Thus Z is represented as a skew matrix ring over f_1Zf_1. Now we construct a J-extension of Z. For each i, j, define a matrix P_{ij} of the same size as R_{ij} by:

$$P_{11} = \begin{pmatrix} A_{11} & \cdots & A_{11} \\ & \ddots & \vdots \\ J(A_{11}) & & A_{11} \end{pmatrix}, \quad P_{12} = \begin{pmatrix} A_{1l} & \cdots & A_{1l} \\ & \cdots & \\ A_{1l} & \cdots & A_{1l} \end{pmatrix},$$

$$P_{21} = \begin{pmatrix} A_{t1} & \cdots & A_{t1} \\ A_{t1} & & \\ & \ddots & \\ A_{t1} & \cdots & A_{t1} \end{pmatrix}, \quad P_{22} = \begin{pmatrix} A_{ll} & \cdots & A_{ll} \\ & \ddots & \vdots \\ J(A_{ll}) & & A_{ll} \end{pmatrix}.$$

Then $P = \begin{pmatrix} P_{11} & P_{12} \\ P_{21} & P_{22} \end{pmatrix}$ is a J-extension ring of Z. Define a map $\tau =$

$(\tau_{ij}): P \to R$ by:

$$\begin{aligned}
\tau_{1j}(\alpha_{11}) &= \alpha_{11}\theta_{1j} & \text{for } 1 \leq j \leq w \\
\tau_{ij}(\alpha_{11}) &= \left[\overline{\theta_{1i}}^{-1}\alpha_{11}\right]\theta_{1j} & \text{for } 1 < i \leq j \leq w \text{ or } 1 \leq j < i \leq w \\
\tau_{ij}(\alpha_{1l}) &= \left[\overline{\theta_{1i}}^{-1}\alpha_{1l}\right]\theta_{lj} & \text{for } 1 \leq i \leq w < j \leq n \\
\tau_{ij}(\alpha_{t1}) &= \left[\overline{\theta_{1i}}^{-1}\alpha_{t1}\right]\theta_{1j} & \text{for } t \leq i, 1 \leq j \leq w \\
\tau_{lj}(\alpha_{ll}) &= \alpha_{ll}\theta_{lj} & \text{for } l \leq j \leq n \\
\tau_{ij}(\alpha_{ll}) &= \left[\overline{\theta_{li}}^{-1}\alpha_{ll}\theta_{l,l+1}\right]\theta_{l+1,j} & \text{for } l < i \leq j \\
\tau_{ij}(\alpha_{ll}) &= \left[\overline{\theta_{li}}^{-1}\alpha_{ll}\right]\theta_{lj} & \text{for } l \leq i < j.
\end{aligned}$$

Then τ is a well-defined ring epimorphism with kernel

$$\begin{pmatrix}
 & 0 & & & & & \\
0 & & \ddots & & & S_1 & \\
 & & & 0 & & & \\
\hline
 & & & & 0 & & \\
 & 0 & & & & \ddots & S_2 \\
 & & & & & & \ddots \\
 & & & & & & 0 \\
\hline
0 & S_3 & & & & & \\
 & \ddots & & & & & 0 \\
0 & 0 & & & & &
\end{pmatrix}.$$

Thus R can be represented as a J/S-extension of Z.

Case (III): $n = sq$, $n \neq s$. Here we consider the partition:

$$\{1, 2, \ldots, s\} \cup \{s+1, \ldots, 2s\} \cup \ldots \cup \{(q-1)s+1, \ldots, n\}.$$

Set $t_1 = 1$, $s_1 = s$, $t_2 = s+1$, $s_2 = 2s$, \ldots, $t_q = (q-1)s+1$, $s_q = n$ and the subring

$$Z = \begin{pmatrix}
A_{t_1 t_1} & A_{t_1 t_2} & \cdots & A_{t_1 t_q} \\
A_{t_2 t_1} & A_{t_2 t_2} & \cdots & A_{t_2 t_q} \\
& & \cdots & \\
A_{t_q t_1} & A_{t_q t_2} & \cdots & A_{t_q t_q}
\end{pmatrix}.$$

Setting $f_i = \langle 1 \rangle_{ii}$ for each i, $\{f_1, \ldots, f_n\}$ is a complete set of orthogonal primitive idempotents of Z and Z is a basic indecomposable QF-Nakayama ring with (KN). For $1 \leq i, j \leq q$, define the $s \times s$ matrices Q_{ij} by:

$$Q_{ii} = \begin{pmatrix}
A_{t_i t_i} & \cdots & A_{t_i t_i} \\
& \ddots & \vdots \\
J(A_{t_i t_i}) & \cdots & A_{t_i t_i}
\end{pmatrix} \text{ and } Q_{ij} = \begin{pmatrix}
A_{t_i t_j} & \cdots & A_{t_i t_j} \\
& \cdots & \\
A_{t_i t_j} & \cdots & A_{t_i t_j}
\end{pmatrix} \quad (i \neq j).$$

Then $P = \begin{pmatrix} Q_{11} & \cdots & Q_{1q} \\ & \cdots & \\ Q_{q1} & \cdots & Q_{qq} \end{pmatrix}$ is a J-extension ring of Z and R is isomorphic to the factor ring of P by the ideal

$$X = \begin{pmatrix} 0 & \cdots\cdots & 0 & X_{1q} \\ X_{21} & 0 & \cdots\cdots & 0 \\ 0 & X_{32} & 0 & \cdots & 0 \\ \vdots & & \ddots & \ddots & \vdots \\ 0 & \cdots & 0 & X_{q,q-1} & 0 \end{pmatrix} \text{ where}$$

$$X_{1q} = \begin{pmatrix} 0 & S(A_{t_1 t_q}) \\ & \ddots & \\ 0 & & 0 \end{pmatrix} \text{ and } X_{i,i-1} = \begin{pmatrix} 0 & S(A_{t_i t_{i-1}}) \\ & \ddots & \\ 0 & & \end{pmatrix}, i > 1.$$

Case (IV): $n = sq + r$, $0 < r < s$. In this case, we consider the partition:

$$\{1, \ldots, s\} \cup \{s+1, \ldots, 2s\} \cup \ldots \cup \{(q-2)s+1, \ldots, (q-1)s\} \cup$$
$$\{(q-1)s+1, \ldots, (q-1)s+r\} \cup \{(q-1)s+r+1, \ldots, n\}.$$

Put $t_1 = 1$, $s_1 = s$, $t_2 = s+1$, $s_2 = 2s, \ldots, t_{q-2} = (q-3)s+1$, $s_{q-2} = (q-2)s$, $t_{q-1} = (q-2)s+1$, $s_{q-1} = (q-2)s+r$, $t_q = (q-2)s+r+1$, $s_q = n$. Then R is the $(q+1) \times (q+1)$ matrix (R_{ij}) where

$$R_{ij} = \begin{pmatrix} A_{t_i t_j} & A_{t_i t_{j+1}} & \cdots & A_{t_i s_j} \\ A_{t_i+1, t_j} & A_{t_i+1, t_j+1} & \cdots & A_{t_i+1, s_j} \\ & & \cdots & \\ A_{s_i t_j} & A_{s_i, t_j+1} & \cdots & A_{s_i s_j} \end{pmatrix}.$$

In particular $\begin{pmatrix} R_{q,q-1} & R_{q,q} & R_{q,q+1} \\ R_{q+1,q-1} & R_{q+1,q} & R_{q+1,q+1} \end{pmatrix}$ is of the form:

$$\begin{pmatrix} A_{t_q, t_{q-1}} & A_{t_q, t_q} & A_{t_q, t_{q+1}} & A_{t_q, t_{q+1}} \\ & & A_{t_{q+1}, t_{q+1}} & \\ & A_{x, t_q} & & \\ A_{t_{q+1}, t_{q-1}} & & A_{t_{q+1}, t_{q+1}} & A_{t_{q+1}, t_{q+1}} \end{pmatrix} \text{ where } x = n - r + 1.$$

Let $Z = \begin{pmatrix} A_{t_1 t_1} & A_{t_1 t_2} & \cdots & A_{t_1 t_{q-1}} & A_{t_1 t_q} & A_{t_1 t_{q+1}} \\ A_{t_2 t_1} & A_{t_2 t_2} & \cdots & A_{t_2 t_{q-1}} & A_{t_2 t_q} & A_{t_2 t_{q+1}} \\ & & \cdots & & & \\ A_{t_q t_1} & A_{t_q t_2} & \cdots & A_{t_q t_{q-1}} & A_{t_q t_q} & A_{t_q t_{q+1}} \\ A_{t_{q+1} t_1} & A_{t_{q+1} t_2} & \cdots & A_{t_{q+1} t_{q-1}} & A_{x t_q} & A_{t_{q+1} t_{q+1}} \end{pmatrix}$ and define

multiplication in Z by the following relations:

a) $\langle\alpha\rangle_{t_it_j}\langle\beta\rangle_{t_jt_k} = \langle\alpha\beta\rangle_{t_it_j}$ $\quad\quad ((t_i,t_j,t_k)\neq(t_{q+1},t_1,t_q))$

b) $\langle\alpha\rangle_{t_{q+1}t_1}\langle\beta\rangle_{t_1t_q} = \langle[\overline{\theta_{t_{q+1}x}}^{-1}\alpha]\beta\rangle_{xt_q}$

c) $\langle\alpha\rangle_{xt_q}\langle\beta\rangle_{t_qt_i} = \begin{cases}\langle\theta_{t_{q+1}x}\alpha\beta\rangle_{t_{q+1}t_i} & (i\neq q)\\ \langle\alpha\beta\rangle_{xt_q} & (i=q)\end{cases}$

d) $\langle\alpha\rangle_{t_it_{q+1}}\langle\beta\rangle_{xt_q} = \begin{cases}\langle\alpha\theta_{t_{q+1}x}\beta\rangle_{t_it_q} & (i\neq q+1)\\ \langle[\overline{\theta_{t_{q+1}x}}^{-1}\alpha]\beta\rangle_{xt_q} & (i=q+1)\end{cases}$

where $\langle\ \rangle_{t_it_j} = \langle\ \rangle_{ij}$ and $\langle\ \rangle_{xt_q} = \langle\ \rangle_{q+1,q}$ in Z. Then, with addition as usual, Z becomes a ring and $\{f_i = \langle 1\rangle_{t_it_j}: i=1,\ldots,q+1\}$ is a complete set of orthogonal primitive idempotents. Moreover, Z is a basic indecomposable QF Nakayama ring with Kupisch series $\{f_{q+1}P, f_qP,\ldots,f_1P\}$ and $f_j \mapsto f_{j-1}$ (with $f_1 \mapsto f_{q+1}$) is a Nakayama permutation of $\{f_1,\ldots,f_{q+1}\}$, i.e., Z is a basic indecomposable QF-ring satisfying (KN). For each i, j, we define a matrix Q_{ij} of the same size as R_{ij} as follows:

$$Q_{ii} = \begin{pmatrix} A_{t_it_i} & \cdots & A_{t_it_i} \\ & & \vdots \\ J(A_{t_it_i}) & \cdots & A_{t_it_i} \end{pmatrix}, Q_{q+1,q} = \begin{pmatrix} A_{xt_q} & \cdots & A_{xt_q} \\ & \cdots & \\ A_{xt_q} & \cdots & A_{xt_q} \end{pmatrix}$$

and $Q_{ij} = \begin{pmatrix} A_{t_it_j} & \cdots & A_{t_it_j} \\ & \cdots & \\ A_{t_it_j} & \cdots & A_{t_it_j} \end{pmatrix}$. Then $P = \begin{pmatrix} Q_{11} & \cdots & Q_{1,q+1} \\ & \cdots & \\ Q_{q+1,1} & \cdots & Q_{q+1,q+1} \end{pmatrix}$

is a J-extension ring of Z and, as in Case (II), there is a ring epimorphism from P to R with kernel

$$\begin{pmatrix} 0 & \cdots\cdots & 0 & X_{1,q+1} \\ X_{21} & 0 & \cdots\cdots & 0 \\ 0 & X_{32} & 0 & \cdots & 0 \\ \vdots & & \ddots & \ddots & \vdots \\ 0 & \cdots & 0 & X_{q+1,q} & 0 \end{pmatrix}$$ where the $X_{ij} \subseteq Q_{ij}$ are given by

$$X_{1,q+1} = \begin{pmatrix} 0 & S(A_{t_1t_{q+1}}) \\ & \ddots \\ 0 & 0 \end{pmatrix} \text{ and } X_{i,i-1} = \begin{pmatrix} 0 & S(A_{t_it_{i-1}}) \\ & \ddots \\ & 0 \end{pmatrix} \text{ for } i \geq 2.$$

Case (V): $s = 1$. $R/S(R)$ is a basic indecomposable QF-Nakayama ring with (KN) so representable as a skew matrix ring over a local Nakayama ring. By the discussion above, the following structure theorem holds:

Theorem 6.3. *Let R be a basic indecomposable QF-Nakayama ring.*

(1) *If the Nakayama permutation of R is non-trivial, R can be represented as a skew matrix ring over a local Nakayama ring or as a J/S-extension ring of a skew matrix ring over a local Nakayama ring. More precisely, there exist Nakayama rings $Z_1, Z_2, \ldots, Z_n$ and ring epimorphisms $\phi_i \colon Z_i \to Z_{i+1}$ where Z_1 is a skew matrix ring over a local Nakayama ring, $Z_n = R$ and $\operatorname{Ker} \phi_i$ is a simple ideal of Z_i for $i = 1, \ldots, n-1$.*

(2) *If the Nakayama permutation of R is the identity, then $R/S(R)$ can be represented as a skew matrix ring over a local Nakayama ring.*

7. Self-duality of Left H-rings

For rings R and S, let $_RM$ and M_S be the categories of all finitely generated left R-modules and right S-modules, respectively. If there exist contravariant functors $F \colon {_RM} \to M_S$ and $G \colon M_S \to {_RM}$ such that GF and FG are isomorphic to the identity functors of $_RM$ and M_S, respectively, then (F, G) is called a *Morita duality* between $_RM$ and M_S and M_S (or $_RM$) is said to be *Morita dual* to $_RM$ (or M_S). We use $_RM \sim M_S$ to indicate that such a duality exists. When $_RM \sim M_R$, R is said to be *self-dual* or to *have self-duality*. Since left H-rings are closely related to QF-rings and Nakayama rings and they are precisely the right co-H-rings, it is natural to ask if left H-rings are self-dual. In this section, we discuss this problem. The results appear in [28]. For later use, we generalize the concept of "Nakayama automorphism" to "Nakayama isomorphism" for a basic artinian ring with a finitely generated injective cogenerator. Let R be a basic indecomposable left artinian ring, E the injective hull of $R/J(R)$ as a left R-module and put $S = \operatorname{End}(E)$. Then it is well-known as Morita-Azumaya's Theorem that $_RM \sim M_S$ if and only if $_RE$ is a finitely generated module, and in this case, S is right artinian. So, if E is finitely generated and $R \simeq S$, then R is self-dual. This criterion is fundamental to the study of Morita duality. Let R be a basic indecomposable artinian ring such that the injective hull $G = E(_RR/J(R))$ is finitely generated. Let $G_i = E(Re_i/J(Re_i))$ where $\{e_1, \ldots, e_n\}$ is a complete set of orthogonal primitive idempotents. Since $G = \oplus_{i=1}^n G_i$, we can write $\operatorname{End}(G)$ as the matrix ring $T = ([G_i, G_j])$ where $[G_i, G_j] = \operatorname{Hom}_R(G_i, G_j)$. If f_i is the matrix with the identity on G_i as the (i,i) entry and other entries zero, then $\{f_1, \ldots, f_n\}$ is a complete set of orthogonal primitive idempotents of T. Then a ring isomorphism $\phi \colon R \to T$ for which $\phi(e_i) = f_i$ for each i is called a *Nakayama isomorphism* of R. Of course, if R is a basic QF-ring, ϕ is just a Nakayama automorphism. Now, let $E = \{e_{11}, \ldots, e_{1n(1)}, \ldots, e_{m1}, \ldots, e_{mn(m)}\}$ be a complete set of orthogonal primitive idempotents of the basic indecomposable left H-ring R such that

(1) each $e_{i1}R_R$ is injective,

(2) $J(e_{i,k-1}R_R)_R \simeq e_{ik}R_R$ for $1 \leq i \leq m$, $1 \leq k \leq n(i)$.

For each e_{i1}, there is a unique i-pair $(e_{i1}R, Rg_i)$. Then Rg_i is injective and

(3) $S_k(_RRg_i) = S(e_{i1}R) + \cdots + S(e_{ik}R_R)$

for $1 \leq i \leq m$, $1 \leq k \leq n(i)$; so $S_k(_RRg_i)$ is a two sided ideal. Moreover

(4) $Rg_i/S_{k-1}(_RRg_i) \simeq E(Re_{ik}/J(Re_{ik}))$ for $1 \leq i \leq m$, $1 \leq k \leq n(i)$.

Put $g_{ik} = g_i + S_{k-1}(_RRg_i) \in Rg_i/S_{k-1}(_RRg_i)$ and

$$G = Rg_{11} \oplus \cdots \oplus Rg_{1n(1)} \oplus \cdots \oplus Rg_{m1} \oplus \cdots \oplus Rg_{mn(m)}.$$

Then G is finitely generated and isomorphic to $E(R/J(R))$. Letting $T = \text{End}(G)$, Morita-Azumaya's Theorem says $_RM \sim M_T$. Thus self-duality of R is equivalent to $R \simeq T$. Represent T as

$$T = \begin{pmatrix} [g_{11}, g_{11}] & \cdots & [g_{11}, g_{mn(m)}] \\ & \cdots & \\ [g_{mn(m)}, g_{11}] & \cdots & [g_{mn(m)}, g_{mn(m)}] \end{pmatrix}$$

where $[g_{ij}, g_{kh}] = \text{Hom}_R(Rg_{ij}, Rg_{kh})$. Let h_{ij} be the matrix such that the (ij, ij) position is unity of $[g_{ij}, g_{ij}]$ and all other entries are zero. Then $\{h_{11}, \ldots, h_{1n(1)}, \ldots, h_{m1}, \ldots, h_{mn(m)}\}$ is a complete set of orthogonal primitive idempotents and

$$T = h_{11}T \oplus \cdots \oplus h_{1n(1)}T \oplus \cdots \oplus h_{m1}T \oplus \cdots \oplus h_{mn(m)}T.$$

Now, R and T are in fact also similar as the following shows:

(5) $(e_{i1}R, Re_{kt})$ is an i-pair if and only if $(h_{i1}T, Th_{kt})$ is an i-pair.

For example, if $m = 2, n(1) = 2, n(2) = 1$ and $(e_{11}R, Re_{21}), (e_{21}R, Re_{11})$ are i-pairs, then $R = \begin{pmatrix} Q & Q & A \\ J & Q & A \\ B & B & P \end{pmatrix}$ and $T = \begin{pmatrix} P & P & B \\ K & P & B \\ A & A & Q \end{pmatrix}$ where $J = J(Q)$, $A = (e_{31}, e_{11})$, $B = (e_{11}, e_{31})$, $K = J(P)$. Then $R = \begin{pmatrix} Q & A \\ B & T \end{pmatrix}$ is a QF-Nakayama ring with (KN) so if $R \simeq T$, there is an automorphism ϕ of the ring $\begin{pmatrix} Q & A \\ B & P \end{pmatrix}$ with $\phi\left(\begin{pmatrix} 1 & 0 \\ 0 & 1 \end{pmatrix}\right) = \begin{pmatrix} 0 & 0 \\ 0 & 1 \end{pmatrix}$ and

$\phi\left(\begin{pmatrix} 0 & 0 \\ 0 & 1 \end{pmatrix}\right) = \begin{pmatrix} 1 & 0 \\ 0 & 0 \end{pmatrix}$ which is simply a Nakayama automorphism of the QF-ring Z. There is thus a close relation between self-duality of left H-rings and existence of Nakayama automorphisms of basic QF-rings. For discussing this situation, we need the following theorem:

Theorem 7.1. *Let S be a simple ideal of the basic indecomposable left H-ring R. If R has a Nakayama isomorphism, so too has R/S.*

Now the following fundamental theorem holds:

Theorem 7.2. *For a basic indecomposable QF-ring R, the following are equivalent:*
  (1) R has a Nakayama automorphism.
  (2) Every J-extension ring of R has a Nakayama isomorphism.
  (3) Every J-extension ring of R is self-dual.
When these conditions hold, by Theorems 5.3 and 6.1, every J/S-extension ring has a Nakayama isomorphism and so is self-dual.

From the theorem we can deduce the well-known fact that a finite dimensional QF-algebra over a field has a Nakayama automorphism, since it is self-dual. We also see that the following problems are equivalent:

(A) Is every left H-ring (resp. finite dimensional H-algebra over a field; resp. Nakayama ring) self-dual?

(B) Does every basic QF-ring (resp. finite dimensional QF-algebra; resp. Nakayama ring) have a Nakayama autmorphism?

Although the author considered (A) at length for left H-rings, Koike [29] has recently noted that the QF-ring in Kraemer [30, Theorem 6.4] has no Nakayama automorphism and so a left H-ring is not self-dual in general.

In the 1980s, Waschbusch [51], Mano [32], Haack [14], Dischinger-Müller [10], and others studied (A) for Nakayama rings. However, as noted in [51], Amdal and Ringdal [1] had earlier settled it affirmatively. Next we confirm self-duality of Nakayama rings by showing that every basic indecomposable QF-Nakayama ring has a Nakayama automorphism.

Let R be a basic indecomposable QF-Nakayama ring and E be a complete set of orthogonal primitive idempotents of R. If the Nakayama permutation on E is trivial, then the identity map is a Nakayama automorphism. If the Nakayama permutation is non-trivial and R satisfies (KN), then, as in Section 6, R can be represented as a skew matrix ring over a local Nakayama ring and so has a Nakayama automorphism by Theorem 5.3. If R does not satisfies (KN) it can be represented as a J/S-extension of a basic indecomposable QF-Nakayama ring Z with (KN). Since Z has a Nakayama automorphism, so too has R by Theorems 7.1 and 7.2.

We note that although Haack [14] was unable to solve self-duality of Nakayama rings, his Proposition 3.2 states that every basic QF-Nakayama ring has a Nakayama automorphism.

8. Kupisch's Work

In [31], Kupisch studied the structure of Nakayama rings and had already considered the skew matrix ring concept. However, his approach is somewhat different from ours and we discuss the interconnection. Let R be a basic indecomposable Nakayama ring with complete set of orthogonal primitive idempotents $E = \{e_1, \ldots, e_n\}$. Let d_i denote the composition length of $e_i R_R$. Then, Kupisch's main result is as follows:

Theorem 8.1. *If $d_i \not\equiv 1 (mod\, n$ for all i, then R can be represented as a factor ring of a skew matrix ring over a local Nakayama ring. Thus any basic indecomposable QF-Nakayama ring with non-trivial Nakayama permutation can be represented as such a factor ring.*

Indecomposable basic QF-Nakayama rings with non-trivial Nakayama permutations are covered in Cases (I)–(IV) of Section 5. By Kupisch's Theorem, these must each be a factor ring of a skew matrix ring over a local Nakayama ring. In (I) the ring is isomorphic to a skew matrix ring over a local Nakayama ring. Otherwise, the rings can be expressed as a J/S-extension ring of a QF-Nakayama ring with (KN). Thus we have:

Theorem 8.2. *For a basic indecomposable QF-Nakayama ring Z with (KN), every J/S-extension ring of Z can be expressed as a factor ring of a skew matrix ring over a local Nakayama ring.*

The theorem is easily proved. For example, if $Z = eZ \oplus fZ$ is a basic indecomposable QF-ring where e, f are orthogonal primitive idempotents, let $Q = eZe$, $J = J(Q)$, $T = fZf$, $A = eZf$ and $B = eZf$, so that $Z = \begin{pmatrix} Q & A \\ B & T \end{pmatrix}$. Since Z satisfies (KN), by Theorem 8.2 it can be expressed as a skew matrix ring over a local Nakayama ring. We now sketch a proof of Theorem 6.2 for this case. First, T can be exchanged with Q:

$$Z = \begin{pmatrix} Q & A \\ B & Q \end{pmatrix}$$

and there exist $\sigma \in \text{Aut}(Q)$, $c \in J$, $\alpha \in A$ and $\beta \in B$ satisfying

$$J = cQ, A = \alpha Q = Q\alpha, B = \beta Q = Q\beta, \alpha\beta = \beta\alpha = c, \sigma(c) = c$$

Theories of Harada in Artinian Rings

and $\sigma(q)c = cq, \alpha q = q\alpha, \beta q = \sigma(q)\beta$ for all $q \in Q$. Then under the map $\begin{pmatrix} x & y \\ z & w \end{pmatrix} \to \begin{pmatrix} x & y\alpha \\ z\beta & w \end{pmatrix}$, Z is isomorphic to $\begin{pmatrix} Q & Q \\ Q & Q \end{pmatrix}_{\sigma,c}$. Here, consider a J/S-extension ring P of Z:

$$\left(\begin{array}{ccc|ccc} Q & \cdots & \cdots & Q & A & \cdots & \cdots & A \\ Qc & \ddots & & \vdots & & \cdots & \cdots & \\ \vdots & \ddots & \ddots & \vdots & & \cdots & \cdots & \\ Qc & \cdots & Qc & Q & A & \cdots & \cdots & A \\ \hline B & \cdots & \cdots & B & Q & \cdots & \cdots & Q \\ & \cdots & \cdots & & Qc & \ddots & & \vdots \\ & \cdots & \cdots & & \vdots & \ddots & \ddots & \vdots \\ B & \cdots & \cdots & B & Qc & \cdots & Qc & Q \end{array} \right).$$

It easily follows that P is isomorphic to a factor of the skew matrix ring

$$\begin{pmatrix} Q & \cdots & Q \\ & \cdots & \\ Q & \cdots & Q \end{pmatrix}_{\sigma,c}$$ by the canonical mapping:

$$\begin{pmatrix} x_{11} & \cdots & & x_{1k} & x_{1l} & \cdots & & x_{1n} \\ x_{21} & \ddots & & \vdots & & \cdots & \cdots & \\ \vdots & \ddots & \ddots & \vdots & & \cdots & \cdots & \\ x_{k1} & \cdots & x_{k,k-1} & x_{kk} & x_{kl} & \cdots & & x_{kn} \\ x_{l1} & \cdots & & x_{lk} & x_{ll} & \cdots & & x_{ln} \\ & \cdots & & & x_{l+1,l} & \ddots & & \vdots \\ & \cdots & & & \vdots & \ddots & \ddots & \vdots \\ x_{n1} & \cdots & & x_{nk} & x_{nl} & \cdots & x_{n,n-1} & x_{nn} \end{pmatrix}$$

$$\mapsto \begin{pmatrix} x_{11} & \cdots & & x_{1k} & x_{1l} & \cdots & & x_{1n} \\ x_{21}c & \ddots & & \vdots & & \cdots & \cdots & \\ \vdots & \ddots & \ddots & \vdots & & \cdots & \cdots & \\ x_{k1}c & \cdots & x_{k,k-1}c & x_{kk} & x_{kl} & \cdots & & x_{kn} \\ x_{l1} & \cdots & & x_{lk} & x_{ll} & \cdots & & x_{ln} \\ & \cdots & & & x_{l+1,l}c & \ddots & & \vdots \\ & \cdots & & & \vdots & \ddots & \ddots & \vdots \\ x_{n1} & \cdots & & x_{nk} & x_{nl}c & \cdots & x_{n,n-1}c & x_{nn} \end{pmatrix}.$$

References

[1] K. Amdal and F. Ringdal, *Categories uniserieles*, C. R. Acad. Sci. Paris, Serie A **267** (1968), 85-86 and 247–249.

[2] G. Azumaya, *A duality theory for injective modules*, Amer. J. Math. **81** (1959), 249–278.

[3] Y. Baba, *Injectivity of quasi-projective modules, projectivity of quasi-injective modules, and projective covers of injective modules*, J. Algebra **155** (1993), 415–434.

[4] Y. Baba and K. Oshiro, *On a theorem of Fuller*, J. Algebra **154** (1993), 86–94.

[5] Y. Baba and K. Iwase, *On quasi-Harada rings*, J. Algebra **185** (1996), 544–570.

[6] H. Bass, *Finitistic dimension and a homological generarization of semiprimary rings*, Trans. Amer. Math. Soc. **95** (1960), 466–488.

[7] J. Clark and R. Wisbauer, Σ-*extending modules*, J. Pure and Applied Algebra **104** (1995), 19–32.

[8] Phan Dan, *Right perfect rings with the extending property on finitely generated free modules*, Osaka J. Math. **26** (1989), 265–273.

[9] N. V. Dung, D. V. Huynh, P. F. Smith and R. Wisbauer, *Extending Modules*, Pitman, Harlow, 1994.

[10] F. Dischinger and W. Müller, *Einreihig zerlegbare artinsche Ringe sind selbstdual*, Arch. Math. **48** (1984), 132–136.

[11] C. Faith, *Algebra II: Ring Theory*, Springer-Verlag, Heidelberg-New York, 1976.

[12] K. R. Fuller, *On indecomposable injectives over artinian rings*, Pacific J. Math. **29** (1969), 115–135.

[13] K. R. Goodearl, *Ring Theory, Nonsingular Rings and Modules*, Pure Appl. Math. Series **33**, Marcel Dekker, New York, 1976.

[14] J. K. Haack, *Self-duality and serial rings*, J. Algebra **59** (1979), 345–363.

[15] M. Harada, *Nonsmall modules and noncosmall modules,* Ring Theory, Proc. 1987 Antwerp Conf., Marcel Dekker, (1979), 669–690.

[16] _____, *On one sided QF-2 rings. I,* Osaka J. Math. **17** (1980), 421–431.

[17] _____, *On one sided QF-2 rings. II,* ibid., 433–438.

[18] _____, *Factor categories with applications to direct decomposition of modules,* Lecture Notes Pure Appl. Math. **88**, Marcel Dekker, New York, 1983.

[19] _____, *On almost relative injectives of finite length,* Preprint.

[20] M. Harada and H. Kanbara, *On categories of projective modules,* Osaka J. Math. **8** (1971), 471–483.

[21] M. Harada and T. Ishii, *On perfect rings and the exchange property,* Osaka J. Math. **12** (1975), 483–491.

[22] M. Harada and K. Oshiro, *On extending property on direct sums of uniform modules,* Osaka J. Math. **18** (1981), 767–785.

[23] M. Hoshino and T. Sumioka, *Injective pairs in perfect rings,* Osaka J. Math. **35** (1998), 501–508.

[24] M. Hoshino, *On strongly QF-rings,* Communications in Algebra **28** (2000), 3585–3600.

[25] L. Jeremy, *Sur les modules et anneaux quasi-continus,* Canad. Math. Bull. **17** (1974), 217–228.

[26] D. V. Huynh and Phan Dan, *Some characterizations of right co-H-rings,* Math. J. Okayama Univ. **34** (1992), 165–174.

[27] J. Kado, *The maximal quotient rings of left H-rings,* Osaka J. Math. **27** (1990), 247–252.

[28] J. Kado and K. Oshiro, *Self-duality and Harada rings,* J. Algebra **211** (1999), 384–408.

[29] T. Koike, *An example of a QF-ring without a Nakayama automorphism and an example of a left H-ring without self-duality,* in: "Proceedings of the 33rd Ring Theory and Representation Theory" vol. 33, Japan, 1999.

[30] J. Kraemer, *Characterizations of the existence of (quasi-) selfduality for complete tensor rings*, Algebra-Berichte **56**, München, 1987.

[31] H. Kupisch, *Über ein Klasse von Artin Ringen*, Arch. Math. **26** (1975), 23–35.

[32] T. Mano, *The invariant system of serial rings and their applications to the theory of self-duality*, Proc. 16th Symp. Ring Theory, Tokyo (1983), 48–53.

[33] S. H. Mohamed and B. J. Müller, *Continuous modules and discrete modules*, London Math. Soc. Lecture Notes **147**, Cambridge Univ. Press, Cambridge, 1990.

[34] K. Morita, *Duality for modules and its applications to the theory of rings with minimal condition*, Sci. Rep. Tokyo Kyoiku Daigaku Sect. A **6** (1958), 83–142.

[35] M. Moromoto and T. Sumioka, *Generalizations of theorems of Fuller*, Osaka J. Math. **34** (1997), 689–701.

[36] T. Nakayama, *Note on uni-serial and generalized uni-serial rings*, Proc. Imp. Acad. Tokyo **16** (1940), 285–289.

[37] B. J. Müller and S. T. Rizvi, *Direct sums of indecomposable modules*, Osaka J. Math. **21** (1984), 365–374.

[38] K. Oshiro, *Semiperfect modules and quasi-semiperfect modules*, Osaka J. Math. **20** (1983), 337–372.

[39] _____, *Continuous modules and quasi-continuous modules*, Osaka J. Math. **20** (1983), 681–694.

[40] _____, *Lifting modules, extending modules and their applications to QF-rings*, Hokkaido Math. J. **13** (1984), 310–338.

[41] _____, *Lifting modules, extending modules and their applications to generalized uniserial rings*, Hokkaido Math. J. **13** (1984), 339–346.

[42] _____, *Structure of Nakayama rings*, Proc. 20th Symp. Ring Theory, Okayama, (1987), 109–133.

[43] _____, *On Harada rings I, II, III*, Math. J. Okayama Univ. **31** (1989), 161–178, 179–188; **32** (1990), 111–118.

[44] K. Oshiro and K. Shigenaga, *On H-rings with homogeneous socles*, Math. J. Okayama Univ. **31** (1989), 189–196.

[45] K. Oshiro and S. H. Rim, *QF-rings with cyclic Nakayama permutation*, Osaka J. Math. **34** (1997), 1–19.

[46] S. T. Rizvi, *Contributions to the theory of continuous modules*, Ph.D. thesis, McMaster University, 1980.

[47] T. Sumioka and S. Tozaki, *On almost QF-rings*, Osaka J. Math. **33** (1986), 649–661.

[48] Y. Utumi, *On continuous regular rings*, Canad. Math. Bull. **4** (1961), 63–69.

[49] N. Vanaja and V. M. Purav, *Characterizations of generalized uniserial rings in terms of factor rings*, Comm. Algebra **20** (1992), 2253–2270.

[50] N. Vanaja, *On Harada modules I, II*, Preprint.

[51] J. Waschbusch, *Self-duality of serial rings*, Comm. Algebra **14** (1986), 581–590.

[52] W. Xue, *On a theorem of Fuller*, J. Pure and Applied Algebra **122** (1997), 159–168.

[53] _____, *Characterization of Morita duality via idempotents for semiperfect rings*, Algebra Colloq. **5** (1998), 99–110.

[54] K. Yamagata, *Nakayama automorphisms and extension rings over algebras*, in Japanese.

[55] Y. Yukimoto, *On decomposition of strongly quasi-Frobenius rings*, Communications in Algebra **28** (2000), 1111–1114.

[56] B. Zimmerman-Huisgen and W. Zimmerman, *Classes of modules with the exchange property*, J. Algebra **88** (1984), 416–434.

Department of Mathematics
Faculty of Science
Yamaguchi University
Yamaguchi 753-8512, Japan
e-mail: oshiro@po.cc.yamaguchi-u.ac.jp

On Some Dimensions of Modular Lattices and Matroids

E. R. Puczyłowski

Abstract

It is shown that methods of matroid theory can be applied in studies of some dimensions of modular lattices and modules. In particular one can obtain some fundamental properties of the Goldie and Kuroš-Ore dimensions of modular lattices.

0. Introduction

In [2] Dawson defined an independence space on the family of uniform submodules of a module and showed that its dimension is equal to the Goldie dimension of the module provided the Goldie dimension is finite. Next in [3] he considered a dual independent space and an independent space connected with the Fleury dimension of modules [4] as well as its dual.

The concept of the Goldie dimension can be extended in a natural way from modules to modular lattices [6, 7]. For each module M the Goldie dimension of M is equal to the Goldie dimension of the lattice L of submodules of M, and the Goldie dimension of the lattice dual to L is equal to the corank dimension (more often called the hollow dimension) of M defined by Varadarajan in [10]. Hence the Goldie dimension of modular lattices gives a common generalization of the Goldie and hollow dimension of modules. In [6] it was proved that the well known concept of Kuroš-Ore dimension of modular lattices is an extension of the Fleury dimension from modules to modular lattices. Thus it is natural to ask whether one can also define independence spaces (or, equivalently, matroids) related to the Goldie and Kuroš-Ore dimensions of modular lattices. We shall show that this is indeed possible. Extending Dawson's ideas we shall define a matroid on the set of uniform elements of a modular lattice L, the dimension of which is equal to the Goldie dimension of L (if the Goldie

---

*The author was supported by KBN Grant 2 PO3A 039 14 and by the organizers of the Third Korea-China-Japan International Symposium on Ring Theory.

dimension of L is defined) and a matroid on the irreducible elements of L the dimension of which is equal to the Kuroš-Ore dimension of L (if the Kuroš-Ore dimension of L is defined). We show that some fundamental properties of all the quoted dimensions follow easily from general results of matroid theory. We also give a new proof of a result of [8] relating the Goldie dimension of modular lattices to the length of some derived lattices.

1. Preliminaries

In this section we shortly recall some general notions and results from matroid theory. For further details see for instance [11].

A *matroid* M is defined as a pair $(S, \mathcal{P})$, where S is a non-empty set and $\mathcal{P}$ is a collection of subsets of S satisfying the following properties:

(I1) $\emptyset \in \mathcal{P}$;

(I2) If $X \in \mathcal{P}$ and $Y \subseteq X$, then $Y \in \mathcal{P}$;

(I3) If $X, Y \in \mathcal{P}$ and $|X| = |Y| + 1$, then there exists $x \in X \setminus Y$ such that $Y \cup \{x\} \in \mathcal{P}$;

(I4) If $A \subseteq S$ and every finite subset of A is in $\mathcal{P}$, then A is in $\mathcal{P}$.

$\mathcal{P}$ is called an *independence space* of M and sets of $\mathcal{P}$ are called independent subsets of S or M. A maximal independent subset of S is called a *basis* of M. Cardinalities of all bases of M are equal. Their common cardinality is called the *dimension* of M.

A subset of S not belonging to $\mathcal{P}$ is called *dependent*, and minimal dependent sets are called *circuits* of M.

Theorem 1. *The set of circuits of a matroid has the following properties*:

(C1) *$\emptyset$ is not a circuit*;

(C2) *If $C_1 \subseteq C_2$ are circuits, then $C_1 = C_2$*;

(C3) *If C_1 and C_2 are distinct circuits and $x \in C_1 \cap C_2$, then $(C_1 \cup C_2) \setminus \{x\}$ contains a circuit*;

(C4) *Circuits are finite.*

Conversely, suppose that S is a non-empty set and $\mathcal{C}$ is a family of subsets of S satisfying conditions (C1) – (C4). *If $\mathcal{P} = \{A \subseteq S \mid$ no subset of A belongs to $\mathcal{C}\}$, then $(S, \mathcal{P})$ is a matroid whose set of circuits coincides with $\mathcal{C}$.*

Let $M = (S, \mathcal{P})$ be a matroid. The *closure* σA of a subset A of S is defined by $\sigma A = A \cup \{x \in S \mid \{x\} \cup T$ is a circuit of M for some $T \subseteq A\}$.

Theorem 2. *The closure operation has the following properties*:

(D1) *For every $A \subseteq S$, $A \subseteq \sigma A$*;

(D2) If $b \in \sigma A$, then $\sigma(A \cup \{b\}) \subseteq \sigma A$;
(D3) If $b \in \sigma(A \cup \{c\})$ but $b \notin \sigma A$, then $c \in \sigma(A \cup \{b\})$;
(D4) If $b \in \sigma A$, then $b \in \sigma A'$ for a finite subset A' of A.

Conversely, suppose that S is a non-empty set, and $\sigma : 2^S \to 2^S$ is a mapping satisfying (D1) – (D4). If $\mathcal{P} = \{A \subseteq S \mid \text{for every } x \in A, x \notin \sigma(A \setminus \{x\})\}$, then the pair $(S, \mathcal{P})$ is a matroid whose closure mapping coincides with σ.

If $\sigma A = A$, then A is called a *flat*. The set of all flats is a semimodular lattice with respect to inclusion. The lattice is modular if and only if for every circuit $\{e_1, e_2, \ldots, e_n\}$ and $2 \leq r \leq n-2$ there exists $e \in S$ such that $\{e_1, \ldots, e_r, e\}$ and $\{e, e_{r+1}, \ldots, e_n\}$ are circuits [9].

2. A Matroid Related to the Goldie Dimension of Modular Lattices

In what follows L is a modular lattice with 0 and 1, i.e., L is a lattice with 0 and 1 such that the following *modular law* is satisfied: if $a, b, c \in L$ and $b \leq a$, then $a \wedge (b \vee c) = b \vee (a \wedge c)$.

For arbitrary $a \leq b$ in L, $[a, b]$ denotes the interval $\{x \in L : a \leq x \leq b\}$. The modular law easily implies that for arbitrary $a, b \in L$, the mapping $x \to b \vee x$ gives an isomorphism of lattices $[a \wedge b, a]$ and $[b, a \vee b]$.

An element $a \in L$ is called *essential* in L if for every $0 \neq x \in L$ we have $x \wedge a \neq 0$. If $a \leq b$ are in L and a is essential in $[0, b]$, then we write $a \leq_e b$. To denote that a is not essential in $[0, b]$ we write $a \not\leq_e b$.

An element $0 \neq u \in L$ is called *uniform* in L if for every $0 \neq a \leq u$, $a \leq_e u$.

Lemma 1. *Let $l \in L$. Every essential element of $[l, 1]$ is an essential element of L.*

Proof. Take $x \in [l, 1]$ essential in $[l, 1]$. Suppose that $x \wedge y = 0$ for some $y \in L$. The modularity law gives: $x \wedge (l \vee y) = l \vee (x \wedge y) = l$. Since $l \vee y \in [l, 1]$ and x is essential in $[l, 1]$, $l \vee y = l$. Consequently $0 = x \wedge y \geq l \wedge y = (l \vee y) \wedge y = y$. This shows that x is essential in L. □

Given a finite subset $S = \{s_1, \ldots, s_n\}$ of L, we put $\bigwedge S = s_1 \wedge \cdots \wedge s_n$ and $\bigvee S = s_1 \vee \cdots \vee s_n$.

The following theorem is well known.

Theorem 3. *For every subset $\{x_1, x_2, \ldots, x_n\}$ of L the following are equivalent:*

(a) For every $1 \leq i \leq n$, $x_i \wedge (\bigvee_{j \neq i} x_j) = 0$;
(b) For every $2 \leq i \leq n$, $x_i \wedge (\bigvee_{j < i} x_j) = 0$;
(c) If $A, B \subseteq \{x_1, x_2, \ldots, x_n\}$ and $A \cap B = \emptyset$, then $(\bigvee A) \wedge (\bigvee B) = 0$.

A set of non-zero elements of L all of whose finite subsets satisfy the conditions of Theorem 3 is called $\vee$-*independent* in L. Subsets of L which are not $\vee$-independent are called $\vee$-*dependent*.

Now we shall define a matroid on the set of uniform elements of L (if it is non-empty).

Let $U(L)$ be the set of all uniform elements of L and let $\mathcal{C}$ be the family of minimal $\vee$-dependent subsets of L contained in $U(L)$.

Theorem 4. *The family $\mathcal{C}$ satisfies conditions* (C1) – (C4) *of Theorem 1.*

Proof. The only not obvious condition is (C3). Let C_1 and C_2 be distinct sets in $\mathcal{C}$ and $x \in C_1 \cap C_2$. Obviously it is enough to prove that the set $(C_1 \cup C_2) \setminus \{x\}$ is $\vee$-dependent. Let $u = \bigvee((C_1 \cap C_2) \setminus \{x\})$, $v = \bigvee(C_1 \setminus C_2)$ and $w = \bigvee(C_2 \setminus C_1)$. It suffices to show that $(u \vee v) \wedge w \neq 0$. Since C_1 and C_2 are circuits, $x \wedge (u \vee v) \neq 0$ and $x \wedge (u \vee w) \neq 0$. This and the fact that x is a uniform element of L imply that $x \wedge (u \vee v) \wedge (u \vee w) \neq 0$. The modular law gives $(u \vee v) \wedge (u \vee w) = u \vee ((u \vee v) \wedge w)$. Hence if $(u \vee v) \wedge w = 0$, then $x \wedge u = x \wedge (u \vee v) \wedge (u \vee w) \neq 0$. However $C_1 \cap C_2$, being a proper subset of C_1, is $\vee$-independent. Consequently $x \wedge u = 0$, a contradiction. □

By Theorems 1 and 4, $\mathcal{C}$ defines a matroid on $U(L)$. We shall shortly denote that matroid by $U(L)$.

Clearly a subset of the set $U(L)$ is $\vee$-independent in L if and only if it is an independent set of the matroid $U(L)$.

For the lattice L of submodules of a module M, $U(L)$ coincides with the matroid defined in [2] and for the lattice L^0 dual to L, $U(L^0)$ coincides with a matroid introduced in [3].

In [7] it was proved that if M is a maximal $\vee$-independent subset of L and all elements of M are uniform, then for any $\vee$-independent subset N of non-zero elements of L, $cardN \leq cardM$. Every such set M was called a basis of L. The cardinality of a basis of L is uniquely determined by L if L has a basis, which is satisfied if and only if for every $x \neq 0$ in L there exists a uniform element u in L such that $u \leq x$. That common cardinality was called the Goldie dimension of L and denoted by $GdimL$. Clearly each basis of L is a basis of the matroid $U(L)$. Thus the Goldie dimension of a lattice having a basis is equal to the dimension of the matroid $U(L)$. There are modular lattices which do not have Goldie dimension but for which the matroid $U(L)$ is defined. Thus with a help of matroids one can extend the Goldie dimension to a wider class of modular lattices. Note however that then some pathology might appear. Let for instance N_1 be a simple module and N_2 be a module which contains no uniform submodules (an example of such a module one can find for instance in [7]). Then the dimension of the matroid $U(L)$, where L is the lattice of submodules of $N_1 \oplus N_2$, is equal 1, which is the Goldie dimension of N_1. Thus the dimension of $U(L)$ is not affected by the presence of N_2.

It is known [6] that $GdimL$ is finite if and only if L does not contain infinite independent subsets and for every non-zero element $a \in L$ there exists a uniform element u of L such that $u \leq a$.

Applying Theorem 3 it is easy to see that if $GdimL = n < \infty$ and $\{u_1, \ldots, u_n\}$ is an independent subset of L, then all elements u_i are uniform and $u_1 \vee \cdots \vee u_n$ is an essential element of L. This implies that $\{u_1, \ldots, u_n\}$ is a basis of the matroid $U(L)$. Using the fact that all bases of the matroid are of the same cardinality, one immediately gets the following fundamental characterization of Goldie dimension of modular lattices [6].

Theorem 5. *The following are equivalent:*
 (i) $GdimL = n < \infty$;
 (ii) *L contains independent uniform elements $u_1, \ldots, u_n$ such that $u_1 \vee \cdots \vee u_n$ is essential in L;*
 (iii) *for arbitrary independent uniform elements $u_1, \ldots, u_n$ of L, $u_1 \vee \cdots \vee u_n$ is an essential element of L.*

This theorem, applied to the lattice of submodules (the dual lattice of submodules) of a module, gives the respective characterization of the Goldie (respectively, hollow) dimension of the module (cf. [6]). These characterizations for modules were obtained with the use of matroids in [2, 3].

3. A Matroid Related to the Kuroš-Ore Dimension of Modular Lattices

An element $a \in L$ is said to be *irreducible* in L if for every $b, c \in L$ we have $a < b \wedge c$ whenever $a < b$ and $a < c$.

Observe that $a \in L$ is irreducible in L if and only if 1 is uniform in $[a, 1]$.

The following theorem is well known (cf. [5, Theorem IV.1.5]).

Theorem 6. (Kuroš-Ore) *If $0 = a_1 \wedge \cdots \wedge a_n = b_1 \wedge \cdots \wedge b_m$ are irredundant representations of 0 and all $a_1, \ldots, a_n, b_1, \ldots, b_m$ are non-redundant elements of L, then $n = m$.*

Thus the number of irreducible elements of L which give such a non-redundant representation of 0 is an invariant of L. It is called the Kuroš-Ore dimension of L.

Now we shall define a matroid connected with this dimension. It in particular will give Theorem 6 as an immediate consequence of general facts from matroid theory.

Lemma 2. *If $n, l \in L$, l is irreducible in L, $n \wedge l \leq_e l$ and $n \wedge l \not\leq_e n$, then n is essential in L.*

Proof. By our assumption there exists $0 \neq x \leq n$ such that $n \wedge l \wedge x = 0$. Since $n \wedge l \leq_e l$, $x \wedge l = 0$. Hence $l < l \vee x$ and since l is irreducible in L, $l \vee x$ is an essential element of $[l, 1]$. Consequently by Lemma 1, $l \vee x$ is essential in L. Now suppose that $y \in L$ and $n \wedge y = 0$. Since $x \leq n$, $n \wedge (x \vee y) = x \vee (n \wedge y) = 0$. Thus $l \wedge n \wedge (x \vee y) = 0$ and since $l \wedge n \leq_e l$, $l \wedge (x \vee y) = 0$. Next, $y \wedge x = 0$ because $x \leq n$ and $y \wedge n = 0$. Applying Theorem 3 one gets that $(l \vee x) \wedge y = 0$. However $l \vee x$ is essential in L, so $y = 0$ and the lemma follows. $\square$

Let $I(L)$ be the set of irreducible elements of L. For a subset A of $I(L)$ put $\sigma A = \{x \in I(L) \mid x \wedge \bigwedge A' \leq_e \bigwedge A'$ for a finite subset A' of $A\}$.

Theorem 7. *The above defined σ satisfies properties (D1) – (D4) of Theorem 4.*

Proof. The condition (D1) is clear. To get (D2) – (D4) it suffices to check (D2) and (D3) for a finite set A.

(D2) Let $x \in \sigma(A \cup \{b\})$, i.e., $x \wedge b \wedge \bigwedge A \leq_e b \bigwedge A$. Since $b \in \sigma A$, we have $b \wedge \bigwedge A \leq_e \bigwedge A$. These imply that $x \wedge b \wedge \bigwedge A \leq_e \bigwedge A$, so $x \wedge \bigwedge A \leq_e \bigwedge A$. Consequently $x \in \sigma A$.

(D3) Assume that $n \in \sigma(A \cup \{l\})$ but $l \notin \sigma(A \cup \{n\})$. We have to show that $n \in \sigma A$. Thus putting $a = \bigwedge A$ we know that $n \wedge l \wedge a \leq_e l \wedge a$, $n \wedge l \wedge a \not\leq_e n \wedge a$ and we have to show that $n \wedge a \leq_e a$. Note that since the lattices $[l \wedge a, a]$ and $[l, l \vee a]$ are isomorphic and l is irreducible in L, $l \wedge a$ is irreducible in $[0, a]$. This allows us to assume that $a = 1$. Now the result is just the conclusion of Lemma 2. $\square$

Now applying Theorem 2 one gets a matroid on $I(L)$ (if $I(L)$ is non-empty). Taking as L the lattice of submodules of a module or the dual lattice one gets some of the matroids introduced in [3].

Observe that if $a_1, \ldots, a_n$ are irreducible elements of L such that $0 = a_1 \wedge \cdots \wedge a_n$ and this representation of 0 is non-redundant then $\{a_1, \ldots, a_n\}$ is a basis of the matroid defined on $I(L)$. Now the fact that all bases of the matroid are of the same cardinality immediately implies Theorem 6.

4. A Relation between the Goldie Dimension and the Length

Let $\overline{L}$ be the lattice of flats of the matroid $U(L)$.

Theorem 8. *The lattice $\overline{L}$ is modular.*

Proof. We have to show that if $\{u_1, \ldots, u_n\}$ is a circuit in $U(L)$ and $2 \leq r \leq n-2$, then there exists a uniform element u in L such that both $\{u_1, \ldots, u_r, u\}$ and $\{u, u_{r+1}, \ldots, u_n\}$ are circuits. Applying Theorem 3 it is easy to observe that since the sets $\{u_1, \ldots, u_r\}$ and $\{u_{r+1}, \ldots, u_n\}$ are independent, whereas the set $\{u_1, \ldots, u_n\}$ is dependent, $(u_1 \vee \cdots \vee u_r) \wedge (u_{r+1} \vee \cdots \vee u_n) \neq 0$. The Goldie dimension of the lattice $[0, u_1 \vee \cdots \vee u_r]$ is finite (equal to r), so there exists a uniform element $u \leq (u_1 \vee \cdots \vee u_r) \wedge (u_{r+1} \vee \cdots \vee u_n)$. We shall show that $\{u_1, \ldots, u_r, u\}$ is a circuit in $U(L)$. Clearly the set $\{u_1, \ldots, u_r, u\}$ is dependent and every subset of $\{u_1, \ldots, u_r\}$ is independent. Take any proper subset S of $\{u_1, \ldots, u_r, u\}$ containing u. Without loss of generality we can assume that $u_1 \notin S$. Since $\{u_1, \ldots, u_n\}$ is a circuit, the set $\{u_2, \ldots, u_r\}$ is independent. Hence, since $u \leq u_{r+1} \vee \cdots \vee u_n$, $(u_2 \vee \cdots \vee u_r) \wedge u \leq (u_2 \vee \cdots \vee u_r) \wedge (u_{r+1} \vee \cdots \vee u_n) = 0$. Applying Theorem 3 we get that the sets $\{u_2, \ldots, u_r, u\}$ and, consequently, S are independent. Similarly one shows that $\{u, u_{r+1}, \ldots, u_n\}$ is a circuit in $U(L)$. □

If A is a flat of $U(L)$ and M is a maximal independent subset of A, then for $u \in U(L) \setminus M$, $u \in A$ if and only if the set $\{u\} \cup M$ is dependent. Moreover if X is an independent subset of $U(L)$, then for every proper subset Y of X, $\sigma(Y)$ is properly contained in $\sigma(X)$. These imply that if the dimension of the matroid $U(L)$ is finite, then it is equal to the length, $l\overline{L}$, of $\overline{L}$. Hence if $GdimL < \infty$, then $GdimL = l\overline{L}$.

Given $x, y \in L$, let $x \sim y$ if and only if $x \wedge y \leq_e x$ and $x \wedge y \leq_e y$. It is easy to see that $\sim$ is a congruence relation in the lower semilattice $(L, \wedge)$ [8]. Let $L^* = L/\sim$. In [8] it was proved that if $GdimL < \infty$, then the semilattice L^* is a modular lattice and $GdimL = lL^*$. It turns out that if $GdimL < \infty$, then the lattice L^* is isomorphic to $\overline{L}$. To get this it obviously suffices to prove that the lower semilattices L^* and $\overline{L}$ are isomorphic. We show this in the next theorem. Note that this theorem combined with Theorem 8 and the above remarks of relations between dimensions and the length give another proof of the quoted result of [8].

Theorem 9. *If $GdimL < \infty$, then the lower semilattices L^* and $\overline{L}$ are isomorphic.*

Proof. Observe that if $x \sim y$, then $\{u \in U(L) \mid x \wedge u \neq 0\} = \{u \in U(L) \mid y \wedge u \neq 0\}$. This implies that putting for every equivalence class $\overline{x}$ of $L/\sim$, $\Phi(\overline{x}) = \{u \in U(L) \mid x \wedge u \neq 0\}$ we get a well defined map $\Phi : L/\sim \to 2^{U(L)}$. We shall show that $\Phi(\overline{x})$ is a flat. Suppose that $u_1, \ldots, u_n \in \Phi(\overline{x})$ and $\{u_1, \ldots, u_n, u\}$ is a circuit. Obviously $Gdim[0, u_1 \vee \cdots \vee u_n] = n$. Hence, since $\{x \wedge u_1, \ldots, x \wedge u_n\}$ is an independent subset of $[0, u_1 \vee \cdots \vee u_n]$, applying Theorem 3 we get that $(x \wedge u_1) \vee \cdots \vee (x \wedge u_n)$ is an essential element of $[0, u_1 \vee \cdots \vee u_n]$. This and the fact that $u \wedge (u_1 \vee \cdots \vee u_n) \neq 0$ imply that $u \wedge [(x \wedge u_1) \vee \cdots \vee (x \wedge u_n)] \neq 0$. Hence $u \wedge x \neq 0$, which means that $u \in \Phi(\overline{x})$. Hence indeed $\Phi(\overline{x}) \in \overline{L}$.

For arbitrary $x, y \in L$ and $u \in U(L)$, $x \wedge y \wedge u \neq 0$ if and only if $x \wedge u \neq 0$ and $y \wedge u \neq 0$. Hence for arbitrary $x, y \in L$, $\Phi(\overline{x} \wedge \overline{y}) = \Phi(\overline{x \wedge y}) = \Phi(\overline{x}) \cap \Phi(\overline{y})$, which implies that Φ is a homomorphism of the lower semilattices.

For a non-empty flat A in $U(L)$ define $\Psi(A) = \{x \in L \mid \text{if } u \in U(L), \text{ then } x \wedge u \neq 0 \text{ iff } u \in A\}$. Note that $\Psi(A) \neq \emptyset$. Indeed, since $GdimL < \infty$, A contains a maximal finite independent subset $\{u_1, \ldots, u_n\}$. Then $u_1 \vee \cdots \vee u_n \in \Psi(A)$. Clearly if $x \sim y$ and $x \in \Psi(A)$, then $y \in \Psi(A)$. Conversely, if $x, y \in \Psi(A)$, then $x \sim y$. Indeed, if for instance $x \wedge y \not\leq_e x$, then since $GdimL < \infty$, there exists $u \in U(L)$ such that $x \wedge y \wedge u = 0$ and $u \leq x$. Now $y \wedge u = x \wedge y \wedge u = 0$, so $u \notin A$. However $x \wedge u = u \neq 0$, so $u \in A$, a contradiction. Hence $\Psi(A) \in L/\sim$. Moreover $\Phi(\Psi(A)) = A$ and $\Psi(\Phi(\overline{x})) = \overline{x}$, which ends the proof. $\square$

The above defined relation $\sim$ on L need not be a lattice congruence even when L is the lattice of submodules of a module. As an example one can take for instance the lattice K of subgroups of the group $\mathbb{Z}_2 \oplus \mathbb{Z}_4$, where $\mathbb{Z}_2$ is a group of order 2 and $\mathbb{Z}_4$ is a cyclic group of order 4. One also can check that $K \setminus \{\mathbb{Z}_2 \oplus S\}$, where S is the cyclic subgroup of order 2 of $\mathbb{Z}_4$, is a sublattice of K on which $\sim$ is not a lattice congruence.

It turns out that the property that $\sim$ is a lattice congruence coincides with a property studied for modules in [1].

A module M is called [1] a *dimension* module if for arbitrary submodules N_1, N_2 of M, $Gdim(N_1 + N_2) + Gdim(N_1 \cap N_2) = GdimN_1 + GdimN_2$.

Suppose that $GdimL < \infty$. We shall say that L is a *dimension* lattice if for arbitrary elements $a, b \in L$, $Gdim[0, a \vee b] + Gdim[0, a \wedge b] = Gdim[0, a] + Gdim[0, b]$.

Theorem 10. *If $GdimL < \infty$, then L is a dimension lattice if and only if $\sim$ is a lattice congruence.*

Proof. Suppose first that $\sim$ is a lattice congruence. Then by the foregoing and properties of the length of modular lattices, for every $x \in L$, $Gdim[0,x] = l[0,\bar{x}]$ and for arbitrary $a, b \in L$, $l[0, \bar{a} \vee \bar{b}] + l[0, \bar{a} \wedge \bar{b}] = l[0,\bar{a}] + l[0,\bar{b}]$. These and the fact that, since $\sim$ is a lattice congruence, $\bar{a} \vee \bar{b} = \overline{a \vee b}$ imply the equality $Gdim[0, a \vee b] + Gdim[0, a \wedge b] = Gdim[0,a] + Gdim[0,b]$.

Suppose now that L is a dimension lattice. In [5, Lemma I.3.8], it was proved that a reflexive relation γ on L is a lattice congruence if and only if for arbitrary $x, y, z, t \in L$

(i) $x\gamma y$ iff $(x \wedge y)\gamma(x \vee y)$;
(ii) $x \leq y \leq z$, $x\gamma y$ and $y\gamma z$ imply that $x\gamma z$;
(iii) $x \leq y$ and $x\gamma y$ imply that $(x \wedge t)\gamma(y \wedge t)$ and $(x \vee t)\gamma(y \vee t)$.

The conditions (i), (ii) and the first part of condition (iii) are always satisfied for $\sim$. To get the second part of condition (iii) for $\sim$ we have to show that if $a, b, c \in L$ and $a \leq_e b$, then $a \vee c \leq_e b \vee c$. Since $a \wedge c \leq_e b \wedge c$, we have that $Gdim[0,a] = Gdim[0,b]$ and $Gdim[0, a \wedge c] = Gdim[0, b \wedge c]$. Applying the assumption that L is a dimension lattice one gets that $Gdim[0, a \vee c] + Gdim[0, a \wedge c] = Gdim[0,a] + Gdim[0,c]$ and $Gdim[0, b \vee c] + Gdim[0, b \wedge c] = Gdim[0,b] + Gdim[0,c]$. These give that $Gdim[0, a \vee c] = Gdim[0, b \vee c]$, so $a \vee c \leq_e b \vee c$. □

References

[1] V. P. Camillo and J. M. Zelmanowitz, *Dimension modules,* Pacific J. Math. **91** (1990), 249–261.

[2] J. E. Dawson, *Independence spaces and uniform modules,* Europ. J. Combinatorics **6** (1985), 29–36.

[3] J. E. Dawson, *Independence structures on the submodules of a module,* Europ. J. Combinatorics **6** (1985), 37–44.

[4] P. Fleury, *A note on dualizing Goldie dimension,* Canad. Math. Bull. **17** (1974), 511–517.

[5] G. Grätzer, *General Lattice Theory,* Birkhäuser, Basel, 1978.

[6] P. Grzeszczuk and E. R. Puczyłowski, *On Goldie and dual Goldie dimension,* J. Pure and Applied Algebra **31** (1984), 47–54.

[7] ———, *On infinite Goldie dimension of modular lattices and modules,* ibid. **35** (1985), 151–155.

[8] P. Grzeszczuk, J. Okniński and E. R. Puczyłowski, *Relations between some dimensions of modular lattices*, Comm. Algebra **17** (1989), 1723–1737.

[9] P. Vamos, *On the representation of independence structures*, 1968, unpublished.

[10] K. Varadarajan, *Dual Goldie dimension*, Comm. Algebra **7** (1979), 565–610.

[11] D. J. A. Welsh, *Matroid Theory*, Academic Press, London, 1976.

Institute of Mathematics
University of Warsaw
02-097 Warsaw, Banacha 2, Poland
e-mail: edmundp@mimuw.edu.pl

On Torsion-free Modules over Valuation Domains

K. M. Rangaswamy

Abstract

In this survey article, we indicate how some of the recent ideas and techniques introduced in the study of infinite rank Butler groups can be successfully used in the investigation of the homological dimensions of torsion-free modules over integral domains and, in particular, over valuation domains.

1. Introduction

Abelian group theory has often been the breeding ground for new concepts and initiatives in the general module theory. R. Baer's idea of an injective module [3], S. Dickson's pioneering work on torsion theories [4] and I. Kaplansky's famous theorem on projective modules (as a byproduct of the theorem on direct summands of completely decomposable abelian groups) [12] are some of the instances where fundamental ideas from abelian group theory have found fruitful applications in module theory. In this paper we survey some of the recent research that utilizes the techniques and concepts from the theory of infinite rank Butler groups in the investigation of torsion-free modules over integral domains. Specifically, the concept of the balanced-projective dimension of modules over an integral domain D is initiated. A new concept of an n-balanced submodule, along with the notion of a pure-essential submodule, plays a fundamental role in this investigation. The n-balanced submodules give rise to a relative homological algebra while the pure-essential submodules help in the construction of balanced extensions and a very special pull-back diagram. This pull-back diagram turns out to be an effective tool in uncovering interesting relations between the projective and the balanced-projective dimensions of torsion-free modules over a valuation domain R. Of course, one has to modify the approach used for abelian groups significantly since R is not noetherian and its prime ideal spectra can differ dramatically from that of the ring $\mathbb{Z}$ of integers. Nevertheless, the idea of a relative balanced projective resolution that L. Fuchs [8] used effectively in his study of Butler groups leads to several characterizations of torsion-free R-modules with finite balanced projective dimension (bpd, for short). These modules are shown to possess a special chain of pure submodules called an n-balanced chain (Theorem 4.3) whose members also have the same bpd, thus providing a converse of the balanced version of the classical Auslander's Lemma on projective dimensions. The n-balanced chain leads to the existence of a special "tight system" of n-balanced submodules in R-modules with finite bpd, indicating their rich internal structure. Necessary and sufficient conditions are given under which a pure submodule of a torsion-free R-module M with $bpd = n \geq 0$ will also have the same bpd. Specialization to the case of modules with finite projective dimension leads to new results. For example, we have (Theorem 4.7): A pure submodule A of a free R-module F is free if and only if there exists a continuous well-ordered ascending chain of pure submodules $A = A_0 < A_1 < \cdots < A_\alpha < \cdots < A_\tau = M$ where, for each $\alpha > 0$, A_α is free and, for each $\alpha \geq 0$, $A_{\alpha+1}/A_\alpha$ is countably generated. The global $bpd(R)$ is calculated and is related to the global dimension of R.

2. Preliminaries

Let D be an integral domain with quotient field Q and let n stand for an integer ≥ 0. The D-modules isomorphic to non-zero submodules of Q are called *rank-1* modules. The direct sums of rank-1 modules are called the *completely decomposable* modules. A pure submodule A of a D-module M is said to be *pure-essential* in M, if A has the property that $A + J$ is not pure in M whenever J is a pure submodule of M with $J \cap A = 0$. A pure-essential completely decomposable submodule B of a torsion-free D-module M is called a *basic* submodule of M. If a D-module M has a generating set of cardinality at most κ, then we say that M is *κ-generated*. We follow the convention under which $\aleph_{-1}$-generated means finitely generated.

An exact sequence of D-modules $0 \to A \to B \to C \to 0$ is called *balanced* if, for every rank-1 D-module J, the induced homomorphism $Hom_D(J, B) \to Hom_D(J, C)$ is surjective. A submodule A of a D-module B is said to be *balanced* in B if the exact sequence $0 \to A \to B \to B/A \to 0$ induced by the inclusion of A in B is balanced. The balanced short exact sequences form a proper class in the sense of MacLane and so the functors $Bext_D^n$ can be defined for all non-negative integers n in such a way that we get the usual long exact sequences. $Bext_D^1$ is a subfunctor of Ext_D^1 consisting of the balanced extensions.

First observe that every torsion-free D-module M fits into a balanced exact sequence $0 \to K \to C \to M \to 0$ where C is completely decomposable. Indeed, if we take $C = \oplus\{\alpha(L)\colon L_D \subseteq Q_D, \alpha \in Hom_D(L,M)\}$, then the inclusion maps $\alpha(L) \to M$ induce an epimorphism $C \to M$ with kernel K which is evidently balanced. By the usual argument one then shows that the balanced-projective torsion-free D-modules are precisely the direct summands of completely decomposable D-modules. One can also define a *balanced-projective resolution* of a torsion-free D-module M to be a long exact sequence $\cdots \to C_n \xrightarrow{\delta_n} C_{n-1} \cdots \to C_1 \xrightarrow{\delta_1} C_0 \xrightarrow{\delta_0} M \to 0$ where the C_i are completely decomposable D-modules and, for each $i \geq 0$, $\ker \delta_i$ is balanced in C_i. The *balanced-projective dimension* of M, in notation $bpd(M)$, is equal to n if n is the smallest index with $Im \delta_n$ balanced-projective; $bpd(M) = \infty$ if no such n exists. The balanced version of Schanuel's lemma shows that this definition is independent of the particular choice of the balanced-projective resolutions. Equivalently, $bpd(M) = n$, if n is the smallest index such that $Bext_D^{n+1}(M,-) = 0$. The projective dimension of M is defined analogously and is denoted by $pd(M)$. The classical Auslander's Lemma and the lemma of Kaplansky on projective dimensions of modules (see [11, Ch.IV, Section 2]) extend easily to the case of balanced-projective dimensions of torsion-free D-modules. In particular, suppose $0 \to A \to B \to C \to 0$ is a balanced exact sequence of torsion-free D-modules. If any two of the three modules A, B, C have finite bpd, then the third one also has finite bpd and $bpd(B) \leq \max\{bpd(A), bpd(C)\}$.

A *valuation domain* R is an integral domain in which all the ideals form a chain under set inclusion. Such an R is also known as a chain domain. R is, in particular, a local ring and so, by Kaplansky's theorem, all the projective R-modules are free. Moreover, a balanced-projective R-module is completely decomposable (see [11]). A well-known result on projective dimension (see [11, Ch.IV, Theorem 5.1]) states that if M is a torsion-free module over a valuation domain R with rank $\leq \aleph_{n-1}$ and $pd(M) \leq n$, then M is $\aleph_{n-1}$-generated. We refer the reader to [11] for general notation, terminology and other results on modules over valuation domains.

3. A Special Pull-back Diagram

Throughout this section R stands for a valuation domain. We introduce a pull-back diagram that was considered in [13]. It comes in handy when establishing connections between the projective and the balanced-projective dimensions of torsion-free R-modules. For instance, if M is a torsion-free R-module which is not completely decomposable, then

$bpd(M) = pd(M/B)$ for any basic submodule B of M. Another result for a torsion-free R-module M states that $bpd(M) \leq n$ if and only if $Bext_R^{n+1}(M, T) = 0$ for all torsion R-modules T.

An important property of a free R-module F [11, Ch.XIV, Theorem 2.2] is that any rank one pure submodule of F is a direct summand and is cyclic. This implies that, for any pure submodule K of F, the projective resolution of K is always a balanced-projective resolution. As a consequence, we have the following lemma.

Lemma 3.1. *Suppose K is a pure submodule of a free R-module. Then $pd(K) = bpd(K)$.*

A significant first step in the construction of balanced extensions using the pure-essential submodules is the next lemma which was proved by L. Fuchs and E. Monari-Martinez [9].

Lemma 3.2. [9] *Consider a commutative diagram of R-modules*

$$\begin{array}{ccccccccc} 0 & \to & K & \to & P & \to & M & \to & 0 \\ & & \| & & \downarrow & & \downarrow & & \\ 0 & \to & K & \to & F & \to & C & \to & 0 \end{array}.$$

If C is torsion-free, P is the pull-back of M and F over C and $B = \ker(M \to C)$ is pure-essential in M, then the top row is balanced exact.

A special case of Lemma 3.2 is particularly useful: Write $C = M/B$ and take F to be a free R-module, so that the bottom row becomes a free resolution of M/B. Then the pull-back P of M and F over C fits into a commutative diagram:

$$\begin{array}{ccccccccc} & & & & 0 & & 0 & & \\ & & & & \downarrow & & \downarrow & & \\ & & & & B & = & B & & \\ & & & & \downarrow & & \downarrow & & \\ 0 & \to & K & \to & P = B \oplus F & \to & M & \to & 0 \\ & & \| & & \downarrow & & \downarrow & & \\ 0 & \to & K & \to & F & \to & M/B & \to & 0 \\ & & & & \downarrow & & \downarrow & & \\ & & & & 0 & & 0 & & \end{array} \quad (D)$$

where the middle column splits and $P = B \oplus F$, since F is free.

We shall now derive an interesting consequence of the pull-back diagram (D) obtained by Nongxa, Rangaswamy and Vinsonhaler [14]. It relates the projective and the balanced projective dimensions of torsion-free R-modules.

Theorem 3.3. [14] *Let M be a torsion-free R-module and B any basic submodule of M with $B \neq M$. If M is not a completely decomposable module, then $bpd(M) = pd(M/B)$. If M is completely decomposable, then $pd(M/B) = 1$ (while, obviously, $bpd(M) = 0$).*

Proof. Being a basic submodule, B is pure-essential in M. With this B, consider the diagram (D) for the module M. If M/B were projective, then B would be a direct summand of M, contradicting the fact that B is pure-essential. So $pd(M/B) \geq 1$. Then $pd(M/B) = 1 + pd(K) = 1 + bpd(K)$, as K is a pure submodule of the free R-module F (Lemma 3.1). Now, by Lemma 3.2, the top row in the diagram (D) is balanced exact and $P = B \oplus F$ is completely decomposable. If M is not completely decomposable, then $bpd(M) = 1 + bpd(K) = pd(M/B)$. If M is completely decomposable, then the top row splits and $bpd(K) = 0$, so $pd(M/B) = 1$. □

Remark. It is interesting to note that for different basic submodules B and B' of M, M/B and M/B' may not be isomorphic but still $pd(M/B) = pd(M/B') = bpd(M)$.

Another consequence of the special pull-back diagram is the following characterization of modules with finite bpd. It uses a deep result, proved in [7], that an R-module K is free if and only if $Ext^1(K,T) = 0$ for all torsion R-modules T.

Proposition 3.4. [14] *Let $n \geq 1$. For a torsion-free R-module M, $bpd(M) \leq n$ if and only if $Bext_R^{n+1}(M,T) = 0$ for all torsion R-modules T.*

Remark. What happens when $n = 0$ in Proposition 3.3? In other words, if $Bext^1(M,T) = 0$ for all torsion modules T, is $bpd(M) \leq 0$ (i.e., M balanced-projective)? This question is considered in Section 7.

4. The n-Balanced Submodules and Modules with Finite bpd

In this section, we introduce the concept of an n-balanced submodule, where n is an integer ≥ 0. Torsion-free modules with finite balanced-projective dimension over a valuation domain R are characterized in terms of the existence of n-balanced chains. A somewhat surprising consequence is that for a torsion-free R-module M, $pd(M) \leq n$ implies that $bpd(M) \leq n$. In particular, $bpd(M) \leq pd(M)$ always holds. Conditions are given under which a pure submodule A of a torsion-free R-module M will have the same balanced-projective dimension as M.

Our first main idea from abelian group theory is the *relative balanced-projective resolution* of a torsion-free module M over an integral domain D with respect to a pure submodule A. This concept, for abelian groups, was introduced in [8]: It is a balanced exact sequence of the form $0 \to K \to A \oplus C \xrightarrow{\varphi} M \to 0$, where C is completely decomposable and φ acts as the identity map on A. Such resolutions always exist. For example, one can choose C to be a completely decomposable module for which there is a balanced epimorphism $\alpha : C \to M$ and then take $\varphi = 1_A \oplus \alpha$. The choice of C is not unique. However, if $0 \to K' \to A \oplus C' \xrightarrow{\varphi'} M \to 0$ is another relative balanced-projective resolution of M with respect to A where C' is completely decomposable and φ' is identity on A, then the balanced version of the argument used to prove the Schanuel's Lemma leads to the isomorphism $A \oplus C' \oplus K \cong A \oplus C \oplus K'$. Since φ and φ' act as the identity map on A, we derive that $C' \oplus K \cong C \oplus K'$. L. Fuchs [8] used the relative balanced-projective resolutions with remarkable success in his study of infinite rank Butler groups and, as we shall see, this tool turns out be equally effective in the study of modules with finite bpd.

Definition. Let D be an integral domain and n an integer ≥ 0. An exact sequence of D-modules $0 \to A \to B \xrightarrow{\eta} C \to 0$ is said to be n-*balanced* if, for any rank-1 module J which is not $\aleph_{n-1}$-generated, the induced map $Hom_D(J, B) \xrightarrow{\eta*} Hom_D(J, C)$ is surjective. A submodule A of a D-module B is said to be n-*balanced* in B, if the exact sequence $0 \to A \xrightarrow{i} B \to B/A \to 0$, where i is the inclusion map, is n-balanced.

Proper Class Properties. The n-balanced short exact sequences of modules over an integral domain D form a proper class in the sense of MacLane. Specifically, suppose C is a D-module and A, B are submodules of C with $A \leq B$:

(i) If A is n-balanced in B and B is n-balanced in C, then A is n-balanced in C.

(ii) If A and B/A are n-balanced respectively in C and C/A, then B is n-balanced in C.

(iii) If B is n-balanced in C, then B/A is n-balanced in C/A.

(iv) If A is n-balanced in C, then A is n-balanced in B.

(v) If A is a direct summand of C, then A is n-balanced in C.

The proofs involve the routine diagram chase arguments.

Since the n-balanced exact sequences form a proper class, the n-balanced extensions of A by C form a subgroup $n\text{-}Bext(C, A)$ of the group $Ext_D^1(C, A)$ of all extensions of A by C.

Suppose $D = R$ is a valuation domain and the module C is torsion-free. Then the required conditions in the above definition of an n-balanced sequence can be somewhat relaxed. Specifically, in order to verify that for each $\alpha : J \to C$ there exists a $\gamma : J \to B$ such that $\eta\gamma = \alpha$ (i.e., to verify that the induced η^* is surjective), we need only confine ourselves to the case when the image $\alpha(J)$ is pure in C. This is because the lifting map $\gamma : J \to B$ satisfying $\eta\gamma = \alpha$ always exists when the image $\alpha(J)$ is not pure in C. To see this, first observe that (because R is a valuation domain) the pure submodule $<\alpha(J)>*$ generated by $\alpha(J)$ in the torsion-free R-module C is uniserial and so for any element $a \in\, <\alpha(J)>*$ with $a \notin \alpha(J)$, we have $\alpha(J) \subset Ra \simeq R$. Clearly, there is a $\beta : Ra \to B$ is such that $\eta\beta$ is the identity on Ra. Then $\gamma = \beta\alpha : J \to B$ satisfies the condition that $\eta\gamma = \alpha$. Also if we identify A with its image in B and consider A as a submodule of B with $C = B/A$, then the n-balanced condition means that for any pure submodule L of B containing A with L/A rank-1, A will be a direct summand of L whenever L/A is not $\aleph_{n-1}$-generated. Thus we are lead to the following reformulation of the above definition:

Over a valuation domain R, a submodule A of an R-module B with B/A torsion-free is n-balanced in B if and only if, for any rank-1 pure submodule L/A of B/A, either A is a direct summand of L or L/A is $\aleph_{n-1}$-generated.

By [11, Ch.IV, Theorems 3.2 and 5.1], the last condition that L/A is $\aleph_{n-1}$-generated is equivalent to having $pd(L/A) \leq n$ when R is a valuation domain.

Observe that the 0-balanced submodules are just the balanced submodules. In [13] 1-balanced submodules are called *pseudo-balanced*.

Examples. It was shown in [11] that every pure submodule of an $\aleph_{n-1}$-generated torsion-free module over a valuation domain R is again $\aleph_{n-1}$-generated. It is then clear that, in an $\aleph_{n-1}$-generated torsion-free R-module, every pure submodule is n-balanced. Moreover, any balanced (more generally, any $(n-1)$-balanced) submodule of a torsion-free module is trivially n-balanced. To see an example of a n-balanced submodule which is not $(n-1)$-balanced, take J to be an ideal of a valuation domain R with a minimal generating set of cardinality $\aleph_{n-1}$ for some $n \geq 1$ and write $J = F/K$ where F is a free R-module. Clearly K is n-balanced in F. Now K cannot be a direct sumand of F. Hence it is not $(n-1)$-balanced and, in particular, not balanced in F.

From this point onwards, R will denote a valuation domain.

The next Proposition gives a criterion for a pure submodule B to be n-balanced in a torsion-free R-module M. It shows how the embedding of B in M is related to the kernel K in a relative balanced-projective resolution of M and was proved in [14].

Proposition 4.1. [14] *Suppose B is a pure submodule of a torsion-free R-module M with M/B having rank one and n is an integer ≥ 1. Then the following are equivalent:*

(a) *B is n-balanced in M;*

(b) *There exists a relative balanced-projective resolution $0 \to K \to B \oplus C \xrightarrow{\varphi} M \to 0$ of M with respect to B where $pd(K) \leq n-1$, C is completely decomposable and C is free whenever B is not a direct summand of M;*

(c) *In one (therefore, every) relative balanced-projective resolution $0 \to K' \to B \oplus C' \to M \to 0$ with C' completely decomposable, we have $bpd(K') \leq n-1$.*

Before we give the next definition, we wish to recall that an ascending chain of modules $A_0 < \cdots < A_\alpha < A_{\alpha+1} < \cdots$, $\alpha < \tau$ is said to be *smooth* if, for all limit ordinals $\sigma \leq \tau$, $A_\sigma = \bigcup_{\alpha < \sigma} A_\alpha$.

Definition. Let A be a pure submodule of a torsion-free R-module M and $n \geq 0$. An *n-balanced chain from A to M* is a smooth well-ordered ascending chain of pure submodules

$$A = A_0 < \cdots < A_\alpha < A_{\alpha+1} < \cdots < A_\tau = \bigcup_{\alpha < \tau} A_\alpha = M$$

where τ is some fixed ordinal and, for each $\alpha < \tau$, A_α is n-balanced in $A_{\alpha+1}$ and $A_{\alpha+1}/A_\alpha$ has rank one. If $A = 0$, the above chain is then called an *n-balanced chain* for M.

Observe that if there is a 0-balanced chain from A to M, then A is simply a direct summand of M whose complementary summand is completely decomposable.

Using transfinite induction and the successive use of Proposition 4.1, Nongxa, Rangaswamy and Vinsonhaler [14] related the existence of n-balanced chains with the kernel of a relative balanced-projective resolution.

Theorem 4.2. [14] *The following are equivalent for a pure submodule A of a torsion-free R-module M and an integer $n \geq 1$:*

(i) *There is an n-balanced chain from A to M;*

(ii) *There exists a relative balanced-projective resolution* $0 \to K \to A \oplus C \to M \to 0$, *with* C *completely decomposable and* $pd(K) \leq n-1$;

(iii) *In one (therefore, every) relative balanced-projective resolution* $0 \to K' \to A \oplus C' \to M \to 0$ *with* C' *completely decomposable, we have* $bpd(K') \leq n-1$.

Taking $A = 0$ in Theorem 4.2, we obtain the following characterization of a torsion-free R-module M with $bpd \leq n$. Here the case $n \geq 1$ follows from Theorem 4.2 while, when $n = 0$, M is completely decomposable, its direct summands can be arranged to form a 0-balanced chain. For convenience of notation, we write $pd(K) \leq -1$ to denote that $K = 0$.

Theorem 4.3. *Let M be a torsion-free R-module and n an integer ≥ 0. Then the following are equivalent:*
(i) $bpd(M) \leq n$;
(ii) M *has an n-balanced chain;*
(iii) *There exists a balanced exact sequence* $0 \to K \to C \to M \to 0$ *where C is completely decomposable and $pd(K) \leq n-1$.*

Corollary. *Let M be a torsion-free R-module with $bpd(M) \leq n$ for some $n \geq 0$. If M is a pure submodule of a torsion-free R-module N, then $bpd(N) \leq n$ if (a) there is an n-balanced chain from M to N or if (b) N/M is $\aleph_{n-1}$-generated.*

Remark. Suppose $bpd(M) \leq n$ so that, by Theorem 4.3, M has an n-balanced chain $0 = A_0 < \cdots < A_\alpha < A_{\alpha+1} < \cdots$, $\alpha < \tau$. Now each member A_α in the n-balanced chain satisfies the hypothesis of Theorem 4.3 and so $bpd(A_\alpha) \leq n$. Moreover, A_β/A_α, for $\alpha < \beta \leq \tau$, and, in particular, M/A_α all have $bpd \leq n$ for the same reason. Suppose further the rank of M is an uncountable regular cardinal τ. Then the above n-balanced chain is actually a τ-filtration since, for each $\alpha < \tau$, A_α will have rank $< \tau$. If $M = \bigcup_{\alpha<\tau} B_\alpha$ is any other τ-filtration of the torsion-free R-module M consisting of pure submodules B_α, then the usual back-and-forth argument [6] implies that there is a closed and unbounded subset C of τ such that $B_\alpha = A_\alpha$ for all $\alpha \in C$. Hence both B_α and B_β/B_α have $bpd \leq n$, for all α, β in C with $\alpha \leq \beta$. This provides the balanced-version of the converse of the classical Auslander Lemma done by P. Eklof [6]. For an alternative proof of the preceding statement using a different approach see [10].

A noteworthy consequence of Theorem 4.3 is the following theorem which is of independent interest and plays a key role in later investigations.

Theorem 4.4. [14] *Let M be a torsion-free R-module and $n \geq 0$. If $pd(M) = n$, then $bpd(M) \leq n$. In particular, $bpd(M) \leq pd(M)$ for every torsion-free R-module M.*

Proof. If $pd(M) = 0$, M is free and clearly $bpd(M) = 0$. Suppose $pd(M) = n \geq 1$. By [11, Ch.IV, Corollary 5.2], M is the union of a smooth ascending chain of pure submodules $0 = M_0 < \cdots < M_\alpha < M_{\alpha+1} < \cdots < M_\tau = M$ where τ is a suitable ordinal and, for each $\alpha < \tau$, $M_{\alpha+1}/M_\alpha$ has rank one and is $\aleph_{n-1}$-generated. This chain is clearly n-balanced. Hence, by Theorem 4.3, $bpd(M) \leq n$. If $pd(M) = \infty$, then trivially $bpd(M) \leq pd(M)$. Thus $bpd(M) \leq pd(M)$ always holds. □

Corollary. *If a torsion-free R-module M can be generated by $\aleph_n$ elements, then $bpd(M) \leq n + 1$.*

Proof. Follows from Theorem 4.4 and the result (see [11, Ch.IV, Proposition 3.2]) that $pd(M) \leq n + 1$. □

Remark. The reverse implication of Theorem 4.4 is false, i.e., $pd(M) \leq bpd(M)$ need not hold for a torsion-free R-module M. If J is an ideal of a valuation domain with a minimal generating set of cardinality $\aleph_1$, then $bpd(J) = 0$ since J has rank one, but $pd(J) = 2$, by Osofsky's Theorem (see [11, Ch.IV, Theorem 2.2]).

Suppose M is a torsion-free R-module with $bpd(M) = n$ for some integer $n \geq 0$. Theorem 4.2 enables us to obtain conditions under which a pure submodule A of M will have $bpd \leq n$. The following theorem points out a curious property that if a torsion-free A with $bpd = n$ is a pure submodule of another torsion-free M with the same bpd, then there is an $(n+1)$-balanced chain from A to M each member of which has $bpd = n$.

Theorem 4.5. *Suppose $n \geq 0$ and A is a pure submodule of a torsion-free R-module M with $bpd(M) \leq n$. Then the following are equivalent:*

(i) $bpd(A) \leq n$;

(ii) *There exists an $(n+1)$-balanced chain from A to M;*

(iii) *There exists a smooth ascending chain of pure submodules $A = A_0 < \cdots < A_\alpha < \cdots < A_\tau = M$, where, for each $\alpha < \tau$, $bpd(A_\alpha) \leq n$ and $A_{\alpha+1}/A_\alpha$ has rank one.*

Proof. (i)⇔(ii). Consider a relative balanced-projective resolution $0 \to K \to A \oplus C \to M \to 0$, where C is completely decomposable. Since $bpd(M) \leq n$, $bpd(A) = bpd(A \oplus C) \leq n$ if and only if $bpd(K) \leq n$. By Theorem 4.2, $bpd(K) \leq n$ exactly when there is an $(n+1)$-balanced chain from A to M.

(ii)⇒(iii). Now each member of the chain in (ii) satisfies the same hypothesis as A and so, by the equivalence of (i) and (ii), its balanced-projective dimension is at most n. Hence (ii) implies (iii).

(iii)⇒(i). Obvious, since $bpd(A_\alpha) \leq n$ for every α and particular for $\alpha = 0$. □

Some of our theorems on modules with finte bpd lead to new results on modules with finite projective dimension. We point out two such results.

Suppose M is a torsion-free R-module with projective dimension n and A is a pure submodule of M. If there is an n-balanced chain from A to M, then, when $n \geq 1$, Theorem 4.2(ii) yields a relative balanced-projective resolution $0 \to K \to A \oplus C \to M \to 0$ where $pd(K) \leq n-1$ and so $pd(A \oplus C) \leq n$. This implies that $pd(A) \leq n$. On the other hand, when $n = 0$, the existence of a 0-balanced (= balanced) chain from A to M implies that A is a direct summand of M and so both A and M will have the same projective as well as the same balanced-projective dimensions. Thus we obtain the following result:

Proposition 4.6. *Suppose M is a torsion-free R-module and there is an n-balanced chain from a pure submodule A to M. If $pd(M) \leq n$, then $pd(A) \leq n$. If $bpd(M) \leq n$, then $bpd(A) \leq n$.*

Fuchs and Salce [11] inquire about the conditions under which a pure submodule of a free R-module is again free. Specialization of our Theorem 4.5 to free R-modules provides an answer as indicated below:

Theorem 4.7. *Let A be a pure submodule of a free R-module F. Then A is free if and only if there is a smooth ascending chain of pure submodules*

$$A = A_0 < \cdots < A_\alpha < A_{\alpha+1} < \cdots < A_\tau = F$$

where τ is a suitable ordinal, A_α is free for each $\alpha > 0$ and $A_{\alpha+1}/A_\alpha$ is countably generated for each $\alpha \geq 0$.

Remark. (i) Observe that the conditions of Theorem 4.7 are sharper than the conditions of Theorem 4.5(iii) since we require that A_α is free only for $\alpha > 0$. Moreover, to prove the "if" part, we need only to assume that A_1/A_0 is countably generated.

(ii) Statements analogous to Theorem 4.7 can be established for torsion-free modules with $pd \leq n$.

5. $\aleph_n$-Families

Let R denote a valuation domain. If M is a torsion-free R-module with $bpd(M) = 0$, then M is completely decomposable and the structure of M is known [11]. So, in this section, we only consider the case when $n = bpd(M)$ is an integer ≥ 1. We begin with a new concept which is an adaptation of the concept of an axiom-3 family introduced by Paul Hill in his investigation of several important classes of abelian groups such as the infinite rank Butler groups [1].

Definition. Let n be an integer ≥ 1. A family $\mathfrak{S}$ of pure submodules of a torsion-free R-module M is said to be an $\aleph_n$-*family* if it satisfies the following conditions:
 (i) $0, M \in \mathfrak{S}$;
 (ii) $\Sigma_{i \in I} N_i \in \mathfrak{S}$, if $N_i \in \mathfrak{S}$, for each $i \in I$;
 (iii) For any $A \in \mathfrak{S}$ and any subset X of M of cardinality $< \aleph_n$, there exists a $B \in \mathfrak{S}$ such that $A \cup X \subset B$ and B/A has rank $< \aleph_n$.

In the presence of conditions (i) and (ii), the condition (iii) of the above definition is equivalent to
 (iii)* Every subset of M of cardinality $\leq \aleph_{n-1}$ is contained in a $B \in \mathfrak{S}$ with rank $B \leq \aleph_{n-1}$.

Nongxa, Rangaswamy and Vinsonhaler [14] obtained the following characterization using the $\aleph_n$-families. It shows how the torsion-free modules with large rank and finite bpd are built up of n-balanced submodules having the same bpd but with smaller rank, indicating the richness of the internal structure of these modules. The proof involves deep and delicate infinite-combinatorics arguments using the idea of closed sets of ordinals considered by Hill (see [1]).

Theorem 5.1. *Let M be a torsion-free R-module and n an integer ≥ 1. Then $bpd(M) \leq n$ if and only if M has an $\aleph_n$-family $\mathfrak{S}$ of n-balanced submodules and every pure submodule of rank $\leq \aleph_{n-1}$ is n-balanced in M. Moreover, for any A in the family $\mathfrak{S}$, both A and M/A have $bpd \leq n$.*

From the proof of Theorem 5.1 in [14] one can derive the following useful proposition.

Proposition 5.2. *Let M be a torsion-free R-module with $bpd(M) \leq n$ and A a pure submodule of rank $\leq \aleph_{n-1}$. Then $bpd(A) \leq n$ and A is a member of an n-balanced chain for M.*

Remark. The $\aleph_n$-family $\mathfrak{S}$ constructed in Theorem 5.1 for an R-module M with $bpd = n$, is the balanced version of a "tight system" as introduced in [11]. A tight sytem $\mathfrak{R}$ is similar to an $\aleph_n$-family $\mathfrak{S}$ whose condition (ii) is weakened by requiring that $\mathfrak{R}$ is closed with respect to taking the unions of ascending chains (instead of closure under module sums) and that in the condition (iii) of $\mathfrak{S}$, B/A has a generating subset of cardinality $< \aleph_n$ instead of having rank $< \aleph_n$.

Summarizing, we give below the different characterizations of a torsion-free R-module whose balanced-projective dimension is finite.

Theorem 5.3. *Let M be a torsion-free R-module and n an integer ≥ 1. Then the following are equivalent:*
(i) $bpd(M) \leq n$;
(ii) $Bext_R^{n+1}(M,T) = 0$ for all torsion R-modules T;
(iii) $pd(M/B) \leq n$ for any basic submodule B of M;
(iv) There exists a balanced exact sequence $0 \to K \to C \to M \to 0$ where C is completely decomposable and $pd(K) \leq n - 1$;
(v) M has an n-balanced chain;
(vi) M has an $\aleph_n$-family of n-balanced submodules and every pure submodule of M of rank $\leq \aleph_{n-1}$ is n-balanced in M.

6. The Global Balanced-Projective Dimension

Let R be a valuation domain with quotient field Q. The results of the preceding section indicate that, for torsion-free R-modules M, $bpd(M) \leq pd(M)$. Hence it is natural to inquire whether the supremum of the $bpd(M)$ for torsion-free R-modules M, is related to the supremum of the $pd(M)$. We start with the following definition.

Definition. The *global balanced-projective dimension* of R, in symbols $gl.bpd(R)$, is defined by $gl.bpd(R) = \{bpd(M) \mid M \text{ torsion-free R-module}\}$.

Observe that if $R = \mathbb{Z}_p$, the localization of the ring $\mathbb{Z}$ of integers at a prime p, then $gl.bpd(\mathbb{Z}_p) = 1$, since (balanced) submodules of free $\mathbb{Z}_p$-modules are themselves free. On the other hand, $gl.bpd(\mathbb{Z}) = \infty$, since for every non-negative integer n, there are even finite rank torsion-free abelian groups whose balanced-projective dimensions are n (cf. [2]).

From Theorem 4.4, we get an upper estimate for $gl.bpd(R)$ if R is a valuation domain: It does not exceed the supremum of the projective dimensions of torsion-free R-modules. In particular, if $gl.\dim(R) = n$ is finite, this supremum is n or $n-1$ according as $pd(Q) = n$ or $pd(Q) \leq n-1$.

Note that by Osofsky [11, Ch IV, Theorem 2.2], the projective dimension of an R-submodule of Q depends on the minimal cardinality of its generating sets. Also, the supremum of the projective dimensions of R-submodules of Q is also the supremum of the projective dimensions of torsion-free (cf. [11]). From these we derive the following:

Proposition 6.1. *Let* $gl.\dim(R) = n$. *Then* $n - 2 \leq gl.bpd(R) \leq n$.

Proof. From Theorem 4.4, $gl.bpd(R) \leq n$. Now $n = 1 + \sup\{pd(L) \mid L$ an ideal of $R\}$. Choose an ideal L of R with $pd(L) = n-1$. If we write $L = F/K$ where F is free, then $pd(K) = n-2$ and so, by Lemma 3.1, $bpd(K) = n - 2$. This implies that $n - 2 \leq gl.bpd(R)$. □

A precise value of the $gl.bpd(R)$ is obtained by Fuchs and Rangaswamy [10]. However, they had to assume the Generalized Continuum Hypothesis (GCH). This is needed to assure that, for arbitrary cardinal κ and λ, $2^\kappa = 2^\lambda$ implies that $\kappa = \lambda$.

Theorem 6.2. [10] *Assume GCH. Let R be a valuation domain with* $gl.\dim(R) = n$. *Then* $gl.bpd(R) = n$ *or* $n-1$ *according as* $pd(Q) = n$ *or* $pd(Q) \leq n - 1$.

7. Butler Modules

A torsion-free R-module M is called a *Butler module* if $Bext^1(M,T) = 0$ for all torsion R-modules T. Can we describe all the Butler modules over a valuation domain R? It is conjectured that every Butler R-module is completely decomposable, i.e., balanced-projective. The first partial answer came from L. Fuchs and E. Monari-Martinez who showed the following:

Theorem 7.1. [9] *A Butler module B over a valuation domain R is completely decomposable provided rank* $B \leq \aleph_1$.

Using the ideas of the n-balanced submodule and the $\aleph_n$-family, the following important result was obtained by the author [15] for modules of arbitrary rank.

Theorem 7.2. [15] *Let R be a valuation domain. Then the following are equivalent for a torsion-free R-module M:*
  (i) *M is a Butler module with $bpd(M) \leq 1$;*
  (ii) *M is completely decomposable;*
  (iii) *$Bext^1(M,T) = Bext^2(M,T) = 0$, for all torsion R-modules T.*

It is not clear if the above theorem holds if we replace the condition that $bpd(M) \leq 1$ by $bpd(M) \leq n$ for an arbitrary positive integer n. This leads to the following

Open Question. Is a Butler module M over a valuation domain balanced-projective (= completely decomposable) if it has finite bpd?

References

[1] U. Albrecht and P. Hill, *Butler groups of infinite rank and axiom 3*, Czech. Math. J. **37** (1987), 293–309.

[2] D. Arnold and C. Vinsonhaler, *Pure subgroups of finite rank completely decomposable groups II*, Lecture Notes in Math. **1006**, Springer-Verlag, Heidelberg, New York (1983), 97–143.

[3] R. Baer, *Abelian groups that are direct summands of every containing abelian group*, Bull. Amer. Math. Soc. **46** (1940), 800–806.

[4] S. Dickson, *Torsion theories for abelian categories*, Trans. Amer. Math. Soc. **121** (1966), 223–235.

[5] R. Dimitric, *Balanced projective dimension of modules*, Acta Math. Inform. Univ. Ostraviensis **5** (1997), 39–51.

[6] P. Eklof, *Homological dimension and stationary sets*, Math. Z. **180** (1982), 1–9.

[7] P. Eklof and L. Fuchs, *Baer modules over valuation domains*, Ann. Mat. Pura Appl. **150** (1988), 363–373.

[8] L. Fuchs, *Infinite rank Butler groups*, J. Pure and Applied Algebra **98** (1995), 25–44.

[9] L. Fuchs and E. Monari-Martinez, *Butler modules over valuation domains*, Canad. J. Math. **43** (1991), 48–59.

[10] L. Fuchs and K. M. Rangaswamy, *On the global balanced-projective dimension of valuation domains*, Periodica Math. Hungarica, to appear.

[11] L. Fuchs and L. Salce, *Modules over Valuation Domains*, Lecture Notes in Pure and Applied Math. **97**, Marcel-Dekker, New York, 1985.

[12] I. Kaplansky, *Projective modules*, Ann. Math. **68** (1958), 372–377.

[13] E. Monari-Martinez and K. M. Rangaswamy, *Torsion-free modules over valuation domains whose balanced projective dimension is at most one*, J. Algebra **205** (1998), 91–104.

[14] L. Nongxa, K. M. Rangaswamy and C. Vinsonhaler, *Torsion-free modules of finite balanced-projective dimension over valuation domains*, J. Pure and Applied Algebra, to appear.

[15] K. M. Rangaswamy, *A criterion for complete decomposability and Butler modules over valuation domains*, J. Algebra **205** (1998), 105–118.

Department of Mathematics
University of Colorado
Colorado Springs, CO 80933-7150, U. S. A.
e-mail: ranga@math.uccs.edu

Hecke Orders, Cellular Orders and Quasi-Hereditary Orders

Klaus W. Roggenkamp

Abstract

The example of the Hecke orders over the integral Laurent polynomials of the dihedral groups of "odd" order are used to explain the notions of *Green orders, quasi-hereditary orders, cellular orders, Hecke-orders and deformations of blocks with cyclic defect.* Green orders, which arise as blocks of cyclic defect of local groups rings and Hecke orders are characterized internally, and their "filtered Cohen-Macaulay modules are described. This gives a local description of the Cohen-Macaulay for the Hecke orders of the "odd" order dihedral groups.

1. Introduction

The aim of this note is to make some propaganda for 2-dimensional orders, e.g. orders over $\mathbb{Z}[q]$ as they occur with integral quantum groups and Hecke orders. We explain the important notions of

GREEN ORDERS, QUASI-HEREDITARY ORDERS, CELLULAR ORDERS, HECKE ORDERS AND DEFORMATIONS OF BLOCKS WITH CYCLIC DEFECT

through a thorough examination of the integral group ring $\mathbb{Z}S_3$ of the symmetric group on three letters and its Hecke order $\mathcal{H}_{S_3}$. This is done in detail in Section 2.

In Section 3 the basic definition of the Hecke order of a BN-pair of rank two is given, and the Hecke order of the dihedral group of order $2 \cdot p^n$ for an odd prime p is described. The structure of these Hecke orders requires a new definition of orders in Section 4.

The cell structure of $\mathcal{H}_{S_3}$ is described in Section 5 and generalized in Section 6. We give definitions which are better suited for the applications

---

1991 *Mathematics Subject Classification:* Primary 16G30
This research was partially supported by the Deutsche Forschungsgemeinschaft.

than the classical ones. For example, with the definition of Graham and Lehrer of cellular algebras [GrLe; 96] the integral group ring of the dihedral groups of order $2 \cdot p^n$ for $p > 3$ is not cellular. With our definition, which coincides with the one of Graham and Lehrer in the splitting case, these rings are cellular. Our description of the Hecke orders $\mathcal{H}_{D_{p^n}}$ is so explicit that we can compute the extension groups of the Weyl modules (Section 7).

Quasi-hereditary orders are introduced as quotients of $\mathcal{H}_{S_3}$ (Section 8 and defined in quite some generality in Section 9). We then turn to separable deformations in Sections 10 and 11. In Sections 13, 14, 15, Green orders – generalizing Brauer tree orders – locally embedded graphs are introduced and an internal description is given. Moreover, their "filtered" Cohen-Macaulay modules are described.

I would like to thank Steffen König for supplying references for the examples of cellular and quasi-hereditary orders.

2. The Hecke Order $\mathcal{H}_{S_3}$ of the Symmetric Group S_3

Definition 2.1. The Hecke order of the symmetric group on three letters is defined as $\mathcal{H}_{S_3} := \mathbb{Z}[q, q^{-1}]\langle a, b \rangle$ subject to the relations:
1. Quadratic relations:

$$x^2 = (q-1) \cdot x + q \cdot 1 \text{ for } x = a, b.$$

2. Homogeneous relations:

$$a \cdot b \cdot a = b \cdot a \cdot b.$$

An irreducible $\mathcal{H}_{S_3}$-representation is given by:
1.

$$a \longrightarrow \begin{pmatrix} -1 & 0 \\ -1 & q \end{pmatrix} \text{ and } b \longrightarrow \begin{pmatrix} -1 & 1+q+q^2 \\ 0 & q \end{pmatrix},$$

2. the index representation: $ind(a) = q$ and $ind(b) = q$, and the sign representation: $sgn(a) = sgn(b) = -1$.
We put

$$[n] = \sum_{i=0}^{n-1} q^i, \ n \in \mathbb{N}.$$

Because of the above representations we may consider

$$\mathcal{H}_{S_3} \subset \begin{pmatrix} \mathbb{Q}(q) & \mathbb{Q}(q) \\ \mathbb{Q}(q) & \mathbb{Q}(q) \end{pmatrix} \oplus \mathbb{Q}(q)^- \oplus \mathbb{Q}(q)^q.$$

Two-Dimensional Orders

The Hecke order of the cyclic group of order 2 is defined as

$$\mathcal{H}_2 := \mathbb{Z}[q, q^{-1}]\langle\, a \,:\, a^2 - (q-1)\cdot a - q = 0 = (a+1)\cdot(a-q)\,\rangle.$$

It can be written as a pull-back

$$\begin{array}{ccccccccc}
& & 0 & & 0 & & 0 & & \\
& & \downarrow & & \downarrow & & \downarrow & & \\
0 \to & & 0 & \to & (a+1)\cdot\mathcal{H}_2^+ & \to & [2]\cdot\mathbb{Z}[q,q^{-1}] & \to & 0 \\
& & \downarrow & & \downarrow & & \downarrow & & \\
0 \to & (a-q)\cdot\mathcal{H}_2^+ & \to & & \mathcal{H}_2 & \to & \mathbb{Z}[q,q^{-1}]^q & \to & 0. \\
& \downarrow & & & \downarrow & & \downarrow & & \\
0 \to & (-[2])\cdot\mathbb{Z}[q,q^{-1}] & \to & & \mathbb{Z}[q,q^{-1}]^- & \to & \mathbb{Z} & \to & 0 \\
& \downarrow & & & \downarrow & & \downarrow & & \\
& 0 & & & 0 & & 0 & &
\end{array}$$

Lemma 2.2. *The $\mathbb{Z}[q, q^{-1}]$-order generated by the two-dimensional representation is given as*

$$\Lambda_3 := \begin{pmatrix} \mathbb{Z}[q,q^{-1}] & [3]\cdot\mathbb{Z}[q,q^{-1}] \\ \mathbb{Z}[q,q^{-1}] & \mathbb{Z}[q,q^{-1}] \end{pmatrix}.$$

This is wrong if q is not invertible. We now describe the amalgamation between Λ_3 and $\mathcal{H}_2$:

Lemma 2.3. *We put*

$$Tr_3 = 1 + q^{-1}\cdot a\cdot b + q^{-2}\cdot(a\cdot b)^2, \quad \text{then}$$

we have a commutative diagram with exact rows and columns

$$\begin{array}{ccccc}
0 & \to & Tr_3\cdot\mathcal{H}_{S_3} & \xrightarrow{\simeq} & [3]\cdot\mathcal{H}_2 \\
\downarrow & & \downarrow & & \downarrow \\
(b-a)\cdot\mathcal{H}_{S_3} & \to & \mathcal{H}_{S_3} & \to & \mathcal{H}_2 \\
\downarrow & & \downarrow & & \downarrow \\
(b-a)\cdot\Lambda_3 & \to & \Lambda_3 & \to & \mathbb{Z}[\theta_3]\times\mathbb{Z}[\theta_3]
\end{array}$$

We have a similar description of the integral group ring of S_3:

$$S_3 :=<\alpha,\,\beta \,:\, \alpha^2,\,\beta^2,\,\alpha\cdot\beta\cdot\alpha = \beta\cdot\alpha\cdot\beta>.$$

We point out that reduction modulo $<q-1>$ is not compatible with taking quotients, so the structure is not a consequence of the structure of the Hecke algebra. For the cyclic group S_2 of order 2 we have a pull-back:

$$\begin{array}{ccc} \mathbb{Z}S_2 & \xrightarrow{mod(a-1)} & \mathbb{Z} \\ {\scriptstyle mod(a+1)}\downarrow & & \downarrow{\scriptstyle mod(2)} \\ \mathbb{Z} & \xrightarrow{mod(2)} & \mathbb{F}_2 \end{array},$$

and if we put

$$\Gamma_3 := \begin{pmatrix} \mathbb{Z} & 3 \cdot \mathbb{Z} \\ \mathbb{Z} & \mathbb{Z} \end{pmatrix},$$

then the group ring $\mathbb{Z}S_3$ is the pull-back:

$$\begin{array}{ccc} \mathbb{Z}S_3 & \xrightarrow{mod(\alpha-\beta)} & \mathbb{Z}S_2 \\ {\scriptstyle mod(1+\alpha\cdot\beta+(\alpha\cdot\beta)^2)}\downarrow & & \downarrow{\scriptstyle mod(3)} \\ \Gamma_3 & \longrightarrow & \mathbb{F}_3 S_2 \simeq \mathbb{F}_3^+ \times \mathbb{F}_3^- \end{array}. \quad (1)$$

We shall use a common description of $\mathbb{Z}S_3$ and of $\mathcal{H}_{S_3}$. To this end we put

$$R = \mathbb{Z} \text{ or } R = \mathbb{Z}[q, q^{-1}] \text{ and } [p] = p \text{ or } [p] = 1 + q + q^2 + \cdots + q^{p-1},$$

and we shall use the *notation*

$$\left(R \xrightarrow{[p]} R\right) = \{(r, r + [p] \cdot s) : r, s \in R\}.$$

Then we may write $\mathcal{H}_{S_3}$ and $\mathbb{Z}S_3$ in the following unified form which we shall use throughout this paper. This shows that both the group ring $\mathbb{Z}S_3$ as well as its Hecke order are "generically" the same. My feeling is that the same should hold for other Hecke orders, in the sense that congruences modulo a prime power p^n in the integral group ring have to be replaced in the Hecke order by congruences modulo $[p^n]$.

$$\mathcal{O} := \begin{pmatrix} R & R \\ <[p_1]> & R \end{pmatrix} \xrightarrow{\quad [3] \quad} (R^-). \qquad (2)$$

with (R^q) above via $[3]$ and $[2]$.

3. Hecke Orders of Rank 2 BN-Pairs

Definition 3.1. Let $m \in \mathbb{N}^+$, then the *Hecke order of the dihedral group of order* $2 \cdot m$ is

$$\mathcal{H}_{D_m} := \mathbb{Z}[q, q^{-1}]\langle a = a_m, b = b_m \rangle \text{ subject to:}$$

1. Quadratic relations:
$$x^2 = (q-1) \cdot x + q \cdot 1 \text{ for } x = a, b.$$

2. Homogeneous relations:
$$(a \cdot b)^k = (b \cdot a)^k \text{ if } m = 2 \cdot k$$

and

$$(a \cdot b)^k \cdot a = (b \cdot a)^k \cdot b \text{ if } m = 2 \cdot k + 1.$$

The Hecke order $\mathcal{H}_{D_m}$ is $\mathbb{Z}^q := \mathbb{Z}[q, q^{-1}]$-free with a semi-simple ring of quotients isomorphic to $\mathbb{Q}(q)D_m$, the group ring of the Dihedral group of order 2 over the rational functions $\mathbb{Q}(q)$.

$\mathbb{Z}[q]$ has *two different types of minimal prime* ideals, height one primes — which are principal, $\mathbb{Z}[q]$ being factorial. *The arithmetic prime ideals*:

$$< f(q) > : f(q) \in \mathbb{Z}[q] \text{ irreducible in } \mathbb{Q}[X], \text{ GGT(coeff)} = 1.$$

As a quotient field of the residue ring, *every algebraic number field* can occur; hence $\mathbb{Z}[q]$ *incorporates all of algebraic number theory*, which provides for the *integral theory*.

The geometric prime ideals:

$$\langle p \rangle, p \text{ a rational prime; then } \mathbb{Z}[q]/\langle p \rangle \simeq \mathbb{F}_p[q],$$

maps onto *every finite field*, thus providing for the *modular theory*.

The integral theory and the modular theory are interrelated via pullbacks

$$\begin{array}{ccc} \mathbb{Z}[q] & \longrightarrow & \mathbb{Z}[q]/<f(q)> \\ \downarrow & & \downarrow \phi \\ \mathbb{F}_p[q] & \longrightarrow & \mathbb{F}_{p^n} \end{array},$$

ϕ is a reduction modulo a maximal ideal above p in $\mathbb{Z}[q]/<f(q)>$. In general, $\mathbb{Z}[q]/<f(q)>$ is not the ring of algebraic integers in $\mathbb{Q}[q]/<f(q)>$.

For the description of the Hecke order we fix some

Notation 3.2. For θ_m a primitive m-th root of unity, with $m = p^n$ for an odd prime p we put $\eta_m := \theta_m + \theta_m^{-1}$.

ring	$R_m := \mathbb{Z}[\eta_m]$	$R_m^q := \mathbb{Z}^q \otimes_\mathbb{Z} \mathbb{Z}[\eta_m]$
field of fractions	K_m	K_m^q
prime element	$\gamma_m := (1 - \theta_m) \cdot (1 - \theta_m^{-1})$	$\gamma_m^q := (q - \theta_m) \cdot (q - \theta_m^{-1})$
prime ideal	$\pi_m := R_m \langle \gamma_m \rangle$	$\pi_m^q := R_m^q \langle \gamma_m^q \rangle$.

Proposition 3.3. *The $\mathbb{Z}^q$-order in $(K_m^q)_2$ generated by the faithful irreducible representation of $\mathcal{H}_{D_m}$ is*

$$\Lambda_m = \begin{pmatrix} R_m^q & \pi_m^q \\ R_m^q & R_m^q \end{pmatrix}.$$

Proposition 3.4. *We put $d_n := q^{-1} \cdot a_m \cdot b_m$, $n > 1$ – this corresponds in the dihedral group to the element of order m – and define*

$$\omega_n^d(d_n) := \Big(\sum_{i=0}^{p^{n-1}-1} d_n^i \Big) \cdot (a_m - b_m)$$

and

$$tr_n^d := \Big(\sum_{i=0}^{p-1} d_n^{i \cdot p^{n-1}} \Big).$$

Then for the ideal generated by $\omega_n^d(d_n)$

$$<\omega_n^d(d_n)>_{\mathcal{H}_{D_m}} = \pi_m^\nu \cdot <a_m - b_m> = \pi_m^\nu \cdot \begin{pmatrix} \pi_m^q \cdot R_m^q & \pi_m^q \cdot R_m^q \\ R_m^q & \pi_m^q \cdot R_m^q \end{pmatrix}.$$

$$\begin{array}{ccccccccc}
 & & 0 & & 0 & & 0 & & \\
 & & \downarrow & & \downarrow & & \downarrow & & \\
0 & \to & 0 & \to & tr_n^d \cdot \mathcal{H}_{D_m} & \xrightarrow{\simeq} & \overline{tr_n^d} \cdot \mathcal{H}_{D_{p^{n-1}}} & \to & 0 \\
 & & \downarrow & & \downarrow & & \downarrow & & \\
0 & \to & \omega_n^d(d_n) \cdot \mathcal{H}_{D_m} & \to & \mathcal{H}_{D_m} & \to & \mathcal{H}_{D_{p^{n-1}}} & \to & 0 \\
 & & \downarrow & & \downarrow & & \downarrow & & \\
0 & \to & \omega_n^d(d_n) \cdot \Lambda_n^q & \to & \Lambda_n^q & \to & \Sigma_n^d & \to & 0 \\
 & & \downarrow & & \downarrow & & \downarrow & & \\
 & & \cup & & 0 & & 0 & &
\end{array}$$

where Σ_n^d is given as a pull-back

$$\begin{array}{ccc}
\Sigma_n^d & \longrightarrow & \mathbb{Z}[\theta_m] \times \mathbb{Z}[\theta_m] \\
\downarrow & & \downarrow \\
\Delta_n^d & \longrightarrow & \mathbb{F}[X]/<X^{2 \cdot \nu}> \times \mathbb{F}[X]/<X^{2 \cdot \nu}>
\end{array}$$

Two-Dimensional Orders   335

with
$$\Delta_n^d = \begin{pmatrix} R_m^q/\pi_m^\nu & \pi_m^q/\pi_m^q \cdot \pi_m^\nu \\ R_m^q/\pi_m^\nu & R_m^q/\pi_m^\nu \end{pmatrix}.$$

4. Orders

We have seen above that in the Hecke orders there occur amalgamations of orders in different characteristics. This leads to the following

Definition 4.1. 1. A commutative ring R is said to be *locally integral* if it has a total ring of quotients K which is a finite product of fields.

2. We say that a finitely – as a module – generated R-algebra Λ is an *R-order in the separable K-algebra* $A := K \otimes_R \Lambda$ provided A is a separable K-algebra and Λ embeds into A.

3. A left Λ-*lattice* M is a finitely generated left Λ-module which embeds into $KM := K \otimes_R M$.

4. The lattice M is said to be *irreducible* provided KM is a simple A-module.

Example 4.2. Define R and Λ as pull-backs

$$\begin{array}{ccc} R & \longrightarrow & \mathbb{Z} \\ \downarrow & & \downarrow \\ \mathbb{F}_p[X] & \longrightarrow & \mathbb{F}_p \end{array} \qquad \begin{array}{ccc} \Lambda & \longrightarrow & \begin{pmatrix} \mathbb{Z} & \mathbb{Z} \\ p \cdot \mathbb{Z} & \mathbb{Z} \end{pmatrix} \\ \downarrow & & \downarrow \\ \mathbb{F}_p[X] \times \mathbb{F}_p[X] & \longrightarrow & \mathbb{F}_p \times \mathbb{F}_p \end{array}.$$

5. The Cell Structure of $\mathbb{Z}S_3$ and $\mathcal{H}_{S_3}$

We shall give a *common description* of $\mathbb{Z}S_3$ and $\mathcal{H}_{S_3}$. To this end we make the following conventions, which we shall use throughout this paper:

$R = \mathbb{Z}$ or $R = \mathbb{Z}[q, q^{-1}]$, $[p] = p$ or $[p] = 1 + q + q^2 + \cdots + q^{p-1}$.

For pull-backs we write

$$\left(R \xrightarrow{[p]} R \right) = \{(r, r + [p] \cdot s) \; : \; r, s \in R\}.$$

The group ring $\mathbb{Z}S_3$ and the Hecke order $\mathcal{H}_{S_3}$ can then be described as (cf. above):

$$\mathcal{O} := \begin{pmatrix} & & (R^q) & & \\ & [3] & \Big| & [2] & \\ & \begin{pmatrix} R & R \\ <[3]> & R \end{pmatrix} & & [3] & \\ & & & & (R^-) . \end{pmatrix} \qquad (3)$$

We now consider some ideals in $\mathcal{O}$:

$$J_0 = \overline{J_0} = P_0 := [2] \cdot [3] \cdot R^q \text{ and if } \mathcal{O}_1 = \mathcal{O}/J_0, \text{ then}$$

$$\mathcal{O}_1 := \begin{pmatrix} R & R \\ <[3]> & R \end{pmatrix} \underline{\quad[3]\quad} (R^-). \qquad (4)$$

$\mathcal{O}_1$ has an ideal

$$\overline{J_1} := \begin{pmatrix} R & <[3]> \\ R & <[3]> \end{pmatrix}. \text{ We put } P_1 := \begin{pmatrix} R \\ R \end{pmatrix}.$$

Let J_1 be the preimage of $\overline{J_1}$ in $\mathcal{O}$. Then we have

$$\mathcal{O}_2 = \overline{J_2} = P_2 := \mathcal{O}_1/\overline{J_1} = \mathcal{O}/J_1 = R^-.$$

What we have constructed is a *chain of cell ideals*

$$J_0 \subset J_1 \subset \mathcal{O}.$$

In addition, $\mathcal{O}$ admits an anti-involution ι, which is the identity on R^q and on R^-, and acts on the 2-dimensional representation by

$$\text{sending } \begin{pmatrix} a & b \\ [3] \cdot c & d \end{pmatrix} \text{ to } \begin{pmatrix} a & c \\ [3] \cdot b & d \end{pmatrix}.$$

For $0 \leq i \leq 2$ we have a very *special structure* of the cell ideals:

$$\overline{J_i} = P_i \otimes_{\text{End}_\mathcal{O}(P_i)} {}^\iota P_i.$$

So, $\mathcal{O}$ is a *cellular order* in the sense of the next section.

6. Cellular Orders

Definition 6.1. (*Cellular Order*) Assume that ι is an R-linear anti-involution on the R-order Λ in the sense of Definition 4.1 in the separable K-algebra A. A two-sided ideal J of Λ is called a *cell ideal*, provided
- $K \cdot J = A \cdot e$ for a central primitive idempotent e of A,
- Λ/J is an $R_J := R \cdot e$-order in $A \cdot e$,
- $\iota(J) = J$ and
- there exists a left Λ-ideal $\Delta := \Delta(J) \subset J$ such that
  - Δ is a Λ-lattice,
  - Δ is irreducible,
  - $E(\Delta) := End_\Lambda(\Delta) \subset \Lambda$ under the natural embedding $End(K \otimes_R \Delta) \subset A$,
  - $\iota(E(\Delta)) = E(\Delta)$;
  - there is an isomorphism of 2-sided Λ-modules

$$\alpha \colon J \to \Delta \otimes_{E(\Delta)} \iota(\Delta),$$

making the following diagram commute

$$\begin{array}{ccc} J & \xrightarrow{\alpha} & \Delta \otimes_{E(\Delta)} \iota(\Delta) \\ \iota \downarrow & & \downarrow x \otimes y \mapsto \iota(y) \otimes \iota(x) \\ J & \xrightarrow{\alpha} & \Delta \otimes_{E(\Delta)} \iota(\Delta) \end{array}.$$

The left ideal Δ is called the *cell module* – sometimes it is also called the *Weyl-module* or in case of the symmetric group the *Specht module*. The pair (Λ, ι) is said to be a *cellular order* provided there exists a chain of two-sided Λ-ideals

$$0 = J_0 \subset J_1 \subset J_2 \subset \cdots \subset J_n = A$$

such that J_i/J_{i-1} is a cell ideal (with respect to ι induced on Λ/J_{i-1}) in Λ/J_{i-1}.

We point out that a cell structure also gives a numbering (ordering) of the simple components of the underlying algebra, and also an ordered set of cell modules $\{\Delta_i = M_i\}$ corresponding to the cell ideals J_i.

A typical example of *mixed characteristics* is given as follows: The congruences "α and β" are given as the pull-backs, where R is an unramified extension of $\mathbb{Z}_p$.

$$\begin{array}{ccccccccc} \mathbb{Z}_{<p>} & \xrightarrow{\alpha} & \mathbb{F}_p[X] & \longrightarrow & \mathbb{Z}_{<p>} & & \mathbb{Z}_{<p>} & \xrightarrow{\beta} & R & \longrightarrow & R \\ \downarrow & & & & \downarrow & & \downarrow & & & & \downarrow \\ \mathbb{F}_p[X] & & & \longrightarrow & \mathbb{F}_p & & \mathbb{Z}_{<p>} & & & \longrightarrow & \mathbb{F}_p \end{array}$$

The Λ is a cellular order of mixed characteristic:

$$\Lambda := \begin{pmatrix} \mathbb{F}_p[X] & \mathbb{F}_p[X] \\ X \cdot \mathbb{F}_p[X] & \mathbb{F}_p[X] \end{pmatrix}$$

$$\diagup \alpha$$

$$\begin{pmatrix} \mathbb{Z}_{<p>} & \mathbb{Z}_{<p>} \\ <p> \cdot \mathbb{Z}_{<p>} & \mathbb{Z}_{<p>} \end{pmatrix}$$

$$\beta \diagdown$$

$$R \ .$$

Example 6.2. 1. Hecke algebras of type A or B or more generally Ariki-Koike algebras (Graham-Lehrer [GrLe; 96]).
  2. Brauer algebras (Graham-Lehrer [GrLe; 96]).
  3. Temperly-Lieb algebras (Graham-Lehrer [GrLe; 96]).
  4. q-Schur algebras (Dipper-James [DiJal; 89]).
  5. Jones' annular algebras (Graham-Lehrer [GrLe; 96]).

Remark 6.3. In the few examples I know explicitly the left ideal Δ is an ideal for a hereditary order $\Lambda(\Delta)$ in a simple algebra, i.e., an order of type

$$\mathbb{H} = \mathbb{H}_{\Omega,\omega,n} := \begin{pmatrix} \Omega & \Omega & \Omega & \cdots & \Omega & \Omega \\ \omega & \Omega & \Omega & \cdots & \Omega & \Omega \\ \omega & \omega & \Omega & \cdots & \Omega & \Omega \\ \cdots & \cdots & \cdots & \ddots & \cdots & \cdots \\ \omega & \omega & \omega & \cdots & \Omega & \Omega \\ \omega & \omega & \omega & \cdots & \omega & \Omega \end{pmatrix}_n,$$

where Ω is a *hereditary* R-order in a skew field, in the sense that R-projective R-lattices are Ω-projective.
  1. $\Lambda(\Delta)$ is R-projective as an R-module.
  2. The anti-involution ι induces an anti-involution on $\Lambda(\Delta)$.
  3. Δ is a projective $\Lambda(\Delta)$-module.

Definition 6.4. If all the cell ideals of a cellular order satisfy the above conditions, we call it a *projectively cellular order*.

Proposition 6.5. [Ro; 97] *The integral group ring of the dihedral group D_m of order $2 \cdot m$ for $m = p^n$ with p an odd prime or integral Hecke order of D_m is a projectively cellular order.*

Two-Dimensional Orders   339

For $m > 3$ these rings are not cellular in the original definition of Graham and Lehrer [GrLe; 96], and the proof here uses totally different techniques.

7. Extensions of the Cell Modules for the Hecke Orders of Dihedral Groups

For the structure of the lattice category of cell orders, the most important category are those lattices which have a filtration by cell modules in the natural order. Thus it is one of the goals to compute the extension groups between the cell modules, which is solved for Hecke orders of dihedral groups in

Theorem 7.1. (1) *For the Hecke order $\mathcal{H}_{D_{p^n}}$ of the dihedral group of order $2 \cdot p^n$ with p an odd prime we have the cell modules, M_0 the index representation, M_{n+1} the sign representation, and M_i a "faithful" irreducible $\mathcal{H}_{D_{p^i}}$-module, $1 \leq i \leq n$, which is the first column of the order Λ_{p^i} from Proposition 3.3.*

(2) (a) *For $i > j \in \{1, \ldots, n\}$ we have*

$$Ext^1_{\mathcal{H}_{D_n}}(M_i, M_j) = Ext^1_{\mathcal{H}_{D_n}}(M_j, M_i) = \mathbb{F}_p[X]/<X^{\mu(j)}>[q, q^{-1}]$$

with $\mu(j) := \frac{p^{j-1} \cdot (p-1)}{2}$.

(b) *For $i \in \{1, \ldots, n\}$ we have*

$$Ext^1_{\mathcal{H}_{D_n}}(M_i, M_0) = \mathbb{Z}[\theta_{p^i}] = Ext^1_{\mathcal{H}_{D_n}}(M_{n+1}, M_i) = \mathbb{Z}[\theta_{p^i}],$$

$$Ext^1_{\mathcal{H}_{D_n}}(M_i, M_{n+1}) = 0 = Ext^1_{\mathcal{H}_{D_n}}(M_0, M_i).$$

(c) *Moreover,*

$$Ext^1_{\mathcal{H}_{D_n}}(M_0, M_{n+1}) = Ext^1_{\mathcal{H}_{D_n}}(M_{n+1}, M_0) = \mathbb{Z}.$$

Note 7.2. We point out that for $n > 1$ the extension groups between the cell modules are partly $\mathbb{Z}$-torsion-free and partly $\mathbb{F}_p[q, q^{-1}]$-torsion-free. So the extension groups are lattices over regular 1-dimensional rings.

8. The Quasi-Hereditary Quotients of $\mathbb{Z}S_3$ and $\mathcal{H}_{S_3}$

Let us look at the order $\mathcal{O}_1$ from Equation 4, which is a quotient of the Hecke order (group ring) of S_3

$$\mathcal{O}_1 := \begin{pmatrix} R & R \\ <[3]> & R \end{pmatrix} \underline{\quad[3]\quad} (R^-).$$

This is a quasi-hereditary order. In fact it has an idempotent

$$e := \left(\begin{pmatrix} 1 & 0 \\ 0 & 0 \end{pmatrix}, 0 \right) \text{ such that } \overline{J_1} = \mathcal{O}_1 \cdot e \cdot \mathcal{O}_1 = \begin{pmatrix} R & \pi \cdot R \\ R & \pi \cdot R \end{pmatrix}$$

is a pure (this means that the quotient is torsion-free) R-free ideal with $\mathcal{O}_2 = \mathcal{O}_1/\overline{J_1} = R$ hereditary. We formulate this as

Lemma 8.1. *Let $R = \mathbb{Z}$ or $R = \mathbb{Z}[q, q^{-1}]$ and $[p] = p$ or $[p] = 1+q+q^2+\cdots+q^{p-1}$. Then $J_0 := [2] \cdot [3] \cdot R^q$ is a pure R-free ideal in $\mathbb{Z}S_3$ and $\mathcal{H}_{S_3}$ resp., such that $\Lambda := \mathbb{Z}S_3/J_0$ and $\Lambda := \mathcal{H}_{S_3}/J_0$ are R-free quasi-hereditary orders. Moreover, in both cases Λ is the Schur order to the direct sum of the trivial and three-dimensional permutation representation; i.e., its endomorphism ring.*

9. Quasi-Hereditary Rings

We shall try here to give a very general definition of quasi-hereditary rings.

Definition 9.1. Let Λ be a two-sided noetherian ring. A two-sided ideal J is said to be a *heredity ideal*, provided
 1. J is projective as a left Λ-module,
 2. J is of the form $J = \Lambda \cdot \epsilon \cdot \Lambda \simeq \Lambda \cdot \epsilon \otimes_{\epsilon \cdot \Lambda \cdot \epsilon} \epsilon \cdot \Lambda$ for an idempotent $\epsilon \in \Lambda$,
 3. the ring $\epsilon \cdot \Lambda \cdot \epsilon$ is a separable algebra over its center.

In case Λ is an R-order in a separable K-algebra A (in the sense of Definition 4.1), we require in addition that $K \cdot J$ is a simple K-algebra. The ring (R-order) Λ is called *quasi-hereditary*, provided there is a proper chain of two-sided ideals $J_0 = 0 \subset J_1 \subset \cdots \subset J_n = \Lambda$ such that for $1 \leq i \leq n$ the quotient J_i/J_{i-1} is a heredity ideal in the ring (R-order) Λ/J_{i-1}.

Note 9.2. We point out that in general the map $\mu : \Lambda \cdot \epsilon \otimes_{\epsilon \cdot \Lambda \cdot \epsilon} \epsilon \cdot \Lambda \to \Lambda \cdot \epsilon \cdot \Lambda$ is not injective. If, however, Λ is an order in a separable K-algebra, then automatically μ is injective, since it is injective on the total ring of quotients.

These quasi-hereditary orders occur in various situations in representation theory:
 • A general reference for integral quasi-hereditary algebras is the basic paper of E. Cline, B. Parshall and L. Scott [CPS; 88] and [CPS; 90].

- J. A. Green gave a combinatorial proof that the classical Schur orders (over $\mathbb{Z}$) are quasi-hereditary [Gr; 93].
- R. Dipper and G. James showed that the q-Schur algebras are quasi-hereditary [DiJa; 89].
- R. Dipper, G. James and A. Mathas defined Schur-algebras to the Ariki-Koike algebras and noted that they are quasi-hereditary [DiJaMa; 98].
- The quotient $\mathcal{H}_{D_{p^n}}/[p^n] \cdot \mathbb{Z}[q, q^{-1}]^q$ is quasi-hereditary [Ro; 97].

In case of an artinian algebra, a quasi-hereditary algebra has finite global dimension. I do not know whether this holds in general.

10. A Separable Deformation of $\mathbb{Z}S_3$

Let us look at the order $\mathcal{D}_3(q) := \mathcal{H}_3^+ := \mathbb{Z}[q] <a, b>$ with $R = \mathbb{Z}[q]$ (cf. Diagram 1).

$$\mathcal{D}_3(q) := \begin{array}{c} (R^q) \\ {}^{[3]}\big| \quad \diagdown {}^{[2]} \\ \begin{pmatrix} R & R \\ <[3]> & R \end{pmatrix} \quad \xrightarrow{[3]} \quad (R^-) \, , \end{array}$$

– note that now q is not invertible anymore, and the structural results from above (cf. Section 2) do not hold any more.

We consider the order $\mathbb{Z}_p \cdot \mathcal{D}_3(q)$ for the various rational primes p and specialize q to various elements in the ring of p-adic integers $\mathbb{Z}_p$.

Case 1: $p = 3$.
- Let $q = 1 + 3 \cdot x \in \mathbb{Z}_3$. Then $<1+q> = \mathbb{Z}_3$ and $<1+q+q^2> = 3 \cdot \mathbb{Z}_3$. In this case

$$\mathbb{Z}_3 \cdot \mathcal{D}_3(q) = \mathbb{Z}_3 S_3 \, .$$

- If $q = 3 \cdot x \in \mathbb{Z}_3$, then both $q+1$ and $1+q+q^2$ are units and so we get the separable order

$$\mathbb{Z}_3 \cdot \mathcal{D}_3(q) = \mathbb{Z}_3 \times (\mathbb{Z}_3)_2 \times \mathbb{Z}_3 \, .$$

- If $q = 2 + 3 \cdot x \in \mathbb{Z}_3$, then $<1+q+q^2> = \mathbb{Z}_3$ but $1 + q = 3 \cdot (1 + x)$. Here we have two different cases:
  - If $x = -1$ then $1 + q = 0$ and we have a *decrease in dimension*:

$$\mathbb{Z}_3 \cdot \mathcal{D}_3(q) = \mathbb{Z}_3 \times (\mathbb{Z}_3)_2 \, .$$

Also this order is separable, but of rank 5.

- If $x \neq -1$, then x can be chosen in such a way that we have

$$<1+q> = <3 \cdot (1+x)> = <3^n> \text{ for any } n \in \mathbb{N}.$$

In this case we obtain a *chain of inseparable orders depending on* n:

$$\mathbb{Z}_3 \cdot \mathcal{D}_3(q) = \{(a, a + 3^n \cdot b) : a, b \in \mathbb{Z}_3\} \times (\mathbb{Z}_3)_2.$$

Case 2: $p = 2$.
- $q = 1 + 2 \cdot x$ in this case $1 + q + q^2$ is a unit and $q + 1 = 2(1+x)$.
- For $x = -1$ we obtain a *decrease in dimension*:

$$\mathbb{Z}_2 \cdot \mathcal{D}_3(q) = (\mathbb{Z}_2)_2 \times \mathbb{Z}_2$$

has dimension 5.

- For $x \neq -1$ the ideal $<2 \cdot (1+x)>$ can be any power $<2^n>$, and we have:

$$\mathbb{Z}_2 \cdot \mathcal{D}_3(q) = (\mathbb{Z}_2)_2 \times \{(a, a + 2^n \cdot b) : a, b \in \mathbb{Z}_2\},$$

which is inseparable, but depends on n.
This includes the group ring.

- For $q = 2 \cdot x$ we get the separable order

$$\mathbb{Z}_2 \cdot \mathcal{D}_3(q) \simeq (\mathbb{Z}_2)_2 \times \mathbb{Z}_2 \times \mathbb{Z}_2.$$

Case 3: $p \neq 2, 3$. In this case it is easily seen that we obtain a separable order, since both $1 + q$ and $1 + q + q^2$ are units.

So we see that apparently we *cannot get a global* $\mathbb{Z}[q]$-*order which upon tensoring with* $\mathbb{Z}_p$ *gives a separable deformation of* $\mathbb{Z}_p S_3$ in the sense of Section 11.

As a matter of fact, we have to do this a *prime at a time*. The proof of the next result follows from the previous calculations:

Lemma 10.1. • For $p = 3$, the $\mathbb{Z}_3[q]$-*order*

$$\{(u, v, \begin{pmatrix} u + [3] \cdot a & [3] \cdot b \\ c & v + [3] \cdot d \end{pmatrix}) : u, v, a, b, c, d \in \mathbb{Z}_3[q]\}$$

is a separable deformation of $\mathbb{Z}S_3$.

• For $p = 2$, the $\mathbb{Z}_2[q]$-*order*

$$(\mathbb{Z}_2[q])_2 \times \{(a, a + (q+1) \cdot b) : a, b \in \mathbb{Z}_2[q]\}$$

is a separable deformation of $\mathbb{Z}S_3$.

Note 10.2. Let us point out that specializing q to a 3rd root of unity, we create a nilpotent ideal.

11. Separable and Maximal Deformations

We now formalize the observations from the previous section.

Definition 11.1. Let R be a complete Dedekind domain with maximal ideal $\mathfrak{m}$ and let Λ be an R-order in a separable algebra over the field of fractions of R. A Cohen-Macaulay $R[q]$-order $\mathbb{H}$ is called a *separable deformation of* Λ, provided
 1. $\mathbb{H} / <q-1> \simeq \Lambda$,
 2. for $r \in R$ not congruent to 1 modulo $\mathfrak{m}$ the order $\mathbb{H}/<q-r>$ is a separable R-order in a separable algebra.

$\mathbb{H}$ is called a *maximal deformation of* Λ, provided
 1. $\mathbb{H} / <q-1> \simeq \Lambda$,
 2. for $r \in R$ not congruent to 1 modulo $\mathfrak{m}$ the order $\mathbb{H} / <q-r>$ is a maximal R-order in a separable algebra.

Note 11.2. We point out that for a separable deformation we have no influence at all over what happens when we specialize q to an element in $\mathfrak{m}$. Moreover, maximal deformations have the advantage that the definition is independent of the ground ring over which we consider the order Λ.

Example 11.3. 1. The Hecke order completed at 3 of the symmetric group S_3 on three letters is a separable deformation of the 3-adic group ring $\mathbb{Z}_3 S_3$.

2. The Hecke order of the dihedral group D_8 of order 16 completed at 2 is a maximal deformation of the 2-adic group ring $\mathbb{Z}D_8$; it is not a separable deformation.

3. More generally, p-adic integral blocks with cyclic defect have maximal deformations which arise by lifting idempotents; they are "Blocks with cyclic defect of Hecke orders" (cf. [Ro; 98, I]).

4. Modular blocks with cyclic defect have semi-simple deformations in a weaker sense (M. Schaps [Sch; 94]).

12. Localizations of $\mathcal{H}_{S_3}$ and $\mathbb{Z}_3 S_3$

$R_\wp$ is the completion of R, a noetherian integral domain, at the prime ideal $\wp$. For our Hecke order $\mathcal{H}_{S_3}$ we are interested in maximal ideals $\wp$ lying above $<q-1>$ – they are of the form $\wp = \langle p, q-1 \rangle$ – since we still want to have an epimorphism

$$\mathbb{Z}[q, q^{-1}]_\wp \otimes_{\mathbb{Z}[q,q^{-1}]} \mathcal{H}_{S_3} \to \mathbb{Z}_p S_3$$

modulo $<q-1>$. Note $q \notin \mathfrak{m}_p$, since $1 \notin \mathfrak{m}_p$. Hence

$$\mathbb{Z}[q,q^{-1}]_{\mathfrak{m}_p} = \mathbb{Z}[q]_{\mathfrak{m}_p}.$$

At the maximal ideal $\mathfrak{m}_p = <q-1, p>$ we put $R_p := \mathbb{Z}[q]_{\mathfrak{m}_p}$.

For $p = 2$ we have $1 = [3] - q \cdot ((q-1) + 2)$ and so $[3]$ is a unit in R_2. Thus

$$R_2 \otimes_R \mathcal{H}_{S_3} = \begin{pmatrix} R_2 & R_2 \\ R_2 & R_2 \end{pmatrix} \oplus R_2^q \xrightarrow{\;[2]\;} R_2^-.$$

It decomposes into two blocks, one of which is a separable order, the other the $\mathbb{Z}[q]$-Hecke order $R_2 \otimes_{\mathbb{Z}[q]} \mathcal{H}_2$ of the cyclic group of order 2.

For $p = 3$ the element $[2]$ is a unit in R_3 and so we have

$$R_3 \otimes_R \mathcal{H}_3 = \begin{array}{c} (R_3^q) \\ [3] \Big| \\ \begin{pmatrix} R_3 & R_3 \\ <[3]>_3 & R_{p_1} \end{pmatrix} \end{array} \xrightarrow{\;[3]\;} (R_3^-).$$

Case 3: For $p \neq 2, 3$ the elements $[2]$ and $[3]$ are units in R_p. Hence the Hecke order is separable,

$$R_p \cdot \mathcal{H}_{S_3} = \begin{pmatrix} R_p & R_p \\ R_p & R_p \end{pmatrix} \oplus R_p^q \oplus R_p^-.$$

The order $\Gamma_3 := R_3 \otimes_R \mathcal{H}_3$ has two non-isomorphic indecomposable projective modules

$$P_1 = \{(u, \begin{pmatrix} u + [3] \cdot a & 0 \\ b & 0 \end{pmatrix}, 0) \,:\, u, a, b \in R_3\} \text{ and}$$

$$P_2 = \{(0, \begin{pmatrix} 0 & c \\ 0 & v + [3] \cdot d \end{pmatrix}, v) \,:\, c, d, v \in R_3\}.$$

Moreover, it has four special non-isomorphic irreducible Cohen-Macaulay-lattices – these are left R_3-free Γ_3-modules, which generate simple modules over the total ring of quotients:

$$M_1 := (R_3^q, 0, 0), \; M_2 := (0, \begin{pmatrix} R_3 & 0 \\ R_3 & 0 \end{pmatrix}, 0),$$

$$M_3 := (0, \begin{pmatrix} 0 & [3] \cdot R_3 \\ 0 & R_3 \end{pmatrix}, 0) \text{ and } M_4 := (0, 0, R_3^-).$$

Over the quotient field K of R_3 the order $K \cdot \Gamma_3$ has three rational components,
$$K \cdot \Gamma = K^q \times (K)_2 \times K^-.$$
The *Brauer tree* of Γ_3 is defined as follows:
- The *vertices* correspond to the rational components of $K \cdot \Gamma$, so let us label them $1^q = K^q$, $2 = (K)_2$, $3^- = K^-$.
- The projective modules P_1 and P_2 have each exactly two simple rational components. We draw an edge between two vertices if there is a projective module which has components in the rational components corresponding to these vertices.

Hence Γ_3 has the Brauer tree
$$A_3 := 1^q \xrightarrow{P_1} 2 \xrightarrow{P_2} 3^-.$$

The projective resolutions – "Green's walk around the Brauer tree" – are obtained as follows

$$\begin{array}{ccccccccc} & P_1 & & P_2 & & P_2 & & P_1 & \\ \nearrow & & \searrow \nearrow & & \searrow \nearrow & & \searrow \nearrow & & \searrow \\ M_1 & & M_2 & & M_4 & & M_3 & & M_1 \end{array},$$

walking clockwise around the tree. We summarize this as

Lemma 12.1. *The Hecke order* $\mathbb{Z}[q]_{<q-1,3>} \otimes_{\mathbb{Z}[q,q^{-1}]} \mathcal{H}_{S_3}$ *and the group ring* $\mathbb{Z}_3 S_3$ *are Brauer tree orders to the graph* A_3. *More generally, the Hecke order* $\mathbb{Z}[q, q^{-1}]_{<p,q-1>} \otimes_{\mathbb{Z}^q} \mathcal{H}_{D_{p^n}}$ *is the Brauer tree order to the Brauer tree* •—•—• *with exceptional vertex of multiplicity* $\mu(n) := \frac{p^{n-1} \cdot (p-1)}{2}$ (*cf.* [Ro; 98, I]) *in the center.*

13. Green Rings – Graph Rings

Definition 13.1. Let Ω be a two-sided noetherian ring (having an identity) with a regular principal ideal
$$\omega := \omega^0 \cdot \Omega = \Omega \cdot \omega^0 \text{ with } \omega \simeq \Omega \text{ on the left and right}.$$
We call the pair (Ω, ω) a *regular pair*.

Definition 13.2. The ring
$$\mathbb{H} = \mathbb{H}_{\Omega, \omega, n} := \begin{pmatrix} \Omega & \Omega & \Omega & \cdots & \Omega & \Omega \\ \omega & \Omega & \Omega & \cdots & \Omega & \Omega \\ \omega & \omega & \Omega & \cdots & \Omega & \Omega \\ \cdots & \cdots & \cdots & \ddots & \cdots & \cdots \\ \omega & \omega & \omega & \cdots & \Omega & \Omega \\ \omega & \omega & \omega & \cdots & \omega & \Omega \end{pmatrix}_n$$

is the *triangular ring of size n* associated to the regular pair (Ω, ω). We denote by P_i the projective left $\mathbb{H}$-module, which corresponds to the i-th column of $\mathbb{H}$. The left ideal ρ generated by

$$\rho^0 := \begin{pmatrix} 0 & 1 & 0 & \cdots & 0 & 0 \\ 0 & 0 & 1 & \cdots & 0 & 0 \\ 0 & 0 & 0 & \cdots & 0 & 0 \\ \vdots & \vdots & \vdots & \ddots & \vdots & \vdots \\ 0 & 0 & 0 & \cdots & 0 & 1 \\ \omega_0 & 0 & 0 & \cdots & 0 & 0 \end{pmatrix}_n$$

is two-sided, and so ρ is isomorphic to $\mathbb{H}$ as a left and as right module; however, not as a bimodule.

Conjugation with ρ^0 induces an automorphism σ of $\mathbb{H}$ which cyclicly permutes the projective modules $\{P_i : 1 \leq i \leq n\}$. The quotient $\mathbb{H}/\rho$ is the product

$$\mathbb{H}/\rho = \prod_1^n \overline{\Omega_i} \text{ with } \overline{\Omega_i} = \Omega/\omega \ .$$

Note 13.3. Let $S_i := P_i/P_{i-1}$, then it has a projective resolution

$$0 \to P_{i-1} \xrightarrow{\phi_i} P_i \to S_i \to 0 ,$$

where ϕ_i is the natural inclusion for $2 \leq i \leq n$ and ϕ_1 is right multiplication by ω_0.

We visualize this as an oriented n-gon, where the vertices correspond to the projective modules P_i and the arrows $P_{i-1} \to P_i$ to the maps ϕ_i.

These rings can be characterized as follows:

Lemma 13.4. *Assume that $\mathcal{H}$ is a basic connected semi-perfect ring which has a two-sided ideal J such that*
- *$J \subset rad(\mathcal{H})$ as a left $\mathcal{H}$-module,*
- *$J \cdot _{\mathcal{H}} \simeq \mathcal{H}$,*
- *$\mathcal{H}/J$ is a product of local rings.*

Then $\mathcal{H}$ is a triangular order of size n, the number of non-isomorphic indecomposable projective left modules P, and the regular pair (Ω, ω) is given as follows: $\Omega = End_\mathbb{H}(P)$ for an indecomposable projective module and ω is generated by the composition $P \simeq J^n \cdot P = P \cdot \omega \subset P$.

Two-Dimensional Orders

Definition 13.5. Let (Ω_i, ω_i) be two regular pairs and $\alpha_i \colon \Omega_i \to \overline{\Omega}$ surjective homomorphisms with kernel ω_i. We can then form the pull-back:

$$\begin{array}{ccc} \Omega_1 \xleftarrow{(\alpha_1,\alpha_2)} \Omega_2 & \longrightarrow & \Omega_1 \\ \downarrow & & \downarrow \alpha_1 \\ \Omega_2 & \xrightarrow{\alpha_2} & \overline{\Omega} \end{array}.$$

Now let $\mathcal{G} = (V, E)$ be a locally embedded graph with vertex set $V = \{v_1, \cdots, v_\nu\}$ and edges $E = E_1 \cup E_2 = \{e_1^1, \cdots, e_\mu^1\} \cup \{e_1^2, \cdots, e_\mu^2\}$, where E_1 are genuine edges and E_2 are truncated edges, which only have one vertex (not to be confused with loops). Given a genuine edge from the vertex v to the vertex w, this edge meets v at the local edge ϵ_i^v and w at the local edge ϵ_j^w. We shall write for this edge $e_{\epsilon_i^v, \epsilon_j^w}$. (Note that this uses heavily that $\mathcal{G}$ is locally embedded into the plane.)

We now construct a ring to $\mathcal{G}$, given the following data:

Data 13.6. 1. $\overline{\Omega}$ is a two-sided noetherian ring.
 2. For a vertex v with valency ν_v the following data are given:
 (a) A regular pair (Ω_v, ω_v).
 (b) A surjective homomorphism $\alpha^v \colon \Omega_v \to \overline{\Omega}$ which has kernel ω_v.
 (c) The triangular ring $\mathbb{H}_v = \mathbb{H}_{\Omega_v, \omega_v, \nu_v}$.
 (d) We number the quotients modulo "ρ" according to the numbering of the local edges at v:

$$\mathbb{H}_v/\rho_v \xrightarrow{(\alpha_1^v, \alpha_2^v, \cdots, \alpha_{\nu_v}^v)} \prod_{i=1}^{\nu_v} \overline{\Omega}_i^v,$$

where each of the rings are equal: $\overline{\Omega}_i^v = \overline{\Omega}$.

Definition 13.7. The *Green ring* $\Lambda_{\mathcal{G}}$ constructed from the above locally embedded graph and our data is defined as a sub-ring of $\mathbb{H} := \prod_{v \in V} \mathbb{H}_v$. The only difference between $\mathcal{G}$ and $\mathbb{H}$ lies in the diagonal entries: We replace the diagonal entry

$$\mathbb{H}_v(i,i) \times \mathbb{H}_w(j,j) \text{ with the pull-back } \Omega_v \xleftarrow{(\alpha_i^v, \alpha_j^w)} \Omega_w.$$

Theorem 13.8. *Let B be a block with cyclic defect of $\mathbb{Z}_p G$ for a finite group G. Then B is a Green order with tree the Brauer tree of B.*

In dimension 2 we have the following examples:

Theorem 13.9. *Let $\mathcal{H}_{D_{p^n}}$ be the Hecke algebra of the dihedral group of order $2 \cdot p^n$ for an odd prime p. Then it is a Green order, corresponding to the tree* •————•————•.

Theorem 13.10. *Maximal and separable deformations of blocks with cyclic defect are Green orders whose tree is again the Brauer tree of the block.*

14. A Characterization of Graph Orders

Let R be a regular complete local integral domain of dimension 2 with field of fractions K. A *Cohen-Macaulay R-order* in a separable K-algebra A is as a module finitely generated projective over R, spanning A. The Cohen-Macaulay Λ-modules are the left Λ-modules, which are R-free of finite rank.

Definition 14.1. We say that a Cohen-Macaulay R-order Λ is an order with a *graph resolution*, provided there exists a graph $\mathcal{G}$ – possibly with *truncated edges* – locally embedded into the plane, subject to:

1. There is a bijection between the edges (including the truncated edges) and the indecomposable projective Λ-modules, $e \longmapsto P_e$. If e is a genuine edge, then P_e is also injective in the category of Cohen-Macaulay Λ-modules.

2. The local edges to each vertex correspond to indecomposable Cohen-Macaulay Λ-modules, $(v, j) \longmapsto M_{v,j}$ for $1 \leq j \leq n_v$, which are rationally isomorphic. There are no homomorphisms between modules corresponding to different vertices.

3. If P_e corresponds to the genuine edge $(v,i) \xrightarrow{e} (w,j)$, then there is a projective resolution
$$0 \to M_{v,i-1} \to P_e \to M_{w,j} \to 0.$$

4. Assume that $e := (v, j)$ is a truncated edge which is neighboring a genuine edge corresponding to the projective module P_e and the injective module Q_e (since $dim(R) = 2$ we have a duality on the Cohen-Macaulay modules, $Hom_R(-, R)$); then we have the projective resolution
$$\begin{array}{ccccccccc} 0 & \to & P_e & \to & P_1 & \to & M_1 & \to & 0 \quad \text{and} \\ 0 & \to & M_2 & \to & P_2 & \to & Q_e & \to & 0 \quad , \end{array}$$
where P_i are projective modules corresponding to genuine edges. Moreover, we have an embedding
$$0 \to Q_e \to P_e \to T_e \to 0,$$

such that $\eta \cdot T = T$ if η is the primitive idempotent corresponding to P_e.

5. If P_1 and P_2 are two neighboring truncated edges, then there is an exact sequence

$$0 \to P_2 \to P_1 \to T \to 0,$$

such that $\eta_1 \cdot T = T$, if η_1 is the primitive idempotent corresponding to P_1.

We point out that the above exact sequences give rise to "Green's" walk around the graph.

Theorem 14.2. *Assume that Λ is a Cohen-Macaulay R-order with a graph resolution to the graph $\mathcal{G}$. Then Λ is a graph order to the graph $\mathcal{G}$, and the associated rings Δ_v are Cohen-Macaulay R-modules, and conversely.*

The origin is Rickard's classification of tree algebras by derived equivalences with its important applications to modular blocks with cyclic defect. The integral version of this result was done by Zimmermann. A detailed analysis of the proofs shows that they carry over to *tree orders* over $\mathcal{O}$, *a commutative complete regular local noetherian integral domain of dimension $d < \infty$*.

Theorem 14.3. [KaRo; 97] *If G is a graph with n edges, then there is a generalized star GSt_n with n edges (i.e., all edges are incident with a fixed vertex v_c,) such that the graph orders $\Lambda_G(\Omega_v, \omega_v, \phi_v)$ and $\Lambda_{GSt_n}(\Omega_v, \omega_v, \phi_v)$ are derived equivalent. (Note that G and GSt_n have the same number over vertices.)*

15. Cohen-Macaulay Modules over Two-dimensional Green Orders

Let $\mathcal{O}$ be a complete regular local integral domain of dimension 2, Λ a Cohen-Macaulay $\mathcal{O}$-order in a separable $K = Frac(\mathcal{O})$-algebra $\mathcal{A}$.

Definition 15.1. 1. A Cohen-Macaulay module M over $\mathcal{O}$ has a *Cohen-Macaulay filtration*, provided it has a filtration

$$0 = M_0 \subset M_1 \subset \cdots \subset M_i \subset M_{i+1} \subset \cdots \subset M_n := M,$$

where the sections are Cohen-Macaulay modules over $\mathcal{O}$.

2. We say that a Cohen-Macaulay order Λ has a *Cohen-Macaulay filtration with respect to* $\mathcal{E}$, provided there exists a complete ordered set of central primitive orthogonal idempotents $\mathcal{E} := (e_1, \cdots, e_s)$ of $\mathcal{A}$ such that with $\epsilon_i := \sum_{j=1}^{i} e_j$ we have a Cohen-Macaulay filtration

$$0 \subset \Lambda \cdot \epsilon_1 \subset \cdots \subset \Lambda \cdot \epsilon_s = \Lambda.$$

3. For $M \in CM(\Lambda)$ we put $M_i = M \cap \epsilon_i \cdot \mathcal{A} \cdot M$. We note that the module M_i is pure in M_{i+1}; but in general, the quotient will not be Cohen-Macaulay. We say that M has a *Cohen-Macaulay filtration with respect to* $\mathcal{E}$, provided

$$0 \subseteq M_1 \subseteq \cdots \subseteq M_s = M$$

is a Cohen-Macaulay filtration.

4. We denote by $CM_{\mathcal{E}}^f(\Lambda)$ or shortly by $CM^f(\Lambda)$ the category of Λ-modules which have a Cohen-Macaulay filtration with respect to $\mathcal{E}$.

In the case of the graph orders one can describe various categories $CM_{\mathcal{E}}^f(\Lambda)$; in those cases, however, I do not know the full category $CM(\Lambda)$ of all Cohen-Macaulay modules for Λ. W. Rump has given an example where a Cohen-Macaulay Λ-module has no Cohen-Macaulay filtration for any numbering of the central primitive idempotents.

Theorem 15.2. [Ro; 99] *Let $\Lambda_{\mathcal{G}}$ be the graph order of a truncated graph with $\Omega_v = \mathcal{O}$ and $\omega_v = \pi$. Let $(e_1, \cdots, e_m)$ be an ordering of the primitive idempotents of the underlying algebra such that genuine edges have neighboring idempotents.*

(1) Apart from the projective irreducible Cohen-Macaulay modules the modules in $CM^f(\Lambda_{\mathcal{G}})$, i.e., filtered with respect to the above sequence of idempotents, decompose into the disjoint union of modules in the categories $CM^f(\Lambda)_e$, one for each genuine edge e.

(2) The categories $CM^f(\Lambda)_e$ are all isomorphic. They are all equivalent to the category $CM^f(\Lambda)$, where Λ is the graph order to $\bullet \bullet$ with respect to the above data.

(3) The objects in $CM^f(\Lambda)$ are given by the homomorphisms

$$\overline{\mathcal{O}}^\mu \to \overline{\mathcal{O}}^\nu \text{ with the bi-action of } (Gl(\mu, \overline{\mathcal{O}}), GL(\nu, \overline{\mathcal{O}})),$$

where $\overline{\mathcal{O}} = \mathcal{O}/\langle \pi \rangle$.

(4) If $\overline{\mathcal{O}}$ has infinite lattice type (note that this is a ring of Krull dimension one), then $CM^f(\Lambda)$ has infinite type.

Remark 15.3. The above results also hold if the rings $\mathcal{O}_v$ at the various vertices are not the same.

Two-Dimensional Orders 351

We shall describe here the filtered Cohen-Macaulay modules for localized Hecke orders of the dihedral groups of order $p \cdot 2$ for an odd prime p. So let $\mathcal{H}_p = \mathcal{H}_{D_p} \otimes_{\mathbb{Z}[q,q^{-1}]} \mathbb{Z}[q, q^{-1}]_{\langle q-1,p \rangle}$ be the completion of the Hecke order $\mathcal{H}_{D_p}$ of the dihedral group of order $2 \cdot p$ for an odd prime p at the maximal ideal $\langle q-1, p \rangle$. Then $\mathcal{H}_p$ is a tree order which is described as follows:

Let $R = \mathbb{Z}[\theta_p]^{C_2}$ be the ring of the integers in the fixed field under the cyclic group of order two of the p-th cyclotomic field and define $\mathcal{O}_2 = R \otimes_{\mathbb{Z}} \mathbb{Z}[q, q^{-1}]_{\langle q-1,p \rangle}$. In $\mathcal{O}_2$ we have the prime ideal $\langle \pi \rangle = \langle (q - \theta_p) \cdot (q - \theta_p^{-1}) \rangle$. We observe that $\overline{\mathcal{O}_2} = \mathcal{O}_2 / \langle \pi \rangle \simeq \mathbb{Z}_{(p)}[\theta_p]$ is regular of dimension one, i.e., a rank one valuation ring. The order Λ_2 is defined as

$$\Lambda_2 =: \begin{pmatrix} \mathcal{O}_2 & \mathcal{O}_2 \\ \langle \pi \rangle & \mathcal{O}_2 \end{pmatrix} .$$

We put $\Lambda_1 = \mathbb{Z}[q, q^{-1}]_{\langle q-1,p \rangle} = \Lambda_3$ and denote by $\Phi_p(q)$ the p-th cyclotomic polynomial. The ring $\mathcal{H}_p$ is then obtained as the pull-back

$$\begin{array}{ccc} \mathcal{H}_p & \longrightarrow & \Lambda_1 \times \Lambda_3 \\ \downarrow & & \downarrow \phi \\ \Lambda_2 & \stackrel{\psi}{\longrightarrow} & \mathbb{Z}_{(p)}[\theta_p] \times \mathbb{Z}_{(p)}[\theta_p] \end{array},$$

where ϕ is reduction modulo $\langle \Phi_p(q) \rangle \times \langle \Phi_p(q) \rangle$, ψ is reduction modulo the ideal

$$\begin{pmatrix} \langle \pi \rangle & \mathcal{O}_2 \\ \langle \pi \rangle & \langle \pi \rangle \end{pmatrix} .$$

We label the rational central primitive idempotents of $\mathcal{A}$, the algebra spanned by $\Lambda = \mathcal{H}_p$ as (e_1, e_2, e_3), where e_i is the unit element in Λ_i. The non-isomorphic irreducible Cohen-Macaulay $\mathcal{H}_p$-modules are the modules:

$$\{ N_1 = \Lambda_1, \; M_1 = \begin{pmatrix} \mathcal{O}_2 \\ \langle \pi \rangle \end{pmatrix}, \; M_2 = \begin{pmatrix} \mathcal{O}_2 \\ \mathcal{O}_2 \end{pmatrix}, \; N_3 = \Lambda_3 \}.$$

Green's walk around the Brauer tree gives the projective resolution

$$N_3 \to P \stackrel{\alpha_3}{\to} Q \stackrel{\alpha_2}{\to} Q \stackrel{\alpha_1}{\to} P \to N_3,$$

which is put together from the projective cover sequences

$$\begin{array}{llllll} \mathbf{E}_1: & 0 & \longrightarrow & M_1 & \longrightarrow P \stackrel{\phi_1}{\longrightarrow} N_3 & \longrightarrow 0 \\ \mathbf{E}_2: & 0 & \longrightarrow & N_1 & \longrightarrow Q \stackrel{\phi_2}{\longrightarrow} M_1 & \longrightarrow 0 \\ \mathbf{E}_3: & 0 & \longrightarrow & M_2 & \longrightarrow Q \stackrel{\phi_3}{\longrightarrow} N_1 & \longrightarrow 0 \\ \mathbf{E}_4: & 0 & \longrightarrow & N_3 & \longrightarrow P \stackrel{\phi_4}{\longrightarrow} M_2 & \longrightarrow 0 \end{array}$$

We now look at the modules X, which have an (e_1, e_2, e_3)-*Cohen-Macaulay filtration*; i.e., $X_1 \subseteq X_2 \subseteq X_3$ such that $X_1 \simeq N_1^{(s_1)}$, $X_2/X_1 \simeq M_1^{(t_1)} \oplus M_2^{(t_2)}$ and $X_3/X_2 \simeq N_3^{(s_3)}$.

Theorem 15.4. *The indecomposable non-isomorphic Cohen-Macaulay $\mathcal{H}_p$-modules having an (e_1, e_2, e_3)-filtration are given as:*
- *The irreducible Cohen-Macaulay $\mathcal{H}_p$-modules*

$$\{N_1, M_1, M_2, N_3\},$$

the modules $\phi_1^{-1}((p^\nu + \pi) \cdot N_3)$, $\nu = 0, 1, \ldots$,
- *the modules* $\phi_2^{-1}(((1 - \theta_p)^\nu + \pi) \cdot M_1)$, $\nu = 0, 1, \ldots$.

If we look at the modules X, which have an (e_2, e_1, e_3)-Cohen-Macaulay filtration, i.e.,

$$X_1 \subseteq X_2 \subseteq X_3$$

such that $X_1 \simeq M_1^{(t_1)} \oplus M_2^{(t_2)}$, $X_2/X_1 \simeq N_1^{(s_1)}$ and $X_3/X_2 \simeq N_3^{(s_3)}$ we obtain

Theorem 15.5. *The indecomposable non-isomorphic Cohen-Macaulay $\mathcal{H}_p$-modules having an (e_2, e_1, e_3)-filtration are given as:*
- *The irreducible Cohen-Macaulay $\mathcal{H}_p$-modules*

$$\{N_1, M_1, M_2, N_3\},$$

- *the modules* $\phi_1^{-1}((p^\nu + \pi) \cdot N_3)$, $\nu = 0, 1, \ldots$,
- *the modules* $\phi_3^{-1}((p^\nu + \pi) \cdot N_1)$, $\nu = 0, 1, \ldots$.

References

[CPS; 88]   Cline, E. – Parshall, B. – Scott, L. L., *Finite dimensional algebras and highest weight categories*, J. Reine Angew. Math. **391** (1988), 85–99.

[CPS; 90]   Cline, E. – Parshall, B. – Scott, L. L., *Integral and graded quasi–hereditary algebras, I.*, J. Algebra **131**, No. 1 (1990), 126–160.

[CR; 87]   Curtis, C. W. – Reiner, I., *Methods of Representation Theory*, Volume II John Wiley & Sons, Interscience Publications, 1987.

[Ca; 85] Carter, R. W., *Finite Groups of Lie Type: Conjugacy Classes and Complex Characters,* John Wiley & Sons, Interscience Publications, 1985.

[Cu; 88] Curtis, C. W., *Representations of Hecke algebras,* Astérisque **168** (1988), 13–60.

[DiJa; 89] Dipper, R. – James, G., *The q–Schur algebra,* Proc. London Math. Soc., III. Ser. **59**, No. 1 (1989), 23–50.

[DiJaMa; 98] Dipper, R. – James, G. – Mathas, A., *Cyclotomic q–Schur algebras,* Math. Z., to appear.

[GrLe; 96] Graham, J. J. – Lehrer, G. I., *Cellular algebras,* Invent. Math. **123** (1996), 1–34.

[Gr; 93] Green, J. A., *Combinatorics and the Schur algebra,* J. Pure and Applied Algebra **88**, No. 1-3 (1993), 89–106.

[KaRo; 97] Kauer, M. – Roggenkamp, K. W., *Higher dimensional orders, graph-orders, and derived equivalences,* J. Pure and Applied Algebra., to appear.

[Ko; 97] König S., *A criterion for quasi–hereditary, and an abstract straightening formula,* Invent. Math. **127**, No. 3 (1997), 481-488.

[KoXi; 96] König, S. – Xi, C. C., *On the Structure of Cellular Algebras,* Preprint 96–113, Sonderforschungsbereich 343, Diskrete Strukturen in der Mathematik, (1996).

[PaSc; 88] Parshal, B. – Scott, L. L., *Derived categories, quasi-hereditary algebras, and algebraic groups,* Proc. of the Ottawa-Moosonee Workshop in Algebra 1987, Math. Lect. Note Series, Carleton University and Université d'Ottawa, (1988).

[Ro; 84] Roggenkamp, K. W., *Automorphisms and isomorphisms of integral group rings of finite groups,* Proceedings of "Groups – Korea 1983", Springer Lecture Notes in Math. **1098** (1984), 118–135.

[Ro; 92] Roggenkamp, K. W., *Blocks with cyclic defect and Green orders,* Comm. Algebra **20**, No. 6 (1992), 1715–1734.

[Ro; 97]   Roggenkamp, K. W., *The structure over* $\mathbb{Z}[q, q^{-1}]$ *of Hecke orders of dihedral groups,* J. Algebra **224** (2000), 346–396.

[Ro; 98, II]   Roggenkamp, K. W., *The cellular structure of integral group rings of dihedral groups,* Manuscript, Stuttgart, (1998).

[Ro; 98, I]   Roggenkamp, K. W., *Blocks with cyclic defect of Hecke orders of Coxeter groups,* Archiv der Math. **74** (2000), 173–182.

[Ro; 99]   Roggenkamp, K. W., *Cohen-Macaulay modules over two-dimensional graph orders,* Colloquium Mathematicum **82**, No. 1 (1999), 25–48.

[Sch; 94]   Schaps, M., *A modular version of Maschke theorem for groups with cyclic p-Sylow subgroups,* J. Algebra **163** (1994), 623–635.

Mathematisches Institut B
Universität Stuttgart
Pfaffenwaldring 57
D-70550 Stuttgart, Germany
e-mail: Roggenkamp@mathematik.uni-stuttgart.de

Some Kind of Duality

Masahisa Sato

Dedicated to Professor Yukio Tsushima on his sixtieth birthday

Abstract

In this paper we generalize the notion of Matlis duality in commutative ring theory to non-commutative ring theory. To do this we analize the essence of Matlis duality and we describe these by the notions in non-commutative ring theory.

1. Preliminaries

In the old paper [7] it is known that a commutative local noetherian ring $(R, \mathfrak{m})$ is complete if and only if the natural homomorphism $R \to \mathrm{End}_R(\mathrm{E}(R/\mathfrak{m}))$ is an isomorphism. Based on the Matlis result, the Matlis duality over a commutative semi-local noetherian ring R with the Jacobson radical $\mathfrak{m}$ was defined as the natural homomorphism $\tau(M) : M \to \mathrm{Hom}_R(\mathrm{Hom}_R(M, \mathrm{E}(R/\mathfrak{m})), \mathrm{E}(R/\mathfrak{m}))$ for an R-module M. In [2, 3, 6, 9] some chain conditions between a module and its Matlis dual module were discussed and the following results were proved;

(1) An R-module M is artinian if and only if there is an exact sequence $0 \to M \to E_0 \to E_1 \to \cdots \to E_i \to \cdots$ such that each E_i is a direct sum of finitely many copies of $\mathrm{E}(R/\mathfrak{m})$.

(2) Particularly $\mathrm{E}(R/\mathfrak{m})$ is artinian.

(3) $\mathrm{E}(R/\mathfrak{m})$ is noetherian if and only if R is artinian.

In this paper we define E-duality for a non-commutative ring, which coincides with the Matlis duality when we take a commutative local ring and $E = \mathrm{E}(R/\mathfrak{m})$ and we will know that the assumption "complete semi-local noetherian" is not necessary to get the above results even in the case of non-commutative rings (Corollary 3.2). Furthermore we will show that noetherian condition implies "semi-local" (Corollary 3.5). The chain conditions between a module and its E-dual module are discussed in Section 2 and we give simple and more general results from the point of view of non-commutative ring theory. (see Theorems 3.1, 3.3 and 3.4).

In the Matlis duality of a commutative ring, the specified module $\mathrm{E}(R/\mathfrak{m})$ is used and $R = \mathrm{End}(\mathrm{E}(R/\mathfrak{m}))$ is assumed. $\mathrm{E}(R/\mathfrak{m})$ satisfies the special properties like *injective*, *cogenerator* and *finitely cogenerated*. Also in case that R is a commutative ring, there is no distinction between endomorphism rings as a left R-module and as a right R-module. So they were not necessary to clarify what properties induced the above results. But in non-commutative ring theory we will not get anything without studying the properties of modules by assuming these properties on a bimodule independently. These observations lead us to understand what is essential to get the above results.

Our purpose in this paper is to explain and reveal its principles which made the above facts hold in the point of view of non-commutative ring theory. After these observations, we will show that the assumptions "complete", "semi-local" and etc. are not necessary to induce the above results and we will get more general results.

Throughout this paper, R and S are associative rings with identity, all modules are unital. Let $_RE_S$ be an R-S bimodule. E is called a **cogenerator** as a left R-module if any left R-module is embedded in a direct product of E as an R-module. Also we call E a **quasi-cogenerator** if every factor module of a finite direct sum E^n of E is cogenerated by E.

We define the E-**dual modules** M_S^* and $_RN^\sharp$ of modules $_RM$ and N_S with respect to E by $\mathrm{Hom}_R(_RM, {}_RE_S)_S$ and $_R\mathrm{Hom}_S(N_S, {}_RE_S)$ respectively.

The E-duality is defined by the canonical homomorphism $\sigma_M : M \to M^{*\sharp}$ (resp. $\sigma_N : N \to N^{\sharp *}$), defined by $\sigma_M(x)(f) = f(x)$ for any $x \in M$ (resp. $x \in N$) and $f \in M^*$ (resp. $f \in N^\sharp$). A module $_RM$ (resp. N_S) is said to be E-**reflexive** if σ_M (resp. σ_N) is an isomorphism. It is easy to see that $_RE$ is a cogenerator if and only if σ_M is a monomorphism for any left R-module M.

Recall that a module M is **finitely cogenerated** if for any set $\mathfrak{A}$ of submodules of M such that $\bigcap_{M_i \in \mathfrak{A}} M_i = 0$, there is some finite subset $\mathfrak{F}$ of $\mathfrak{A}$ such that $\bigcap_{M_i \in \mathfrak{F}} M_i = 0$. Also this is equivalent to the condition that *the injective hull $\mathrm{E}(M)$ of M is isomorphic to a direct sum of injective hulls of finitely many simple modules, that is, M has the finitely generated essential socle* [1, Proposition 18.18].

We define a notion of an E-finitely cogenerated module as a generalization of a finitely cogenerated module. A module M is called an E-**finitely cogenerated module** if M is embedded into a direct sum of a finite number of copies of E.

By [1, Proposition 10.10], *every factor module of M is finitely cogenerated if and only if M is artinian*. If a module E is a finitely cogenerated cogenerator, then *a module M is finitely cogenerated if and only if M is embedded into a direct sum of a finite number of copies of E* [1, Proposition 10.2].

2. Reflexive Modules

In this section, we investigate the conditions for a module to become reflexive under the E-duality. Also results in [3, 6] given for commutative rings will be stated more generally for non-commutative rings.

Lemma 2.1. *For an R-S bimodule ${}_RE_S$ it holds that:*

(1-1) *A left R-module ${}_RM$ is E-reflexive implies that a right S-module M_S^* is E-reflexive;*

(1-2) *Assume that ${}_RE$ is a cogenerator. Then a left R-module ${}_RM$ is E-reflexive if and only if a right S-module M_S^* is E-reflexive.*

Dually it holds that:

(2-1) *A right S-module N_S is E-reflexive implies that a left R-module ${}_RN^\sharp$ is E-reflexive;*

(2-2) *Assume that E_S is a cogenerator. Then a right S-module N_S is E-reflexive if and only if a left R-module ${}_RN^\sharp$ is E-reflexive.*

Proof. These are proved from the fact $(\sigma_M)^* \sigma_{M^*} = 1_{M^*}$ and $(\sigma_N)^\sharp \sigma_{N^\sharp} = 1_{N^\sharp}$. □

A module Q is called a **quasi-injective module** if $\text{Ext}^1(Q/Q', Q) = 0$ for any submodule Q' of Q. For details, refer to [1].

Lemma 2.2. (1) *The following statements are equivalent for a cogenerator ${}_RE$:*

(1-1) *R is an E-reflexive left R-module (i.e., $R = \text{End}(E_S) = R^{*\sharp}$);*

(1-2) *E_S is an E-reflexive right S-module.*

If E_S is quasi-injective, then the following property is equivalent to the above properties;

(1-3) *Every finitely generated left R-module is E-reflexive.*

If E_S is a quasi-injective quasi-cogenerator, then the following property is equivalent to the above properties;

(1-4) *Every E-finitely cogenerated right S-module is E-reflexive.*

(2) *Dually the following statements are equivalent for a cogenerator E_S:*

(2-1) *S is an E-reflexive right S-module (i.e., $S = \text{End}({}_RE) = S^{\sharp*}$);*

(2-2) *${}_RE$ is an E-reflexive left R-module.*

If ${}_RE$ is quasi-injective, then the following property is equivalent to the above properties;

(2-3) *Every finitely generated right S-module is E-reflexive.*

If ${}_RE$ is a quasi-injective quasi-cogenerator, then the following property is equivalent to the above properties;

(2-4) *Every E-finitely cogenerated left R-module is E-reflexive.*

Proof. We apply Lemma 2.1(1-2) to $_RM = R$ and $E_S = R^*$ to get the equivalence of (1-1) and (1-2).

(1-1)⇔(1-3). (1-3) implies (1-1) is trivial. Now suppose $_RR$ is E-reflexive. By assumption, there is some positive integer n such that $_RR^n \to {}_RM \to 0$ is exact. Remarking that $R^* = E$ and E_S is quasi-injective, we obtain the following commutative diagram with exact rows;

$$\begin{array}{ccccc} _RR^n & \longrightarrow & _RM & \longrightarrow & 0 \\ \downarrow \sigma_{R^n} & & \downarrow \sigma_M & & \\ (_RR^n)^{*\sharp} & \longrightarrow & _RM^{*\sharp} & \longrightarrow & 0. \end{array}$$

Since σ_{R^n} is an isomorphism, σ_M is an epimorphism. Also σ_M is a monomorphism since $_RE$ is a cogenerator. Hence σ_M is an isomorphism and $_RM$ is E-reflexive.

(1-2)⇔(1-4). (1-4) implies (1-2) is trivial since E_S is E-finitely cogenerated. Now suppose E_S is an E-reflexive right S-module. From the assumption, there is some positive integer n such that $0 \to N_S \to E_S^n \to E_S^n/N_S \to 0$ is exact. Because E_S is quasi-injective, we obtain the following commutative diagram with exact rows;

$$\begin{array}{ccccccccc} 0 & \longrightarrow & N_S & \longrightarrow & E_S^n & \longrightarrow & E_S^n/N_S & \longrightarrow & 0 \\ & & \downarrow \sigma_N & & \downarrow \sigma_{E^n} & & \downarrow \sigma_{E^n/N} & & \\ 0 & \longrightarrow & N_S^{\sharp *} & \longrightarrow & (E_S^n)^{\sharp *} & \longrightarrow & (E_S^n/N_S)^{\sharp *} & & . \end{array}$$

Since σ_{E^n} is an isomorphism, $\text{Cok}(\sigma_N) \cong \text{Ker}(\sigma_{E^n/N})$. But σ_N and $\sigma_{E^n/N}$ are monomorphisms, hence σ_N is an isomorphism. Therefore N_S is E-reflexive. □

Theorem 2.3. *For a left R-module $_RM$ and a right S-module N_S, we have:*

(1) We assume that $R = \text{End}(E_S)$, $_RE$ is a cogenerator and E_S is a quasi-injective quasi-cogenerator. Then it holds that;

(1-1) $_RM$ is finitely generated if and only if M_S^* is E-finitely cogenerated.

(1-2) If N_S is E-finitely cogenerated, then $_RN^\sharp$ is finitely generated.

(1-3) Assume E_S is a cogenerator. If $_RN^\sharp$ is finitely generated, then N_S is E-finitely cogenerated, i.e., there is a duality R-mod $\sim$ f.cog.E_S, where R-mod is the category of finitely generated left R-modules and f.cog.E_S is the category of E_S-finitely cogenerated modules.

(2) We assume that $S = \text{End}(_RE)$, E_S is a cogenerator and $_RE$ is a quasi-injective quasi-cogenerator. Then it holds that;

Some Kind of Duality   359

(2-1) N_S is finitely generated if and only if $_RN^\sharp$ is E-finitely cogenerated.

(2-2) If $_RM$ is E-finitely cogenerated, then M_S^* is finitely generated.

(2-3) Assume $_RE$ is a cogenerator. If M_S^* is finitely generated, then $_RM$ is E-finitely cogenerated, i.e., there is a duality $f.cog._RE \sim$ mod-S, where mod-S is the category of finitely generated right S-modules and $f.cog._RE$ is the category of $_RE$-finitely cogenerated modules.

Proof. (1-1) We only need to prove that there is an exact sequence $_RR^n \to {_RM} \to 0$ if there is an exact sequence $0 \to M_S^* \to E_S^n$, where n is a positive integer. Since E_S is quasi-injective, we have an exact sequence $_R(E^n)^\sharp \to {_RM^{*\sharp}} \to 0$. On the other hand, M_S^* is E-reflexive by Lemma 2.2, so $_RM$ is E-reflexive by Lemma 2.1. Since $_RR$ is E-reflexive, $_R(E^n)^\sharp \cong {_RR^n}$. Hence $_RR^n \to {_RM} \to 0$ is exact.

(1-2) Let $f : N_S \to E_S^{(n)}$ be a monomorphism. Then $f^\sharp : {_R(E_S^{(n)})^\sharp} \to {_RN^\sharp}$ is an epimorphism. Remarking that $_RR^n \cong {_R(E_S^{(n)})^\sharp}$, $_RN^\sharp$ is finitely generated.

(1-3) We take an epimorphism $f : R^n \to {_RN^\sharp}$. We have a monomorphism $f^* : N_S^{\sharp*} \to (R^n)_S^*$. We remark that $(R^n)_S^* \cong E_S^n$ and $\sigma_N : N_S \to N_S^{\sharp*}$ is a monomorphism since E_S is a cogenerator. Hence N_S is E-finitely cogenerated.

The remaining part is similar to the above.   □

3. Chain Conditions

In this section, we study chain conditions between a module and its E-dual module under the E-duality.

Theorem 3.1. *Let $_RE_S$ be an R-S bimodule. It holds that:*

(1) *If $_RE$ is a cogenerator, then it holds that;*

   (a) *An R-module $_RM$ is noetherian (resp. artinian) if M_S^* is artinian (resp. noetherian).*

   (b) *Assume that E_S is quasi-injective and $R = \mathrm{End}(E_S)$. Then $_RM^{*\sharp}$ is noetherian if and only if $_RM$ is noetherian for an R-module $_RM$.*

   (c) *Assume that E_S is a quasi-injective quasi-cogenerator and $R = \mathrm{End}(E_S)$. Then $_RM^{*\sharp}$ is artinian if and only if $_RM$ is artinian for an R-module $_RM$.*

(2) *If E_S is a cogenerator, then it holds that;*

   (a) *An S-module N_S is noetherian (resp. artinian) if $_RN^\sharp$ is artinian (resp. noetherian).*

   (b) *Assume that $_RE$ is quasi-injective and $S = \mathrm{End}(_RE)$. Then $N_S^{\sharp*}$ is noetherian if and only if N_S is noetherian for an S-module N_S.*

(c) *Assume that $_RE$ is a quasi-injective quasi-cogenerator and $S = \text{End}(_RE)$. Then $N_S^{\sharp*}$ is artinian if and only if N_S is artinian for an S-module N_S.*

Proof. (1)(a) Assume $M_S^* = \text{Hom}(_RM, {_RE_S})$ is artinian. For an ascending chain of submodules of $_RM$;

$$M_1 \subset M_2 \subset \cdots,$$

we have the chain of epimorphisms;

$$M \to M/M_1 \to M/M_2 \to \cdots.$$

We apply $\text{Hom}_R(-, {_RE_S})$ to these epimorphisms, we have the descending chain;

$$M_S^* \supset (M/M_1)_S^* \supset (M/M_2)_S^* \supset \cdots.$$

Since M_S^* is artinian, there is some i such that $(M/M_i)_S^* = (M/M_{i+1})_S^* = \cdots$. By assumption $_RE$ is a cogenerator, $M/M_i = M/M_{i+1} = \cdots$. Hence $M_i = M_{i+1} = \cdots$. Thus $_RM$ is noetherian.

(b) We claim $_RM = {_RM^{*\sharp}}$ if $_RM$ or $_RM^{*\sharp}$ is noetherian. Since $_RM$ is a submodule of $_RM^{*\sharp}$, $_RM$ is finitely generated in both cases. Hence $_RM$ is reflexive by Lemma 2.2(1-3).

(c) We claim $_RM = {_RM^{*\sharp}}$ if $_RM$ or $_RM^{*\sharp}$ is artinian. Since $_RM$ is a submodule of $_RM^{*\sharp}$, $_RM$ is finitely cogenerated in both cases. Hence $_RM$ is reflexive by Lemma 2.2(1-4).

(2) is a dual argument. □

Remark 3.1. We do not know if $_RM$ is noetherian (resp. artinian) implies that $_RM^*$ is artinian (resp. noetherian) and that $_RM^{*\sharp}$ is noetherian (resp. artinian) under the assumption that $_RE$ is a cogenerator.

Corollary 3.2. *Assume that R is a commutative ring and $_RE$ is a cogenerator. Then $_RE$ is noetherian if and only if $_RE$ has a composition series. In this case, R is an artinian ring.*

Proof. We consider E as an R-R bimodule. Since $R_R^* = E_R = {_RE}$, E_R is noetherian. Hence $_RR$ is artinian by Theorem 3.1(1)(a). Thus $_RE$ is artinian since $_RE$ is finitely generated. □

Theorem 3.3. *Let $_RE_S$ be a bimodule.*

(1) *If $_RE$ is a cogenerator and E_S is a quasi-injective quasi-cogenerator with $R = \text{End}(E_S)$, then the following statements are equivalent for an R-module $_RM$ and an S-module N_S;*

(a) $_RM$ *is noetherian.*

Some Kind of Duality

(b) $_RM^{*\sharp}$ is noetherian.

(c) M_S^* is artinian.

(2) If $_RE$ is a cogenerator, then it holds that;

(a) E_S is noetherian implies $_RR$ is left artinian.

(b) Assume E_S is a quasi-injective quasi-cogenerator and $R = \text{End}(E_S)$. Then E_S is noetherian implies E_S has a composition series.

Proof. (1)(b)⇒(c). Assume $_RM^{*\sharp}$ is noetherian. Let $M_S^* \supset N_1 \supset N_2 \supset \cdots$ be a descending chain of M_S^*. This induces R-homomorphims

$$_RM^{*\sharp} = \text{Hom}_S(M_S^*, {}_RE_S) \xrightarrow{f_i} \text{Hom}_S(N_i, {}_RE_S) = N_i^\sharp$$

for each i. We put $M_i = \text{Ker}(f_i)$. Then we have an ascending chain $M_1 \subset M_2 \subset \cdots$ of submodules of $_RM^{*\sharp}$. Hence there is some i such that $M_i = M_{i+1} = \cdots$. On the other hand, $\sigma_M : {}_RM \to {}_RM^{*\sharp}$ is a monomorphism by assumption that $_RE$ is a cogenerator. Thus $_RM$ is finitely generated by assumption that $_RM^{*\sharp}$ is noetherian. Let $f : {}_RR^n \to {}_RM$ be an epimorphism. Then this induces a monomorphism $f^* : M_S^* \to E_S^n$. Hence M_S^* is E_S-finitely cogenerated. Thus $f_i, f_{i+1}, \ldots$ are epimorphisms since E_S is quasi-injective. This means $N_i^\sharp = N_{i+1}^\sharp = \cdots$. Since N_i's are submodules of E_S^n, they are E-reflexive by Lemma 2.2. Thus $N_i = N_{i+1} = \cdots$. Hence M_S^* is artinian.

The equivalence of (a) and (b) holds by Theorem 3.1(1)(b).

(c)⇒(a) holds by Theorem 3.1(1)(a).

(2) Remark that $R_S^* = \text{Hom}(_RR, {}_RE_S)_S = E_S$. Hence (a) holds by Theorem 3.1(1)(a). Also it holds $R = \text{End}(E_S) = R^{*\sharp}$ if $R = \text{End}(E_S)$. Then $E_S = R^*$ is artinian since $_RR$ is noetherian from (1). □

By assuming that $_RE_S$ is a cogenerator both as an R-module and as an S-module, the properties for chain conditions under the E-duality are simplified in the following theorem and we can prove some dual conditions in the above theorem.

Theorem 3.4. *Assume $_RE_S$ is a cogenerator both as an R-module and as an S-module.*

(1) *It holds that;*

(a) *If $_R\text{End}(E_S)$ is an artinian (resp. noetherian) R-module, then $\text{End}(E_S)$ is a left artinian (resp. noetherian) ring.*

(b) *If $\text{End}(E_S)$ is a left artinian (resp. noetherian) ring, then E_S is noetherian (resp. artinian).*

(c) *If E_S is noetherian (resp. artinian), then R is a left artinian (resp. noetherian) ring.*

(2) *If E_S is quasi-injective and $R = \text{End}(E_S)$, then the following statements are equivalent for an R-module $_RM$ and an S-module N_S;*

(a) $_RM$ is artinian (resp. noetherian).
  (b) $_RM^{*\sharp}$ is artinian (resp. noetherian).
  (c) M_S^* is noetherian (resp. artinian).

(3) If $R = \text{End}(E_S)$ and E_S is quasi-injective, then the following statements are equivalent;
  (a) E_S is noetherian.
  (b) R is a left artinian ring.
  (c) E_S has a composition series.

Proof. (1) (a) is clear since left $\text{End}(E_S)$-ideals are left R-modules.

To prove (b), we can apply Theorem 3.1(2)(a) for the bimodule $_{\text{End}(E_S)}E_S$ since $E_S = \text{Hom}(\text{End}(E_S), {}_{\text{End}(E_S)}E_S) = \text{End}(E_S)^*$ and $_{\text{End}(E_S)}\text{End}(E_S) = \text{End}(E_S)^{*\sharp}$.

Since $E_S = R^*$, (c) is a direct consequence of Theorem 3.1(1)(a).

(2) The equivalence of (a) and (b) holds by Theorem 3.1(1)(c). Theorem 3.1(2)(a) implies (b)$\Rightarrow$(c) and Theorem 3.1(1)(a) implies (c)$\Rightarrow$(a).

The respective case is already proved in Theorem 3.3(1).

(3) A cogenerator is a quasi-cogenerator, so this holds by Theorem 3.1(2). □

Corollary 3.5. *Let R be a commutative ring and $_RE$ a cogenerator with $R = \text{End}(E_R)$.*

  (1) The following statements are equivalent for an R-module;
  (a) M^{**} is noetherian (resp. artinian).
  (b) M^* is artinian (resp. noetherian).
  (c) M is noetherian (resp. artinian).

  (2) The following statements are equivalent;
  (a) $_RE$ is noetherian (resp. artinian).
  (b) R is an artinian (resp. noetherian) ring.

In this case, R is semi-local.

  (3) If $_RE$ is noetherian, then $_RE$ has a composition series. Particularly R noetherian implies $_RE$ artinian.

Proof. We remark $R = \text{End}(E_R) = \text{End}(_RE)$ and E is a cogenerator as both a left and right R-module by assumption that R is commutative and $R = \text{End}(E_R)$. Hence E_R is injective by [4, Corollary 7]. Thus (1) holds by Theorem 3.4(2). Also (2) holds by Theorem 3.4(3).

E_R is an injective cogenerator, so any simple module is isomorphic to a submodule of E. On the other hand, the socle of E_R is finitely generated in the case E_R is noetherian or artinian. Thus there are finitely many simple modules, that is, R is semi-local. □

Remark 3.2. In Corollary 3.5, if E is artinian or noetherian, then E becomes finitely cogenerated since the socle of E is finitely generated.

Some Kind of Duality

Hence the fact that R is semi-local is shown by the fact R is semi-perfect [1, §27, Exercise 3]. This was pointed out by the referee. The above proof provides another way of direct proof.

A module M is called E-dom.dim $M \geq n$ if there is an exact sequence $0 \to M \to E_1 \to E_2 \to \cdots \to E_n$ such that each E_i is a direct summand of a finite direct sum of copies of E. We denote E-dom.dim $M = \infty$ if E-dom.dim $M \geq n$ for any $n \geq 1$.

Proposition 3.6. *The following two statements are equivalent for an R-module $_RM$:*

(1) M_S^* *is noetherian (resp. artinian).*

(2) $\mathrm{Hom}_R(_RM, _RN_S)_S$ *is noetherian (resp. artinian) for any bimodule $_RN_S$ with $_RE_S$-dom.dim $N \geq 1$;*

If $_RE$ is injective, then the following is also equivalent to the above.

(3) *For each n, $\mathrm{Ext}_R^n(_RM, _RN_S)_S$ is noetherian (resp. artinian) for any bimodule $_RN_S$ with $_RE_S$-dom.dim $N \geq n$.*

Proof. (1)$\Rightarrow$(2). Suppose $_RN_S$ is a bimodule with $_RE_S$-dom.dim $N \geq 1$. Then $_RN_S$ can be embedded into $_RE_S^k$ for some positive integer k. We have

$$\mathrm{Hom}_R(_RM, _RN_S) \subset \mathrm{Hom}_R(_RM, _RE_S^k) \cong \bigoplus_{i=1}^k \mathrm{Hom}_R(_RM, _RE_S) = \bigoplus_{i=1}^k M^*.$$

Hence $\bigoplus_{i=1}^k M^*$ and $\mathrm{Hom}_R(_RM, _RN_S)$ are noetherian right S-modules. (3)$\Rightarrow$(1) and (2)$\Rightarrow$(1) are clear.

(2)$\Rightarrow$(3). It holds clearly in the case $n = 0$. Now suppose $n \geq 1$. By assumption, $_RN_S$ is a bimodule which has the following exact sequence;

$$0 \to {}_RN_S \to E_0 \xrightarrow{d_0} E_1 \xrightarrow{d_1} \cdots \to E_i \xrightarrow{d_i} \cdots \xrightarrow{d_{n-1}} E_n.$$

Here each E_i is a direct summand of a direct sum of a finite number of E. The exact sequence $0 \to {}_RN_S \to E_0 \to \mathrm{Im} d_0 \to 0$ induces an epimorphism $\mathrm{Hom}_R(_RM, \mathrm{Im} d_0) \to \mathrm{Ext}_R^1(_RM, _RN_S) \to 0$ and isomorphisms $\mathrm{Ext}_R^{n+1}(_RM, _RN_S) \cong \mathrm{Ext}_R^n(_RM, \mathrm{Im} d_0)$ for any $n \geq 1$. Then $\mathrm{Hom}_R(_RM, \mathrm{Im} d_0)$ is a noetherian right S-module by (2). So $\mathrm{Ext}_R^1(_RM, _RN_S)$ is also a noetherian right S-module. Since $_RE_S$-dom.dim $\mathrm{Im} d_0 \geq n-1$, $\mathrm{Ext}_R^1(_RM, \mathrm{Im} d_0)$ is a noetherian right S-module from the above argument. Hence $\mathrm{Ext}_R^2(_RM, _RN_S)$ is a noetherian right S-module. We have the desired result by repeating this procedure. □

Acknowledgements. The author would like to express great thanks to the referee for useful suggestions and corrections.

References

[1] F. W. Anderson and K. R. Fuller, *Rings and Categories of Modules*, (Second Edition), Springer-Verlag, Heidelberg-New York, 1992.

[2] R. G. Belshoff, *Matlis reflexive modules*, Comm. Algebra **19** (1991), 1099–1118.

[3] F. C. Cheng and M. Y. Wang, *Homological dimension of G-Matlis dual modules over semilocal rings*, Comm. Algebra **21** (1993), 1215–1220.

[4] R. Colby and K. R. Fuller, *QF-3' rings and Morita duality*, Tsukuba J. Math. **8**(1) (1981), 183–188.

[5] C. Faith, *Algebra II, Ring Theory*, Springer-Verlag, Heidelberg-New York, 1976.

[6] Z. Y. Huang, *Some notes on G-Matlis reflexive modules*, J. Math. Research and Exposition **16** (1996), 493–496.

[7] E. Matlis, *Injective modules over Noetherian rings*, Pacific J. Math. **8** (1958), 511–528.

[8] B. Stenström, *Rings of Quotients*, Springer-Verlag, Heidelberg-New York, 1975.

[9] J. R. Strooker, *Homological Questions in Local Algebra*, London Math. Soc. Lecture Note **145**, Cambridge Univ. Press, Cambridge, 1990.

Department of Civil and Environmental Engineering
Faculty of Engineering
Yamanashi University
Kofu, Yamanashi 400-8511, Japan
e-mail:sato@mail.yamanashi.ac.jp

On Inertial Subalgebras of Certain Rings

Takao Sumiyama

Abstract

R. Raghavendran, W. E. Clark and R. S. Wilson have proved that finite rings have inertial subalgebras. The author will show that this result is naturally extended to certain infinite rings, and generalize Wedderburn's theorem concerning commutativity of finite division rings.

Let A be a finite-dimensional algebra over a field K, and $J(A)$ be the Jacobson radical of A. If $A/J(A)$ is separable over K, then by the Wedderburn-Malcev theorem [3, Theorem 72.19, p.491], there exists a semi-simple subalgebra S of A such that $A = S \oplus J(A)$ (as K-vector spaces). In this case, we see that S is naturally isomorphic to $A/J(A)$ in the sense that the mapping $a \mapsto a + J(A)$ ($a \in S$) gives an isomorphism of S onto $A/J(A)$. If S' is another subalgebra of A such that $A = S' \oplus J(A)$, then S and S' are conjugate in A, that is, there exists $x \in J(A)$ such that $S' = (1+x)^{-1} S (1+x)$.

From this, in [5], E. Ingraham made the following definition: Let R be a commutative ring, and A be a module finite R-algebra. A subalgebra S of A is called an *inertial subalgebra* of A if S is separable over R and $A = S + J(A)$. The commutative ring R is called an *inertial coefficient ring* if every module finite R-algebra A such that $A/J(A)$ is separable over R has an inertial subalgebra. We shall say that the *uniqueness statement holds* for R if, for any module finite R-algebra A and for any two inertial subalgebras S and S' of A, there exists $x \in J(A)$ such that $S' = (1+x)^{-1} S (1+x)$.

The results of R. Raghavendran [9], W. E. Clark [2] and R. S. Wilson [11] imply that, if p is a prime, then $\mathbb{Z}_{p^k} = \mathbb{Z}/(p^k)$ is an inertial coefficient ring. They also determined the structure of inertial subalgebras of finite rings. In fact, a Henselian ring is an inertial coefficient ring [5, p.89].

It is clear that a module finite $\mathbb{Z}_{p^k}$-algebra is a finite ring. In this paper, the author will show that these results can be extended naturally to certain infinite rings.

In what follows, by a ring we mean an associative ring with 1. When we say that S is a subring of a ring R, S must contain 1 of R. When R is a ring, $(R)_{n\times n}$ denotes the ring of all $n \times n$ matrices having entries in R. By a local ring we mean a ring (not necessarily commutative) such that $R/J(R)$ is a division ring. A finite ring is a ring consisting of only finitely many elements. When R is a finite ring, the number of elements of R is called the order of R. If R is a finite ring of order p^n (p a prime), then the characteristic of R must be a power of p. It is easy to see that a finite ring is a direct sum of finite rings of prime-power order. The order of a finite local ring is a prime-power. When R is a finite local ring, though $R/J(R)$ is a commutative field by Wedderburn's theorem, R itself is not necessarily commutative.

Let p^k be a fixed prime-power. For any positive integer r, there exists uniquely (up to $\mathbb{Z}_{p^k}$-algebra automorphism) an r-dimensional local separable extension of $\mathbb{Z}_{p^k}$ (see [8, Chapters XV and XVI]). This ring is called the *Galois ring of characteristic p^k and rank r*, and is denoted by $GR(p^k, r)$. The ring $GR(p^k, r)$ is a finite commutative local ring consisting of p^{kr} elements. The radical of $R = GR(p^k, r)$ is pR, and $R/pR = GF(p^r)$. Galois rings were first noticed by W. Krull [7], and later rediscovered by G. T. Janusz [6] and R. Raghavendran [9].

If r and r' are positive integers such that $r'|r$, then the Galois ring $GR(p^k, r)$ contains a unique copy of $GR(p^k, r')$ (see [8, Lemma XVI.7, p.319]). So, if $1 = r_0 \leq r_1 \leq r_2 \leq \cdots$ is an infinite sequence of positive integers such that $r_i | r_{i+1}$ ($i = 0, 1, 2, \cdots$), then we can construct the inductive limit $R = \bigcup_{i=0}^{\infty} GR(p^k, r_i)$, which will be called an IG-ring (inductive limit of Galois rings). If $R = \bigcup_{i=0}^{\infty} GR(p^k, r_i)$ is an IG-ring, then $J(R) = pR$ and $R/pR = \bigcup_{i=0}^{\infty} GF(p^{r_i})$.

The results of R. Raghavendran, W. E. Clark and R. S. Wilson can be stated as follows.

Theorem 1. [2, Theorem] and [8, Theorem XIX.5, p.376] *Let R be a finite ring with characteristic p^k (p a prime). Then there exists a subring S of R such that*
  (1) $S \cong \bigoplus_{i=1}^{d} (S_i)_{n_i \times n_i}$, *where each S_i ($1 \leq i \leq d$) is a Galois ring,*
  (2) $J(S) = S \cap J(R) = pS$, *and*
  (3) S/pS *is naturally isomorphic to* $R/J(R)$.
If S' is another subring of R which satisfies (1)–(3), then $S \cong S'$.

Viewing R as an algebra over $\mathbb{Z}_{p^k}$, such subrings S of R described in Theorem 1 are inertial subalgebras of R. It is unknown whether, in Theorem 1, the uniqueness statements holds or not (see [2]).

On Inertial Subalgebras   367

In what follows, we shall show that Theorem 1 can be extended naturally to certain infinite rings.

Theorem 2. *Let R be a ring whose characteristic is a power of a prime p. Assume that $J(R)$ is nilpotent, $R/J(R)$ is Artinian, and for each $\bar{a} \in R/J(R)$, there exist positive integers $m > n$ such that $\bar{a}^m = \bar{a}^n$. Then there exists a subring S of R such that*
  (1) $S \cong \bigoplus_{i=1}^{d} (S_i)_{n_i \times n_i}$, *where each S_i ($1 \leq i \leq d$) is an IG-ring,*
  (2) $J(S) = S \cap J(R) = pS$, *and*
  (3) S/pS *is naturally isomorphic to $R/J(R)$.*
If S' is another subring of R which satisfies (1)–(3), then $S \cong S'$.

Proof. Let $\bar{R} = R/J(R) = \bar{R}\bar{e}_1 \oplus \bar{R}\bar{e}_2 \oplus \cdots \oplus \bar{R}\bar{e}_d$, where $\bar{R}\bar{e}_i$ ($1 \leq i \leq d$) is a simple component of $\bar{R}$, and $\bar{e}_i$ is a central idempotent of $\bar{R}$. Let $\bar{R}\bar{e}_i \cong (K_i)_{n_i \times n_i}$, where K_i ($1 \leq i \leq d$) is a division ring. By our assumption and Jacobson's commutativity theorem (see [4, Theorem 3.1.2, p.73]), K_i ($1 \leq i \leq d$) is a commutative field which is algebraic over $GF(p)$. Let $\pi : R \longrightarrow \bar{R}$ be the natural homomorphism. There are mutually orthogonal idempotents $e_1, e_2, \cdots, e_d$ of R such that $e_1 + e_2 + \cdots + e_d = 1$ and $\pi(e_i) = \bar{e}_i$ ($1 \leq i \leq d$). Then, $R = (\bigoplus_{i=1}^{d} e_i R e_i) \bigoplus (\bigoplus_{i \neq j} e_i R e_j)$ as abelian groups. Since each $e_i R e_i$ is semiperfect and $e_i R e_i / J(e_i R e_i) \cong \bar{R}\bar{e}_i \cong (K_i)_{n_i \times n_i}$, by [1, Theorem 21, p.160], there exist a local ring A_i and an isomorphism φ_i of $e_i R e_i$ onto $(A_i)_{n_i \times n_i}$. Let $U = \bigoplus_{i=1}^{d} (A_i)_{n_i \times n_i}$, and $\varphi = \varphi_1 + \varphi_2 + \cdots + \varphi_d : \bigoplus_{i=1}^{d} e_i R e_i \longrightarrow U$ be the isomorphism. Since $A_i / J(A_i) \cong K_i$, by [10, Theorem 2.2 and Theorem 2.3(I)], there exist a subring S_i of A_i and an S_i-submodule N_i of A_i such that S_i is an IG-ring, $A_i = S_i \bigoplus N_i$ (as abelian groups), and S_i / pS_i is naturally isomorphic to $A_i / J(A_i)$. Then $G = \bigoplus_{i=1}^{d} (S_i)_{n_i \times n_i}$ is a subring of U. Let $S = \varphi^{-1}(G)$. As $J(e_i R e_i) \cap \varphi^{-1}((S_i)_{n_i \times n_i}) = J(\varphi^{-1}((S_i)_{n_i \times n_i}))$, we see $J(S) = S \cap J(R) = pS$ and that S/pS is naturally isomorphic to $(\bigoplus_{i=1}^{d} e_i R e_i)/J(\bigoplus_{i=1}^{d} e_i R e_i) = R/J(R)$.

Now, let us suppose that S' is another subring of R which satisfies (1) – (3). Let e and f be primitive idempotents of S'. We claim that $Re \cong Rf$ (as left R-modules) if and only if $S'e \cong S'f$ (as left S'-modules). Let $\pi(e) = \bar{e}$ and $\pi(f) = \bar{f}$. Assume that $Re \cong Rf$. Then $\bar{R}\bar{e} \cong \bar{R}\bar{f}$ as left $\bar{R}$-modules. Both $\bar{R}\bar{e}$ and $\bar{R}\bar{f}$ are minimal left ideals of $\bar{R}$, so they are contained in the same simple component of $\bar{R}$, which implies that $J(R)$ does not include eRf. Conversely, if $J(R)$ does not include eRf, then $\bar{R}\bar{e} \cong \bar{R}\bar{f}$, which means $Re \cong Rf$ by [1, Theorem 16, p.158]. Thus we see that $Re \cong Rf$ (as left R-modules) if and only if $J(R)$ does not include eRf. Similarly, $S'e \cong S'f$ (as left S'-modules) if and only if $J(S') = pS'$ does not include $eS'f$. Since S'/pS' is naturally isomorphic to $R/J(R)$, $J(R)$

includes eRf if and only if pS' includes $eS'f$. So we see that $Re \cong Rf$ (as left R-modules) if and only if $S'e \cong S'f$ (as left S'-modules).

By making use of matrix units, 1 of R is written in S as $1 = (e_{11} + e_{12} + \cdots + e_{1n_1}) + (e_{21} + e_{22} + \cdots + e_{2n_2}) + \cdots + (e_{d1} + e_{d2} + \cdots + e_{dn_d})$, where e_{ki} are mutually orthogonal primitive idempotents of S, and $Se_{ki} \cong Se_{\ell j}$ (as left S-modules) if and only if $k = \ell$. Similarly, $1 = (f_{11} + f_{12} + \cdots + f_{1m_1}) + (f_{21} + f_{22} + \cdots + f_{2m_2}) + \cdots + (f_{d1} + f_{d2} + \cdots + f_{dm_d})$, where f_{ki} are mutually orthogonal primitive idempotents of S', and $S'f_{ki} \cong S'f_{\ell j}$ (as left S'-modules) if and only if $k = \ell$.

As $e_{ki}Se_{ki}/pe_{ki}Se_{ki} \cong e_{ki}Re_{ki}/e_{ki}J(R)e_{ki}$, we see that e_{ki} and $f_{\ell j}$ are primitive idempotents of R. Then $R = \bigoplus Re_{ki} = \bigoplus Rf_{\ell j}$ are indecomposable decompositions.

By what was stated above, the Krull-Schmidt theorem tells us that there exists a permutation σ of $\{1, 2, \ldots, d\}$ such that $n_i = m_{\sigma(i)}$ and $Re_{ik} \cong Rf_{\sigma(i)\ell}$ as left R-modules $(1 \leq i \leq d, 1 \leq k, \ell \leq n_i)$. By renumbering, we may assume $n_i = m_i$ and $Re_{ik} \cong Rf_{i\ell}$ $(1 \leq i \leq d, 1 \leq k, \ell \leq n_i)$. Now,

$$S \cong (e_{11}Se_{11})_{n_1 \times n_1} \bigoplus (e_{21}Se_{21})_{n_2 \times n_2} \bigoplus \cdots \bigoplus (e_{d1}Se_{d1})_{n_d \times n_d}$$

and

$$S' \cong (f_{11}S'f_{11})_{n_1 \times n_1} \bigoplus (f_{21}S'f_{21})_{n_2 \times n_2} \bigoplus \cdots \bigoplus (f_{d1}S'f_{d1})_{n_d \times n_d},$$

where $e_{i1}Se_{i1}$ and $f_{j1}S'f_{j1}$ are IG-rings. Hence, to complete the proof, it will suffice to show $e_{i1}Se_{i1} \cong f_{i1}S'f_{i1}$.

We see $e_{i1}Re_{i1} \cong \text{End}(_RRe_{i1}) \cong \text{End}(_RRf_{i1}) \cong f_{i1}Rf_{i1}$. As $e_{i1}Se_{i1}/e_{i1}J(S)e_{i1}$ is naturally isomorphic to $e_{i1}Re_{i1}/e_{i1}J(R)e_{i1}$, and $f_{i1}S'f_{i1}/f_{i1}J(S')f_{i1}$ is naturally isomorphic to $f_{i1}Rf_{i1}/f_{i1}J(R)f_{i1}$, by [10, Theorem 2.3(II)], we see $e_{i1}Se_{i1} \cong f_{i1}S'f_{i1}$. □

Corollary. *Let R be a ring whose characteristic is a power of a prime p. Assume that $J(R) = pR$, $R/J(R)$ is Artinian, and for each $\bar{a} \in R/J(R)$, there exist positive integers $m > n$ such that $\bar{a}^m = \bar{a}^n$. Then*

$$R \cong (S_1)_{n_1 \times n_1} \bigoplus \cdots \bigoplus (S_d)_{n_d \times n_d},$$

where each S_i $(1 \leq i \leq d)$ is an IG-ring.

Proof. Let p^t be the characteristic of R. By Theorem 2, R contains a subring S such that $S \cong \bigoplus_{i=1}^d (S_i)_{n_i \times n_i}$, where S_i $(1 \leq i \leq d)$ is an IG-ring, and $R = S + pR$. Then we have that $R = S + p(S + pR) = S + p^2R = \cdots = S + p^tR = S$. □

Note that this corollary is a generalization of Wedderburn's theorem concerning commutativity of finite division rings (see, for instance, [4, Theorem 3.1.1, p.70]).

References

[1] E. A. Behrens, *Ring Theory,* Academic Press, New York, 1972.

[2] W. E. Clark, *A coefficient ring for finite non-commutative rings,* Proc. Amer. Math. Soc. **33** (1972), 25–28.

[3] C. W. Curtis and I. Reiner, *Representation Theory of Finite Groups and Associative Algebras,* Pure and Applied Math. Series **11**, Interscience, New York, 1962.

[4] I. N. Herstein, *Noncommutative Rings,* Math. Assoc. Amer., Carus Monograph **15**, 1968.

[5] E. C. Ingraham, *Inertial subalgebras of algebras over commutative rings,* Trans. Amer. Math. Soc. **124** (1966), 77–93.

[6] G. T. Janusz, *Separable algebras over commutative rings,* Trans. Amer. Math. Soc. **122** (1966), 461–479.

[7] W. Krull, *Algebraische Theorie der Ringe II,* Math. Ann. **91** (1924), 1–46.

[8] B. R. McDonald, *Finite Rings with Identity,* Pure and Applied Math. Series **28**, Marcel Dekker, New York, 1974.

[9] R. Raghavendran, *Finite associative rings,* Compositio Math. **21** (1969), 195–229.

[10] T. Sumiyama, *Coefficient subrings of certain local rings with prime-power characteristic,* International J. Math. Math. Sci. **18** (1995), 451–462.

[11] R. S. Wilson, *On the structure of finite rings,* Compositio Math. **26** (1973), 79–93.

Department of Mathematics
Aichi Institute of Technology
Toyota 470-0392, Japan
email: sumiyama@ge.aitech.ac.jp

Some Recent Results on Hopficity, Co-hopficity and Related Properties

K. Varadarajan

Abstract

In this paper we give a survey of some recent results on hopficity, co-hopficity, residual finiteness etc., in many familiar categories, after briefly describing the origin of the subject. At the end of the paper we mention many open problems.

1. Introduction

This is an expository article dealing with some aspects of the development of an area of mathematics which has its origin in geometry, namely in the study of surfaces, and which has led in a natural way to many interesting and deep results in algebra. We will be presenting some of those results. No proofs will be given. They can be found in the articles appearing in the list of references.

It is customary to refer to a compact connected 2-dimensional manifold M (with or without boundary) as a surface. M will be called closed if the boundary ∂M of M is empty. A celebrated result of H. Hopf [31] asserts that a degree 1 self map $\varphi : M \to M$ of a closed orientable surface M is a homotopy equivalence.

One of the crucial steps in his proof is the following algebraic property of the fundamental group $\pi_1(M)$ of such an M :

Any surjective endomorphism $f : \pi_1(M) \to \pi_1(M)$ is an automorphism.

This acts as the motivation for the following:

Definition 1.1. A group G is said to be hopfian if every surjective endomorphism $f : G \to G$ is an automorphism.

Any group G possessing the dual property that any injective endomorphism $f : G \to G$ is an automorphism will be referred to as a co-hopfian group.

Let G be a group. It is easily seen that G is hopfian (resp. co-hopfian) $\Leftrightarrow G$ is not isomorphic to a proper quotient (resp. a proper subgroup) of G.

The proof of the hopficity of $\pi_1(M)$ for any closed orientable surface M given by H. Hopf in [31] is very highly geometric. A purely algebraic proof for the same result was later given by K. N. Frederick [20]. This naturally led Hopf to raise the question whether every finitely generated (in future abbreviated as f.g.) group is hopfian.

The first example of a f.g. non-hopfian group was given by B. H. Neumann [36]. Let
$$G = \langle a, b \ ; \ e_2 = e_3 = \cdots = 1 \rangle$$
and
$$H = \langle a, b \ ; \ e_1 = e_2 = e_3 = \cdots = 1 \rangle$$
where
$$e_i = a^{-1}b^{-1}ab^{-i}ab^{-1}a^{-1}b^i a^{-1}bab^{-i}aba^{-1}b^i$$
for $i \geq 1$. Then H is a proper quotient of G. Let F be the free group on two letters a, b. The isomorphism $\alpha : F \to F$ satisfying $\alpha(a) = b^{-1}a$, $\alpha(b) = b$ carries e_i to e_{i+1} for all $i \geq 1$, hence there is an induced isomorphism $\bar{\alpha} : H \to G$. The first example of a finitely presented non-hopfian group was given by G. Higman [28]. Let
$$G = \langle a, b, c \ ; \ a^{-1}ca = b^{-1}cb = c^2 \rangle \text{ and}$$
$$H = \langle \bar{a}, \bar{b}, \bar{c} \ ; \ (\bar{a})^{-1}\bar{c}\bar{a} = (\bar{b})^{-1}\bar{c}\bar{b} = \bar{c}^2, \ \bar{a}\bar{c}(\bar{a})^{-1} = \bar{b}\bar{c}(\bar{b})^{-1} \rangle.$$
Then H is a proper quotient of G and there exists an isomorphism $\gamma : H \to G$ satisfying $\gamma(\bar{a}) = a$, $\gamma(\bar{b}) = b$ and $\gamma(\bar{c}) = c^2$.

$\mathbb{Z}$, $\mathbb{Q}$, $\mathbb{Z}_{p^\infty}$ for any prime p will have their usual meanings.

Example 1.2. (i) Any finite group is simultaneously hopfian and co-hopfian.

(ii) Any f.g. abelian group is hopfian. The only f.g. abelian groups which are co-hopfian are finite.

(iii) For any group G and any indexing set J, let G^J denote the direct product $\prod_{j \in J} G_j$ where $G_j = G$ for each $j \in J$ and $G^{(J)} = \{(g_j) \in G^J \mid g_j = e \text{ for almost all } j \in J\}$, e being the identity element of G. If $G \neq \{e\}$ and J any infinite set, then G^J as well as $G^{(J)}$ will neither be hopfian nor co-hopfian.

(iv) Let A be a torsion free abelian group. Then A is co-hopfian $\Leftrightarrow$ $A \simeq \mathbb{Q}^k$ where k is an integer ≥ 0.

(v) Let A be a torsion abelian group and $t_p(A)$ the p-primary part of A. Then A is hopfian (resp. co-hopfian) $\Leftrightarrow$ $t_p(A)$ is hopfian (resp. co-hopfian) for each prime p.

(vi) Let A be a divisible abelian group. Then A is hopfian $\Leftrightarrow A \simeq \mathbb{Q}^k$; A is co-hopfian $\Leftrightarrow A \simeq \mathbb{Q}^k \oplus \left(\bigoplus_{p \in P} \mathbb{Z}_{p^\infty}^{k(p)} \right)$ where P denotes the set of prime numbers and k, $k(p)$ are all integers ≥ 0. In particular $\mathbb{Z}_{p^\infty}$ is co-hopfian but not hopfian.

One of the earlier papers dealing with hopfian groups as well as co-hopfian groups is [3] by R. Baer. He refers to hopfian groups as Q-groups and co-hopfian groups as S-groups.

Definition 1.3. A group G is said to be residually finite if given any $x \neq e$ in G, there exist a finite group L_x and a homomorphism $f_x : G \to L_x$ with $f_x(x) \neq e$ in L_x.

This is equivalent to requiring that if $\{N_\alpha\}_{\alpha \in J}$ is the family of normal subgroups of G with $[G : N_\alpha] < \infty$, then $\bigcap_{\alpha \in J} N_\alpha = \{e\}$.

A result of Malcev asserts that any f.g. residually finite group is hopfian [33, p.197]. In [27] J. Hempel gives a very quick geometric proof for the residual finiteness of $\pi_1(N)$ for any surface N (N need neither be closed nor orientable). Since $\pi_1(N)$ is finitely presented it follows that $\pi_1(N)$ is hopfian for any surface N.

2. Hopficity and co-Hopficity in Some Familiar Categories

To my knowledge it is J. Lewin [32] who first introduced the concepts of residual finiteness, hopficity etc., in rings and obtained the analogue of Malcev's theorem for rings. A basic step in his proof is that if A is a f.g. ring and n any integer ≥ 1, then the number of subrings B of A satisfying $[A : B] = n$ is finite. This is the analogue for rings of a result of M. Hall [26] for groups. Later M. Orzech and L. Ribes [39] carried out a further study of residual finiteness and hopficity in rings based on the work of Lewin. All of them dealt with rings not necessarily possessing an identity element. While studying cancellation of quasi-injective modules, G. F. Birkenmeier [7] has introduced the dual notions of hopficity and co-hopficity for (unital) modules over a ring with identity $1 \neq 0$. Later V. A. Hiremath [30] obtained some results on hopfian rings and hopfian modules.

In [48], I carried out a systematic study of hopficity and co-hopficity in rings, modules and topological spaces.

Unless otherwise mentioned all the rings considered in Section 2 will be associative rings with identity, all the ring homomorphisms will be assumed to preserve the identity elements and all the modules will be unital. For any ring A, the category of left (resp. right) A-modules will be denoted by A-mod (resp. mod-A). Given a commutative ring K, we denote the category of K-algebras by K-alg. The category of topological spaces and continuous maps will be denoted by Top.

Definition 2.1. (i) A ring A is said to be hopfian (resp. co-hopfian) if every surjective (resp. injective) ring homomorphism $f : A \to A$ is an automorphism of A. Similarly a K-algebra A is said to be hopfian (co-hopfian) if every surjective (resp. injective) K-algebra homomorphism $f : A \to A$ is an automorphism.

(ii) Let $M \in A$-mod. We say that M is hopfian (resp. co-hopfian) in A-mod if every surjective (resp. injective) map $f : M \to M$ in A-mod is an automorphism. We have similar definitions in mod-A.

(iii) Let $X \in$ Top. We say that X is hopfian (resp. co-hopfian) if every surjective (resp. injective) map $f : X \to X$ in Top is a homeomorphism.

We list below some elementary results and examples as well as sources where proofs could be found. Before listing these we explain the terminology we use. An element x of a ring A will be called left regular if $\ell_A(x) = 0$ where $\ell_A(x) = \{\lambda \in A \mid \lambda x = 0\}$. A will be called directly finite if $x \in A$, $y \in A$, $xy = 1 \Rightarrow yx = 1$.

2.2. Some Elementary Results and Examples

(a) Any noetherian (resp. artinian) module is hopfian (resp. co-hopfian) [41, p.41].

(b) Let A be a ring. If A satisfies a.c.c for two sided ideals then A is hopfian. If A satisfies d.c.c for subrings then A is co-hopfian [48, Proposition 1.9]. A is hopfian in A-mod $\Leftrightarrow A$ is directly finite $\Leftrightarrow A$ is hopfian in mod-A [7, p.102]. A is co-hopfian in A-mod $\Leftrightarrow$ every left regular element in A is a unit in A [7, p.102]. A co-hopfian in A-mod $\Rightarrow A$ hopfian in A-mod [48, Proposition 1.10].

(c) Let A be a commutative ring. Then every f.g. A-module is hopfian [47], [55]. A is co-hopfian in A-mod $\Leftrightarrow A$ is its own total quotient ring. Every f.g. A-module is co-hopfian $\Leftrightarrow$ every prime ideal in A is maximal [56].

(d) Let $A = \begin{pmatrix} \mathbb{Z}/2\mathbb{Z} & \mathbb{Z}/2\mathbb{Z} \\ 0 & \mathbb{Z}_{(2)} \end{pmatrix}$ where $\mathbb{Z}_{(2)} = \left\{ \dfrac{m}{n} \in \mathbb{Q} \mid n \text{ odd} \right\}$.

Then A is co-hopfian in A-mod, but not co-hopfian in mod-A. Also, A is left as well as right noetherian but is neither left nor right artinian [48, Example 1.5].

(e) Let $A = K\left[(X_\alpha)_{\alpha \in J}\right]$ where K is a commutative ring and J an infinite set. Then A is hopfian as an A-module, but not hopfian as a ring.

Let K be a field, V a countable infinite dimensional vector space over K. Then $A = \text{End}\,(_K V)$ is hopfian as a ring because it has only three 2-sided ideals. Since A is not directly finite it is not hopfian as an A-module [48, Example 1.8(h)].

(f) Let $L = K\left((X_\alpha)_{\alpha \in J}\right)$ the field of rational functions in an infinite number of indeterminates over a field K. Then L is not co-hopfian as a ring, thus illustrating that even a field need not be co-hopfian as a ring [48, Example 1.8(e)].

For the rest of this section, A denotes a ring. Let $M \in A$-mod, X an indeterminate over A, n any integer ≥ 1. Let $M[X]$, $M[X]/(X^n)$ and $M[[X]]$ denote respectively the polynomial, truncated polynomial and power series module in X with coefficients from M. We have $M[X] \in A[X]$-mod, $M[X]/(X^n) \in A[X]/(X^n)$-mod and $M[[X]] \in A[[X]]$-mod. The following is the analogue of Hilbert's basis theorem for module hopficity.

Theorem 2.3. [48] *Let $M \in A$-mod. Then the following are equivalent.*
(i) M is hopfian in A-mod.
(ii) $M[X]$ is hopfian in $A[X]$-mod.
(iii) $M[X]/(X^n)$ is hopfian in $A[X]/(X^n)$-mod.
(iv) $M[[X]]$ is hopfian in $A[[X]]$-mod.

Remarks 2.4. (a) For any $0 \neq M \in A$-mod, the modules $M[X]$ in $A[X]$-mod and $M[[X]]$ in $A[[X]]$-mod are never co-hopfian. In fact multiplication by X is an injective non-surjective map in both cases.

(b) Let $A[X, X^{-1}]$ denote the ring of Laurent polynomials over A in one indeterminate X. The rings $A[X]$, $A[X, X^{-1}]$, $A[[X]]$ and $A[[X, X^{-1}]]$ are never co-hopfian as rings. In each case the unique ring homomorphism which is identity on A and which carries X to X^2 is injective but not surjective.

(c) Let $\{A_\alpha\}_{\alpha \in J}$ be any family of rings and $A = \prod_{\alpha \in J} A_\alpha$ their direct product. Let $M_\alpha \in A_\alpha$-mod and $M = \prod_{\alpha \in J} M_\alpha$ with the obvious A-action. Then M is hopfian (resp. co-hopfian) in A-mod $\Leftrightarrow M_\alpha$ is hopfian (resp. co-hopfian) in A_α-mod for every $\alpha \in J$ [48, Proposition 1.17].

Concerning group rings we have the following results:

Proposition 2.5. [48] *Let $A[G]$ denote the group ring of a group G over the ring A.*

(i) If $A[G]$ is hopfian (resp. co-hopfian) as a ring, then A is hopfian (resp. co-hopfian) as a ring and G is hopfian (resp. co-hopfian) as a group.

(ii) If $A[G]$ is hopfian (resp. co-hopfian) in $A[G]$-mod, then A is hopfian (resp. co-hopfian) in A-mod.

If A is commutative and G abelian, then $A[G]$ is commutative, hence directly finite. Thus $A[G]$ is hopfian in $A[G]$-mod. G need not be hopfian.

Concerning matrix rings the following are always valid. We denote the $n \times n$ matrix ring over A by $M_n(A)$.

Proposition 2.6. [48] *(i) $M_n(A)$ hopfian (resp. co-hopfian) as a ring $\Rightarrow$ A hopfian (co-hopfian) as a ring.*

(ii) $M_n(A)$ hopfian (resp. co-hopfian) in $M_n(A)$-mod $\Rightarrow$ A hopfian (resp. co-hopfian) in A-mod.

(iii) $M_n(A)$ hopfian (resp. co-hopfian) in A-mod $\Rightarrow$ A hopfian (resp. co-hopfian) in A-mod.

The following is a well-known result of N. H. McCoy.

Lemma 2.7. [34, Theorem 51] *Let A be a commutative ring and $P \in M_n(A)$. Let $d = \det P$. Then there exists a non-zero column $u \in A^n$ with $Pu = 0 \Leftrightarrow \mathrm{ann}_A(d) \neq 0$ where $\mathrm{ann}_A(d)$ denotes the annihilator of d in A.* The theory of determinants shows that $M_n(A)$ is directly finite whenever A is commutative. Thus $M_n(A)$ is hopfian in $M_n(A)$-mod (or mod-$M_n(A)$) for a commutative ring A.

Using Lemma 2.7 the following can easily be proved.

Proposition 2.8. [48] *Let A be a commutative ring with the property that A is co-hopfian as an A-module. Then for any integer $n \geq 1$, A^n is co-hopfian as an A-module.*

Proposition 2.9. [48] *Let A be a commutative ring which is co-hopfian in A-mod. Then $M_n(A)$ is co-hopfian in both $M_n(A)$-mod and mod-$M_n(A)$.*

Remarks 2.10. (a) Given any integer $n > 1$, G. M. Bergmann [6] constructs a ring A with the property that each of its regular (= both left and right regular) elements is invertible, but $M_n(A)$ is not its own classical ring of quotients. A careful examination shows that in A the set of left regular elements is the same as the set of right regular elements. Thus A is co-hopfian in both A-mod and mod-A but $M_n(A)$ is neither co-hopfian in $M_n(A)$-mod nor co-hopfian in mod-$M_n(A)$.

(b) Clearly $M \in A$-mod hopfian $\Rightarrow$ End$(_AM)$ directly finite. In [44] J. C. Shepherdson gives examples of directly finite A with $M_n(A)$ not directly finite for some $n \geq 2$. Any such A is hopfian in A-mod, but A^n is not hopfian in A-mod.

(c) Let B be an abelian group with the property $t_p(B) \neq 0$ for every prime p where $t_p(B)$ is the p-primary torsion of B. Let $A = B \oplus \mathbb{Z}$ with multiplication given by $(b, m)(b', m') = (mb' + m'b, mm')$ for all $b \in B, m \in \mathbb{Z}$. Then A is a commutative ring with the property that A^n is hopfian as well as co-hopfian in A-mod for every integer $n \geq 1$. However, the cyclic A-module A/B is not co-hopfian. If for some prime p_0 we choose $t_{p_0}(B)$ to be an infinite direct sum of copies of $\mathbb{Z}/p_0\mathbb{Z}$, then the A-submodule B of A is neither hopfian nor co-hopfian [48, Example 3.3].

(d) For any prime p, $\mathbb{Z}_{p^\infty} \oplus \mathbb{Z}$ is neither hopfian nor co-hopfian in $\mathbb{Z}$-mod. Its localization at the prime ideal 0 of $\mathbb{Z}$ (i.e., rationalization) is $\mathbb{Q}$ in $\mathbb{Q}$-mod which is simultaneously hopfian and co-hopfian. Let $\mathbb{Z}_{(p)} = \left\{\dfrac{m}{n} \in \mathbb{Q} \mid (p, n) = 1\right\}$. Then $\bigoplus_{p \in P} \mathbb{Z}_{(p)}$ is hopfian in $\mathbb{Z}$-mod, but its rationalization is a countable infinite direct sum of copies of $\mathbb{Q}$. This is not hopfian in $\mathbb{Q}$-mod [48, Examples 3.6].

Recall that a ring A is said to be boolean if $a^2 = a$ for all $a \in A$. It is well known that any boolean ring A is commutative and that $2a = 0$ for all $a \in A$. Given a compact totally disconnected Hausdorff space X, let $B(X)$ denote the set of clopen subsets of X. Then $B(X)$ is a boolean ring under $C + D = C \nabla D$, the symmetric difference of C and D, and $CD = C \cap D$. Let $H = \mathbb{Z}/2\mathbb{Z} = \{0, 1\}$ with the discrete topology. For any set S, the direct product $H^S = \prod_{s \in S} H_s$ where $H_s = H$ for all $s \in S$ is a compact totally disconnected Hausdorff space. Given a boolean ring A let $X_A = \{f \in H^A \mid f : A \to \mathbb{Z}/2\mathbb{Z}$ a ring homomorphism$\} = \{f \in H^A \mid f(a+b) = f(a) + f(b), f(ab) = f(a)f(b)$ for all $a, b \in A$, $f(1) = 1\}$. Then X_A is a closed subspace of H^A, as such X_A is compact, Hausdorff and totally disconnected. Let $T : A \to B(X_A)$ be defined by $T(a) = \{f \in X_A \mid f(a) = 1\}$. Then M. H. Stone's representation theorem [45], [46] asserts that T is a ring isomorphism. X_A will be referred to as the Stone space of A. Then we have the following:

Theorem 2.11. [48] *A boolean ring A is hopfian (resp. co-hopfian) if and only if its Stone space X_A is co-hopfian (resp. hopfian) in Top.*

For the remainder of this section X will denote a compact Hausdorff space, $C_\mathbb{R}(X)$ (resp. $C_\mathbb{C}(X)$) the $\mathbb{R}$-algebra (resp. $\mathbb{C}$-algebra) of real valued (resp. complex valued) continuous functions on X.

Theorem 2.12. [48] $C_\mathbb{R}(X)$ is hopfian (resp. co-hopfian) in $\mathbb{R}$-alg $\Leftrightarrow$ X is co-hopfian (resp. hopfian) in Top $\Leftrightarrow$ $C_\mathbb{C}(X)$ is hopfian (resp. co-hopfian) in $\mathbb{C}$-alg.

The proof of Theorem 2.12 uses the Gelfand representation theorem identifying X with the maximal ideal space of $C_\mathbb{R}(X)$ (or $C_\mathbb{C}(X)$) with a suitable topology. A nice treatment of Gelfand's representation theorem can be found in [45, pp. 327–330].

An important class of spaces is the class of compact topological manifolds. In [48] we have completely characterized the hopfian and co-hopfian objects in Top belonging to this class. The main result is:

Theorem 2.13. (i) Any compact manifold M without boundary is co-hopfian in Top.

(ii) If M is a compact manifold with a non-empty boundary ∂M, then M is neither hopfian nor co-hopfian.

(iii) The only compact manifolds which are hopfian are finite discrete spaces.

Corollary 2.14. Let M be a compact manifold with or without boundary. Then:

(i) If $\partial M = \phi$, $C_\mathbb{R}(M)$ (resp. $C_\mathbb{C}(M)$) is hopfian in $\mathbb{R}$-alg (resp. $\mathbb{C}$-alg);

(ii) If $\partial M \neq \phi$, $C_\mathbb{R}(M)$ (resp. $C_\mathbb{C}(M)$) is neither hopfian nor co-hopfian in $\mathbb{R}$-alg (resp. $\mathbb{C}$-alg);

(iii) $C_\mathbb{R}(M)$ is co-hopfian in $\mathbb{R}$-alg $\Leftrightarrow M$ is a finite discrete space $\Leftrightarrow$ $C_\mathbb{C}(M)$ is co-hopfian in $\mathbb{C}$-alg.

The proof for Theorem 2.13 can be found in §6 of [48]. Recall that a ring A is said to be left (resp. right) π-regular if given any $a \in A$, there exist an element $b \in A$ and an integer $n \geq 1$ satisfying $a^n = ba^{n+1}$ (resp. $a^n = a^{n+1}b$). G. Azumaya [2] referred to a ring which is both left and right π-regular as strongly π-regular. In [17], [18] F. Dischinger has shown that A is left π-regular $\Leftrightarrow A$ is right π-regular. Hence any such A is strongly π-regular. In the above mentioned papers Dischinger also proved the following:

Theorem 2.15. Let A be a ring.

(i) A is strongly π-regular $\Leftrightarrow$ every cyclic left or right A-module is co-hopfian.

(ii) The following conditions are equivalent:

(a) Every f.g. left A-module is co-hopfian.

(b) Every f.g. right A-module is co-hopfian.

(c) $M_n(A)$ is strongly π-regular for all integers $n \geq 1$.

Complete proofs of slightly more general results can be found in [1]. In [23] K. R. Goodearl introduced the concept of a left repetitive ring. A is said to be left repetitive if given any f.g. left ideal I of A and any $a \in A$, the left ideal $\sum_{n \geq 0} I a^n$ of A is also f.g. One of the main results proved by Goodearl in [23] is the following:

Theorem 2.16. *Every f.g. $M \in A$-mod is hopfian $\Leftrightarrow M_n(A)$ is left repetitive for all $n \geq 1$.*

3. Hopfian and co-Hopfian Zero Dimensional Spaces

In Section 2 we have already stated that a boolean ring A is hopfian (resp. co-hopfian) $\Leftrightarrow$ the associated Stone space X_A is co-hopfian (resp. hopfian) [Theorem 2.11]. Thus the problem of finding infinite hopfian (resp. co-hopfian) boolean rings translates into finding infinite compact, totally disconnected, Hausdorff spaces which are co-hopfian (resp. hopfian). In [12] S. Deo and I obtained a complete solution for the problem in the co-hopfian case, using powerful techniques from set theoretic topology and functional analysis. Unfortunately our techniques do not yield any infinite compact Hausdorff totally disconnected space X which is hopfian. Our method of attack consists of two steps:

Step 1. Obtain some useful conditions which will allow us to conclude from the co-hopficity (resp. hopficity) of a completely regular space X that the Stone-Čech compactification βX will inherit the same property.

Step 2. Construct an infinite completely regular co-hopfian (resp. hopfian) space X satisfying all these conditions and further possessing the property that βX is totally disconnected.

Observing that βX is totally disconnected $\Leftrightarrow \dim \beta X = 0$ and that $\dim X = \dim \beta X$ (when we use Katetov's dimension theory), for tackling Step 2 we need to construct infinite completely regular 0-dimensional co-hopfian (resp. hopfian) spaces satisfying "the useful conditions" in Step 1. For co-hopficity we do succeed in getting such useful conditions. Then techniques similar to those used by J. deGroot [9], J. deGroot and R. H. McDowell [10] in the construction of "rigid" spaces help us complete Step 2. Unfortunately for hopficity our methods do not succeed.

Throughout this section X will denote a completely regular (i.e., a T_3) space and βX will denote its Stone-Čech compactification. By a zero set in X we mean a set of the form $z_f = f^{-1}(0) = \{x \in X \mid f(x) = 0\}$ for some $f \in C_{\mathbb{R}}(X)$. By a co-zero set we mean the complement of a zero set. We define the concept of X having dimension 0 (in the sense of Katetov).

Definition 3.1. X will be called 0-dimensional if for any two disjoint zero sets E, F of X we can find a clopen set C of X with $E \subset C$ and $F \cap C = \emptyset$.

Definition 3.2. X will be called weakly 0-dimensional if clopen subsets of X form a base for the topology of X.

In general one has the implications X 0-dimensional $\Rightarrow$ X weakly 0-dimensional $\Rightarrow$ X totally disconnected. When X is compact we have X totally disconnected $\Leftrightarrow$ X weakly 0-dimensional $\Leftrightarrow$ X 0-dimensional. When we talk of maps between topological spaces they will be assumed to be continuous.

If for every injective map $f : \beta X \to \beta X$ we have $f(X) \subseteq X$, then it is easily seen that the co-hopficity of X implies the co-hopficity of βX. Thus we are led to investigate conditions which will ensure that any injective map $f : \beta X \to \beta X$ automatically satisfies $f(X) \subseteq X$. We actually tackle a more general question. For this purpose we introduce the following definition. Let A, B denote T_3-spaces.

Definition 3.3. A map $\varphi : A \to B$ is said to be locally injective if for any $a \in A$ we can find an open set U_a of A with $a \in U_a$ and $\varphi_{|U_a} : U_a \to B$ injective.

Any injective map is clearly locally injective. We investigate conditions which will ensure that any locally injective map $f : \beta A \to \beta B$ will automatically satisfy $f(A) \subseteq B$.

Definition 3.4. Let $\varphi : A \to B$ and $a \in A$. We say that φ is nice at a if there exists a sequence $\{a_n\}_{n \geq 1}$ of distinct elements in A converging to a in A and satisfying the condition that $\{\varphi(a_n) \mid n \geq 1\}$ is an infinite set. By passing to a subsequence if necessary we may require $\{\varphi(a_n)\}_{n \geq 1}$ to be all distinct.

Definition 3.5. X is said to be real compact if for any $u \in \beta X \setminus X$, there exists an $f \in C_\mathbb{R}(X)$ depending on u satisfying the condition that the unique continuous extension $\hat{f} : \beta X \to S^1 = \mathbb{R} \cup \{\infty\}$ of f carries u to ∞.

All Lindelöf spaces, hence all second countable spaces are real compact [21].

Proposition 3.6. [12] *Let Y be real compact. Suppose $f : \beta X \to \beta Y$ satisfies the condition that $f_{|X} : X \to \beta Y$ is nice at all $x \in X$. Then $f(X) \subseteq Y$.*

Definition 3.7. $\varphi : A \to B$ is said to be non-constant on every open set of A if for any non-empty open set U of A we have $|\varphi(U)| \geq 2$, where $|\varphi(U)|$ denotes the cardinality of the set $\varphi(U)$.

Proposition 3.8. [12] *Let A be 1st countable and $\varphi : A \to B$ be non-constant on every open set of A. Then φ is nice at all $a \in A$.*

Corollary 3.9. [12] *Let X be 1st countable and Y real compact. Let $f : \beta X \to \beta Y$ satisfy the condition that $f_{|X} : X \to \beta Y$ is non-constant on every open set of X. Then $f(X) \subseteq Y$.*

Observing that any locally injective map $\varphi : A \to B$ with A perfect and 1st countable is nice at all $a \in A$ we get the following:

Corollary 3.10. [12] *Let X be 1st countable and perfect, Y real compact. Suppose $f : \beta X \to \beta Y$ satisfies the condition that $f_{|X} : X \to \beta Y$ is locally injective. Then $f(X) \subseteq Y$.*

Definition 3.11. X is said to be rigid if the only self homeomorphism of X is Id_X. We call X rigid for injective maps if the only injective map $X \xrightarrow{f} X$ is $f = Id_X$. A similar definition of rigidity holds for locally injective maps or surjective maps. X is said to be strongly rigid if the only non-constant map $X \to X$ is Id_X.

Remark 3.12. Any X which is rigid for injective (resp. surjective) maps is clearly co-hopfian (resp. hopfian).

Theorem 3.13. [12] *Let X be perfect, 1st countable and real compact. Then:*
  (i) *X co-hopfian $\Rightarrow \beta X$ co-hopfian;*
  (ii) *X rigid for injective maps $\Rightarrow \beta X$ rigid for injective maps;*
  (iii) *X rigid for locally injective maps $\Rightarrow \beta X$ rigid for locally injective maps.*

Remark 3.14. If $f : X \to Y$ satisfies the condition that $f(X)$ is dense in Y, then $\beta(f) : \beta(X) \to \beta(Y)$ is surjective.

The following is an easy consequence:

Proposition 3.15. [12] (i) *βX hopfian $\Rightarrow X$ hopfian.*
  (ii) *βX rigid for surjective maps $\Rightarrow X$ rigid for surjective maps.*

Example 3.16. Let $X = \mathcal{I} \cup \mathbb{N}$ where $\mathcal{I}$ is the set of irrationals. We regard X as a subspace of $\mathbb{R}$. Let $\varphi : X \to X$ be given by $\varphi(x) = x + 1$. Then $\varphi(X) = X \setminus \{1\}$ is dense in X. Hence $\beta(\varphi) : \beta(X) \to \beta(X)$ is surjective. Clearly $\varphi : X \to X$ is injective. But $\beta(\varphi) : \beta(X) \to \beta(X)$ is not injective [12, Example 6.1].

Definition 3.17. X is said to be pseudo-compact if every real valued continuous function on X is bounded.

Proposition 3.18. [12] *Let X be pseudo-compact and Y be 1st countable. Suppose $f : X \to Y$ satisfies the condition that $\beta f : \beta X \to \beta Y$ is surjective. Then $f : X \to Y$ is surjective.*

Definition 3.19. [12] $f : X \to X$ is called a continuous displacement if $|\{x \in X \mid f(x) \neq x\}| \geq c = |\mathbb{R}|$.

Lemma 3.20. [12] *Let X be a separable metric space which is perfect. Let $f : X \to X$ be either surjective or a locally injective map. Then either $f = Id_X$ or is a continuous displacement.*

With Lemma 3.20 in place of Lemma 2 in [9] the same construction as in [9] yields the following:

Theorem 3.21. [12] *There is a family $\{F_\gamma\}$ of 2^c perfect, 0-dimensional subspaces of $\mathbb{R}$ satisfying the following conditions:*
  (a) *The only surjective or locally injective map $F_\gamma \to F_\gamma$ is Id_{F_γ}.*
  (b) *For $\gamma \neq \gamma'$ there is no surjective or locally injective map $F_\gamma \to F_{\gamma'}$.*

Each F_γ occurring in Theorem 3.21 is rigid for locally injective (hence for injective) maps as well as for surjective maps. Hence each F_γ is simultaneously co-hopfian and hopfian. Each F_γ is perfect and 2nd countable. From Theorem 3.13(iii) we see that each βF_γ is rigid for locally injective maps. Also from Corollary 3.10 we see that for $\gamma \neq \gamma'$ there exists no map $\beta F_\gamma \xrightarrow{\theta} \beta F_{\gamma'}$ with $\theta_{|F_\gamma} : F_\gamma \to \beta F_{\gamma'}$ one-one. In particular βF_γ is not homeomorphic to $\beta F_{\gamma'}$. Hence we see that there exist at least 2^c compact 0-dimensional co-hopfian spaces. Each βF_γ is rigid for locally injective maps, hence rigid. Naturally we would like to construct infinite compact 0-dimensional co-hopfian spaces which are not rigid. Let $X_\gamma = F_\gamma \dot\cup F_\gamma$ be the disjoint union of two copies of F_γ. Then we have the following:

Theorem 3.22. [12] (i) *Each βX_γ is a compact, 0-dimensional non-rigid co-hopfian space.*
  (ii) *βX_γ is not homeomorphic to $\beta X_{\gamma'}$, whenever $\gamma \neq \gamma'$.*

Defining a boolean ring B to be rigid if the only ring automorphism of B is Id_B from Theorem 3.22 we immediately get the following:

Theorem 3.23. *There are at least 2^c mutually non-isomorphic hopfian non-rigid boolean rings each of which is of infinite dimension as a vector space over $\mathbb{Z}/2\mathbb{Z}$.*

4. Residual Finiteness in Rings and Modules

In this section by a ring we mean an associative ring not necessarily possessing an identity element. We will generally state results for left modules. Analogous results are valid for right modules. Thus unless otherwise mentioned by an A-module we mean a left A-module.

In what follows A denotes a ring.

Definition 4.1. A is said to be residually finite as a ring if for any $0 \neq x \in A$ we can find a finite ring A_x and a ring homomorphism $\theta_x : A \to A_x$ with $\theta_x(x) \neq 0$.

This is equivalent to requiring that $\bigcap_{\alpha \in J} I_\alpha = 0$ where $\{I_\alpha\}_{\alpha \in J}$ is the family of two-sided ideals of A with A/I_α finite.

Definition 4.2. An A-module M is said to be residually finite if for any $0 \neq x \in M$ we can find an A-module M_x with M_x (actually) finite and an A-homomorphism $\theta_x : M \to M_x$ with $\theta_x(x) \neq 0$.

A is hopfian as a ring if every surjective ring homomorphism $\varphi : A \to A$ is an automorphism of A. Similarly an A-module M is hopfian if every surjective A-module homomorphism $f : M \to M$ is an automorphism of M. The following analogues of Malcev's result are true.

Theorem 4.3. [32] *Any f.g. residually finite ring is hopfian.*

Theorem 4.4. [51] *Let A be any ring, M any f.g. residually finite A-module. Then M is hopfian.*

A well-known result of G. Baumslag [5] asserts that if G is a f.g. residually finite group, the group $Aut(G)$ of automorphisms of G is also residually finite. The following analogues of this result of Baumslag are true.

Theorem 4.5. [51] *Let A be an f.g. residually finite ring. Then the group $Aut(A)$ of ring automorphisms of A is residually finite.*

Theorem 4.6. [51] *Let A be an f.g. ring and M an f.g. residually finite A-module. Then the group $Aut(M)$ of A-automorphisms of M is residually finite.*

The following are some other significant results on residual finiteness of rings and modules.

Theorem 4.7. [51] *Let A be any ring and M be any A-module. Then the following are equivalent.*
   (a) *M is a residually finite A-module.*
   (b) *$M[X]$ is a residually finite $A[X]$-module.*

(c) For any given integer $n \geq 1$, $M[X]/(X^n)$ is a residually finite $A[X]/(X^n)$-module.

(d) $M[[X]]$ is a residually finite $A[[X]]$-module.

Proposition 4.8. [51] *The following conditions are equivalent for a ring A.*

(a) *For any $0 \neq x \in A$, there exists a subring B of A with $x \notin B$ and $[A:B] < \infty$.*

(b) *A is residually finite as a ring.*

(c) *A is residually finite as a left A-module.*

(d) *A is residually finite as a right A-module.*

(e) *All the matrix rings $M_n(A)$ are residually finite as rings (for all $n \geq 1$).*

Proposition 4.9. [51] *Let A be a f.g. residually finite ring. Then the group $U(A)$ of units of A is residually finite.*

Proposition 4.10. [51] *Let A be a ring and $A[G]$ the group ring of a group G over A. Then $A[G]$ residually finite as a ring $\Rightarrow$ A residually finite as a ring and any f.g. subgroup of G is residually finite.*

Corollary 4.11. [51] *Let G be an f.g. group. Then $\mathbb{Z}[G]$ is residually finite as a ring $\Leftrightarrow$ G is residually finite as a group.*

Example 4.12. [51] Let X be the Cantor set and $A = B(X)$ the boolean ring of clopen subsets of X. From M. H. Stone's representation theorem, the group $Aut(A)$ of ring automorphisms of A is isomorphic to the group $G(X)$ of homeomorphisms of X. It is known that $G(X)$ is an infinite simple group. Hence $Aut(A)$ is not residually finite. In this example, A is a residually finite ring which is not f.g. as a ring. This shows that in Theorem 4.5 we can not dispense with the assumption that A is f.g. as a ring.

5. Determining Co-hopfian Surface Groups and Algebraic Consequences

In [13] S. Deo and I completely determined the surfaces M with $\pi_1(M)$ co-hopfian.

Theorem 5.1. [13] (i) *The only closed surfaces M with $\pi_1(M)$ non-co-hopfian are $S^1 \times S^1$ and K where K is the Klein bottle.*

(ii) *The only surface M with $\partial M \neq \phi$ and $\pi_1(M)$ co-hopfian is D^2.*

Denoting the closed orientable surface of genus g by Σ_g we have $\pi_1(\Sigma_g) = \langle a_1, b_1, \ldots, a_g, b_g; \prod_{i=1}^{g} [a_i, b_i] = 1 \rangle$. Also $\Sigma_0 = S^2, \Sigma_1 = S^1 \times S^1$. The surfaces Σ_g for $g \geq 2$ are aspherical with non-vanishing Euler-characteristic. By a result of D. H. Gottlieb, we immediately see that $\pi_1(\Sigma_g)$ is centerless for $g \geq 2$. Clearly, the free group F_g of rank g is a quotient of $\pi_1(\Sigma_g)$. Clearly any finitely generated group is a quotient of F_g for some $g \geq 2$. Combining this with Theorem 5.1 stated above and the hopficity of $\pi_1(\Sigma_g)$ we obtain the following:

Theorem 5.2. [13] *Any finitely generated group can be expressed as a quotient of a finitely presented centerless group which is simultaneously hopfian and co-hopfian.*

The above result is a partial dual to a deep result of Miller and Schupp [35] which we state now.

Theorem 5.3. [35] *Let C_m denote the cyclic group of order m. Let $m \geq 2$ and $n \geq 3$ and $C_m * C_n$ denote the free product of C_m and C_n. Let G be any given countable group. Then there exists an imbedding of G in a quotient H of $C_m * C_n$ satisfying the following conditions:*

(i) H *is a complete hopfian group.*
(ii) *If G is finitely presented, H can be chosen to be finitely presented.*
(iii) *If G has no elements of order m or n, H can be chosen to be further co-hopfian.*

Recall that a group H is said to be complete if H is centerless and the only automorphisms of H are inner automorphisms. An important consequence of Theorem 5.1 is the following result which complements the celebrated result of H. Hopf stated in §1.

Theorem 5.4. *Let M be a closed surface different from $S^1 \times S^1$ and K. Then any $f : M \to M$ with $f_* : \pi_1(M) \to \pi_1(M)$ injective is a homotopy equivalence.*

A result of Goryushkin strengthens Theorem 5.3 in some sense.

Theorem 5.5. [24] *Let $m \geq 2$, $n \geq 3$ and G any given countable group. Then there exists a simple quotient S of $C_m * C_n$ into which G can be imbedded.*

Actually every group can be imbedded in an algebraically closed group [43] and all algebraically closed groups are simple [37]. Thus every group can be imbedded in a simple (hence hopfian) group. However in [13] we show that there is no functorial imbedding of groups (resp. f.g. groups) into hopfian groups.

6. Hopficity of Polynomial Rings and Product Rings

The following are important questions concerning hopfian rings.

1. If A is a hopfian ring, is the polynomial ring $A[T_1,\ldots,T_n]$ in a finite number of indeterminates $T_1,\ldots,T_n$ over A hopfian?

2. If A_1, A_2 are hopfian rings, is $A_1 \times A_2$ a hopfian ring?

In [52], [53] we have obtained some positive results concerning these problems though they remain unsolved. We state our main results.

Theorem 6.1. [52] *Let A_1, A_2 be hopfian rings. Suppose the only central idempotents in A_2 are 0 and 1_{A_2}, and A_2 is not a homomorphic image of A_1. Then $A_1 \times A_2$ is a hopfian ring.*

Corollary 6.2. [52] *Let A_1 be a boolean hopfian ring and A_2 a hopfian ring $\not\simeq \mathbb{Z}/2\mathbb{Z}$ and further satisfying the condition that 0 and 1_{A_2} are the only central idempotents in A_2. Then $A_1 \times A_2$ is hopfian.*

However if A_1 is a boolean hopfian ring, it is true that $A_1 \times \mathbb{Z}/2\mathbb{Z}$ is hopfian. This is an immediate consequence of the following easy

Lemma 6.3. *Let X be a compact, Hausdorff 0-dimensional co-hopfian space and $a \notin X$. Then the disjoint union $X \,\dot\cup\, \{a\}$ with the union topology is co-hopfian.*

Theorem 6.4. [53] *Let A be a reduced exchange ring. If A is hopfian so is the polynomial ring $A[T]$.*

Theorem 6.5. [53] *Let A be a reduced commutative exchange ring. If A is hopfian so is the polynomial ring $A[T_1,\ldots,T_n]$ in n commuting indeterminates for every integer $n \geq 1$.*

In §7 we will list many open problems. Before doing so we want to remind the readers that there is an extensive literature on hopfian groups and groups satisfying various residual properties. Obviously our list of references would become too long if we were to include all the articles dealing with these topics. We will include all the papers referred to in our present survey article along with some which are needed to understand the present status of the open problems.

7. Some Open Problems

In [4] G. Baumslag attempts to construct an abelian hopfian group of any given cardinality. Unfortunately there is a gap in his proof. In [11] S. Deo, P. Sankaran and I, assuming the continuum hypothesis, have shown that for any cardinal α with $1 \leq \alpha \leq c = |\mathbb{R}|$, there exists an abelian group H which is simultaneously hopfian and co-hopfian and satisfying $|H| = \alpha$.

Problem 7.1. Given an arbitrary infinite cardinal α, is there a hopfian abelian group (resp. hopfian torsion-free abelian group) H with $|H| = \alpha$?

Taking the clue from P. Vamos' definition of a finitely co-generated module M, we define a group G to be finitely co-generated if for any family $\{G_\alpha\}_{\alpha \in J}$ of subgroups of G satisfying $\bigcap_{\alpha \in J_0} G_\alpha = \{e\}$, there exists a finite subset J_0 of J with $\bigcap_{\alpha \in J_0} G_\alpha = \{e\}$.

Problem 7.2. Construct a finitely co-generated group G which is not co-hopfian.

Any finitely co-generated abelian group is artinian and hence co-hopfian. Thus in Problem 7.2, G is necessarily non-abelian.

Problem 7.3. Find co-hopfian abelian groups A, B with $A \times B$ not co-hopfian.

In [8] A. L. S. Corner has given an example of a torsion free hopfian abelian group A with $A \times A$ not hopfian.

Problem 7.4. Find an f.g. hopfian group G with $Aut(G)$ not hopfian.

Problem 7.5. Let A be a ring with identity $1 \neq 0$ and $M \in A$-mod be hopfian. Is $M[X, X^{-1}]$ hopfian in $A[X, X^{-1}]$-mod?

Problem 7.6. Let A be a hopfian (resp. co-hopfian) ring and n an integer ≥ 1. Is the matrix ring $M_n(A)$ hopfian (resp. co-hopfian)?

In [25] A. Haghany studies hopficity and co-hopficity for Morita contexts. Defining a ring A to have property (P) if $M_n(A) \simeq M_n(B)$ for some $n \Rightarrow A \simeq B$, he shows that Problem 7.6 has a positive answer if A has property (P).

Define a completely regular space X to be extremally disconnected (abbreviates as e.d.) if the closure in X of any open set of X is open in X. According to a result of A. M. Gleason [22], e.d. spaces are the projective objects among compact Hausdorff spaces. In [7] G. F. Birkenmeier observes that a quasi-projective co-hopfian module is hopfian. In [12] S. Deo and I made a similar observation that a compact, Hausdorff co-hopfian e.d. space is automatically hopfian. One way of constructing infinite compact Hausdorff 0-dimensional hopfian spaces is to answer

Problem 7.7. Are there infinite compact e.d. spaces which are co-hopfian?

The analogue of Problem 7.7 with "co-hopfian" replaced by "rigid" has only recently been answered by A. Dow, A. V. Gubbi and A. Szymansky in [19]. Using I. I. Parovicenko's result [40] that $\mathbb{N}^* = \beta \mathbb{N} \setminus \mathbb{N}$ is universal for compact spaces of weight $\leq \aleph_1$, we have shown in [12] that there are no infinite compact co-hopfian e.d. spaces of weight $\leq \aleph_1$ ($\aleph_1$ is the first uncountable cardinal).

In [38] K. C. O'Meara, C. I. Vinsonhaler and W. J. Wickless have shown that any countable ring with identity is a unital subring of a 2-generated ring. This is the analogue of a famous result of G. Higman, B. H. Neumann and Hanna Neumann [29] in group theory.

It is well known that a ring R is a subring of a simple ring if and only if the additive group $(R, +)$ is either torsion free or an elementary p-group for some prime p.

Problem 7.8. Let R be a countable subring of a simple ring. Is R a subring of a 2-generated simple ring?

I. M. S. Dey [14] and then I. M. S. Dey and Hanna Neumann [15] have studied the problem of hopficity for free products of groups. There are many papers on residual finiteness of free products of groups. Similar questions are being studied for HNN extensions of groups. Free products of rings have been studied by G. M. Bergmann, P. M. Cohn, etc.. Warren Dicks has introduced HNN constructions for rings [16]. A challenging program, in my view, is the study of hopficity, residual finiteness etc. for free products of rings as well as HNN extensions of rings.

Our final problem in this section is:

Problem 7.9. Is there a compact Hausdorff hopfian space X which is not finite? Note that we do not require X to be totally disconnected.

Acknowledgements. The author would like to thank the referee for informing him about rings that occur as subrings of simple rings (comment preceding Problem 7.8).

References

[1] E. P. Armendariz, J. W. Fisher and R. L. Snider, *On injective and surjective endomorphisms of finitely generated modules*, Comm. Algebra **6** (1978), 659–72.

[2] G. Azumaya, *Strongly π-regular rings*, J. Fac. Sci. Hokkaido Univ. **13** (1954), 34–39.

[3] R. Baer, *Groups without proper isomorphic quotient groups*, Bull. Amer. Math. Soc. **50** (1944), 267–278.

[4] G. Baumslag, *On abelian hopfian groups*, Math. Z. **78** (1962), 53–54.

[5] ———, *Automorphism groups of residually finite groups*, J. London Math. Soc. **38** (1963), 117–118.

[6] G. M. Bergmann, *Some examples in PI ring theory*, Israel J. Math. **18** (1974), 257–277.

[7] G. F. Birkenmeier, *On the cancellation of quasi-injective modules*, Comm. Algebra **4** (1976), 101–109.

[8] A. L. S. Corner, *Three examples on hopficity in torsion-free abelian groups*, Acta Math. Hungarica **16** (1965), 303–310.

[9] J. deGroot, *Groups represented by homeomorphism groups I*, Math. Ann. **138** (1959), 80–102.

[10] J. deGroot and R. H. McDowell, *Auto homeomorphism groups of 0-dimensional spaces*, Compositio Math. **15** (1963), 203–209.

[11] S. Deo, P. Sankaran and K. Varadarajan, *Some finiteness properties of groups and their automorphism groups*, Algebra Colloq., to appear.

[12] S. Deo and K. Varadarajan, *Hopfian and co-hopfian zero dimensional spaces*, J. Ramanujan Math. Soc. **9** (1994), 177–202.

[13] ———, *Hopfian and co-hopfian groups*, Bull. Australian Math. Soc. **56** (1997), 17–24.

[14] I. M. S. Dey, *Free products of hopf groups*, Math. Z. **85** (1964), 274–284.

[15] I. M. S. Dey and Hanna Neumann, *The hopf property of free products*, Math. Z. **117** (1970), 325–339.

[16] W. Dicks, *The HNN construction for rings*, J. Algebra **81** (1983), 434–487.

[17] F. Dischinger, *Sur les anneaux fortement π-reguliers*, C. R. Acad. Sci. Paris Ser. A **283** (1976), 571–573.

[18] _____, *Stark π-regular Ringe*, Dissertation, Ludwig-Maximilians-Universität, München, 1977.

[19] A. Dow, A. V. Gubbi and A. Szymansky, *Rigid Stone spaces within ZFC*, Proc. Amer. Soc. **102** (1988), 745–748.

[20] K. N. Frederick, *The Hopfian property for a class of fundamental Groups*, Comm. Pure and Applied Math. **16** (1963), 1–8.

[21] L. Gillman and M. Jerison, *Rings of Continuous Functions*, Van Nostrand, New York, 1960.

[22] A. M. Gleason, *Projective topological spaces*, Illinois J. Math. **2** (1958), 482–489.

[23] K. R. Goodearl, *Surjective endomorphisms of finitely generated modules*, Comm. Algebra **15** (1987), 589–609.

[24] Goryushkin, *Imbedding of countable groups in two generator groups*, Math. Zametki **16** (1974), 231–235.

[25] A. Haghany, *Hopficity and co-hopficity of Morita contexts*, Comm. Algebra **27** (1999), 477–492.

[26] M. Hall, *Subgroups of finite index in free groups*, Canad. J. Math. **1** (1949), 187–190.

[27] J. Hempel, *Residual finiteness of surface groups*, Proc. Amer. Soc. **32** (1972), 323.

[28] G. Higman, *A finitely related group with an isomorphic proper factor group*, J. London Math. Soc. **26** (1951), 59–61.

[29] G. Higman, B. H. Neumann and Hanna Neumann, *Embedding theorems for groups*, J. London Math. Soc. **24** (1949), 247–254.

[30] V. A. Hiremath, *Hopfian rings and hopfian modules*, Indian J. Pure Applied Math. **17** (1986), 895–900.

[31] H. Hopf, *Beiträge zur klassifizierung der Flä chenabbildungen*, J. Reine Angew. Math. **165** (1931), 225–236.

[32] J. Lewin, *Subrings of finite index in finitely generated rings*, J. Algebra **5** (1967), 84–88.

[33] R. C. Lyndon and P. E. Schupp, *Combinatorial Group Theory*, Ergebnisse Math. **89**, Springer-Verlag, Heidelberg, New York, 1977.

[34] N. H. McCoy, *Rings and Ideals*, Carus Math. Monograph **8**, Math. Assoc. Amer., 1948.

[35] C. F. Miller and P. E. Schupp, *Embeddings into hopfian groups*, J. Algebra **17** (1971), 171–176.

[36] B. H. Neumann, *A two generator group isomorphic to a factor group*, J. London Math. Soc. **25** (1950), 247–248.

[37] _____, *The isomorphism problem for algebraically closed groups, word problems*, North Holland (1973), 553–562.

[38] K. C. O'Meara, C. I. Vinsonhaler and W. J. Wickless, *Identity preserving embeddings of countable rings into 2-generator rings*, Rocky Mountain J. Math. **19** (1989), 1095–1105.

[39] M. Orzech and L. Ribes, *Residual finiteness and the hopfian property in rings*, J. Algebra **15** (1970), 81–88.

[40] I. I. Parovicenko, *A universal bicompact of weight $\aleph_1$*, Soviet Math. Doklady **4** (1963), 592–595.

[41] P. Ribenboim, *Rings and Modules*, Tracts in Math. **24**, Interscience Publ., New York, 1969.

[42] P. E. Schupp, *Embeddings into simple groups*, J. London Math. Soc. **13** (1976), 90–94.

[43] W. R. Scott, *Algebraically closed groups*, Proc. Amer. Math. Soc. **2** (1951), 118–121.

[44] J. C. Shepherdson, *Inverses and zero divisors in matrix rings*, Proc. London Math. Soc. **1** (1951), 71–85.

[45] G. F. Simmons, *Introduction to Topology and Modern Analysis*, McGraw-Hill, New York, 1963.

[46] M. H. Stone, *Applications of the theory of boolean rings to general topology*, Trans. Amer. Math. Soc. **41** (1937), 375–481.

[47] J. R. Strooker, *Lifting projectives*, Nagoya Math. J. **27** (1966), 747–751.

[48] K. Varadarajan, *Hopfian and co-hopfian objects*, Publ. Mat. **36** (1992), 293–317.

[49] _____, *A note on the hopficity of $M[X]$ or $M[[X]]$*, National Acad. Sci. Letters **15** (1992), 53–56.

[50] _____, *Hopficity of cyclic modules*, ibid. **17** (1992), 217–221.

[51] _____, *Residual finiteness in rings and modules*, J. Ramanujan Math. Soc. **18** (1993), 29–48.

[52] _____, *On hopfian rings*, Acta Math. Hungarica **83** (1999), 17–26.

[53] _____, *Study of hopficity in certain classes of rings*, Comm. Algebra, to appear.

[54] _____, *Rings with all modules residually finite*, Proc. Indian Acad. Sci., to appear.

[55] W. Vasconcelos, *On finitely generated flat modules*, Trans. Amer. Math. Soc. **138** (1969), 505–512.

[56] _____, *Injective endomorphisms of finitely generated modules*, Proc. Amer. Math. Soc. **25** (1970), 900–901.

[57] W. Xue, *Hopfian and co-hopfian modules*, Comm. Algebra **23** (1995), 1219–1229.

Department of Mathematics
The University of Calgary
Calgary, Alberta T2N 1N4, Canada
e-mail: varadara@math.ucalgary.ca

Some Studies on QcF-coalgebras

Mingyi Wang

Abstract

In this paper, we first give some characterizations for left QcF-coalgebras, one of them is very natural since we use the concept of left semi-perfect coalgebras; we also give some new characterizations for cosemisimple coalgebras in this section. Then we generalize some important results about co-Frobenius coalgebras in [4] to QcF-coalgebras. In the last part of this paper, we discuss some other interesting properties which are related to QcF-coalgebras.

1. Introduction

Throughout this paper C will be a coalgebra over a field k and $\mathcal{M}^C$ (resp. $^C\mathcal{M}$) will denote the Grothendieck category of all right (resp. left) C-comodules. The dual space $C^* = Hom_k(C, k)$ is naturally endowed with the structure of a k-algebra such that every right C-comodule can be viewed as a left C^*-module. The coalgebra C is said to be left semi-perfect [3] if $^C\mathcal{M}$ has enough projectives. A sufficient condition for C to be left semi-perfect is that C is projective as a left C^*-module [3, Theorem 23]. A coalgebra C is said to be *co-Frobenius* if there is a left C^*-monomorphism from C to C^*. The finite-dimensional left and right co-Frobenius coalgebras are precisely the dual coalgebras of the Frobenius algebras. In [6], Torrecillas and Nastasescu extend the notion of co-Frobenius coalgebras in a very natural sense to the concept of left quasi-co-Frobenius (denoted by QcF). A coalgebra C is said to be left QcF if and only if there exists a monomorphism of left C^*-modules from $_{C^*}C$ to a free left C^*-module. The notion of QcF-coalgebra is more general than the notion of co-Frobenius [6, Remark 1.5]. The finite-dimensional QcF-coalgebras are precisely the dual of finite-dimensional QF-algebras. In fact, the behaviour of QcF-coalgebras is in many aspects similar to that of QF-rings. In [6], the following equivalences are proved: C is left QcF if and only if C is a torsionless left C^*-module if and only if there exists a family of C-balanced bilinear forms $\{b_i : C \times C \to k \mid i \in I\}$ such that for every nonzero $x \in C$ there is $i \in I$ such that $b_i(C, x) \neq 0$ if and only if every injective right C-comodule is projective if and only if C is a projective right

C-comodule if and only if C is a projective left C^*-module. Naturally, one could ask: is every projective right (resp. left) C-comodule injective? We give an affirmative answer to it, and we give some natural characterizations for QcF-coalgebras. Also several characterizations of co-semisimple coalgebras are obtained including one in terms of QcF-coalgebras.

In [4], D. Stefen investigates the dimension of the space $\int_M$ of comodule morphisms from C to M. He proves that $dim(\int_M) \leq dim(M)$, for any finite-dimensional C-comodule M and any coalgebra C (with some additional properties). In this paper, we generalize some results in [4] to the case of QcF-coalgebras. Finally, we discuss some properties which are related to QcF-coalgebras. For example, some properties which are preserved by the equivalence of comodules categories. In [7], it was pointed out that the property "finitely cogenerated" is not invariant. We show that it must be invariant if C is almost connected (i.e., the coradical of C is finite-dimensional). In this case, we show that QcF-coalgebras are invariant under the equivalence of comodules categories.

2. Some Characterizations

Theorem 2.1. *The following conditions are equivalent*:
  (i) C *is left QcF*;
  (ii) C *is left semi-perfect and every projective left C-comodule is injective, and there is no left C-comodule N with $inj.dim N = 1$*;
  (iii) *Every finitely cogenerated right injective C-comodule is projective*;
  (iv) *Every right C-comodule can be imbedded in a free left C^*-module*.

Proof. (i)$\Rightarrow$(ii). By [6, Theorem 1.3] we know that C is left semi-perfect. Now there are enough projective left C-comodules for $^C\mathcal{M}$. Let P be any projective left C-comodule; we have an exact sequence $C^{(I)} \to P \to 0$ by C being a generator for $^C\mathcal{M}$ [6, Proposition 2.5]. But P is projective, so P is a summand of $C^{(I)}$. It follows that P is injective. Suppose that N is a C-comodule with $inj.dim N = 1$. Then there exists a simple left C-comodule S such that $Ext^1_C(S, N) \neq 0$ by [8, Proposition 8]. For the exact sequence $0 \to S \to C \to C/S \to 0$ we can obtain that $Ext^2_C(C/S, N) \simeq Ext^1_C(S, N) \neq 0$. It follows from this that $inj.dim N \neq 1$. This is a contradiction.

(ii)$\Rightarrow$(i). Let S be a minimal right coideal of C. We are going to prove that the injective envelope $E(S)$ of S is projective. Since C is left semi-perfect, there exists a projective cover P such that $0 \to ker f \to P \xrightarrow{f} E(S) \to 0$ is exact. By (ii) we know that $ker f$ must be injective. It follows from this that $ker f = 0$ and hence $E(S)$ is projective. So C is left QcF by [6, Theorem 1.3] for $C = \oplus_{i \in I} E(S_i)$.

(i)⇒(iii). Let M be any finitely cogenerated injective right C-comodule, so the exact sequence $0 \to M \to C^{(I)}$ must split. It follows from [6, Theorem 1.3] that M is projective.

(i)⇒(iv). Since there exists an exact sequence $0 \to M \to C^{(I)}$ of right C-comodules, it must be exact as left C^*- modules. But C is left QcF, so M can be imbedded in a free left C^*-module.

(iii)⇒(i) and (iv)⇒(i) follows easily. □

Remark. (iv) is similar to a characterization of QF-rings.

Proposition 2.2. (i) *Assume C is a left semi-perfect coalgebra and every projective left C-comodule is injective. Then C is a generator for $^C\mathcal{M}$.*

(ii) *If C is a generator for $^C\mathcal{M}$ and every flat left C^*-module is torsionless, then C is left QcF.*

Proof. (i) For any left C-comodule M, since C is left semi-perfect, we have an epimorphism $P \to M \to 0$, where P is a projective left C-comodule. For P we have a monomorphism $0 \to P \to C^{(I)}$. By our assumption, we must have an epimorphism $C^{(I)} \to P \to 0$. It follows that C is a generator for $^C\mathcal{M}$.

(ii) Since C is a generator for $^C\mathcal{M}$, we can prove that every finite-dimensional right C-comodule is C^*-torsionless. In fact, let M be any finite-dimensional right C-comodule, then there exists a finite-dimensional left C-comodule N such that $M \simeq N^*$. Since C is a generator, we have an exact sequence $C^{(I)} \to N \to 0$. It follows from this that M is C^*-torsionless. By the locally finite property we have $C = \lim C_i$, where each C_i is a finite-dimensional right coideal. For each C_i, we have an exact sequence $0 \to C_i \to C^{*(m_i)}$, where m_i is a natural number. It follows from the exactness of a co-limit functor that $0 \to C \to \lim C_i^{(m_i)}$ is exact. It is easy to prove that $\lim C_i^{(m_i)}$ is a flat left C^*-module. So C is left QcF by [6, Theorem 1.3]. □

Now let us give a list of the characterizations for co-semisimple coalgebras, some of them are well known.

Proposition 2.3. *The following statements are equivalent:*
(i) *C is co-semisimple;*
(ii) *C is a direct sum of simple subcoalgebras;*
(iii) *Every rational left C^*-module is completely reducible;*
(iv) *Every simple right C-comodule is injective;*
(v) *Every finite-dimensional right C-comodule is injective;*
(vi) *Every finitely cogenerated right C-comodule is injective;*
(vii) *Every quasi-finite right C-comodule is injective;*

(viii) C is left QcF and left hereditary;
(ix) C is a generator for $^C\mathcal{M}$ and is left hereditary.

Proof. (i)⇔(ii)⇔(iii) follow by [5, p.289]. (i)⇒(vii)⇒(vi)⇒(v)⇒(iv) ⇒(i) follow easily by [8, Theorem 1] and every finitely cogenerated comodule is quasi-finite. (viii)⇒(ix) follows by [8, Proposition. 2.5].

(ix)⇒(i). For any $M \in {}^C\mathcal{M}$ since C is a generator, we have an exact sequence $C^{(I)} \to M \to 0$. But C is left hereditary and $C^{(I)}$ is injective, so M is injective.

(i)⇒(vii) follows easily by [8]. □

3. The Uniqueness of Integrals for QcF-coalgebras

Now let us investigate the dimension of the space $\int_M$ of comodule morphisms from C to M. Recall that $C^*_{rat} = \{c^* \in C^* \mid kerc^*$ contains a right coideal of finite codimension$\}$. Now we give the following.

Theorem 3.1. *Let C be a left and right QcF-coalgebra and M a finite-dimensional right comodule. If $Ext^1_{C^*}(C^*_{rat}/C, M) = 0$, then $dim(\int_M) \le dim(M)$.*

Proof. From $0 \to C \to {}_{C^*}C^*_{rat} \to {}_{C^*}C^*_{rat}/C \to 0$, we obtain an exact sequence $0 \to Hom_{C^*}({}_{C^*}C^*_{rat}/C, M) \to Hom_{C^*}({}_{C^*}C^*_{rat}, M) \to Hom_{C^*}(C, M) \to 0$ by $Ext^1_{C^*}({}_{C^*}C^*_{rat}/C, M) = 0$. Now let us show that $Hom_{C^*}({}_{C^*}C^*_{rat}, M) \simeq M$ (as k-spaces). Since C is left QcF, C must be left semi-perfect. It follows that $E(M)$ is finite-dimensional and is an injective C^*-module by [1]. Since C is right QcF, ${}_{C^*}C^*_{rat}$ must be dense in ${}_{C^*}C^*$ by [4]. So if we use the same proof of Proposition 2 in [4], we can obtain that $Ext^1_{C^*}(C^*/{}_{C^*}C^*_{rat}, M) = 0$. From the exact sequence $0 \to {}_{C^*}C^*_{rat} \to C^* \to C^*/{}_{C^*}C^*_{rat} \to 0$, we have an exact sequence $0 \to Hom_{C^*}(C^*/{}_{C^*}C^*_{rat}, M) \to Hom_{C^*}(C^*, M) \to Hom_{C^*}({}_{C^*}C^*_{rat}, M) \to 0$. By the same discussion as in [4, Theorem 3], we have that $Hom_{C^*}({}_{C^*}C^*_{rat}, M) \simeq M$ (as k-spaces). It follows from the exact sequence $0 \to Hom_{C^*}({}_{C^*}C^*_{rat}/C, M) \to Hom_{C^*}({}_{C^*}C^*_{rat}, M) \to Hom_{C^*}(C, M) \to 0$ that $dim(\int_M) \le dim(M)$. □

Remark. Note that Theorem 3.1 also holds for any finite-dimensional left C-comodule M.

Lemma 3.2. *The following statements concerning a Hopf algebra H are equivalent:*
(i) *H has a nonzero left integral;*
(ii) *H is left QcF;*

(iii) H has a nonzero right integral;
(iv) H is right QcF.

Proof. (i)$\Rightarrow$(ii) follows from the fact any left Frobenius coalgebra is left QcF.

(ii)$\Rightarrow$(iii). Since H is left QcF, there exists a family of H-balanced bilinear forms $\{b_i : H \times H \to k\}$ such that $b_i(H, x) \neq 0$ by [6, Theorem 1.3]. Now let I be a minimal left coideal of H, so it is finite-dimensional. $I^\perp$ is a right coideal of H which is cofinite dimensional. So we obtain that H contains a proper right coideal of finite codimension. It follows from [5, 2.14] that H has a nonzero right ideal.

(iii)$\Rightarrow$(iv) and (iv)$\Rightarrow$(i) are obviously confirmed by the above proof. □

Now we can obtain the following result by careful reading of the proof for Theorem 5 in [4] and by Lemma 3.2.

Corollary 3.3. *Let H be a left QcF Hopf algebra and let M be a finite dimensional right H-comodule. Then $dim(\int_M) \leq dim(M)$. In particular, for any Hopf algebra H, $dim(\int_k) \leq 1$.*

4. QcF-coalgebras and Equivalences

Let C and D be any two coalgebras (over the same field k). Let $\mathcal{M}^C$ and $\mathcal{M}^D$ be their categories of right comodules, and let 1_C and 1_D be their indentity functors, respectively. We say that C and D are equivalent if the categories $\mathcal{M}^C$ and $\mathcal{M}^D$ are equivalent, i.e., in case there exist a pair of covariant additive functors

$$F : \mathcal{M}^C \rightleftarrows \mathcal{M}^D : G$$

such that $G \circ F$ is naturally equivalent to 1_C and $F \circ G$ is naturally equivalent to 1_D (see [2]). In [2], it has been pointed out that any property of M in $\mathcal{M}^C$ that is categorical is a property of its correspondent in $\mathcal{M}^D$. Examples of such properties include being projective, being a generator. Now let us consider what happens about QcF coalgebras. By the way, we know that the property "finitely cogenerated" is not invariant in general (see [7]). Now we give the following.

Theorem 4.1. *Let C and D be equivalent by $F : \mathcal{M}^C \rightleftarrows \mathcal{M}^D : G$. Then we have the following:*

(i) C is almost connected if and only if D is almost connected.

(ii) C is right QcF if and only if D is right QcF.

(iii) If C is almost connected, then N_C is finitely cogenerated if and only if $F(N_C)$ is finitely cogenerated.

Proof. We first show (i) that D also is almost connected. By the equivalence we can show that $Corad(D) = F(Corad(C))$. But being "finite-dimensional" is invariant under the equivalence. So D also is almost connected.

(ii) Let C be right QcF and M_D be any injective right D-comodule. For any N_D we have an isomorphism:

$$Com_D(M_D, N_D) \simeq Com_C(G(M_D), G(N_D)).$$

Since C is Morita-Takeuchi equivalent to D, $G(M_D)$ is an injective C-comodule. But C is right QcF, so $G(M_D)$ is projective and hence $Com_D(M_D, -)$ is exact. It follows that D is right QcF.

(iii) Since C is equivalent to D, we obtain that $\mathcal{M}^C$ is strongly equivalent to $\mathcal{M}^D$ by the remark in [2]. So $F(C)$ is an ingenerator in $\mathcal{M}^D$ and $G(D)$ is an ingenerator in $\mathcal{M}^C$. Now let X_C be a finitely cogenerated C-comodule; we have an exact sequence $0 \to X_C \to C^{(n)}$. Acting by the functor F we obtain an exact sequence $0 \to F(X_C) \to F(C)^{(n)}$. But $F(C) \oplus N_D = D^{(m)}$ for some $N_D \in \mathcal{M}^D$. So $F(X_C)$ must be finitely cogenerated. The converse can be proved in a similar way. □

The following lemma is obvious.

Lemma 4.2. *Let $C \times_\alpha H$ be a crossed coproduct. If $dim H < \infty$, then $(C \times_\alpha H)^* \simeq C^* \#_{\alpha^*} H^*$ as algebras.*

Proposition 4.3. *Let $C \times_\alpha H$ be a crossed coproduct. If H and C are finite-dimensional, α is an invertible 2-cocycle, then $C \times_\alpha H$ is QcF if and only if C is QcF.*

Proof. Since C is finite-dimensional, C is QcF if and only if C^* is QF if and only if $C^* \#_{\alpha^*} H^*$ is QF (since $C^* \#_{\alpha^*} H^*$ is an excellent extension of C^*) if and only if $(C \times_\alpha H)^*$ is QF if and only if $C \times_\alpha H$ is QcF. □

Proposition 4.4. *Let $C = \oplus_{i \in I} C_i$. Then C is left QcF if and only if each C_i is left QcF.*

Proof. First, each C_i is an injective right C-comodule, and hence it is a summand of a right free C-comodule. If C is left QcF, then each C_i is a projective C-comodule. Of course, each C_i is projective as a right C_i-comodule, i.e., each C_i is left QcF. Now let each C_i be left QcF. Then we have an exact sequence $0 \to {}_{C^*}C_i \to C_i^{*(k_i)}$ for each i. It follows that $0 \to {}_{C^*}C \to \oplus_{i \in I} C_i^{*(k_i)} \hookrightarrow C^{*(I)}$ is exact. So C is left QcF. □

Questions. 1. Does the converse of the Proposition 2.2(ii) hold? That is, does the condition that C is a QcF-coalgebra imply that every flat left

C^*- module is torsionless? Clearly, it holds in the finite-dimensional case, since C^* is a QF-ring in this case.

2. Is C^{*o} QcF if C^* is QF?

References

[1] Y. Doi, *Homological coalgebra*, J. Math. Soc. Japan **33**(1) (1981), 32–50.

[2] I. Peng Lin, *Morita's theorem for coalgebras*, Comm. Algebra **1**(4) (1974), 311–344.

[3] _____, *Semi-perfect coalgebras*, J. Algebra **49** (1977), 357–373.

[4] D. Stefen, *The uniqueness of integrals*, Comm. Algebra **23**(5) (1995), 1657–1662.

[5] M. E. Sweedler, *Integrals for Hopf algebras*, Ann. Math. **89** (1969), 323–335.

[6] J. G. Torrecillas and C. Nastasescu, *Quasi-Frobenius coalgebras*, J. Algebra **174** (1995), 909–923.

[7] B. Torrecillas, F. Van Oystaeyen and Y. H. Zhang, *The Brauer group of a cocommutative coalgebra*, J. Algebra **177** (1995), 536–568.

[8] _____, *Hereditary coalgebras*, Comm. Algebra **21**(4) (1996), 1521–1528.

Institute of Mathematics,
Southwest Jiaotong University,
Chengdu 610031, P. R. China
e-mail: mywang@center2.swjtu.edu.cn

Finitely Pseudo-Frobenius Rings

Hiroshi Yoshimura

Dedicated to Professor Yoshiki Kurata on his seventieth birthday

Abstract

We make a survey of progress in the study of finitely pseudo-Frobenius rings during the last fifteen years. We also study when a triangular matric ring by a module of finite length and its endomorphism ring is a finitely pseudo-Frobenius ring.

Introduction

Rings R whose every finitely generated faithful right R-module is a generator for Mod-R, the category of right R-modules, are called *right finitely pseudo-Frobenius (right FPF) rings*; the study was initiated in connection with quasi-Frobenius (QF) rings or pseudo-Frobenius (PF) rings, i.e., rings whose every faithful module is a generator [E] and [Ta]. The class of FPF rings thus includes QF or PF rings and also bounded Dedekind (noetherian) prime rings, self-injective (von Neumann) regular rings of bounded index and commutative self-injective or Prüfer rings. The monograph, *FPF Ring Theory* [FaiPa] by C. Faith and S. Page, published in 1984, is the first one that provides a systematic study of FPF rings and overviews of its background; it includes most of the known results related to the subject up to that time.

The aim of this note is to make a brief survey of progress in the study of FPF rings since 1984 up to the present; and in addition, to study when a triangular matrix ring is FPF.

In Section 1 we shall make a survey of known results on FPF rings during the last fifteen years. To this end, included in the References at the end of this note are the articles on FPF rings appearing in this period; however, we note that the list may be modified by the availability of the

---

The author was financially supported in part by Research Grant of the Japan Society for the Promotion of Science for attending the Second Korea-Japan Joint Seminar combined with the Third Korea-China-Japan International Symposium on Ring Theory and for preparing the article.

resources. In Section 2 we shall present a necessary and sufficient condition for a triangular matrix ring by a ring S, an S-module M of finite length and $T = \mathrm{End}_S(M)$ to be an FPF ring.

Notation and Terminology

Throughout this note, all rings considered are associative rings with identity and all modules are unitary.

Let R be a ring. We denote by $Q^r_{max}(R)$ the maximal right quotient ring of R and by $M_n(R)$ the ring of all $n \times n$ matrices over R for a positive integer n. The notation M_R means that M is viewed as a right R-module. We denote by $E(M)$, $J(M)$ and $Z(M)$ the injective hull, the Jacobson radical and the singular submodule of M respectively. For $A \subset R$ and $X \subset M$, we denote by $l_M(A)$ and $r_R(X)$ the left annihilator of A in M and the right annihilator of X in R respectively.

A ring R is said to be *right bounded* if every essential right ideal of R contains a (two-sided) ideal that is essential as a right ideal.

1. Results on FPF rings after Faith-Page [FaiPa]

1.1. Semiperfect FPF Rings

Tachikawa [Ta] in 1969 showed that a one-sided artinian ring is right FPF if and only if it is QF (cf. Endo [E] in 1967 also showed this for commutative rings). Faith [Fai1] [Fai2] in 1976, 1977 investigated the FPF condition more generally over semiperfect rings and improved the result of Tachikawa, where he showed the basic results of semiperfect right FPF rings R, e.g., every indecomposable projective right R-module is uniform; the basic ring R_0 of R is *strongly right bounded*, i.e., every nonzero right ideal of R_0 contains a nonzero ideal of R_0.

After the initial work of Faith and [FaiPa], further studies of semiperfect FPF rings were made in [Ca], [Fat2], [Fat3], [P6], [P8], [Yos4], [You1] and [You2]. Among others, Faticoni [Fat2] investigated the structure of those rings in detail and showed the following result:

Theorem 1.1. [Fat2, Corollaries 2.7 and 4.2] (1) *Let R be a semiperfect right FPF ring. Then,*

$$R/Z(R_R) \cong M_{n_1}(S_1) \times \cdots \times M_{n_k}(S_k)$$

where $S_1, \ldots, S_k$ are serial domains.

(2) *Let R be a semiperfect ring satisfying the condition $(*)$ that every right regular element of R is left regular. Then R is right FPF if and only if R satisfies the following three conditions:*

(a) The basic ring R_0 of R is strongly right bounded;
(b) Every finitely generated faithful right ideal of R is a generator for Mod-R;
(c) R has the classical quotient ring that is right self-injective.

Camillo [Ca] presented another characterization of (basic) semiperfect FPF rings, which shows a close relation between FPF rings and self-injective rings.

Semiperfect FPF rings (with the condition (∗) in the theorem above) as well as noetherian FPF rings have finite Goldie dimension and possess the classical quotient ring [Fat1], [P7]; hence Faticoni [Fat3] provided a general approach to localizations in FPF rings with finite Goldie dimension and showed the following relation between those rings and their quotient rings:

Theorem 1.2. [Fat2, Theorem 4.3 and Proposition 4.9] *Let R be a right FPF ring.*
(1) *The following conditions are equivalent*:
  (a) $Q^r_{max}(R)$ *is a semiperfect right self-injective ring*;
  (b) R *has finite right Goldie dimension and satisfies* (∗).
(2) *If R is a semiperfect ring, then $Q^r_{max}(R)$ is the classical left quotient ring of R.*

Remark. It is not known, as noted in [Fat2] [Fat3], whether the condition (∗) in Theorems 1.1 and 1.2 may be dropped.

It was shown in [Fai1] that a semiperfect right FPF ring with the Jacobson radical nil is right self-injective. Page [P6] and Yousif [You1] generally studied self-injective semiperfect FPF rings and showed the following results respectively.

Theorem 1.3. [P6, Theorem 9] *Let R be a semiperfect right self-injective right FPF ring. Then, $R \cong S \times T$, where S is a semisimple artinian ring and T is a ring with $Z(T_T)$ essential in T_T.*

Theorem 1.4. [You1, Theorem 1] *Let R be a semiperfect right FPF ring. Then R is right self-injective if and only if $J(R) = Z(R_R)$.*

In [P8] [You2] they also studied generalized semiperfect FPF rings.

1.2. Noetherian FPF Rings

Endo [E] in 1967 made a study of QF algebras and determined the structure of commutative noetherian FPF rings, viz., they are precisely the finite direct products of local QF rings and Dedekind domains.

Faticoni [Fat1] and Page [P7] extended the result of Endo to the noncommutative case.

Theorem 1.5. [Fat1, Theorem 2.4] and [P7, Theorem 18] *Let R be a left and right noetherian ring. Then R is left and right FPF if and only if*

$$R \cong Q \times D_1 \times \cdots \times D_k$$

where Q is a QF ring and $D_1, \ldots, D_k$ are bounded Dedekind prime rings (where a ring is said to be Dedekind if it is a hereditary noetherian ring with no non-trivial idempotent ideals).

Remark. We do not know any characterization of one-sided noetherian and one-sided FPF rings.

1.3. (Von Neumann) Regular FPF Rings

Page [P1] in 1978 determined the structure of regular right FPF rings, viz., they are precisely the self-injective regular rings of bounded index, or equivalently the finite direct products of matrix rings over self-injective abelian regular rings.

In relation to the result above of Page, Birkenmeier [Bi1] [Bi2], Kobayashi [Ko3] and the author [Yos2] considered the following condition on a ring R.

(GFC) *Every cyclic faithful right R-module is a generator for* Mod -R.

For regular rings with (GFC), they showed the following:

Theorem 1.6. [Yos2, Theorem 4.3], cf. [Bi2, Theorem 3.5] and [Ko3, Theorem 1] *Let R be a regular ring. Then R satisfies* (GFC) *if and only if*

$$R \cong S_1 \times M_{n_2}(S_2) \times \cdots \times M_{n_k}(S_k)$$

where S_1 is abelian regular and $S_2, \ldots, S_k$ are self-injective abelian regular.

The following is another generalization of Page's theorem for regular FPF rings.

Theorem 1.7. [Yos3, Theorem B] *Let R be a regular ring. Then R satisfies the condition that every finitely generated faithful right R-module contains a submodule that is a generator for* Mod -R *if and only if*

$$R \cong M_{n_1}(S_1) \times \cdots \times M_{n_k}(S_k)$$

where each S_i ($i = 1, \ldots, k$) is abelian regular such that every finitely generated faithful right S_i-submodule of $Q^r_{max}(S_i)$ contains a unit in $Q^r_{max}(S_i)$.

Remark. It was shown in [P1] and [Yos2] that over regular rings, both the conditions FPF and GFC are left-right symmetric. Similarly, this is the case for the condition of the theorem above [Yos6].

1.4. Nonsingular FPF Rings

Page [P2] [P3] in 1982, 1983 showed the basic results of nonsingular or semihereditary FPF rings, e.g., every right nonsingular right FPF ring R is a semiprime ring with $Q^r_{max}(R)$ an FPF ring; every right semihereditary right FPF ring is Morita equivalent to a right FPF ring whose maximal right quotient ring is abelian regular.

Remark. It was shown in [Yos2] that the results above hold also for nonsingular rings with (GFC).

As a general characterization of nonsingular FPF rings, Kobayashi [Ko1] showed the following:

Theorem 1.8. [Ko1, Theorem 1] *A ring R is right nonsingular and right FPF if and only if R satisfies the following three conditions:*
  (a) *R is right bounded;*
  (b) *The inclusion map $R \to Q := Q^r_{max}(R)$ is a left flat epimorphism, i.e., the multiplication map $Q \otimes_R Q \to Q$ is an isomorphism and Q is flat as a right R-module;*
  (c) *For every finitely generated right ideal I of R,*
$$R = r_R(I) \oplus Tr_R(I)$$
where $Tr_R(I)$ is the trace ideal of I, i.e., $Tr_R(I) = \sum \operatorname{Im} f$ (where f ranges over all homomorphisms in $\operatorname{Hom}_R(I, R)$).

All FPF rings as seen in §§1.1–1.3 have the classical quotient rings. Similarly, this is the case for nonsingular FPF rings, which was shown in Burgess [Bu] by using a technique of (Pierce) sheaf theory. By the method, he also obtained the following further results on nonsingular or semihereditary FPF rings:

Theorem 1.9. [Bu, Theorem 1.8 and Theorem 1.10] *(1) Let R be a right semihereditary ring. Then R is right FPF if and only if R satisfies the following five conditions:*
  (a) *The inclusion map $R \to Q := Q^r_{max}(R)$ is a left and right flat epimorphism;*
  (b) *Q is a self-injective regular ring of bounded index;*
  (c) *$B(R) = B(Q)$, the sets of central idempotents of R and Q respectively;*

    (d) *Every nonzero finitely generated right ideal of the stalk, i.e., $R/R\boldsymbol{x}$ ($\boldsymbol{x} \in \operatorname{Spec} B(R)$), is a generator for* Mod-$R/R\boldsymbol{x}$;
    (e) *R is right bounded.*

   (2) *Let R be a semiprime ring that is module finite over its center C. Then R is FPF if and only if R is a semihereditary maximal C-order in a left and right self-injective ring.*

Remark. Kobayashi [Ko1] [Ko2] showed another characterization of semihereditary FPF rings and also studied matrix representations of nonsingular FPF rings.

It is known that every right nonsingular right FPF ring is a semiprime quasi-Baer ring with (GFC), where a ring R is said to be *quasi-Baer* (resp. *Baer*) if the annihilator of every ideal (resp. non-empty subset) of R is a direct summand of R_R. For such rings, Birkenmeier [Bi4] showed the following decomposition.

Theorem 1.10. [Bi4, Theorem 1.4] *Let R be a semiprime quasi-Baer ring with* (GFC). *Then R has a decomposition*:

$$R = A \oplus B \oplus C$$

where

   (i) *A is an abelian Baer ring;*
   (ii) *B is a ring whose every idempotent is central;*
   (iii) *C is a ring that is an essential extension of $\langle N_E(R) \rangle$, the subring (without unity) of R generated by $\bigcup eR(1-e)$, where e ranges over all idempotents of R;*
   (iv) *$B \oplus C$ is the densely nil MDSN (the minimal direct summand containing the nilpotent elements) of R;*
   (v) *If R contains no infinite set of orthogonal central idempotents, then R is a finite direct product of prime right Goldie rings and $C = \langle N_E(R) \rangle$.*

Remark. More generally, Birkenmeier [Bi3] [Bi5] showed the splitting theorems for GFC or FPF rings in detail (also cf. Birkenmeier-Kim-Park [BiKP]).

1.5. Commutative FPF Rings

Faith [Fai3] [Fai4] in 1979, 1982 provided an account of the results on commutative FPF rings, where he showed that a commutative ring R is FPF if and only if the classical quotient ring of R is self-injective and every finitely generated faithful ideal of R is projective.

Further studies of commutative FPF rings were made in [Fai5], [Fai6], [Fai7], [FaiPi], [Yos4] and [Yos5]. In particular, Faith [Fai6] [Fai7] made a study of the valuations (defined by Manis [M1] [M2]) of commutative rings and showed the following relation between commutative FPF rings and Manis valuation rings.

Theorem 1.11. [Fai6, Theorem, §12 and Theorem, Appendix B] *Let R be a commutative local ring with $Q_{cl}(R)$ the classical quotient ring.*
  (1) *If R is an FPF ring, then R is a valuation ring for $Q_{cl}(R)$.*
  (2) *If R is a valuation ring for $Q_{cl}(R)$ with $Q_{cl}(R)$ self-injective, then R is an FPF ring.*

The related results in detail may be found in Faith-Pillay [FaiPi]. By virtue of the results of Faith, the author [Yos4] [Yos5] could show the following.

Theorem 1.12. [Yos4, Theorem 2.1 and Remark, §3] and [Yos5, Corollary 1.6 and Remark 1.7] *Let R be a commutative ring. Then R is FPF if and only if R satisfies any of the following equivalent five conditions:*
  (a) *There exists a one-to-one correspondence*
$$I \mapsto I^* \ (= \{x \in E(R) \mid xI \subset R\})$$
from Δ onto Δ^*, where
$$\Delta = \{I \mid I \text{ an ideal of } R \text{ s.t. } I^*I = R\} \text{ and}$$
$$\Delta^* = \{J \mid J \text{ a finitely generated } R\text{-submodule of } E(R) \text{ s.t. } R \subset J\};$$
  (b) *For every $x \in E(R)$, there exist $a, b \in (R : x) \ (= \{r \in R \mid xr \in R\})$ such that $a + xb = 1$;*
  (c) (i) *$Q := Q_{max}(R)$ is self-injective;*
     (ii) *Every subring of Q containing R is integrally closed in Q;*
  (d) (i) *$Q := Q_{max}(R)$ is self-injective;*
     (ii) *For every $p \in \operatorname{Spec} R$, $(R_{[p]}, [p])$ is a valuation pair on Q, where*
$$R_{[p]} = \{x \in Q \mid (R : x) \not\subset p\} \text{ and } [p] = \{x \in Q \mid (p : x) \not\subset p\};$$
  (e) *For every $p \in \operatorname{Spec} R$, $R_{[p]}$ is an FPF ring.*

1.6. Problems in [FaiPa] and Others

In this subsection, we shall expose the results to date of (some of) the problems posed in [FaiPa] and other topics.

1.6.1. When is a Trivial Extension $R \ltimes M$ of a Ring R by an R-bimodule M FPF?

In the commutative case, Faith [Fai5] showed the following.

Theorem 1.13. [Fai5, Theorem 2] *Let R be a commutative ring and M a faithful R-module. Then $R \ltimes M$ is an FPF ring if and only if the following three conditions hold*:

(a) *M is injective*;

(b) $\text{End}_R(M) \cong RS^{-1}$ *canonically, where RS^{-1} is the localization of R at $S = \{a \in R \mid l_M(a) = 0\}$*;

(c) *Every finitely generated ideal I of R with $l_M(I) = 0$ is invertible in RS^{-1}*.

In the non-commutative case, we do not know any general result on this problem. However, Page [P4] constructed an FPF ring that is a triangular matrix ring, which is a special case of trivial extensions. In §2, more generally we shall characterize triangular matrix FPF rings under some condition.

1.6.2. Is the Center of an FPF Ring FPF?

Clark [Cl1] presented an example of a QF group ring whose center is not FPF. Herbera-Menal [HM] also presented such a PF ring and furthermore showed several interesting results on the center of FPF rings, one of which is the following.

Theorem 1.14. [HM, Theorem 2.8] *Every integrally closed commutative domain is isomorphic to the center of a (not necessarily commutative) bounded Bezout (FPF) domain.*

1.6.3. Characterize FPF Group Rings

Herbera-Menal [HM] (cf. Kitamura [Ki]) showed the following.

Theorem 1.15. [HM, Theorem 2.11] (cf. [Ki, Corollary 10]) *Let R be a commutative ring and G a finite group whose order is invertible in R. Then the group ring RG is FPF if and only if R is FPF.*

1.6.4. Is a Nonsingular FPF Ring Semihereditary?

Herbera-Menal [HM] constructed a reduced (and hence nonsingular) FPF ring that is not semihereditary.

1.6.5. Is there an FPF Ring R with G a Finite Group of Automorphisms of R whose Galois Subring $R^G (= \{r \in R \mid g(r) = r \, (g \in G)\})$ is not FPF?

In fact, Clark [Cl2] constructed two FPF rings R with a finite group G such that R^G is not FPF, while Herbera-Menal [HM] also constructed such rings and conversely showed the following.

Theorem 1.16. [HM, Theorem 2.2] *Let R be a reduced commutative FPF ring and G a finite group of automorphisms of R. Then R^G is FPF.*

1.6.6. Other Topics

Johns rings: Faith-Menal [FaiM1] [FaiM2] studied *right Johns rings*, i.e., right noetherian rings R whose every right ideal is the right annihilator of a subset of R, and *strongly right Johns rings*, i.e., rings R such that $M_n(R)$ is a right Johns ring for all positive integers n. As a result, they showed the following:

Theorem 1.17. [FaiM2, Theorem 4.2] *Every strongly right Johns ring is right FPF.*

FPF matrix rings: It was shown in [Ko3] that a ring R is right FPF if and only if $M_n(R)$ satisfies (GFC) for all positive integers n. As a related result, Birkenmeier [Bi1], Kobayashi [Ko3] and the author [Yos4] showed the following.

Theorem 1.18. [Bi1, Theorem 2], [Ko3, Introduction and §2] and [Yos4, Corollaries 1.2 and 2.2] *Let R be a ring satisfying any of the following four conditions*:

(i) *right self-injective*; (ii) *regular*; (iii) *semiperfect*; (iv) *commutative*.

Then R is right FPF if and only if $M_2(R)$ satisfies (GFC) (or equivalently every faithful right R-module generated by two elements is a generator for Mod-R).

Remark. We do not know whether this is true in general; in particular, is it true for nonsingular FPF rings?

FPF polynomial rings: Herbera-Pillay [HP] studied when a polynomial ring has the classical quotient ring that is self-injective or QF. As a result, they showed the following.

Theorem 1.19. [HP, Corollaries 3.5 and 3.7] *Let R be a ring that is either commutative or right nonsingular.*

(1) *A polynomial ring $R[X]$ in indeterminates x ($x \in X \neq \emptyset$) over R is right FPF if and only if $X = \{x\}$ and R is semisimple artinian.*

(2) *A group ring RG of a free abelian group G over R is right FPF if and only if $G \cong \mathbb{Z}$, the group of integers, and R is semisimple artinian.*

Azumaya or separable algebras: Page [P5] investigated Azumaya algebras (= central separable algebras) over a commutative FPF ring and showed that they are also FPF rings, while Kitamura [Ki] showed the following general result.

Theorem 1.20. [Ki, Theorem 9] and [P5, Theorem 2.1] *Let A be a separable algebra over a commutative ring R such that the center of A is free as an R-module. Then A is FPF if and only if R is FPF.*

FPF endomorphism rings: Page [P6] showed a sufficient condition for projective modules to have FPF endomorphism rings.

Theorem 1.21. [P6, Theorem 4] *Let P be a finitely generated distinguished projective right module over a ring R.*

(1) If P is an FPF module, i.e., every P-finitely generated module cogenerating P generates P, then $\text{End}_R(P)$ is a right FPF ring.

(2) If $Tr_R(P)$ is pure in $_RR$, i.e., $R/Tr_R(P)$ is flat as a left R-module, then the converse of (1) holds.

García Hernández-Gómez Pardo [GG] studied PF or FPF endomorphism rings of modules with some projectivity and showed more general results concerning the theorem above. Thuyet [Th1] [Th2] defined co-FPF modules and showed a similar relation between co-FPF modules and its endomorphism rings and studied rings with a related condition.

Maximal quotient rings of FPF rings: It is known that the maximal quotient ring $Q_{max}(R)$ of 'almost all' FPF rings R (as seen in §§1.1–1.5) is the classical quotient ring of R and is a self-injective FPF ring. However, we do not know whether this is true in general.

2. Triangular Matrix FPF Rings

Throughout this section, let S and T be rings and M a (T, S)-bimodule and consider the triangular matrix ring

$$R = \begin{pmatrix} S & 0 \\ M & T \end{pmatrix}$$

and set

$$e_1 = \begin{pmatrix} 1 & 0 \\ 0 & 0 \end{pmatrix}, \quad e_2 = \begin{pmatrix} 0 & 0 \\ 0 & 1 \end{pmatrix} \quad (\in R).$$

In this section, we shall study when R is right FPF in a special case: M_S is of finite length and $T = \text{End}_S(M)$.

Since the triangular matrix ring R may be considered as a trivial extension $(S \times T) \ltimes M$ of the ring $S \times T$ by the $S \times T$-bimodule M naturally, it follows from [FoGR, Corollary 4.36] (or a direct proof by the Baer Criterion) that we have the following.

Lemma 2.1. *Let* $R = \begin{pmatrix} S & 0 \\ M & T \end{pmatrix}$.

(1) $e_1 R$ *is injective if and only if the following two conditions hold:*
  (a) S *is right self-injective;*
  (b) $\operatorname{Hom}_S(M, S) = 0$.

(2) $e_2 R$ *is injective if and only if the following three conditions hold:*
  (a) M_S *is injective;*
  (b) *The canonical map* $\theta : T \to \operatorname{End}_S(M)$ *is an epimorphism;*
  (c) $\operatorname{Ker} \theta \, (= \{ t \in T \mid tM = 0 \})$ *is injective as a right T-module.*

Lemma 2.2. *Let R be a ring with $J = J(R)$ and $Z = Z(R_R)$ and let e be an idempotent of R such that eRe/eJe is an artinian ring and $eJe = eZe$. Then every element $x \in eRe$ with $r_{eR}(x) = 0$ is a unit in eRe.*

Proof. Let $x \in eRe$ with $r_{eR}(x) = 0$. Then, x is right regular modulo eZe, i.e., $xy \in eZe$ $(y \in eRe)$ implies $y \in eZe$. Thus, noting that every unit modulo Jacobson radical of a ring S is a unit in S, we see from the assumption on e that x is a unit in eRe. □

For R-modules M and N, the notation $N \lesssim_\oplus M$ means that N is isomorphic to a direct summand of M. For a positive integer n, we denote by $M^{(n)}$ the direct sum of n copies of M.

The following lemma is well known (e.g., see [Ba, Lemma 3.5, Ch. I] and [AF, Proposition 18.20 and Exercises §27, 3]).

Lemma 2.3. (1) *Let M and M_i $(i = 1, \ldots, n)$ be right R-modules such that $\operatorname{End}_R(M)$ is a local ring. If $M \lesssim_\oplus M_1 \oplus \cdots \oplus M_n$, then $M \lesssim_\oplus M_i$ for some i.*

(2) *Let E be an injective right R-module. Then*
$$J(\operatorname{End}_R(E)) = \{ f \in \operatorname{End}_R(E) \mid \operatorname{Ker} f \text{ is essential in } E \}.$$

(3) *Every injective right R-module with finite Goldie dimension has a semiperfect endomorphism ring.*

For the sake of completeness we provide a proof of the following lemma, (1) and (2), (a)⇔(b) of which are essentially in [Fa1, Theorem 1] and [You1, Theorem 1].

Lemma 2.4. *Let R be a right FPF ring with $J = J(R)$ and $Z = Z(R_R)$ and let e be an idempotent of R such that $(1-e)Re = 0$.*

(1) *If eRe is a semiperfect ring, then eR has finite Goldie dimension.*

(2) *The following conditions are equivalent:*
  (a) eR *is an injective right R-module with finite Goldie dimension;*

(b) eRe is a semiperfect ring such that $eJe = eZe$;

(c) eRe is a semiperfect ring such that every $x \in eRe$ with $r_{eR}(x) = 0$ is a unit in eRe.

Proof. Assume that eRe is a semiperfect ring. Then we may take orthogonal idempotents $e_1, \ldots, e_k, e_{k+1}, \ldots, e_n$ of R such that

(i) $e = e_1 + \cdots + e_n$;

(ii) Each $e_i R$ ($i = 1, \ldots, n$) is a local R-module, i.e., $e_i R$ contains $e_i J$ as the unique maximal R-submodule;

(iii) $\{e_1 R, \ldots, e_k R\}$ is a non-redundant set of pairwise non-isomorphic modules in $\{e_1 R, \ldots, e_n R\}$.

Throughout the proof, we shall consider $e_1, \ldots, e_k, e_{k+1}, \ldots, e_n$ as the idempotents above, when eRe is semiperfect.

(1) Assume that eRe is semiperfect. It then suffices to show that each $e_j R$ ($j = 1, \ldots, k$) is uniform. To prove this in case $j = 1$, let I and H be proper submodules of $e_1 R$ for which $I \cap H = 0$. Then, the R-module $(e_1 R/I) \oplus (e_1 R/H) \oplus e_2 R \oplus \cdots \oplus e_k R \oplus (1-e)R$ is finitely generated faithful and hence a generator for Mod-R. Thus, in particular,

$$e_1 R \lesssim_\oplus ((e_1 R/I) \oplus (e_1 R/H) \oplus e_2 R \oplus \cdots \oplus e_k R \oplus (1-e)R)^{(m)}$$

for some positive integer m. Since $e_1 R, \ldots, e_k R$ are pairwise non-isomorphic local modules and $(1-e)Re = (1-e)Re_1 = 0$, it follows from Lemma 2.3(1) that either $e_1 R \lesssim_\oplus e_1 R/I$ or $e_1 R \lesssim_\oplus e_1 R/H$. If $e_1 R \lesssim_\oplus e_1 R/I$, then $I = 0$, because $e_1 R$, and hence $e_1 R/I$, has the unique maximal submodule. Thus, $e_1 R$ is uniform, as desired.

(2) (a)$\Rightarrow$(b). Assume (a). It then follows from Lemma 2.3(2) and (3) that eRe is semiperfect and $eJe \subset eZe$. On the other hand, since each $e_i R$ ($i = 1, \ldots, n$) is a local module, it follows that $e_i Z \subset e_i J$; hence $eZe \subset eJe$. Thus we have $eJe = eZe$.

(b)$\Rightarrow$(c). This follows from Lemma 2.2.

(c)$\Rightarrow$(a). Assume (c). Then, by (1) one only needs to show that eR is injective. Suppose, to the contrary, that eR is not injective. Then, renumbering e_j ($j = 1, \ldots, k$) if necessary, we may take an $x \in E(e_1 R) - e_1 R$. Now, applying the same argument as that in the proof of (1) to the finitely generated faithful R-module $(e_1 R + xR) \oplus e_2 R \oplus \cdots \oplus e_k R \oplus (1-e)R$, we have $e_1 R \lesssim_\oplus e_1 R + xR$. This implies an isomorphism

$$\varphi : e_1 R + xR \xrightarrow{\cong} e_1 R,$$

because $e_1 R$, and hence $e_1 R + xR$, is a uniform module as shown in the proof of (1). Since $r_{e_1 R}(\varphi(e_1)) = 0$ and hence $r_{eR}(\varphi(e_1) + e - e_1) = 0$, it follows from (c) that $\varphi(e_1)+e-e_1$ is a unit in eRe, from which $e_1 \in \varphi(e_1 R)$. On the other hand, since $e_1 R + xR \cong e_1 R$ is a local module and $x \notin e_1 R$, it follows that $e_1 R \subsetneq xR \cong e_1 R$, from which $\varphi(e_1 R) \subset e_1 J$. Thus we have $e_1 \in e_1 J$, which is a contradiction. □

Corollary 2.5. (cf. [You1, Theorem 1]) *Let R be a right FPF ring. Then the following conditions are equivalent:*
  (1) *R is a right self-injective ring with finite Goldie dimension;*
  (2) *R is a semiperfect ring such that $J(R) = Z(R_R)$;*
  (3) *R is a semiperfect ring such that every right regular element of R is a unit.*

The following lemma follows from [Yos1, Propositions 3 and 4].

Lemma 2.6. *Let R be a right FPF ring.*
  (1) *If I is a nilpotent ideal of R, then $r_R(I)$ is essential in R_R.*
  (2) *R is right bounded.*

Now, we can prove the following theorem.

Theorem 2.7. *Let $R = \begin{pmatrix} S & 0 \\ M & T \end{pmatrix}$, where $T = \mathrm{End}_S(M)$. Assume that M_S is artinian and T is a semiperfect ring, and consider the following two conditions:*
  (1) *R is right FPF.*
  (2) *S is right FPF and M_S is injective with $r_S(M)$ essential in S_S.*
Then, (1) implies (2). If $l_S r_S(M/S(M_S)) = 0$ (where $S(M_S)$ is the socle of M_S), then (2) implies (1).

Proof. Set
$$I = e_2 R = \begin{pmatrix} 0 & 0 \\ M & T \end{pmatrix}.$$

Assume (1). Since $S \cong R/I$ as rings, we shall first show that R/I is right FPF. To prove this, let X be a finitely generated faithful right R/I-module. Then, $X \oplus I$ is a finitely generated faithful right R-module and hence a generator for Mod-R. Thus there exists an exact sequence
$$(X \oplus I)^{(n)} \to R \to 0$$
for some positive integer n. Applying $- \otimes_R R/I$ to the sequence above, we obtain the exact sequence

$$
\begin{array}{ccccc}
(X \oplus I)^{(n)} \otimes_R R/I & \to & R \otimes_R R/I & \to & 0 \\
\| \wr & & \| \wr & & \\
X^{(n)} & \to & R/I & \to & 0,
\end{array}
$$

from which X is a generator for Mod-R/I. Therefore, R/I is right FPF.

Applying Lemma 2.6(1) to the nilpotent ideal
$$\begin{pmatrix} 0 & 0 \\ M & 0 \end{pmatrix}$$
of R, we see that $r_S(M)$ is essential in S_S. On the other hand, since $T = \text{End}_S(M) \cong e_2 R e_2$ and M_S is artinian, it follows from the Fitting Lemma that every $x \in e_2 R e_2$ with $r_{e_2 R}(x) = 0$ is a unit in $e_2 R e_2$. Thus, Lemmas 2.1(2) and 2.4(2) imply that M_S is injective.

Conversely, assume (2) and $l_S r_S(M/S(M_S)) = 0$. Since M_S is injective and artinian, we may assume
$$M = E(V_1)^{(n_1)} \oplus \cdots \oplus E(V_t)^{(n_t)}$$
where $V_1, \ldots, V_t$ are pairwise non-isomorphic simple S-modules and $n_1, \ldots, n_t$ are positive integers. Set
$$K_i = \begin{pmatrix} 0 & 0 \\ V_i & 0 \end{pmatrix} \ (i = 1, \ldots, t) \ \text{ and } \ L = \begin{pmatrix} S & 0 \\ M & J(T) \end{pmatrix}.$$
Note from Lemma 2.1(2) that I_R is injective and hence
$$I = E(K_1)^{(n_1)} \oplus \cdots \oplus E(K_t)^{(n_t)} \quad (*)$$

Now, to prove that R is right FPF, let X be a finitely generated faithful right R-module. By the faithfulness of X, for each i ($i = 1, \ldots, t$) there exists an $x_i \in X$ such that $x_i K_i \neq 0$. Since K_i is simple and essential in $E(K_i)$, the left multiplication map $E(K_i) \to X$ of x_i is a monomorphism; hence $E(K_i) \lesssim_\oplus X$. Thus we may take a positive integer m_i that is largest with respect to $E(K_i)^{(m_i)} \lesssim_\oplus X$. (Indeed, $m_i \leq$ the length of X/XL, a finitely generated module over the semisimple artinian ring $R/L \cong T/J(T)$). Then, by the (pairwise non-isomorphic) simplicity of K_i's and the choice of m_i's we have an R-module Y such that
$$X \cong E(K_1)^{(m_1)} \oplus \cdots \oplus E(K_t)^{(m_t)} \oplus Y \text{ and}$$
$$Y \begin{pmatrix} 0 & 0 \\ S(M_S) & 0 \end{pmatrix} \ (= Y(K_1^{(n_1)} \oplus \cdots \oplus K_t^{(n_t)})) = 0. \quad (*')$$
In particular, it follows from $(*)$ that X generates I. On the other hand, note from $(*)$ and $(*')$ that
$$(r_R(Y/YI) \cap \begin{pmatrix} r_S(M) & 0 \\ 0 & 0 \end{pmatrix}) \begin{pmatrix} r_S(M/S(M_S)) & 0 \\ 0 & 0 \end{pmatrix}$$
$$= r_R(I) \cap r_R(Y) = r_R(X) = 0.$$

It then follows from $l_S r_S(M/S(M_S)) = 0$ and the essentiality of $r_S(M)$ that $r_R(Y/YI) = I$. Thus, Y/YI is a finitely generated faithful module over the right FPF ring R/I ($\cong S$), whence Y/YI, and hence X, generates R/I. Therefore, X generates $I \oplus (R/I) \cong R$, i.e., X is a generator for Mod-R, as desired. □

Note that if S is a right bounded ring, then the right annihilator $r_S(M)$ of every finitely generated singular right S-module M is essential in S_S and that every module of finite length has a semiprimary endomorphism ring (e.g., see [AF, Corollary 29.3]). Then, Theorem 2.7 combined with Lemmas 2.1 and 2.6(2) immediately implies the following.

Corollary 2.8. *Let* $R = \begin{pmatrix} S & 0 \\ M & T \end{pmatrix}$, *where* $T = \mathrm{End}_S(M)$. *Assume any of the following conditions*:
  (i) M_S *is finitely generated semisimple*;
  (ii) S *is right nonsingular and* M_S *is of finite length*.
Then the following conditions are equivalent:
  (1) R *is right FPF*;
  (2) S *is right FPF and* M_S *is singular and injective*.
In this case, R *is right self-injective if and only if* S *is right self-injective*.

Let S be a regular right FPF ring and P a prime ideal of S. It then follows from Page's theorem [P1, Theorem 9] that S is a right self-injective regular ring of bounded index; hence S/P is a simple artinian ring and every right S-module annihilating P is semisimple and injective. Thus, the corollary above implies the following, which is obtained by Page [P4] in case S is commutative and M_S is simple.

Example 2.9. Let S be a regular right FPF ring and M a finitely generated singular right S-module annihilating a prime ideal of S. Then,

$$R = \begin{pmatrix} S & 0 \\ M & \mathrm{End}_S(M) \end{pmatrix}$$

is a right self-injective right FPF ring.

Acknowledgements. The author would like to thank the organizers of the Third Korea-China-Japan International Symposium and the 2nd Korea-Japan Joint Seminar on Ring Theory for inviting him to the conferences and for recommending that he present this survey in this volume; in particular, the author acknowledges Professors Jae Keol Park and Masahisa Sato. The author would also like to thank the staff of the conferences for the kind hospitality.

References

[AF] F. W. Anderson and K. R. Fuller, *Rings and Categories of Modules*, Graduate Texts in Math. **13**, 2nd Edition, Springer-Verlag, Heidelberg, New York, 1992.

[Ba] H. Bass, *Algebraic K-Theory*, Benjamin, New York, 1968.

[Bi1] G. F. Birkenmeier, *Quotient rings of rings generated by faithful cyclic modules*, Proc. Amer. Math. Soc. **100** (1987), 8–10.

[Bi2] _____, *A generalization of FPF rings*, Comm. Algebra **17** (1989), 855–884.

[Bi3] _____, *Reduced right ideals which are strongly essential in direct summands*, Arch. Math. **52** (1989), 223–225.

[Bi4] _____, *A decomposition of rings generated by faithful cyclic modules*, Canad. Math. Bull. **32** (1989), 333–339.

[Bi5] _____, *When does a supernilpotent radical essentially split off?*, J. Algebra **172** (1995), 49–60.

[BiKP] G. F. Birkenmeier, J. Y. Kim and J. K. Park, *Splitting theorems and a problem of Müller*, Advances in Ring Theory, eds.: S. K. Jain and S. Tariq Rizvi, Trends in Math., Birkhäuser, Boston, Basel, 1997, 39–47.

[Bu] W. D. Burgess, *On nonsingular right FPF rings*, Comm. Algebra **12** (1984), 1729–1750.

[Ca] V. Camillo, *How injective is a semiperfect F.P.F ring?*, Comm. Algebra **17** (1989), 1951–1954.

[Cl1] J. Clark, *The center of an FPF ring need not be FPF*, Bull. Austral. Math. Soc. **38** (1988), 235–236.

[Cl2] _____, *A note on the fixed subring of an FPF ring*, Bull. Austral. Math. Soc. **40** (1989), 109–111.

[E] S. Endo, *Completely faithful modules and quasi-Frobenius algebras*, J. Math. Soc. Japan **19** (1967), 437–456.

[Fai1] C. Faith, *Injective cogenerator rings and a theorem of Tachikawa*, Proc. Amer. Math. Soc. **60** (1976), 25–30.

[Fai2] _____, *Injective cogenerator rings and a theorem of Tachikawa II*, Proc. Amer. Math. Soc. **62** (1977), 15–18.

[Fai3] _____, *Injective quotient rings of commutative rings*, Module Theory, eds.: C. Faith and S. Wiegand, Lecture Notes in Math. **700**, Springer-Verlag, Heidelberg, New York, 1979, 151–203.

[Fai4] _____, *Injective Modules and Injective Quotient Rings*, Lecture Notes in Pure and Applied Math. **72**, Marcel Dekker, New York, 1982.

[Fai5] _____, *Commutative FPF rings arising as split-null extensions*, Proc. Amer. Math. Soc. **90** (1984), 181–185.

[Fai6] _____, *The structure of valuation rings*, J. Pure and Applied Algebra **31** (1984), 7–27.

[Fai7] _____, *The structure of valuation rings II*, J. Pure and Applied Algebra **42** (1986), 37–43.

[FaiM1] C. Faith and P. Menal, *A counter example to a conjecture of Johns*, Proc. Amer. Math. Soc. **116** (1992), 21–26.

[FaiM2] _____, *The structure of Johns rings*, ibid. **120** (1994), 1071–1081.

[FaiPa] C. Faith and S. Page, *FPF Ring Theory: Faithful modules and generators of mod-R*, London Math. Soc., Lecture Note Series **88**, Cambridge Univ. Press, Cambridge, 1984.

[FaiPi] C. Faith and P. Pillay, *Classification of Commutative FPF Rings*, Notas de Matematica, Vol. 4, Universidad de Murcia, 1990.

[Fat1] T. G. Faticoni, *FPF rings I: The Noetherian case*, Comm. Algebra **13** (1985), 2119–2136.

[Fat2] _____, *Semi-perfect FPF rings and applications*, J. Algebra **107** (1987), 297–315.

[Fat3] _____, *Localization in finite dimensional FPF rings*, Pacific J. Math. **134** (1988), 79–99.

[FoGR] R. M. Fossum, P. A. Griffith and I. Reiten, *Trivial Extensions of Abelian Categories,* Lecture Notes in Math. **456**, Springer-Verlag, Heidelberg, New York, 1975.

[GG] J. L. García Hernández and J. L. Gómez Pardo, *Self-injective and PF endomorphism rings,* Israel J. Math. **58** (1987), 324–350.

[HM] D. Herbera and P. Menal, *On rings whose finitely generated faithful modules are generators,* J. Algebra **122** (1989), 425–438.

[HP] D. Herbera and P. Pillay, *Injective classical quotient rings of polynomial rings are quasi-Frobenius,* J. Pure and Applied Algebra **86** (1993), 51–63.

[Ki] Y. Kitamura, *Inheritance of FPF rings,* Comm. Algebra **19** (1991), 157–165.

[Ko1] S. Kobayashi, *On non-singular FPF-rings I,* Osaka J. Math. **22** (1985), 787–795.

[Ko2] _____, *On non-singular FPF-rings II,* Osaka J. Math. **22** (1985), 797–803.

[Ko3] _____, *On regular rings whose cyclic faithful modules are generators,* Math. J. Okayama Univ. **30** (1988), 45–52.

[M1] M. E. Manis, *Extension of valuation theory,* Bull. Amer. Math. Soc. **73** (1967), 735–736.

[M2] _____, *Valuations on a commutative ring,* Proc. Amer. Math. Soc. **20** (1969), 193–198.

[P1] S. Page, *Regular FPF rings,* Pacific J. Math. **79** (1978), 169–176; *Correction,* Proc. Amer. Math. Soc. **97** (1981), 488–490.

[P2] _____, *Semi-prime and non-singular FPF rings,* Comm. Algebra **10** (1982), 2253–2259.

[P3] _____, *Semihereditary and fully idempotent FPF rings,* Comm. Algebra **11** (1983), 227–242.

[P4] _____, *FPF rings and some conjectures of C. Faith,* Canad. Math. Bull. **26** (1983), 257–259.

[P5] _____, *Azumaya algebras and the Brauer group of FPF rings*, Comm. Algebra **13** (1985), 329–336.

[P6] _____, *FPF endomorphism rings with applications to QF-3 rings*, Comm. Algebra **14** (1986), 423–435.

[P7] _____, *FQF-3 rings*, Math. J. Okayama Univ. **30** (1988), 79–91.

[P8] _____, *Large FPF rings*, Comm. Algebra **23** (1995), 2125–2134.

[Ta] H. Tachikawa, *A generalization of quasi-Frobenius rings*, Proc. Amer. Math. Soc. **20** (1969), 471–476.

[Th1] L. V. Thuyet, *On co-FPF modules*, Bull. Austral. Math. Soc. **48** (1993), 257–264.

[Th2] _____, *On ring extensions of FSG rings*, Bull. Austral. Math. Soc. **49** (1994), 365–371.

[Yos1] H. Yoshimura, *On finitely pseudo-Frobenius rings*, Osaka J. Math. **28** (1991), 285–294.

[Yos2] _____, *On rings whose cyclic faithful modules are generators*, Osaka J. Math. **32** (1995), 591–611.

[Yos3] _____, *On regular rings whose cyclic faithful modules contain generators*, Osaka J. Math. **34** (1997), 363–380.

[Yos4] _____, *FPF rings characterized by two-generated faithful modules*, Osaka J. Math. **35** (1998), 855–871.

[Yos5] _____, *Rings whose invertible ideals correspond to finitely generated overmodules*, Comm. Algebra **26** (1998), 997–1004.

[Yos6] _____, *Rings whose finitely generated faithful modules contain generators*, preprint.

[You1] M. Yousif, *On semiperfect FPF-rings*, Canad. Math. Bull. **37** (1994), 287–288.

[You2] ———, *On large FPF-rings*, Comm. Algebra **26** (1998), 221–224.

Department of Mathematics
Faculty of Science
Yamaguchi University
Yamaguchi 753-8512, Japan
e-mail: yoshi@po.cc.yamaguchi-u.ac.jp

Surjectivity of Linkage Maps

Yuji Yoshino

Abstract

In this paper we give necessary and sufficient conditions for the linkage maps Φ_r ($r = 1, 2$) to be surjective onto the set of isomorphism classes of Cohen-Macaulay modules. More precisely, for a Gorenstein complete local ring R, we prove that Φ_1 is surjective iff R is an integral domain, and that Φ_2 is surjective iff R is a unique factorization domain.

1. Introduction

In the previous paper [YI], we defined the notion of linkage for Cohen-Macaulay modules over a Gorenstein complete local ring. In the present paper we shall give some necessary and sufficient conditions for these linkage maps to be surjective.

To be more precise, let $(R, \mathfrak{m}, k)$ be a commutative Gorenstein complete local ring of Krull dimension d. We denote the category of finitely generated R-modules by R-mod and denote the category of maximal Cohen-Macaulay modules (resp. the category of Cohen-Macaulay modules of codimension r) as a full subcategory of R-mod by CM(R) (resp. CM$^r(R)$). Hence the objects of CM$^r(R)$ are finitely generated R-modules M with depth$M = \dim M = d - r$. Note that CM$^0(R) = $ CM(R).

We also denote the stable category by $\underline{\text{CM}}(R)$ (resp. $\underline{\text{CM}^r}(R)$) that is defined in such a way that the objects are the same as that of CM(R) (resp. CM$^r(R)$), while the morphisms from M to N are the elements of $\underline{\text{Hom}}_R(M, N) = \text{Hom}_R(M, N)/P(M, N)$ where $P(M, N)$ is the set of morphisms which factor through free R-modules.

We recall the definition of Cohen-Macaulay approximations from the paper of Auslander and Buchweitz. It is shown in [AB] that for any $M \in R$-mod, there is an exact sequence

$$0 \longrightarrow \text{Y}_R(M) \longrightarrow \text{X}_R(M) \longrightarrow M \longrightarrow 0$$

where $\text{X}_R(M) \in $ CM(R) and $\text{Y}_R(M)$ is of finite projective dimension. Such a sequence is not unique, but $\text{X}_R(M)$ is known to be unique up to free summand, and hence it gives rise to the functor

$$X_R : \underline{R\text{-mod}} \longrightarrow \underline{CM}(R)$$

which we call the Cohen-Macaulay approximation functor.

As in the paper [YI], we define the linkage functor $L_R : \underline{CM}(R) \longrightarrow \underline{CM}(R)$ by $L_R = D_R \circ \Omega^1_R$, where $D_R = \operatorname{Hom}_R(\ ,R)$ and Ω^1_R is the first syzygy functor. Let N_1, N_2 be two Cohen-Macaulay modules of codimension r. Then we can find a regular sequence $\underline{\lambda}$ (resp. $\underline{\mu}$) on R such that N_1 (resp. N_2) is a maximal Cohen-Macaulay module over $R/\underline{\lambda}R$ (resp. $R/\underline{\mu}R$). If there exists a module $N \in \underline{CM}^r(R)$ that belongs to both $\underline{CM}(R/\underline{\lambda}R)$ and $\underline{CM}(R/\underline{\mu}R)$ satisfying

$$N_1 \cong L_{R/\underline{\lambda}R}(N) \text{ in } \underline{CM}(R/\underline{\lambda}R) \quad \& \quad N_2 \cong L_{R/\underline{\mu}R}(N) \text{ in } \underline{CM}(R/\underline{\mu}R),$$

then we say N_1 (resp. N_2) is linked to N through the regular sequence $\underline{\lambda}$ (resp. $\underline{\mu}$). We also say in this case that N_1 and N_2 are doubly linked through $(\underline{\lambda}, \underline{\mu})$. If there is a sequence of modules $N_1, N_2, \ldots, N_s$ in $\underline{CM}^r(R)$ such that N_i and N_{i+1} are doubly linked for $1 \leq i < s$, then we say that N_1 and N_s are evenly linked, or N_1 and N_2 are in the same even linkage class.

In the previous paper [YI] we have shown that if two modules in $\underline{CM}^r(R)$ are in the same even linkage class, then the functor X_R takes the same value for these modules. Hence it turns out that the Cohen-Macaulay approximation functor X_R yields the map from the set of even linkage classes in $\underline{CM}^r(R)$ to the set of isomorphism classes of objects in $\underline{CM}(R)$, which we denote by Φ_r.

$$\Phi_r : \underline{CM}^r(R)/(\text{even linkage}) \longrightarrow \underline{CM}(R)/(\text{isomorphism}).$$

In this paper we are interested in studying when the map Φ_r is surjective. Our main results say that Φ_1 is surjective if and only if R is an integral domain, and Φ_2 is surjective if and only if R is a unique factorization domain. And we conjecture that R would be a regular local ring if Φ_r is surjective for some $r \geq 3$.

Remark. As the referee pointed out in his report, it is known that minimal Cohen-Macaulay approximations exist and are unique up to isomorphism, since R is Gorenstein. See E. E. Enochs, O. M. G. Jenda and X. Jinzhong; A generalization of Auslander's last theorem, Algebra and Representation Theory vol.2 (1999), 259–268. Hence there is a functor X from R-mod to CM(R) and the above theory might be modified to use minimal approximation. However, the nature of the above mentioned surjectivity problem is unchanged. In fact, if $\underline{\lambda}$ is an R-regular sequence of length r, then $R/\underline{\lambda}R$ is a Cohen-Macaulay module of codimension r and $X(R/\underline{\lambda}R) = R$, hence every free R-module is in the image of minimal Cohen-Macaulay approximation functor.

2. Main Results

As in the introduction, R always denotes a commutative Gorenstein complete local ring of dimension d. We first remark the following which is obvious from the definition of Φ_r. See [YI] for more detail.

Lemma 1. *The following conditions are equivalent for an integer $r \geqq 1$.*

(1) Φ_r *is surjective.*

(2) *For any $X \in \underline{CM}(R)$, there is an $M \in \underline{CM}^r(R)$ such that there exists an exact sequence;*

$$0 \longrightarrow Y \longrightarrow X \oplus F \longrightarrow M \longrightarrow 0$$

where F is a free R-module and Y is a module of finite projective dimension.

The following is one of the key points when we consider the surjectivity of Φ_r.

Lemma 2. *Suppose Φ_r is surjective. Then the ring R satisfies the Serre (R_{r-1})-condition, i.e., $R_{\mathfrak{p}}$ is a regular local ring for $\mathfrak{p} \in \mathrm{Spec} R$ with $\mathrm{ht}(\mathfrak{p}) \leqq r - 1$.*

Proof. Let $\mathfrak{p}$ be in $\mathrm{Spec} R$ with $\mathrm{ht}(\mathfrak{p}) \leqq r - 1$. And suppose $R_{\mathfrak{p}}$ is a nonregular local ring. Then, putting X to be a dth syzygy module of $R/\mathfrak{p}$, we can show that X is in $\underline{CM}(R)$. Actually, every dth syzygy module is a maximal Cohen-Macaulay module over R. See [Y, Proposition 1.16].

Now, since Φ_r is surjective, we can find a module $M \in \underline{CM}^r(R)$ and an exact sequence;

$$0 \longrightarrow Y \longrightarrow X \oplus F \longrightarrow M \longrightarrow 0$$

where F is a free R-module and Y is a module of finite projective dimension. Since M has codimension r, we see that $M_{\mathfrak{p}} = 0$, hence it follows from the exact sequence that $X_{\mathfrak{p}}$ is a direct summand of $Y_{\mathfrak{p}}$, in particular, $X_{\mathfrak{p}}$ is of finite projective dimension as an $R_{\mathfrak{p}}$-module. On the other hand, since $X_{\mathfrak{p}}$ is a maximal Cohen-Macaulay module over $R_{\mathfrak{p}}$, we conclude that $X_{\mathfrak{p}}$ is a free $R_{\mathfrak{p}}$-module. Then $(R/\mathfrak{p})_{\mathfrak{p}}$ must have finite projective dimension, and $R_{\mathfrak{p}}$ should be a regular local ring. □

As we remarked in the beginning of the section 2 in [YI], if R is an integral domain, then Φ_1 is obviously surjective. We can show the converse is also true.

Theorem 3. *The following conditions are equivalent.*
  (1) *Φ_1 is surjective.*
  (2) *R is an integral domain.*

Proof. For the convenience of the reader we include here a proof of the implication (2)$\Rightarrow$(1) that was given in [YI, Section 2].

If R is an integral domain, then every maximal Cohen-Macaulay module $X \in \underline{\mathrm{CM}}(R)$ is a torsion free module, hence it has a well-defined rank, say s, and a free module of rank s can be embedded in X:

$$0 \longrightarrow R^s \longrightarrow X \longrightarrow M \longrightarrow 0 \quad (\text{exact})$$

where one can easily see that $M \in \underline{\mathrm{CM}}^1(R)$. Therefore Φ_1 is surjective by Lemma 1.

Now we prove (1)$\Rightarrow$(2). (Part of this proof was obtained through discussions with Kiriko Kato. The author would like to thank her.)

Suppose Φ_1 is surjective. Then, by Lemma 2, we see that R satisfies the (R_0)-condition, hence R is a reduced ring (i.e., without nilpotent elements). Therefore, to show that R is an integral domain, we have only to prove that R has a unique minimal prime ideal.

Suppose we have distinct minimal prime ideals $\mathfrak{p}$ and $\mathfrak{q}$. Let X be the dth syzygy module of $R/\mathfrak{p}$ and

$$0 \longrightarrow X \longrightarrow F_{d-1} \longrightarrow \cdots \longrightarrow F_0 \longrightarrow R/\mathfrak{p} \longrightarrow 0$$

is exact such that every F_i is a free R-module. Since $R_\mathfrak{p}$ is a field (hence $R_\mathfrak{p}/\mathfrak{p}R_\mathfrak{p} = R_\mathfrak{p}$), we see that $X_\mathfrak{p}$ is a free $R_\mathfrak{p}$-module whose rank is $\sum_{i=1}^{d}(-1)^{i-1}\mathrm{rank}F_{d-i} + (-1)^d$. On the other hand, noting that $(R/\mathfrak{p})_\mathfrak{q} = 0$, we see that $X_\mathfrak{q}$ is also a free $R_\mathfrak{q}$-module, but its rank is $\sum_{i=1}^{d}(-1)^{i-1}\mathrm{rank}F_{d-i}$. Therefore $\mathrm{rank}_{R_\mathfrak{p}}X_\mathfrak{p} \neq \mathrm{rank}_{R_\mathfrak{q}}X_\mathfrak{q}$.

Now, since Φ_1 is surjective, we have an exact sequence $0 \to R^r \to X \oplus R^s \to M \to 0$, where $M \in \underline{\mathrm{CM}}^1(R)$. Then, since $M_\mathfrak{p} = 0$ and $M_\mathfrak{q} = 0$, we have that $\mathrm{rank}_{R_\mathfrak{p}}X_\mathfrak{p} = r - s = \mathrm{rank}_{R_\mathfrak{q}}X_\mathfrak{q}$. This contradiction completes the proof. □

In the previous paper [YI] we have shown that when R has Krull dimension two, then Φ_2 is surjective if and only if R is a unique factorization domain. Here we are going to show that this is true for any dimension.

Theorem 4. *The following conditions are equivalent.*
  (1) *Φ_2 is surjective.*
  (2) *R is a unique factorization domain.*

Proof. (2)⇒(1). The proof (a)⇒(b) of [YI, Theorem 2.2] is valid for any dimension as it is.

(1)⇒(2). Note from Lemma 2 that R is a normal domain. Therefore we have only to show that the divisor class group $\mathrm{Cl}(R)$ is trivial. To show this, let $\mathfrak{p}$ be a prime ideal of height one and take the Cohen-Macaulay approximation of $\mathfrak{p}$.

$$0 \longrightarrow Y(\mathfrak{p}) \longrightarrow X(\mathfrak{p}) \longrightarrow \mathfrak{p} \longrightarrow 0 \qquad (i)$$

where $X(\mathfrak{p}) \in \underline{\mathrm{CM}}(R)$ and $Y(\mathfrak{p})$ is of finite projective dimension. Now, since Φ_2 is surjective, there is an exact sequence

$$0 \longrightarrow W \longrightarrow X(\mathfrak{p}) \oplus F \longrightarrow L \longrightarrow 0 \qquad (ii)$$

where F is a free R-module, W is a module of finite projective dimension and $L \in \underline{\mathrm{CM}}^2(R)$. Taking the divisor classes of modules in the sense of Bourbaki (or first Chern classes), we see that $c(\mathfrak{p}) = c(X(\mathfrak{p}))$ from the exact sequence (i), and $c(X(\mathfrak{p})) = c(L)$ from (ii). It hence follows that $c(\mathfrak{p}) = c(L)$. For the definition of c, see [B, §4, $n°7$]. Note here that $c(Y) = 0$ if Y has finite projective dimension. Also note that $c(L) = 0$ since L is of codimension > 1. Therefore we have $c(\mathfrak{p}) = 0$. Since this is true for any prime ideal $\mathfrak{p}$ of height one, we conclude that $\mathrm{Cl}(R) = \{0\}$ as desired. □

Remark 5. Assume that R is a regular local ring. Then it is well known that every maximal Cohen-Macaulay module over R is free, hence $\underline{\mathrm{CM}}(R)$ consists of only the zero object. Since Φ_r is a map whose image is a subset of the set of isomorphism classes of objects in $\underline{\mathrm{CM}}(R)$, we see that Φ_r is surjective for any $r \geq 1$.

Hence Theorem 4 implies the well-known theorem of Auslander and Buchsbaum which says that every regular local ring is a unique factorization domain.

When we consider the surjectivity of Φ_r for $r \geq 3$, we naturally expect the following conjecture is true.

Conjecture 6. The following three conditions would be equivalent:
(1) Φ_r is surjective for some $r \geq 3$.
(2) Φ_r is surjective for any $r \geq 3$.
(3) R is a regular local ring.

References

[AB] M. Auslander and R.-O. Buchweitz, *The homological theory of maximal Cohen-Macaulay approximations,* Soc. Math. France, Mém. **38** (1989), 5–37.

[B] N. Bourbaki, *Commutative Algebra: Chapters 1–7,* Springer-Verlag, Heidelberg, New York, 1988.

[Y] Y. Yoshino, *Cohen-Macaulay modules over Cohen-Macaulay rings,* London Math. Soc. Lecture Note Series **146**, Cambridge University Press, Cambridge, 1990.

[YI] Y. Yoshino and S. Isogawa, *Linkage of Cohen-Macaulay modules over a Gorenstein ring,* J. Pure and Applied Algebra **149** (2000), 305–318.

Department of Mathematics
Faculty of Science
Okayama University
Okayama 700-8530, Japan
e-mail: yoshino@math.okayama-u.ac.jp

Infinite Quivers and Cohomology Groups

Pu Zhang

Abstract

The aim of this paper is to report some results on the cohomology groups $Ext_{A^e}^n$ and the Hochschild cohomology groups $H^n(A)$, where A is an algebra given by arbitrary quiver (not necessarily a finite quiver) and relations. In this case, the two cohomologies no longer coincide. In particular, we prove that $Ext_{A^e}^1(A, A) = 0$ if and only if Q is a tree, where $A = kQ$ with Q an arbitrary quiver; and that $H^1(A) = 0$ if and only if Q is a finite tree, where $A = kQ/I$ is a monomial algebra. This generalizes the corresponding results of Happel, Bardzell and Marcos to the infinite-dimensional case.

By a theorem of Gabriel (cf. [G]), any finite-dimensional basic algebra over an algebraically closed field k is of the form kQ/I, where Q is a finite quiver, I is an ideal of the path algebra kQ satisfying $J^N \subseteq I \subseteq J^2$ for some $N \geq 2$, and J is the ideal of kQ generated by arrows of Q. In dealing with infinite-dimensional algebras, it is also natural to consider the algebras given by infinite quivers and relations. On the other hand, in recent years, there have been many papers studying the Hochschild cohomology groups of finite-dimensional algebras by using some methods in the representation theory of finite-dimensional algebras. This inspires us to consider the Hochschild cohomology groups of some infinite-dimensional algebras given by infinite quivers and relations.

The aim of this paper is to report some results on cohomology groups of algebras given by arbitrary quivers (not necessarily finite quivers) and relations. Different from the finite-dimensional case, the algebras A given by infinite quivers and relations have no units, and hence the Hochschild cohomology groups $H^n(A)$ are not isomorphic to the cohomology groups $Ext_{A^e}^n(A, A)$ (see e.g., 1.5 below), thus, we should distinguish the two cohomology groups for algebras given by infinite quivers and relations. The

---

Supported by a grant from the Volkswagen-Stiftung, Germany, and the National Natural Science Foundation of China.

main results reported here are Theorems 2.2 and 3.1, and some other vanishing conditions on the first Hochschild cohomology group, see Theorems 4.1–4.3.

1. Preliminaries

1.1. A quiver is a datum $Q = (Q_0, Q_1, h, t)$, where Q_0, Q_1 are two sets, which are respectively called the set of vertices and the set of arrows of Q, and $h, t : Q_1 \longrightarrow Q_0$ are two maps, for which $h(\alpha)$ and $t(\alpha)$ are respectively called the head and the tail of an arrow α. Thus for $\alpha \in Q_1$, we write $\alpha : h(\alpha) \longrightarrow t(\alpha)$. We emphasize that the quiver Q considered here may be an infinite quiver, i.e., at least one of Q_0 and Q_1 is an infinite set.

A path p in Q of length l means a sequence of arrows $p = \alpha_1 \cdots \alpha_l$ with $t(\alpha_i) = h(\alpha_{i+1})$ for $1 \leq i \leq l-1$. Call $h(p) = h(\alpha_1)$, $t(p) = t(\alpha_l)$, and $l(p) = l$ respectively the head, the tail, and the length of p. If we regard a vertex $i \in Q_0$ as a path of length 0, it will be denoted by e_i. A path p with $l(p) \geq 1$ is called an oriented cycle provided $h(p) = t(p)$. Denote by Q_p the set of all paths in Q. We emphasize that the quiver Q considered here may contain an oriented cycle.

A quiver is called locally finite provided that for any $i \in Q_0$ there are only finitely many $\alpha \in Q_1$ with $h(\alpha) = i$, and there are only finitely many $\beta \in Q_1$ with $t(\beta) = i$.

For an arrow $\alpha \in Q_1$, consider the formal inverse α^{-1}. Define $h(\alpha^{-1}) = t(\alpha)$ and $t(\alpha^{-1}) = h(\alpha)$. A walk w is an "unoriented path", i.e., a sequence $w = \beta_1 \cdots \beta_l$, where β_i is an arrow α_i or a formal inverse α_i^{-1} of an arrow α_i such that $t(\beta_i) = h(\beta_{i+1})$, $1 \leq i \leq l-1$, and that w contains no subsequences of the form $\alpha \alpha^{-1}$ and the form $\alpha^{-1}\alpha$, where α is an arrow. Define the head $h(w)$ of w to be $h(\beta_1)$, and the tail $t(w)$ of w to be $t(\beta_l)$. A walk of length l is denoted by $i_1 - i_2 - \cdots - i_{l+1}$, where $i_j - i_{j+1}$ means the arrow can go in either direction. In particular, we regard a vertex as a walk of length 0. Two walks w_1 and w_2 are said to be parallel provided that they have the same head and the same tail. A walk $w : i_1 - i_2 - \cdots - i_{l+1}$ is called an unoriented cycle provided that $l > 0$, $h(w) = i_1 = i_{l+1} = t(w)$, and that the vertices $i_1, \cdots, i_l$ are pairwise distinct. Thus, by definition, for any arrow α, at most one of α and α^{-1} occurs in an unoriented cycle. A quiver Q is called a tree provided that Q contains no unoriented cycles. A quiver is called connected provided that for any two vertices i and j, there is a walk connecting i and j.

For a field k and a quiver Q, let $A = kQ$ be the k-space with basis of all paths in Q. For $p = \alpha_1 \cdots \alpha_m$, $q = \beta_1 \cdots \beta_n \in Q_p$, define the multiplication

$$pq = \begin{cases} \alpha_1 \cdots \alpha_m \beta_1 \cdots \beta_n, & t(p) = h(q) \\ 0, & t(p) \neq h(q). \end{cases}$$

Then $A = kQ$ becomes a k-algebra, which is called the path algebra of Q. Note that A may have no unit, however $A^2 = A$; and that A has the unit if and only if Q_0 is a finite set, and in this case $1 = \sum_{i \in Q_0} e_i$; also note that A is finite-dimensional if and only if Q is a finite quiver (i.e., both Q_0 and Q_1 are finite sets) and Q contains no oriented cycles.

We are interested in considering the monomial algebras, which are by definition of the form $A = kQ/I$, where Q is an arbitrary quiver, and I is an ideal of kQ generated by a set of paths of length longer than 1. In particular, path algebras are monomial algebras. Note that if $A = kQ/I$ is a monomial algebra, then $I \subseteq J^2$, where J is the ideal of kQ generated by all arrows of Q. We emphasize that monomial algebras considered here may be infinite-dimensional.

1.2. Let R be a ring and $e^2 = e \in R$. Then Re and eR are respectively left and right projective R-modules. We do not assume that R has a unit, but assume that R has a set of orthogonal idempotents $\{e_i \mid i \in I\}$ such that $R = \oplus_{i \in I} Re_i = \oplus_{i \in I} e_i R$. Consider the category R-Mod of all left R-modules X with $RX = X$, or equivalently, $X = \oplus_{i \in I} e_i X$ (the morphisms in this category are just the R-module homomorphisms). Clearly, Re_i and R are objects of R-Mod; and R-Mod is an extension closed abelian category. Such a ring R is called (left) hereditary provided that every submodule $X \in R$-Mod of a projective module $P \in R$-Mod is also projective.

Now, let $Q = (Q_0, Q_1, h, t)$ be a quiver. Then $A = kQ = \oplus_{i \in Q_0} Ae_i = \oplus_{i \in Q_0} e_i A$. Note that Ae_i is the k-space with basis the set of all paths in Q with a tail i. Let $X \in A$-Mod. The following construction of a projective resolution of X is the explicit form of Happel's resolution in [Ha, 1.6], and its proof is due to Crawley-Boevey [CB]; both are stated for A being finite-dimensional. Fortunately, it is also valid for infinite quivers. For the convenience of the reader we include the proof here. The tensor product $\otimes$ will mean $\otimes_k$.

Lemma. [Happel and Crawley-Boevey] *We have the projective resolution of $X \in A$-Mod,*

$$0 \to \bigoplus_{\alpha \in Q_1} Ae_{h(\alpha)} \otimes e_{t(\alpha)} X \xrightarrow{f} \bigoplus_{i \in Q_0} Ae_i \otimes e_i X \xrightarrow{g} X \to 0, \tag{1}$$

where g and f are A-homomorphisms defined by $g(a \otimes x) = ax$ for $a \in Ae_i$, $x \in e_i X$ and $f(a \otimes x) = a\alpha \otimes x - a \otimes \alpha x$ for $a \in Ae_{h(\alpha)}$, $x \in e_{t(\alpha)} X$.

Proof. Note that $Ae_i \otimes V$ is always A-projective for any k-space V. Since $X \in A$-Mod, it follows that g is epic. We claim that f is mono. In fact, let $0 \neq \xi \in \bigoplus_{\alpha \in Q_1} Ae_{h(\alpha)} \otimes e_{t(\alpha)} X$. Then ξ can be written as a finite sum

$$\xi = \sum_{\alpha \in Q_1} \sum_{p \in Q, t(p) = h(\alpha)} p \otimes x_{\alpha,p} \quad \text{with} \quad x_{\alpha,p} \in e_{t(\alpha)} X.$$

Let p be a path of maximal length such that $x_{\alpha,p} \neq 0$ for some α. Then

$$f(\xi) = \sum_{\alpha \in Q_1} \sum_{p \in Q_p, t(p) = h(\alpha)} (p\alpha \otimes x_{\alpha,p} - p \otimes \alpha x_{\alpha,p}).$$

Since $p\alpha \otimes x_{\alpha,p} \neq 0$ by assumption, it follows that $f(\xi) \neq 0$.

Clearly, $fg = 0$. Let $\eta \in \bigoplus_{i \in Q_0} Ae_i \otimes e_i X$. Then η can be uniquely written as a finite sum

$$\eta = \sum_{i \in Q_0} \sum_{p \in Q_p, t(p) = i} p \otimes x_p \quad \text{with} \quad x_p \in e_i X.$$

Define $deg(\eta)$ to be the maximal length of the paths p with $x_p \neq 0$. Write $p \in Q_p$ with $t(p) = i$ and $l(p) \geq 1$ as $p = p'\alpha$ with $\alpha \in Q_1$. Then

$$f(p' \otimes x_p) = p \otimes x_p - p' \otimes \alpha x_p.$$

Now we can claim that $\eta + Im(f)$ contains an element of degree 0; if $deg(\eta) = d > 0$, then $\eta - f(\sum_{i \in Q_0} \sum_{p \in Q_p, t(p) = i, l(p) = d} p' \otimes x_p)$ is of degree less than d, and the assertion follows from induction.

Let $\eta \in Ker(g)$, and $\eta' \in \eta + Im(f)$ with $deg(\eta') = 0$. Then we can write $\eta' = \sum_{i \in Q_0} e_i \otimes x'_{e_i}$ with $x'_{e_i} \in e_i X$, and then

$$0 = g(\eta) = g(\eta') = g(\sum_{i \in Q_0} e_i \otimes x'_{e_i}) = \sum_{i \in Q_0} x'_{e_i} \in \bigoplus_{i \in Q_0} e_i X;$$

it follows that every component $x'_{e_i} = 0$, and hence $\eta' = 0$ and $\eta \in Im(f)$. □

Corollary. *Let $A = kQ$ with Q a quiver. Then*
(i) *For $X \in A$-Mod, the projective dimension $p.d.X \leq 1$;*
(ii) *A is hereditary.*

1.3. Let Λ be an algebra over a field k. We will not insist that Λ have a unit. Let $\Lambda^e = \Lambda \otimes \Lambda^*$ be the enveloping algebra of Λ, where Λ^* is the opposite algebra of Λ. For $a \in \Lambda$, the corresponding element in Λ^* is denoted by a'. Thus, in Λ^e we have $(a \otimes b')(c \otimes d') = ac \otimes (db)'$. Regard Λ as a left Λ^e-module in a natural way: $(a \otimes b')x = axb$ for $a \otimes b' \in \Lambda^e, x \in \Lambda$.

Now, we consider the path algebra $A = kQ$, where Q is an arbitrary quiver. Then $A^e = \bigoplus_{i,j \in Q_0} A^e(e_i \otimes e'_j) = \bigoplus_{i,j \in Q_0} Ae_i \otimes (e_j A)'$.

Corollary. *Let $A = kQ$ with Q a quiver. Then we have the projective resolution of A over A^e,*

$$0 \to \bigoplus_{\alpha \in Q_1} Ae_{h(\alpha)} \otimes (e_{t(\alpha)} A)' \xrightarrow{f} \bigoplus_{i \in Q_0} Ae_i \otimes (e_i A)' \xrightarrow{g} A \to 0, \qquad (2)$$

where g and f are A^e-homomorphisms defined by $g(a \otimes b') = ab$ for $a \in Ae_i$, $b \in e_i A$ and $f(a \otimes b') = a\alpha \otimes b - a \otimes \alpha b$ for $a \in Ae_{h(\alpha)}$, $b \in e_{t(\alpha)} A$.

Proof. Take X in (1) in Lemma 1.2 to be $_{A^e} A$, then the exact sequence (1) gives a projective resolution of the A^e-module $_{A^e} A$, and hence the assertion follows. $\square$

1.4. Let Λ be an algebra over a field k. As in 1.3, also we will not insist on Λ having a unit. Let $\Lambda^{\otimes n}$ denote the n-fold tensor product of Λ with itself over k, and X a Λ^e-module. Regard X as a Λ^e-bimodule by $axb =: (a \otimes b')x$ for $a, b \in \Lambda$. Recall that the Hochschild complex (C^n, d^n) introduced in [Ho] is defined as follows:

$C^n = 0$ for $n < 0$; $C^0 = X$; $C^n = \text{Hom}_k(\Lambda^{\otimes n}, X)$ for $n > 0$;
$d^0 : X \longrightarrow \text{Hom}_k(\Lambda, X)$ with $d^0 x(a) = ax - xa$ for $x \in X$, $a \in \Lambda$;

and $d^n : \text{Hom}_k(\Lambda^{\otimes n}, X) \longrightarrow \text{Hom}_k(\Lambda^{\otimes(n+1)}, X)$ with

$$\begin{aligned} d^n f(a_1 \otimes \cdots \otimes a_{n+1}) &= a_1 f(a_2 \otimes \cdots \otimes a_{n+1}) \\ &+ \sum_{1 \le j \le n} (-1)^j f(a_1 \otimes \cdots \otimes a_j a_{j+1} \otimes \cdots \otimes a_{n+1}) \\ &+ (-1)^{n+1} f(a_1 \otimes \cdots \otimes a_n) a_{n+1} \end{aligned}$$

for $f \in C^n$ and $a_1, \cdots, a_{n+1} \in \Lambda$; and the n-th Hochschild cohomology of Λ with coefficients in X is the k-space $H^n(\Lambda, X) = Ker(d^n)/Im(d^{n-1})$ by definition. In particular, $H^n(\Lambda) = H^n(\Lambda, \Lambda)$ is called the n-th Hochschild cohomology of Λ.

It is clear that $H^0(\Lambda, X) = \{x \in X \mid xa = ax, \forall a \in \Lambda\}$; in particular, $H^0(\Lambda)$ is the center $Z(\Lambda)$ of Λ. Let

$$Der(\Lambda, X) = \{\delta \in \text{Hom}_k(\Lambda, X) \mid \delta(ab) = \delta(a)b + a\delta(b)\},$$

and
$$Der^0(\Lambda, X) = \{\delta_x \in \text{Hom}_k(\Lambda, X) \mid x \in X\}$$
with $\delta_x(a) = xa - ax$, for all $a \in \Lambda$. Then we have that $H^1(\Lambda, X) = Der(\Lambda, X)/Der^0(\Lambda, X)$; in particular, $H^1(\Lambda) = Der(\Lambda)/Der^0(\Lambda)$, where $Der(\Lambda) = Der(\Lambda, \Lambda)$ and $Der^0(\Lambda) = Der^0(\Lambda, \Lambda)$ are respectively the k-spaces of the derivations and the inner derivations of Λ.

1.5. Note that in standard literature (see e.g., [CE, W]) the Hochschild cohomology is defined for algebras with unit, and then we have the isomorphism of k-spaces $H^n(A) \cong \text{Ext}^n_{A^e}(A, A)$ for $n \geq 0$. But for an algebra without unit, this isomorphism is no longer valid. We have the following

Lemma. *Let $A = kQ$ with Q a connected quiver. Then the following are equivalent:*
  (i) *A has a unit.*
  (ii) *Q_0 is a finite set.*
  (iii) *$Z(A) \neq 0$.*
  (iv) *We have an isomorphism of k-spaces $H^n(A) \cong \text{Ext}^n_{A^e}(A, A)$ for $n \geq 0$.*

Proof. It is clear that (i) is equivalent to (ii), and (i) implies (iii). The implication of (i) to (iv) is well known, see e.g., [CE]; and the implication of (iv) to (iii) follows from the fact that
$$Z(A) = H^0(A) \cong \text{Hom}_{A^e}(A, A) \neq 0.$$
It remains to prove that if Q_0 is infinite, then $Z(A) = 0$. Otherwise, let $0 \neq a = \sum c_i p_i \in Z(A)$ with $p_i \in Q_p$ and $c_i \in k^*$. Let p be the path among those p_i which is of maximal length. Then $h(p) = t(p) = j$.

Since Q_0 is infinite and Q is connected, it follows that there exists an arrow α, such that $h(\alpha) = j \neq t(\alpha)$; or $t(\alpha) = j \neq h(\alpha)$. Without loss of generality, we may assume that $h(\alpha) = j \neq t(\alpha)$. Then we get the contradiction $a\alpha \neq \alpha a$. □

2. Cohomology Groups $\text{Ext}^n_{A^e}(A, A)$

We are interested in the cohomology groups $\text{Ext}^n_{A^e}(A, A)$, where $A = kQ$ with Q an arbitrary quiver.

We need the following observation.

Lemma 2.1. Let $Q = (Q_0, Q_1, h, t)$ be a quiver. Then Q is a tree if and only if for any $(d_\alpha)_{\alpha \in Q_1} \in \prod_{\alpha \in Q_1} k_\alpha$, where $k_\alpha = k$ for any $\alpha \in Q_1$, the system of linear equations

$$x_{t(\alpha)} - x_{h(\alpha)} = d_\alpha, \quad \forall \alpha \in Q_1 \tag{3}$$

has a solution $x_i = c_i$, $\forall i \in Q_0$.

Proof. Assume that the system of linear equations (3) has a solution. If Q is not a tree, then there is an unoriented cycle $c = \beta_1 \cdots \beta_n$ in Q, where β_i is an arrow α_i, or a formal inverse of α_i. Choose $(d_\alpha) \in \prod_{\alpha \in Q_1} k_\alpha$ with $d_\alpha = 0$ for all $\alpha \neq \alpha_1$, and $d_{\alpha_1} = 1$. Then from (3) we have the contradiction

$$0 = \sum_{1 \leq i \leq n} (-1)^{\sigma(\beta_i)} d_{\alpha_i} = (-1)^{\sigma(\beta_1)},$$

where $\sigma(\beta_i) = 1$ if $\beta_i = \alpha_i$, and $\sigma(\beta_i) = -1$ if $\beta_i = \alpha_i^{-1}$.

Conversely, without loss of generality, we may assume that Q is a connected tree. Start from an arbitrary vertex i, and take a fixed value of x_i, say $x_i = c_i$. Then for any vertex j we have a walk w starting at i and ending at j, since Q is connected. Then by (3) we obtain the value of $x_j = c_j$. Since Q is a tree, it follows that such a walk w is unique; in this way we get a solution $x_j = c_j, \forall j \in Q_0$, of the system of linear equations (3). $\square$

Theorem 2.2. Let $A = kQ$, where Q is a quiver. Then:
(i) we have $\mathrm{Ext}^n_{A^e}(A, A) = 0$ for $n \geq 2$;
(ii) $\mathrm{Ext}^1_{A^e}(A, A) = 0$ if and only if Q is a tree.

Proof. The assertion (i) follows from Corollary in 1.3. In order to prove (ii), applying $\mathrm{Hom}_{A^e}(-, A)$ to the exact sequence (2) in Corollary of 1.3, we see that $\mathrm{Ext}^1_{A^e}(A, A) = 0$ if and only if the map

$$f^* : \mathrm{Hom}_{A^e}(\bigoplus_{i \in Q_0} Ae_i \otimes (e_i A)', A) \longrightarrow \mathrm{Hom}_{A^e}(\bigoplus_{\alpha \in Q_1} Ae_{h(\alpha)} \otimes (e_{t(\alpha)} A)', A)$$

induced by f is epic.

Note that

$$\mathrm{Hom}_{A^e}(\bigoplus_{i \in Q_0} Ae_i \otimes (e_i A)', A) \cong \prod_{i \in Q_0} \mathrm{Hom}_{A^e}(Ae_i \otimes (e_i A)', A);$$

$$\mathrm{Hom}_{A^e}(\bigoplus_{\alpha \in Q_1} Ae_{h(\alpha)} \otimes (e_{t(\alpha)} A)', A) \cong \prod_{\alpha \in Q_1} \mathrm{Hom}_{A^e}(Ae_{h(\alpha)} \otimes (e_{t(\alpha)} A)', A).$$

If Q is a tree, then
$$\mathrm{Hom}_{A^e}(\bigoplus_{i\in Q_0} Ae_i \otimes (e_iA)', A) \cong \prod_{i\in Q_0} kf_i,$$
where $f_i \in \mathrm{Hom}_{A^e}(Ae_i \otimes (e_iA)', A)$ is given by $f_i(e_i \otimes e'_i) = e_i$; and
$$\mathrm{Hom}_{A^e}(\bigoplus_{\alpha\in Q_1} Ae_{h(\alpha)} \otimes (e_{t(\alpha)}A)', A) \cong \prod_{\alpha\in Q_1} kf_\alpha,$$
where $f_\alpha \in \mathrm{Hom}_{A^e}(Ae_{h(\alpha)} \otimes (e_{t(\alpha)}A)', A)$ is given by $f_\alpha(e_{h(\alpha)} \otimes e'_{t(\alpha)}) = \alpha$. Note that
$$f^*((c_if_i)_{i\in Q_0}) = ((c_{t(\alpha)} - c_{h(\alpha)})f_\alpha)_{\alpha\in Q_1}. \qquad (4)$$
Take an element $(d_\alpha f_\alpha)_{\alpha\in Q_1}$ in $\mathrm{Hom}_{A^e}(\bigoplus_{\alpha\in Q_1} Ae_{h(\alpha)} \otimes (e_{t(\alpha)}A)', A)$. Then by Lemma 2.1 the system of linear equations
$$x_{t(\alpha)} - x_{h(\alpha)} = d_\alpha, \quad \forall \alpha \in Q_1$$
has a solution $x_i = c_i$, $\forall i \in Q_0$. By (4) this means
$$f^*((c_if_i)_{i\in Q_0}) = (d_\alpha f_\alpha)_{\alpha\in Q_1},$$
that is, f^* is epic.

Conversely, if f^* is epic, then for any $(d_\alpha)_{\alpha\in Q_1} \in \prod_{\alpha\in Q_1} k_\alpha$, where $k_\alpha = k$ for any $\alpha \in Q_1$, there exists $\delta = (\delta_i)_{i\in Q_0} \in \prod_{i\in Q_0} \mathrm{Hom}_{A^e}(Ae_i \otimes (e_iA)', A)$, such that $f^*(\delta) = (d_\alpha f_\alpha)_{\alpha\in Q_1}$. Let $\delta_i(e_i \otimes e'_i) = \alpha_i \in e_iAe_i$. Then
$$\begin{aligned}
f^*(\delta)(e_{h(\alpha)} \otimes e'_{t(\alpha)}) &= \delta(\alpha \otimes e'_{t(\alpha)} - e_{h(\alpha)} \otimes \alpha') \\
&= (\alpha \otimes e'_{t(\alpha)})\delta_{e_{t(\alpha)}}(e_{t(\alpha)} \otimes e'_{t(\alpha)}) \\
&\quad - (e_{h(\alpha)} \otimes \alpha')\delta_{e_{h(\alpha)}}(e_{h(\alpha)} \otimes e'_{h(\alpha)}) \\
&= \alpha\alpha_{t(\alpha)} - \alpha_{h(\alpha)}\alpha.
\end{aligned}$$
It follows that
$$\alpha\alpha_{t(\alpha)} - \alpha_{h(\alpha)}\alpha = d_\alpha\alpha \text{ for } \alpha \in Q_1. \qquad (5)$$
Let $\alpha_i = c_ie_i + x_i$ with $x_i \in e_iAe_i \cap J$, where J is the ideal of A generated by all arrows in Q. Then (5) forces $x_i = 0$ and
$$c_{t(\alpha)} - c_{h(\alpha)} = d_\alpha \text{ for } \alpha \in Q_1;$$
it follows from Lemma 2.1 that Q is a tree. □

3. Monomial Algebras

The following result generalizes and unifies several well-known corresponding results in [BM, 2.2] and [Ha, 1.6, 2.2, 2.3, 3.2].

Theorem 3.1. *Let $A = kQ/I$ be a monomial algebra with Q connected. Then the following are equivalent:*
  (i) $H^1(A) = 0$;
  (ii) Q *is a finite tree*;
  (iii) $H^n(A) = 0$ *for* $n \geq 1$.

Note that if Q is a tree, then any ideal of kQ is generated by a set of paths in Q. It follows that we have

Corollary 3.2. *Let $A = kQ/I$ with Q a connected tree, and $I \subseteq J^2$, where J is the ideal of kQ generated by all arrows of Q. Then $H^1(A) = 0$ if and only if Q_0 is a finite set.*

In order to prove the theorem, we need the following construction of derivations, which is introduced by Happel [Ha, 3.2] and developed by Bardzell and Marcos [BM, 2.1].

Given any function $w : Q_1 \longrightarrow k$, one can easily see that w can be extended to a function from Q_p to k, again denoted by w, such that $w(e_i) = 0$ for $i \in Q_0$ and $w(pq) = w(p) + w(q)$ for $p, q \in Q_p$.

Define a k-map $d(w) : kQ \longrightarrow kQ$ by $d(w)(p) = w(p)p$ for $p \in Q_p$. Then it is clear that $d(w) \in Der(kQ)$.

Further, the function w can be extended to the set of all walks in Q by $w(\alpha^{-1}) = -w(\alpha)$ for $\alpha \in Q_1$, again denoted by w.

Lemma 3.3. *Let $A = kQ/I$ with Q a quiver and I an ideal of kQ with $I \subseteq J^2$. Let $w : Q_1 \longrightarrow k$ be a function such that $d(w)(I) \subseteq I$. If $H^1(A) = 0$, and q_1, q_2 are two parallel walks in Q, then $w(q_1) = w(q_2)$.*

Proof. The proof is the same as the one in [BM, 2.1], which is stated only for the finite-dimensional case. For the convenience of the reader, we include the proof.

For any element $q \in kQ$, denote the canonical image of q in A by $\bar{q}$. Since $d(w)(I) \subseteq I$, it follows that $d(w)$ induces a derivation of A, again denoted by $d(w)$. Thus, $d(w)(\bar{p}) = w(p)\bar{p}$ for $p \in Q_p$. Since $H^1(A) = 0$, it follows that $d(w) = \delta_{\bar{a}}$ for some $a = \sum_{i \in Q_0} c_i e_i + x$ (a finite sum) with $x \in J$. Then for any $\alpha \in Q_1$ we have $w(\alpha)\bar{\alpha} = \delta_{\bar{a}}(\bar{\alpha}) = \bar{a}\bar{\alpha} - \bar{\alpha}\bar{a} = (c_{h(\alpha)} - c_{t(\alpha)})\bar{\alpha} + \bar{y}$ with $y \in J^2$. Since $I \subseteq J^2$, it follows that $(w(\alpha) - c_{h(\alpha)} + c_{t(\alpha)})\alpha \in J^2$. This forces

$$c_{h(\alpha)} - c_{t(\alpha)} = w(\alpha), \qquad (6)$$

which means that for any walk γ we have $w(\gamma) = w(h(\gamma)) - w(t(\gamma))$. Thus the assertion follows. $\square$

Corollary 3.4. *Let A and w be as in Lemma 3.3. If $H^1(A) = 0$, then $w(c) = 0$ for any unoriented cycle c.*

For the general case, we have the following similar result as in the finite-dimensional case due to Bardzell and Marcos [BM, 2.2].

Lemma 3.5. *Let $A = kQ/I$ be a monomial algebra. If $H^1(A) = 0$, then Q is a tree.*

Proof. Otherwise, let c be an unoriented cycle containing an arrow α. Choose a function $w : Q_1 \longrightarrow k$ to be $w(\alpha) = 1$ and $w(\beta) = 0$ for $\beta \, (\neq \alpha)$. Since A is monomial, it follows that $d(w)(I) \subseteq I$, and hence $w(c) = 0$. But $w(c) = 1$ by the choice of w. □

Lemma 3.6. *Let $A = kQ/I$ be a monomial algebra. If $H^1(A) = 0$, then Q is locally finite.*

Proof. Otherwise, since Q contains no loops (i.e., arrows α with $h(\alpha) = t(\alpha)$) by Lemma 3.5, it follows that Q must contain either a subquiver $D = (D_0, D_1)$ with $D_0 = \mathbb{N}_0$ and $D_1 = \{a_i : 0 \to i \mid i \in \mathbb{N}_1\}$, or a subquiver $D' = (D'_0, D'_1)$ with $D'_0 = \mathbb{N}_0$ and $D'_1 = \{a_i : i \to 0 \mid i \in \mathbb{N}_1\}$; here $\mathbb{N}_i = \{n \in \mathbb{Z} \mid n \geq i\}$. By the dual argument, it suffices to assume that Q contains a subquiver D. Consider a function $w : Q_1 \longrightarrow k$ defined by

$$w(\alpha_{2n}) = 1 \text{ for } n \geq 1 \text{ and } w(\beta) = 0, \forall \beta \in Q_1 - \{a_{2n} \mid n \in \mathbb{N}_1\}.$$

Since A is a monomial algebra, it follows that $d(w)(I) \subseteq I$, and hence we have the derivation $d(w)$ of A given by $d(w)(\bar{p}) = w(p)\bar{p}$ for $p \in Q_p$, where $\bar{p}$ denotes the canonical image in A of p. Let $\delta = \delta_{-\bar{\alpha}_1} + d(w)$. Then $\delta \in Der(A)$. We claim that $\delta \notin Der^0(A)$, and hence we get a desired contradiction to the assumption $H^1(A) = 0$.

In fact, if $\delta = \delta_{\bar{a}}$, where $a \in kQ$ is a finite sum of the form

$$a = \sum_{i \in \mathbb{N}_0} c_i e_i + \sum_{i \in \mathbb{N}_1} d_i \alpha_i + \sum_{\alpha \in \Omega} t_\alpha \alpha,$$

with $\Omega = Q_p - \{e_0, e_i, \alpha_i \mid i \in \mathbb{N}_1\}$, then we have

$$\bar{\alpha}_{2n} = \delta(\bar{\alpha}_{2n}) = \delta_{\bar{a}}(\bar{\alpha}_{2n}) = \bar{a}\bar{\alpha}_{2n} - \bar{\alpha}_{2n}\bar{a} = (c_0 - c_{2n})\bar{\alpha}_{2n} + \bar{x}$$

with $x \in J^2$. Since $I \subseteq J^2$, it follows that

$$c_0 - c_{2n} = 1. \tag{6}$$

Similarly, we have

$$c_0 - c_{2n-1} = 0. \tag{7}$$

Since a is a finite sum, it follows that almost all c_i are zero, which contradicts (7) and (8). □

Lemma 3.7. *Let $A = kQ/I$ be a monomial algebra with Q connected and $H^1(A) = 0$. Then Q is a finite tree.*

Proof. Since Q is connected, by Lemmas 3.5 and 3.6 it is enough to prove Q_1 is a finite set.

Consider the k-map $\delta : A \longrightarrow A$ given by $\delta(\bar{p}) = l(p)\bar{p}$ for $p \in Q_p$, where $l(p)$ is the length of p. Then $\delta \in Der(A)$ since A is monomial, and hence $\delta = \delta_{\bar{a}}$ for some $a = \sum_{i \in Q_0} c_i e_i + x$ (a finite sum) with $x \in J$. Since $I \subseteq J^2$, it follows that

$$c_{h(\alpha)} - c_{t(\alpha)} = 1, \quad \text{for} \quad \alpha \in Q_1. \tag{8}$$

Let Q_0' be the set of vertices i with $c_i \neq 0$. Since a is a finite sum, it follows that Q_0' is a finite set. For every arrow α we see from (4) that at least one of $h(\alpha)$ and $t(\alpha)$ belongs to Q_0'.

If Q_1 is an infinite set, then there exists a vertex $i \in Q_0'$ and infinitely many arrows α_j such that i is the head or the tail of α_j. This is impossible since Q is locally finite by Lemma 3.6. □

Lemma 3.8. [Ha, 2.2] *Let $A = kQ/I$ with Q a finite tree and $I \subseteq J^2$. Then $H^n(A) = 0$ for all $n \geq 1$.*

Proof of Theorem 3.1. The implication of (i) to (ii) follows from Lemma 3.7 and the implication of (ii) to (iii) follows from Lemma 3.8. □

Remark. For a non-monomial algebra $A = kQ/I$, even if Q is a finite quiver, Theorem 3.1 is no longer valid. For example, let Q be the quiver with $Q_0 = \{1, 2, 3, 4\}$ and $Q_1 = \{\alpha : 1 \longrightarrow 2; \beta : 2 \longrightarrow 4; \gamma : 1 \longrightarrow 3; \delta : 3 \longrightarrow 4\}$; and $I = \langle \alpha\beta - \gamma\delta \rangle$. Then $H^1(kQ/I) = 0$, but Q is not a tree.

4. Vanishing of the First Hochschild Cohomology

Let $A = kQ/I$ with Q a finite quiver and I an ideal of kQ with $J^N \subseteq I \subseteq J^2$ for some positive integer N. Recently, Buchweitz and Liu [BL] have constructed an algebra A with a loop with $H^1(A) = 0$. However, in many cases, $H^1(A) = 0$ implies that Q is directed, that is, Q contains no oriented cycles (cf. [Ha]). We include several results towards this direction.

For a given quiver Q, recall that a relation ρ in Q is a finite combination $\sum c_i p_i$ of paths p_i of length longer than 1, such that all p_i have the same head, and have the same tail, see [R, p.43]. Note that any ideal I of kQ with $I \subseteq J^2$ is generated by a set of relations in Q. An ideal $I = \langle \rho_i \rangle$ with all $\rho_i = \sum c_{i,j} p_{i,j}$ being relations is called a homogeneous ideal provided that $l(p_{i,j})$ does not depend on j.

Denote by $\tilde{A}_{p,q}$ the quiver with $Q_0 = \{1, \cdots, p+q\}$ and $Q_1 = \{\alpha_i : i \longrightarrow i+1$ for $1 \leq i \leq p-1$; $\alpha_p : p \longrightarrow p+q$; $\beta_1 : 1 \longrightarrow p+1$; $\beta_j : p+j-1 \longrightarrow p+j$ for $2 \leq j \leq q\}$.

Theorem 4.1. *Let $A = kQ/I$ with Q a quiver and I a homogeneous ideal of kQ. If $H^1(A) = 0$, then Q does not contain a subquiver $\tilde{A}_{p,q}$ with $p \neq q$.*

Proof. Otherwise, denote c by the unoriented cycle given by a subquiver $\tilde{A}_{p,q}$. Consider the length function l, that is, $l(\alpha) = 1$ for $\alpha \in Q_1$. Since I is homogeneous, it follows that $d(w)(I) \subseteq I$ and $l(c) = 0$ by Corollary 3.4. On the other hand, $l(c) \neq 0$ since $p \neq q$. □

Theorem 4.2. *Let $A = kQ/I$ with Q a quiver and I an ideal of kQ with $I \subseteq J^2$. If $H^1(A) = 0$, then either Q is a tree, or for any unoriented cycle c and any arrow α occurring in c, there exist a generating relation $g = \sum_{1 \leq i \leq n} c_i p_i$ of I with $n \geq 2$, and i, j such that $m_\alpha(p_i) \neq m_\alpha(p_j)$, where $m_\alpha(p_i) =$ the times of the occurrences of α in p_i.*

Proof. Assume that Q is not a tree. Then by Theorem 3.1 we see that A is not monomial. Let $I = \langle \rho_i \mid i \rangle$ with all ρ_i being relations. Consider the set S of all generators of I which are not paths. If there exist an oriented cycle c and an arrow α occurring in c, such that $m_\alpha(p_i) = m_\alpha(p_j)$ for any relation $g = \sum_{1 \leq i \leq n} c_i p_i \in S$ and all $1 \leq i, j \leq n$, then define $w : Q_1 \longrightarrow k$ by $w(\alpha) = 1$ and $w(\beta) = 0$ for $\beta (\neq \alpha)$. Then $d(w)(I) \subseteq I$ and by Corollary 3.4 we have $w(c) = 0$. But $w(c) = 1$ by the choice of w. □

Theorem 4.3. *Let Q be a cyclic quiver, that is, $Q_0 = \{1, 2, \cdots, l\}$ and $Q_1 = \{\alpha_i : i \longrightarrow i+1, \forall 1 \leq i \leq l-1$; $\alpha_l : l \longrightarrow 1\}$; and let I be an ideal of kQ with $J^N \subseteq I \subseteq J^2$ for some positive integer N. Then I is generated by some paths, and $H^1(kQ/I) \neq 0$.*

Proof. By Theorem 3.1, it is enough to prove the first assertion. Otherwise, assume that $g = \sum_{1 \leq i \leq n} c_i p_i \in I$ is a relation with $\prod_{1 \leq i \leq n} c_i \neq 0$, $n \geq 2$, $p_i \in Q_p$, and $p_i \notin I$ for $1 \leq i \leq n$, such that n is minimal among all such relations in I. Since Q is a cyclic quiver, we see $l(p_1) > \cdots > l(p_n)$.

Since $J^N \subseteq I \subseteq J^2$ for some positive integer N, it follows that kQ/I is finite-dimensional, and hence there exists a unique path q such that $l(p_1 q)$ is minimal among $p_i q \in I$. Denote $p_1 q$ by p. Then

$$p = p_1 q = c_1^{-1}(gq - \sum_{2 \leq i \leq n} c_i p_i q),$$

it follows that $\sum_{2 \leq i \leq n} c_i p_i q \in I$. By the minimality of n, we see that $p_i q \in I$ for $2 \leq i \leq n$.

Let c be an oriented cycle with $l(c) = l$, $h(c) = t$, and $t(c) = t$, where $t = t(p_i)$. Then $p_1 = p_n c^m$ for some $m \geq 1$. Write q as $q = c^r q_1$ with $r \geq 0$ and $h(q_1) = t$, $l(q_1) < l$.

If $r \geq m$, then we have

$$p_n q = p_n c^r q_1 = (p_n c^m)(c^{r-m} q_1) = p_1(c^{r-m} q_1) \in I,$$

which contradicts the assumption of q since $l(c^{r-m} q_1) < l(q)$.

If $m > r$, then there exists a unique $p' \in Q_p$ such that $c^{m-r} = q_1 p'$, and hence we have

$$p_1 = p_n c^m = (p_n c^r) c^{m-r} = (p_n c^r)(q_1 p') = (p_n q) p' \in I,$$

a contradiction. This completes the proof. $\square$

References

[BL] R-O. Buchweitz and S. P. Liu, *Artin algebras with loops but no outer derivations*, preprint.

[BM] M. J. Bardzell and E. Marcos, *Induced boundary maps for the cohomology of monomial and Auslander algebras*, Canad. Math. Soc. Conf. Proc. **24** (1998), 47–54.

[CB] W. Crawley-Boevey, *Lectures on representations of quivers*, preprint.

[CE] H. Cartan and S. Eilenberg, *Homological Algebra*, Princeton University Press, 1956.

[G] P. Gabriel, *Auslander-Reiten sequences and representation-finite algebras*, Lecture Notes in Math. **832**, Springer-Verlag, Heidelberg, New York, (1980), 1–71.

[Ha] D. Happel, *Hochschild cohomology of finite-dimensional algebras*, Lecture Notes in Math. **1404**, Springer-Verlag, Heidelberg, New York, (1989), 108–126.

[Ho] G. Hochschild, *On the cohomology groups of an associative algebra*, Ann. Math. **46** (1945), 58–67.

[R] C. M. Ringel, *Tame algebras and integral quadratic forms*, Lecture Notes in Math. **1099**, Springer-Verlag, Heidelberg, New York, 1984.

[W] C. A. Weibel, *An Introduction to Homological Algebra*, Cambridge Studies in Advanced Math. **38**, Cambridge Univ. Press, Cambridge, 1995.

Department of Mathematics
University of Science and Technology of China
Hefei 230026, P. R. China
e-mail: pzhang@ustc.edu.cn

Open Problems

Pere Ara

Problem. Determine the kernel of the natural map $GL_1(R) \to K_1(R)$ when R is a separative exchange ring. This has been solved for unit-regular rings and regular right self-injective rings by Menal and Moncasi [P. Menal and J. Moncasi, K_1 of von Neumann regular rings, J. Pure and Applied Algebra **33** (1984), 295–312], and for exchange rings with primitive factor rings Artinian by Chen and Li [H. Chen and F.-U. Li, Whitehead groups of exchange rings with primitive factor rings artinian, Preprint]. In all cases, one gets $K_1(R) = GL_1(R)^{ab}$ provided that $1/2 \in R$.

Yoshitomo Baba

Problem. For given modules M and N, we say that M is N-injective if there exists an extension homomorphism f' from N to M for each submodule N' of N and each homomorphism f from N' to M. Also we say M is simple-N-injective if there exists an extension homomorphism f' from N to M for each submodule N' of N and each homomorphism f from N' to M with the image of f simple.

Let R be a perfect ring and M, N given R-modules with the socle of M simple. If M is simple-N-injective, then is M an N-injective module?

If R is a semiprimary ring, then we know that any simple-N-injective module is N-injective for given R-modules M and N with the socle of M simple by Proposition 2 in [Y. Baba and K. Oshiro, On a theorem of Fuller, J. Algebra **154** (1993), 86–94]. This open problem was first given by M. Hoshino of the University of Tsukuba to Y. Baba.

Gary F. Birkenmeier, Jae Keol Park and Jin Yong Kim

Problem. In [G. F. Birkenmeier, Idempotents and completely semiprime ideals, Comm. Algebra **11** (1983), 567–580], an idempotent $e \in R$ is called left (right) semicentral if $Re = eRe$ ($eR = eRe$). We use $\mathcal{S}_\ell(R)$ and $\mathcal{S}_r(R)$ for the sets of all left and right semicentral idempotents, respectively. For an idempotent $e \in R$, observe that $\mathcal{S}_\ell(eRe) = \{0, e\}$ if and only if $\mathcal{S}_r(eRe) = \{0, e\}$; when this occurs e is said to be semicentral reduced. If 1, the unity of R, is semicentral reduced, then R is said to be semicentral reduced [G. F. Birkenmeier, H. E. Heatherly, J. Y. Kim and J. K. Park,

Triangular matrix representations, To appear in J. Algebra]. In the paper "Semicentral reduced algebras" by G. F. Birkenmeier, J. Y. Kim and J. K. Park in this Proceedings, it is shown that for any positive integer n, R is semicentral reduced if and only if the n-by-n full matrix ring $M_n(R)$ is semicentral reduced.

Now we ask: Is the semicentral reduced property a Morita invariant property?

Yasuyuki Hirano

Problem. A ring is called a quasi-Baer ring, if the left annihilator of every left ideal is generated by an idempotent. Let R be a quasi-Baer ring, G a finite group and suppose that the order of G is invertible in R. Then is the group ring $R[G]$ quasi-Baer?

Hidetoshi Marubayashi

Problem 1. Find necessary and sufficient conditions for the generalized crossed product algebras $R * G$ to be semi-hereditary and Prüfer in terms of G and R.

Problem 2. Let P be a prime ideal of a fully bounded Prüfer order in a simple Artinian ring. Is R/P a Goldie ring? (If R is a Dubrovin valuation ring, then the answer is affirmative.)

Problem 3. Does the approximation theorem hold in the case that the order is a semi-local Bezout order without the assumption that the order is a PI-ring? We may assume that R_i are all fully bounded Dubrovin valuation rings.

For definitions and relevant references, see the paper "Non-Commutative valuation rings and their global theories" by H. Marubayashi in this Proceedings.

Kaoru Motose

Problem. Let $\Phi_n(x)$ be the cyclotomic polynomial of order n. Factorize the next number (repunit of 97 digits):

$$\Phi_{97}(10) = \frac{10^{97} - 1}{10 - 1} = \overbrace{111\cdots 1}^{97}.$$

Recently it has been shown that $\Phi_n(a)$ for $2 \le n \le 100$ and $2 \le a \le 10$, up to the above case $n = 97$ and $a = 10$, can be factorized.

W. K. Nicholson and M. F. Yousif

Problem. Following Harada, an associative ring R is called right simple-injective if, for each right ideal T of R, every R-linear map from T into R with simple-image extends to R, i.e., the map is given by left multiplication by an element c of R. In [W. K. Nicholson and M. F. Yousif, On perfect simple-injective rings, Proc. Amer. Math. Soc. **125** (1997), 979–985], it was shown that every semiprimary right simple-injective ring R is right self-injective, every left perfect left and right simple-injective ring R is a quasi-Frobenius ring, and an example of a left perfect left simple-injective ring was provided.

It is an open question whether every left perfect right simple-injective ring is right self-injective.

K. M. Rangaswamy

Problem. Is a Butler module over a valuation domain completely decomposable? (For the definition and relevant references, see the paper "On torsion-free modules over valuation domains" by K. M. Rangaswamy in this Proceedings.)

S. Tariq Rizvi

If x, y are elements in a modular lattice L with $x \leq y$, then y/x will denote the interval $[x, y]$, i.e., $y/x = \{a \in L \mid x \leq a \leq y\}$.

The *Krull dimension* of a partially ordered set $(P, \leq)$, denoted by $\kappa(P)$, is an ordinal number defined recursively as follows:

$\kappa(P) = -1$ if and only if P is a trivial poset, where -1 is assumed to be the predecessor of zero. Let $\alpha \geq 0$ be an ordinal number, and assume that we have already defined which posets have Krull dimension β for any ordinal $\beta < \alpha$. Then we define: $\kappa(P) = \alpha$ if and only if $\kappa(P) \neq \beta$ for all ordinals β with $\beta < \alpha$, and for any descending chain $x_1 \geq x_2 \geq \cdots \geq x_n \geq \cdots$ of elements of P, there exists a positive integer n_0 such that $\kappa(x_n/x_{n+1}) < \alpha$ for all $n \geq n_0$, i.e., for all $n \geq n_0$, $\kappa(x_n/x_{n+1})$ has previously been defined and it is an ordinal $< \alpha$. If no ordinal α exists such that $\kappa(P) = \alpha$, we say that P does not have Krull dimension.

The *dual Krull dimension* of the poset P, denoted by $\kappa^0(P)$, is defined as being the Krull dimension $\kappa(P^0)$ of the opposite poset P^0 of P (if it exists).

Recall that a module M_R is called *quotient finite dimensional* (or *qfd*), if every factor module of M has finite uniform (or Goldie) dimension. For any module M_R, we denote:

$$C(M) := \{X \leq M \mid X \text{ is cyclic}\};$$

$$F(M) := \{X \leq M \mid X \text{ is finitely generated}\}.$$

Recently, Albu and Rizvi in [1, Theorem 1.12] have shown that

$$M_R \text{ is Artinian} \Leftrightarrow M \text{ is qfd and } \kappa(C(M)) \leq 0.$$

(Here, $\kappa(C(M))$ means the Krull dimension of the poset $C(M)$ of all cyclic submodules of M, ordered by inclusion.)

It is natural to ask whether this result can be extended to an arbitrary Krull dimension α? Hence prove or disprove:

Problem 1. For any right R-module M, $\kappa(M_R) \leq \alpha \Leftrightarrow M$ is qfd and $\kappa(C(M)) \leq \alpha$.

Remark. We have shown that a slightly weaker version of this problem holds true in case one replaces $C(M)$ by $F(M)$, see [1, Theorem 1.17]:

$$\kappa(M_R) \leq \alpha \Leftrightarrow M \text{ is qfd and } \kappa(F(M)) \leq \alpha.$$

In [6, Lemma 6] (also [5, Theorem 6.3]), Huynh, Dung and Smith have shown the following interesting and useful result:
The following are equivalent for a module M_R and an ordinal $\alpha \geq 0$.
  (1) $\kappa(M) \leq \alpha$.
  (2) *Every homomorphic image of M has an essential submodule E with $\kappa(E) \leq \alpha$.*

It would be very nice to obtain a dual of this result. Hence prove or disprove:

Problem 2. The following are equivalent for a module M_R and an ordinal $\alpha \geq 0$.
  (1) $\kappa^0(M) \leq \alpha$.
  (2) Every submodule X of M has a small submodule S with $\kappa^0(X/S) \leq \alpha$.

Remarks. We suspect that this equivalence may hold true. Note that the equivalence holds true for $\alpha = 0$ since $\kappa^0(X/S) \leq 0$ for a small submodule S of X implies that X is finitely generated. In order to prove this result in general, we only need to show that condition (2) implies that M is qfd, and then use [1, Proposition 2.2].

A module M is called *extending* if every submodule of M is essential in a direct summand. We call a module *FI-extending* if every fully invariant submodule is essential in a direct summand. In a recent paper, Birkenmeier, Müller and Rizvi [2] have shown that in contrast to the extending modules, arbitrary direct sums of FI-extending modules are always FI-extending. (The FI-extending property is also shown to carry over to matrix rings, unlike the extending property.) It is presently unknown if the FI-extending property is inherited by direct summands of arbitrary FI-extending modules.

Problem 3. Let M be an FI-extending R-module. Is every direct summand of M also FI-extending? If not, then find necessary and sufficient conditions for this to hold true.

Remarks. Birkenmeier, Călugăreanu, Fuchs and Goeters in their investigations have proved that for FI-extending abelian groups, the answer to this question is yes [3, Theorem 3.2]. Recently Birkenmeier, Park and Rizvi [4] have shown that the answer is also affirmative if the FI-extending module is nonsingular.

References

[1] T. Albu and S. T. Rizvi, *Chain conditions on quotients dimensional modules,* Comm. Algebra, to appear.

[2] G. F. Birkenmeier, B. J. Müller and S, T. Rizvi, *Modules in which every fully invariant submodule is essential in a direct summand,* preprint.

[3] G. F. Birkenmeier, G. Călugăreanu, L. Fuchs and H. P. Goeters, *The fully-invariant-extension property for abelian groups,* Comm. Algebra, to appear.

[4] G. F. Birkenmeier, J. K. Park and S. T. Rizvi, *Modules with fully invariant submodules essential in fully invariant summands,* preprint.

[5] D. V. Huynh, N. V. Dung and P. F. Smith, *A characterization of rings with Krull dimension,* J. Algebra **32** (1990), 104–112.

Mingyi Wang

A coalgebra C is called right (resp. left) conoetherian if every quotient comodule C/I of C_C (resp. $_CC$) is finitely cogenerated. A coalgebra C is called right (resp. left) Co-Noetherian if every quotient module C/I of C_C (resp. $_CC$) is quasi-finite.

Problem 1. Is every right conoetherian coalgebra right Co-Noetherian? Does there exist a one-sided conoetherian or Co-Noetherian coalgebra?

Problem 2. Let T_C be a classical tilting comodule over a right conoetherian right semiperfect coalgebra and D the coendomorphism coalgebra of T_C. Is D a left conoetherian left semiperfect coalgebra? Is $_D T$ a classical tilting comodule?

References

[1] M. Wang and Z. X. Wu, *Conoetherian coalgebras*, Algebra Colloq. **5** (1998), 117–120.

[2] M. Wang, *Some co-hom functors and classical tilting comodules*, SEA Bull. Math. **22** (1998), 455–468.

[3] M. Wang, *Tilting comodules over semi-perfect coalgebras*, Algebra Colloq. **6** (1999), 461–472.

Alexander Zimmermann

Problem. Given a finite group G, is there a tilting complex in the derived category of the integral group ring $D^b(\mathbb{Z}G)$ which is not a Morita bimodule?